Best Practices in Manufacturing Processes

Jorge Luis García Alcaraz
Leonardo Rivera Cadavid
Rosa Guadalupe González-Ramírez
George Leal Jamil · Mario Gustavo Chong Chong
Editors

Best Practices in Manufacturing Processes

Experiences from Latin America

Springer

Editors
Jorge Luis García Alcaraz
Universidad Autónoma de Ciudad Juárez
Ciudad Juárez, Chihuahua, Mexico

Leonardo Rivera Cadavid
Ciudad Universitaria Meléndez
Universidad del Valle
Valle del Cauca, Colombia

Rosa Guadalupe González-Ramírez
Universidad de Los Andes
Santiago, Chile

George Leal Jamil
Informações em Rede Consultoria e
Treinamento Ltda
Belo Horizonte, Minas Gerais, Brazil

Mario Gustavo Chong Chong
Universidad del Pacífico
Lima—Jesús María, Peru

ISBN 978-3-030-07574-3 ISBN 978-3-319-99190-0 (eBook)
https://doi.org/10.1007/978-3-319-99190-0

This Springer imprint is published by the registered company Springer Nature Switzerland AG
The registered company address is: Gewerbestrasse 11, 6330 Cham, Switzerland

Preface

Nowadays, the globalization environment and fierce competition have forced companies worldwide to place significant efforts to improve their performance and be more competitive to stay in the market. Since the Industrial Revolution, several methodologies and concepts have been introduced to support manufacturing companies to reduce costs and increase their productivity and efficiency in their operations. As an example of such methodologies, we can mention mass customization, just in time, lean manufacturing, additive manufacture, and concurrent engineering, among others. Another important trend is related to the automation of the manufacturing processes and industrial robotics. According to the International Federation for Robotics, since 2010, the technical improvement in industrial robots and their demand has been considerably accelerated.

Furthermore, the emergence of new technologies such as Internet of Things and Sensors, Cloud Computing, Big Data, Blockchain, etc., presents challenges and opportunities for manufacturing companies on how to adopt such technologies to support their decision-making and improve their performance.

Latin America and the Caribbean is a region recognized by a strong manufacturing sector, being Mexico one of the main manufacturing countries. According to the Global Manufacturing Index (GMI) computed by Deloite, in 2016, Mexico was positioned among the top 10 countries in the ranking (out of 40), with the 8th place. Other countries from the region were Brazil, Colombia, and Argentina that occupied the 29th, 36th, and 39th positions, respectively. As indicated in the GMI report, there are three dominant regional clusters: North America, Europe, and Asia Pacific, with Mexico being part of the first one.

In this book, we aim to report the best practices that companies established in Latin America and the Caribbean are implementing in their manufacturing processes to generate high-quality products and stay in the market, as well as new technologies and operational and managerial philosophies that are being proposed by researchers and implemented by practitioners. The book has 24 chapters, from Brazil (2), Chile (1), Colombia (3), Ecuador (1), Spain (1), Mexico (12), Peru (3), and Trinidad and Tobago (1).

The content of this book is structured into four parts: (I) Techniques, Tools and Methodologies, (II) Production Management, (III) Manufacturing and Technology, and (IV) Human Factors.

Part I: Techniques, Tools and Methodologies comprises eight chapters, which are further described below:

Chapter 1 titled "Digital Transformation: Digital Maturity Applied to Study Brazilian Perspective for Industry 4.0" is carried out by Ferreira-Braga et al. and provides an analysis of the digital transformation efforts in Brazilian companies, aiming to identify opportunities and limitations.

Chapter 2 titled "Modelling and Analysis of the Apples Export Supply Chain Business Processes: Experiences from Chile" presented by López-Campos et al. presents an analysis of the fresh apple export supply chain in Chile, analyzing the shipment business process and providing a redesign proposal that considers the implementation of electronic data interchange procedures.

Chapter 3 titled "Considerations for the Latin American and Caribbean Region in Light of the Global Move Towards Low Carbon Shipping" is proposed by Singh and Rambarath-Parasram. The chapter presents some realities and developments on climate action in the international maritime industry that may impact the economies of the region and highlights the opportunities to overcome the barriers for the implementation and adoption of maritime technological advances to mitigating climate change.

Chapter 4 titled "Risk-based Strategic Inventory Supply Model for New Products" is carried out by Sánchez-Vega et al. They propose a risk-based strategic inventory supply model to make more efficient the supply chain of new products. For this, they perform a comprehensive study to determine the most important factors that may affect the supply of new products.

Chapter 5 titled "Fuzzy QFD and TOPSIS for Dispatching Prioritization in Maritime Transportation Considering Operational Risk" presented by Osorio-Gómez et al. presents a multi-criteria approach to prioritize container shipments dispatching according to the availability of resources (vessels) and the associated risks, considering a case study, the Port of Buenaventura in Colombia.

Chapter 6 titled "Logistics Planning for the Synchronization of Key Functional Areas of a Latin American Bottling Company" is carried out by Sánchez-Partida et al. and, based on a case study real bottling company in Mexico, proposes a forecast and inventory policy with the aim of reducing their forecast variability and ensuring the availability of materials for production lines.

Chapter 7 titled "Demand Management to Optimize the Supply Chain of an Agribusiness Company" is presented by Pérez-Pérez et al., which analyzes the quinoa supply chain of the agribusiness company BAMSA and proposes a redesign of its supply chain strategy with the aim to reduce lost sales and improve their inventory levels (safety stock).

Chapter 8 titled "Courier Logistics Strategy to Create Commercial Impact in Peru. Application of Correlation and Regression Analysis, Lean Manufacturing" presents a research carried out by Núñez-Pauca et al. They analyze the Express services in Peru with the aim to determine if imports of international express shipments explain the external trade volume of the country, to which extent they do it and which is the future field of action of projected foreign trade based on Express shipment growth.

Part II: Production Management consists of six chapters regarding techniques and method used in industry for support decision-making; those charters are described as follows:

Chapter 9 titled "Implementation of the S&OP Process in Textile Company Case Study: Ecuadorian Textile ABA" by Bofill-Altamirano and Avilés-Sacoto reports a case study from textile industry for improving communication and coordination among different departments involved in a production order, and they propose an S&OP as a solution that gives results fast and with high reliability, making easy the production control.

Chapter 10 is entitled "Reflecting on Industrial Business Models: A History of Tradition, Challenges, and Potential Innovations" by Jamil et al. They analyze and evaluate market options for business models to be adopted by industrial agents and after a deep literature review, and they propose a methodological approach aligned to production strategy. Their report is interesting because they integrate the experience of several successful companies.

Chapter 11 is entitled "The Constrained Joint Replenishment Problem Using Direct and Indirect Grouping Strategies with Genetic Algorithms" by Zapata-Cortes et al.; a model for inventory optimization for solving the joint replenishment problem (JRP) when ordering multiple products is required. The model allows reducing total inventory costs compared to the traditional practice of performing individual optimization of each product.

Chapter 12 is entitled "Improvement of the Demand Planning of Imported Seeds in the Company Agro Perú SA" by Coba et al. They analyze the current structure of the demand planning in a Peruvian enterprise and the potential of rethinking the demand planning structure and decisions on the whole supply chain performance. In their case study, they combine the CSAR framework for rethinking supply chain strategies and a multi-criteria method.

Chapter 13 is entitled "Numerical Analysis in a Beverage Can Utilizing Tube Hydroforming Process" and is presented by León-Anaya et al. They are reporting a numerical analysis for Tube Hydroforming (THF) and predict the behavior of extruded aluminum tube in a forming die for beverage can applications. They are using simulation and experimental determination of the mechanical properties of aluminum 6061-T5 for test specimens.

Chapter 14 is entitled "The E-Strategy for Lean-Sigma Solutions, Latin American Case Study in a New Product Validation Process" by Estrada-Orantes et al. They integrate lean manufacturing and six sigma concepts for proposing an E-strategy for solving problems in production systems and report a case study or proof of the new methodology.

Part III: Manufacturing and Technology presents studies that relate traditional, classical, and innovative applied research for industrial contexts, such as finite elements analysis, fused deposition modeling, and manufacturing/operations planning and systems implementations, among others, resulting in a coherent evaluation on how theoretical contents can decisively be considered to implement real solutions, the new environment of "best practices" in times of fast-changing technologies. With these researches, authors from several countries aim to improve the diffusion of consistent scientific methodological principles, resulting in a more reliable level of comprehension of critical aspects regarding emerging technologies and scenarios for industrial complexes.

Chapter 15 entitled "Manufacturing's Strategic Role and Management Practices: Evidence from Colombian Companies" by Vivares and Sarache is aimed to improve the understanding of manufacturing's strategic role (MSR) applying a survey in a group of Colombian companies of the coffee sector, where they explored MSR's potential as one fundamental component for manufacturing/operations system (MOS), taken as an integrative part of the organizational operational strategy (OS). For this, a consistent literature review and modeling relationship were provided by the authors. With their contribution, practices inserted in the MSR context are related to corporative strategies, as elements of the OS and their study also helps to configure the basics of MSR research variables to apply in further studies and designing practical, applicable solution in real organizational systems, with strategic evaluation.

Chapter 16 is entitled "Additive Manufacturing: Fused Deposition Modeling Advances"; through an application of systematic literature review, Aguilar-Duque et al. approached the advances of fused deposition modeling (FDM) as one of the most used technologies for additive manufacturing (AM) in the last years in this chapter. FDM was considered as an alternative for limitations of the classical manufacturing methods. For instance, those shown by environmental issues, costs and overall economic impacts, as well as the production of complex shapes in new assemblies. As the modern integration of manufacturing systems is reached every day, additive methodologies, such as FDM, can become more applied, with versatility and flexibility, resulting in appreciable opportunity to customize design and production, implicating in a space for new set of best practices for industrial complexes and associated productive arrays.

Chapter 17 entitled "Analysis of the Productivity of a Shoe Production Line—Application of Queueing Theory and Lean Manufacturing" by Hernández-González et al. studies one of the most competitive and complex production lines with its associated systems and processes, reporting the case of footwear industry. They are focused to propose decision-making to allow future planning and strategic implementation, nowadays expressively demanded by this type of industrial complex. To accomplish their intent, the authors applied a two-step methodology: First, a physical analysis was conducted through a Factory Physics approach, and then, the results were studied from lean manufacturing principles. As an outcome of the case approached, it was eventually reached a significant cost of production cost.

Chapter 18 is entitled "Performance Evaluation of a Commercial 3D Printer that Uses Fused Filament Deposition Technology" by Ramos-Lozano et al., which promotes a study on an opportune market demand: the performance evaluation of 3D printing resources. Through a design of experiment approach, Ramos-Lozano et al. applied an FDM—filament deposition technique—procedure to determine parameters for this analysis, such as those related to dimensional accuracy, resulting in a first proposition for an analytical criteria. As 3D printing is expanding continuously, eventually posing as an element of industrial operational planning, this study contributes to initiate a comprehension toward best practices not only for its immediate usage but also for its implementation as parts of optimal new industrial arrangements.

Chapter 19 is entitled "Organizational Systems Convergence with the Industry 4.0 Challenge" by Pérez-Lara et al. They evaluated the issue of Industry 4.0 conceptualization, and although largely announced, debated, and treated by commercial players, Industry 4.0 still needs a stable conceptualization, as this trend is supported by many different technological fronts, emerging resources and solutions, under the market pressures to answer economical competitive demands. In their work, authors, inspired by supply chain principles, propose guidance toward a measurement system, as a fundamental to understand the 4.0 market evidence, conducting to a more predictable comprehension for these phenomena. This recommendation, proposed by the study, was supported by a methodological approach from a literature review, measurement proposition, and a final comparison with supply chain alternatives, allowing it to be developed for application.

Chapter 20 is entitled "Modeling by Finite Element of a Turning Process with Chip Detachment" by Lucero-Velázquez et al. It concludes the third section of our book reporting the application of finite element modeling techniques. After a modeling proposition and correspondent simulation, the results obtained by authors were validated through a real case study, producing a solid case study about finite element analysis, improving its comprehension as a best practice for both management, trend analysis, design, and implementation, with clear application implications.

Part IV: Human Factors, as its name declares, reports four chapter regarding the human factors in production and manufacturing process, a concept or factor forgotten usually, and that can make the difference between the success and failure of the company.

Chapter 21 is entitled "The Role of Knowledge Transfer in Supply Chain Flexibility and Performance" by Díaz-Reza et al. They propose a structural equation model to analyze four latent variables: internal knowledge transfer, external knowledge transfer, supply chain flexibility, and supply chain performance. Those variables are interrelated through three research hypotheses. The model is validated with information from 269 Mexican manufacturing companies. The results indicate that internal knowledge transfer has the most important direct effect on supply chain flexibility and the highest indirect effect on supply chain performance.

Chapter 22 is entitled "Magnitude of Low Back Pain, Occupation, Education, and Economic Level in Mexican Workers" by Prado-León and Rosales-Cinco. They report an investigation to describe the magnitude of low back pain in workers with lumbar spondyloarthrosis and determine occupational and sociodemographic factors related to the disorder. They compare two groups using information obtained by the Nordic Questionnaire. The cases demonstrated that the disorder had presented itself at a young age.

Chapter 23 is entitled "The Knowledge-Based Maintenance: An Approach for Reusing Experiences in Industrial Systems" by López-Ramos et al. They report the importance of maintenance in any production process. They indicate that frequently, the common vision about the maintenance process takes only into account the probability of an event and the risk associated with some expected effects with the goal to reduce the effective cost. They proposed a learning process in the maintenance process for reducing the loss of knowledge and increased the dependency on some experts and presented a knowledge-based maintenance approach that aims to create a framework to manage a maintenance program but also to preserve the valuable experience deployed in the maintenance activities.

Chapter 24 is entitled "Industry 4.0 and Engineering Education: An Analysis of Nine Technological Pillars Inclusion in Higher Educational Curriculum" by Hernández-Muñoz et al. They present an analysis of engineering curriculum and the relation with nine technological pillars of the Industry 4.0, considering the case of four universities in the state of Nuevo Leon, in Mexico. Their results show that the engineering curriculum is contributing directly to those pillars, but some are not entirely defined or missed, such as: Big data, Cloud, Augmented Virtual Reality, Internet of Things and Additive Manufacturing. Hence, they provide some recommendations for public policies and incentives that enhance the revision and actualization of the engineering curriculum to prepare professionals that meet the requirements of the digital era.

Ciudad Juárez, Mexico — Jorge Luis García Alcaraz
Cali, Colombia — Leonardo Rivera Cadavid
Santiago, Chile — Rosa Guadalupe González-Ramírez
Rio de Janeiro, Brazil — George Leal Jamil
Lima, Peru — Mario Gustavo Chong Chong

Acknowledgements

This project is the sum of efforts from many colleagues, friends, and institutions, and therefore, the invited editors wish to thank to:

- The talented technical reviewers who helped review and improve every chapter in this book. The knowledge and enthusiasm they brought to the project was simply amazing.
- The contributors or authors that during one-year support with their research works and suggestions.
- We equally and especially wish to thank Springer International Publishing AG and associate editors for granting us the opportunity to edit this book and providing valuable comments to improve the selection of research works.
- Guest editors are grateful to the Autonomous University of Ciudad Juárez (Mexico), University of El Valle (Colombia), University of Los Andes (Chile), University of El Pacifico (Peru), and Fundação Dom Cabral (Brazil), our affiliation institutions.
- We want to express our gratitude to the Mexican National Council of Science and Technology (CONACYT) for supporting the project named Thematic Network in Industrial Process Optimization, as well as by the Public Education Secretary (SEP).

Contents

Part III Manufacturing and Technology

Part IV Human Factors

List of Figures

List of Tables

Contributors

Mario Acosta-Flores Faculty of Chemical Sciences and Engineering, Autonomous University of the State of Morelos, Cuernavaca, Morelos, Mexico

Julián Israel Aguilar-Duque Faculty of Engineering, Architecture and Design, Autonomous University of Baja California, Ensenada, Baja California, Mexico

Noé G. Alba-Baena Department of Industrial Engineering and Manufacturing, Universidad Autónoma de Ciudad Juárez, Juárez, Chihuahua, Mexico

Giner Alor-Hernández Division of Research and Postgraduate Studies, Tecnologico Nacional de México/Instituto Tecnológico de Orizaba, Orizaba, Veracruz, Mexico

Guillermo Amaya-Parra Faculty of Engineering, Architecture and Design, Autonomous University of Baja California, Ensenada, Baja California, Mexico

Martín Darío Arango-Serna Departamento de Ingeniería de la Organización, Universidad Nacional de Colombia, Medellín, Antioquia, Colombia

Dulce Arroyo-García Department of Logistics and Supply Chain Management, Universidad Popular Autónoma del Estado de Puebla, Puebla, Puebla, Mexico

David Atayde-Campos Department of Industrial Engineering and Manufacturing, Universidad Autónoma de Ciudad Juárez, Juárez, Chihuahua, Mexico

Liliana Avelar-Sosa Department of Industrial Engineering and Manufacturing, Autonomous University of Ciudad Juarez, Ciudad Juárez, Chihuahua, Mexico

Sonia Valeria Avilés-Sacoto Department of Industrial Engineering, Universidad San Francisco de Quito, Quito, Ecuador

Lorena Bearzotti School of Transport Engineering, Pontificia Universidad Católica de Valparaíso, Valparaíso, Chile

María Angeline Bofill-Altamirano Department of Industrial Engineering, Universidad San Francisco de Quito, Quito, Ecuador

Hugo Ferreira Braga Tadeu Fundação Dom Cabral, Belo Horizonte, Brazil

Antonio Juan Brioñes-Peñalver Department of Economy and Business, Universidad Politecnica de Cartagena, Cartagena, Spain

Santiago-Omar Caballero-Morales Universidad Popular Autónoma del Estado de Puebla A.C., Puebla, Mexico

Ismael Canales-Valdiviezo Department of Electrical Engineering and Computational, Universidad Autónoma de Ciudad Juárez, Juárez, Chihuahua, Mexico

Salvatore Cannella Università Degli Studi Di Catania, Catania, Italy

Mario Gustavo Chong Chong Graduate School, Universidad del Pacífico, Jesús María, Lima, Peru

Juan Carlos Cisneros-Ortega Universidad Popular Autónoma del Estado de Puebla, Puebla, Mexico

Leonilda Coba Department of Supply Chain Management, Universidad Del Pacifico, Lima, Peru

Guillermo Cortés-Robles Division of Research and Postgraduate Studies, Tecnologico Nacional de México/Instituto Tecnológico de Orizaba, Orizaba, Veracruz, Mexico

André Luis de Castro Moura Duarte Insper Institute, São Paulo, Brazil

Juan José Delfín-Vázquez Department of Mechanical Engineering, Instituto Tecnológico Superior de Cajeme, Ciudad Obregón, Sonora, Mexico

José Roberto Díaz-Reza Department of Electrical Engineering and Computational, Universidad Autónoma de Ciudad Juárez, Juárez, Chihuahua, Mexico

Eloy Dimas-Celestino División de Ingeniería Industrial, Universidad Politécnica de Valle de México, Tultitlán, Estado de México, Mexico

Francisco J. Estrada-Orantes Department of Industrial Engineering and Manufacturing, Universidad Autónoma de Ciudad Juárez, Juárez, Chihuahua, Mexico

Jorge Luis García Alcaraz Department of Industrial Engineering and Manufacturing, Universidad Autónoma de Ciudad Juárez, Juárez, Chihuahua, Mexico

Flor Araceli García-Castillo Universidad Autónoma de Nuevo León, Facultad de Ciencias Físico Matemáticas, San Nicolás Los Garza, Nuevo León, Mexico

Abimael H. García-Pérez Department of Industrial Engineering and Manufacturing, Universidad Autónoma de Ciudad Juárez, Juárez, Chihuahua, Mexico

Domingo García-Perez de Lema Financial Economics and Accounting, Universidad Politécnica de Cartagena, Cartagena, Spain

Jesús González-Feliu Institut Henri Fayol, Saint-Etienne Cedex, France

Rosa Guadalupe González-Ramírez Faculty of Engineering and Applied Sciences, Universidad de Los Andes, Chile, Santiago, Chile

Lizbeth Habib-Mireles Universidad Autónoma de Nuevo León, Facultad de Ingeniería Mecánica Y Eléctrica. Centro de Investigación E Innovación En Ingeniería Aeronáutica, San Nicolás de Los Garza, Nuevo León, Mexico

Juan Luis Hernández-Arellano Department of Design, Autonomous University of Ciudad Juarez, Ciudad Juárez, Mexico

Salvador Hernández-González Department of Industrial Engineering, Tecnológico Nacional de México en Celaya, Celaya, Guanajuato, Mexico

Guadalupe Maribel Hernández-Muñoz Universidad Autónoma de Nuevo León, Facultad de Ingeniería Mecánica Y Eléctrica. Centro de Investigación E Innovación En Ingeniería Aeronáutica, San Nicolás de Los Garza, Nuevo León, Mexico

George Leal Jamil Fundação Dom Cabral, Belo Horizonte, Brazil

José Alfredo Jiménez-García Department of Industrial Engineering, Tecnológico Nacional de México en Celaya, Celaya, Guanajuato, Mexico

Eusebio Jiménez-López Department of Research and Development, Technological University of Southern Sonora-ULSA NOROESTE-IIMM, Ciudad Obregón, Sonora, Mexico

José Antonio Juanico-Lorán División de Ingeniería Industrial, Universidad Politécnica de Valle de México, Tultitlán, Estado de México, Mexico

Emmanuel Juárez-García Department of Logistics and Supply Chain Management, Universidad Popular Autónoma del Estado de Puebla, Puebla, Puebla, Mexico

Gisela Landa-Gálvez Escuela de Post Grado, Universidad del Pacífico, Jesús María, Lima, Peru

Jorge Carlos León-Anaya División de Ingeniería Industrial, Universidad Politécnica de Valle de México, Tultitlán, Estado de México, Mexico

Mónica López-Campos Industrial Engineering Department, Universidad Tecnológica Federico Santa María, Valparaíso, Chile

Luis Alberto López-Ramos Division of Research and Postgraduate Studies, Tecnologico Nacional de México/Instituto Tecnológico de Orizaba, Orizaba, Veracruz, Mexico

César López-Rosas Graduate School, Universidad del Pacífico, Jesús María, Lima, Peru

Baldomero Lucero-Velázquez Department of Mechanical Engineering, Instituto Tecnológico Superior de Cajeme, Ciudad Obregón, Sonora, Mexico

Diego Fernando Manotas-Duque Grupo de Investigación en Logística y Producción, Universidad del Valle, Colombia, Cali, Valle, Colombia

José Antonio Marmolejo-Saucedo Facultad de Ingeniería, Universidad Panamericana, Ciudad de México, Mexico

José-Luis Martínez-Flores Department of Logistics and Supply Chain Management, Universidad Popular Autónoma del Estado de Puebla, Puebla, Puebla, Mexico

Valeria Martínez-Loya Department of Industrial Engineering and Manufacturing, Universidad Autónoma de Ciudad Juárez, Juárez, Chihuahua, Mexico

Javier Molina-Salazar Department of Industrial Engineering and Manufacturing, Universidad Autónoma de Ciudad Juárez, Juárez, Chihuahua, Mexico

Fernando Montemayor-Ibarra Universidad Autónoma de Nuevo León, Facultad de Ingeniería Mecánica Y Eléctrica. Centro de Investigación E Innovación En Ingeniería Aeronáutica, San Nicolás de Los Garza, Nuevo León, Mexico

Daniel Núñez-Pauca Escuela de Post Grado, Universidad del Pacífico, Jesús María, Lima, Peru

Juan Carlos Osorio-Gómez Grupo de Investigación en Logística y Producción, Universidad del Valle, Colombia, Cali, Valle, Colombia

Roberto Pérez-Franco Centre for Supply Chain and Logistics, Deakin University, Melbourne, Australia

Magdiel Pérez-Lara Department of Logistic and Supply Chain, Universidad Autónoma de Nuevo León, San Nicolás de Los Garza, Nuevo León, Mexico

Alicia Pérez-Pérez Graduate School, Universidad del Pacífico, Jesús María, Lima, Peru

Lilia Roselia Prado-León University of Guadalajara, Guadalajara, Jalisco, Mexico

Vivian Rambarath-Parasram Centre for Maritime and Ocean Studies, The University of Trinidad and Tobago, Chaguaramas, Trinidad and Tobago

Ricardo Ramírez-Tapia Department of Industrial Engineering, Tecnológico Nacional de México en Celaya, Celaya, Guanajuato, Mexico

Secundino Ramos-Lozano Department of Industrial Engineering and Manufacturing, Universidad Autónoma de Ciudad Juárez, Juárez, Chihuahua, Mexico

Lázaro Rico-Pérez Department of Industrial Engineering and Manufacturing, Universidad Autónoma de Ciudad Juárez, Juárez, Chihuahua, Mexico

Eduardo Roldán-Reyes Division of Research and Postgraduate Studies, Tecnologico Nacional de México/Instituto Tecnológico de Orizaba, Orizaba, Veracruz, Mexico

Rosa Amelia Rosales-Cinco University of Guadalajara, Guadalajara, Jalisco, Mexico

José Efrén Ruelas-Ruíz Department of Mechanical Engineering, Instituto Tecnológico Superior de Cajeme, Ciudad Obregón, Sonora, Mexico

Tomás Eloy Salais-Fierro Department of Logistic and Supply Chain, Universidad Autónoma de Nuevo León, San Nicolás de Los Garza, Nuevo León, Mexico

Víctor Jaime Saldarriaga-Romero Escuela de Negocios, Institución Universitaria CEIPA, Sabaneta, Antioquia, Colombia

Diana Sánchez-Partida Department of Logistics and Supply Chain Management, Universidad Popular Autónoma del Estado de Puebla, Puebla, Puebla, Mexico

Cuauhtémoc Sánchez-Ramírez Division of Research and Postgraduate Studies, Tecnologico Nacional de México/Instituto Tecnológico de Orizaba, Orizaba, Veracruz, Mexico

María-del-Rosario Sánchez-Vega Universidad Popular Autónoma del Estado de Puebla A.C., Puebla, Mexico

William Sarache Facultad de Ingeniería y Arquitectura, Departamento de Ingeniería Industrial, Universidad Nacional de Colombia, Manizales, Colombia

Jania Astrid Saucedo-Martínez Department of Logistic and Supply Chain, Universidad Autónoma de Nuevo León, San Nicolás de Los Garza, Nuevo León, Mexico

Cesar Shimabuku Department of Supply Chain Management, Universidad Del Pacifico, Lima, Peru

Sukhjit Singh Centre for Maritime and Ocean Studies, The University of Trinidad and Tobago, Chaguaramas, Trinidad and Tobago

Ulises Jesús Tamayo-Pérez Faculty of Engineering, Architecture and Design, Autonomous University of Baja California, Ensenada, Baja California, Mexico

Cezar Taurion I2a2 (Applied Artificial Intelligence Institute), Brasilia, Brazil; IBM, Rio de Janeiro, Brazil; PWC, Rio de Janeiro, Brazil

José Claudio Villamonte-Cornejo Graduate School, Universidad del Pacífico, Jesús María, Lima, Peru

Jorge A. Vivares Facultad de Ingeniería y Arquitectura, Departamento de Ingeniería Industrial, Universidad Nacional de Colombia, Manizales, Colombia

Julian Andres Zapata-Cortes Escuela de Negocios, Institución Universitaria CEIPA, Sabaneta, Antioquia, Colombia

Part I
Techniques, Tools and Methodologies

Chapter 1
Digital Transformation: Digital Maturity Applied to Study Brazilian Perspective for Industry 4.0

Hugo Ferreira Braga Tadeu, André Luis de Castro Moura Duarte, Cezar Taurion and George Leal Jamil

Abstract This paper aims to analyze digital transformation in a Brazilian perspective, aiming to understand it for Industry 4.0. Starting from academic researches, from Brazilian companies, a model based on dimensions to define a construct—digital maturity—was suggested, allowing a definite starting point to analyze it for industry 4.0 future scenario. To strengthen these definitions, results of a research about digitalization efforts in Brazilian companies, held by the innovation and entrepreneurship research group of Fundação Dom Cabral, a major Brazilian player on executive education, is presented, where opportunities and limitations for digital transformation in Brazilian companies were analyzed.

Keywords Digital transformation · Digital maturity · Brazilian organizations

H. F. Braga Tadeu · G. L. Jamil (✉)
Fundação Dom Cabral, Belo Horizonte, Brazil
e-mail: gljamil@gmail.com

H. F. Braga Tadeu
e-mail: hugo.tadeu@fdc.org.br

A. L. de Castro Moura Duarte
Insper Institute, São Paulo, Brazil

C. Taurion
I2a2 (Applied Artificial Intelligence Institute), Brasilia, Brazil

C. Taurion
IBM, Rio de Janeiro, Brazil

C. Taurion
PWC, Rio de Janeiro, Brazil

J. L. García Alcaraz et al. (eds.), *Best Practices in Manufacturing Processes*,
https://doi.org/10.1007/978-3-319-99190-0_1

1.1 Introduction

Organizational transformation need is a challenge for managers and always called their attention and interest in various levels of corporative decisions. Transformation abilities allow that organizations grow and obtain sustainable competitive advantages. Those which show better conditions in transforming themselves, generally overcome those which do not develop this dynamic competence (Karimi and Walter 2015).

External factors can ignite organizational transformation like emerging technologies or market changes. But these changes can also start from the inside its borders, by an internal restructuration or strategic decision. Although transformation is not a new topic in management studies, recent advances in new digital technologies are demanding more studies, to clarify essential aspects for organizational survival in markets progressively more dynamic and intensive on new these technologies potentialities (Andal-Ancion et al. 2003). Facing these new digital era challenges, managers of different sectors shall develop abilities to transform their organizations.

Digital transformation concept arises as a propeller of the so called "Industrial fourth revolution" (or Industry 4.0). From various opportunities, related to advances of those technologies, it is possible to automatize processes, increase operational efficiency and obtain progressive gains of productivity. Moreover, it is possible to develop more proximity with customers, producing new experiences in this relationship. These actions are supported by systems for data and information intelligent capture, collection, management, and analysis. The ability to develop substantially innovative and customized offers produce an opportunity for business model reviews for these organizations.

Digital transformation (DT) becomes a new kind of change, one theme to be researched, as it imposes new challenges for markets and organizations (Matt et al. 2015). However, DT shows complex and varied perspectives, which make this a tough work, even defiant and passionate, to allow new ways towards the growth.

Digital transformation provocations are more difficult in developing countries, with dynamic and unstable political and economic environments, such in Brazil, deserving a specific observation. This way, this study was proposed with the central objective to provide results for managers enabling better comprehension of digital transformation process and perspectives, in its various dimensions. With this purpose, initially, a wide theoretical review about DT was conducted followed by one survey with 246 managers of competitive Brazilian companies, from several sectors, to produce an understanding about limitations and risks they face towards digital transformation.

1.2 Industry 4.0

Before approaching the main research model, adopted for study development, two important conceptual contexts must be, at least in an initial way, discussed to level intended basis: Industry 4.0 and innovation.

Industry 4.0 became one of the main buzzwords adopted in the world, because, mainly, it represents the totality of innovative propositions regarding industrial processes and their integration inside manufacturing environments and external connections, through the implementation of new technologies that can promote production optimization and fast adjustments to several business models.

Several authors describe Industry 4.0 in terms of expressive information technology introduction to implement, optimize, and redesign industrial processes (Maier 2017; Schwab 2017; Ford 2018). Citations from these works, for instance, approach Internet of things, big data, communications resources (such as cloud services, RFID tags), automation infrastructure and several other active and passive components. Arranged in configurations, these machinery-oriented arrangements, theoretically, can optimize production indexes, cost management, process monitoring, redefinition, and supervision. There are many opportunities from these and other features to an initial provocation for complete understanding:

- "Industry" is to be restricted just to industrial focus, industrial origin, to comprehend this massive evolutionary trend?
- What are the main reasons, in a stable consideration, to understand this as different from the previous degrees of industrial evolution (from 0 to 1.0, 1.0 to 2.0 and so for)?
- Is there any kind of paradigmatic definition already being executed by main competitive players of these markets (potentially establishing some type of best practices)?
- What are expected impacts of this new technological development?
- Among many others.

This section aims to discuss Industry 4.0 to address these main topics.

First, a comprehension of the context. Why 4.0? What really produced this new scenario and its related propositions? As pointed out by West (2017), industrial four main phases were characterized by:

- Industry 1.0 was related to water steam-based engines and its application in replacement of hand production, enabling the configuration and building to what became known as "the factory". This happened about the decade of 1780.
- Industry 2.0 phase remarkably happened based on two main factors and opportunities: (1) Electricity was introduced as a type of energy, and (2) Labor was organized, starting the association with managerial practices and propositions. The decade of 1850 shown these events and, for many, this period, from 1780 to 1850, is regarded as the First Industrial revolution.
- The third phase, of what can be cited as Industry 3.0, was referred to the final moments of the 1960 decade and the beginning of the 1970s, where automation

started to be approached in industrial plants. Solutions for this effort were based on machines, nowadays considered simple, that simply were configured to repeat human interventions in production lines.

- Finally, we reach the considered 4.0 era, where, cited by the author, we face the dissemination of "cyberphysical systems" with potential manufacturing "democratization" and mass customization.

Following the main conception for these definitions, it is opportune to infer that both—technological devices implementation and business models associated with their usage for industrial strategies—present a remarkable potential of change of the previous situation. As it can be seen from WEF (2018), a potential trend for industry 4.0 reached the "digital transformation" (DT) status, signaling exactly the wave of potential changes that reached not only industrial installations but their productive chains, implicating in severe modifications, from basic and repetitive tasks, to possible automation (decided by sophisticated machinery with fair connectivity) of decision-making in production lines. Digital transformation can be considered, this way, a top-level case, where industry 4.0 encompasses several alternatives, for a variety of depth, towards its implementations (Bleicher and Stanley 2016).

An attempt usually approached by authors is to appreciate how strategies, tactics, and operational planning activities were impacted by this industrial evolution. Along with intensive operational jobs creation, during the expansion of new factories, regarding the start of the manufacturing era, new urbanization was demanded, or even forced, creating several implications for the years to come. Organizational impacts were remarkable, as planning abilities, communication, coordination, human resources training, development and remuneration were introduced as factors of administration. Works from Frederick Taylor and Jules Henri Fayol, both in the end of Nineteenth to the beginning of the Twentieth Century, focused on statistically based control of production equipment, mainly human operational taskforce management. Also, the introduction of electricity produced a scenario where additional services were demanded by industries, resulting in a situation of implementation of facilities, like powerhouses, dams, transmission lines and complex electricity supply management. These facts brought so impact, reaching the level of transforming and permanently affecting ecosystems, with huge engineering projects, showing typical undeniable and remarkable signals to what is called as the "second industrial revolution".

The "third" wave introduced controllers and programmable features in the industrial plant, allowing a progressive automation. Here, ideas previously discussed of "artificial intelligence" started to be thought in the workplace, however demanding computing resources—processing, storage and communication—not available at that time. Automation spread as a solution to automatize what human operators did, producing the first wave of replacement of workers by machines, today considered a limited aspect for this process.

New methods, tools, and approaches for management were studied, proposed, and practiced, introducing a modern-fashion for productive arrangements. But, on the contrary to an unthinkable path of expansion, economic, political, and social crisis posed several obstacles to industrial expansions, composing a difficult picture to formulate and propose strategies in this context (Utterback 1971; Ansoff 1990; Porter 2008; Mintzberg et al. 2008). With these new demands for more strategic thinking around industrial complexes and its competitive unfolding, together with movements by their productive chains, a new wave of management studies was produced, resulting in a dynamic arena where organizations are still being thought nowadays.

This superficial, brief approach to what happened with industry in the previous "revolutions", or remarkable evolutionary steps and phases, illustrate what is reasonable to expect from the actual panorama, as changes are being introduced in a fast pace, with emerging technologies, new economic arrangements, social and political processes and dynamics and, increased competition.

With the new step, proposed as Industry 4.0, it is possible to reach the complete automation of a plant, with potentials such as inter-related interventions, two-way communications, risk management and auto-correction and intervention systems (Celaschi 2017; Colquhoun 2017). Undoubtfully, this topic—industry 4.0—still demand studies, as the one here promoted, to understand its implications, but offering real potential on appreciating multidisciplinary contributions, exactly performing the desired associations with other fields and deepening the conceptual understanding. For instance, in Celaschi (2017), it is possible to comprehend both the view of several conceptualizations for this new technological wave and the associated managerial thunderstorm, by the view of Design field, adding a contributive point of view. For him, relations such as "business-to-business" and "business-to-customer" decisions and production indexes must be approached together with the consideration that we face digital intangible resources to improve (supposing its successful implementations) the sometimes classical, traditional problems that arise in the tangible manufacturing environment. This set of solutions, with its related descriptions for processes and decisions—management indeed—results in the Industry 4.0 focus.

From Bishop (2018) and Ford (2018), we learn several trends of new apparatus, implements, interconnected components that can serve for Industry 4.0. Some of these features, according to these authors, are: Collaborative robots (not only automated, but with connections and communications services that allow inter programming, among other features), cloud storage application (as to store data, information and knowledge regarding production-level events and their associated decisions), big data and analytics servers and interpreters (allowing its fast adjustment, management and some degree of potential autonomy by interconnected machines, with decision-making based on facts) and, mainly, Internet of things.

Celaschi (2017) developed a picture about management, social and customers-side issues regarding Industry 4.0 introduction, bringing special focus to important facts, also not related to strict technological development, but to its usage and implications. For this author, industrial "revolution" takes place in function of information, reaching the possibility of continuous changes to product supplies, according to customers' reaction and its processing through analytics and big data resources.

According to Bishop (2018), supporting ideas already found in the works of Maier (2017), resistances, uncertainties, fear of risk and need to face tough systems adaptation are among the main problems to be faced by interested investors, technical staff, and managers who intend to develop and deliver Industry 4.0. Lack of investments follow-up techniques, communication principles, internal (aligned to overall strategy) marketing actions, towards people integration and integrated planning actions are to be considered in this delicate, but irreversible path.

Internet of things deserve a specific observation as, for some authors, it is considered the automation source or even Industry 4.0 instance, by itself. Chui et al. (2010) classify Internet of things, or IoT, as the myriad of devices and related installations which can communicate—commanded or autonomously—via Internet services and protocols. Not only humans, but also “things”, like our domestic apparatus, car components, usual physical artifacts which we deal informally, like bottles for any kind of beverage, packaging material, etc., can connect and communicate through Internet. This is one of the main platforms for Industry 4.0 implementations—Boyd (2017)—allowing to design compositions, from basic daily usage devices to complex integrated manufacturing machinery. This author remarks the comparison between the traditional and advanced, modern manufacturing, just observing how IoT is installed, configured and managed in modern installations. IoT also enables an opportune integration, which is the organizational network that implements a potential array of productive plants (maybe not owned by the same corporation, defining a scheme for its inter-connection) that can, for example, in one relevant managerial (maybe automated) decision, distribute production over several different plants, dynamically, optimizing production demands and loads.

For Kocsi and Oláh (2017), industry 4.0 is defined as a new proposal where machines are interconnected as a “collaborative community”, making it possible to automatically receive data and information from any environment—internal or external—of a manufacturing context. This allows its integration to processes like productivity monitoring, control, prediction and adaptation to market demands. From these authors’ study, it is possible to infer changes about technology introduction and its impacts on management for organizations.

As seen from these sources, the industrial revolution, by several ways, arrived at the market as a real phenomenon, with implications still unknown. Industry 4.0 became a new paradigm for several marketing, commercial and strategic actions which can develop industrial sectors and almost everything that surround them, as we live in a globalized economy, producing substantive changes even for services, governmental and other places.

Some reflections, presumptions and even afflictions are approached regarding this fast-moving, powerful event, to be addressed in a specific further study, which can consider this paper as a starting point. For example, contextual limitations of Industry 4.0 phenomena—if it is strict to one type of industrial complex or it is spread for any kind of industrial park (Serban 2018); does it demand any special training, skills development and change learning plans and platforms of actual organizational installations (Saturno et al. 2018; Meyer and Han 2017; Rose et al. 2017)? How new ways to offer and negotiate value can be considered in productive chains (Roblek et al. 2017)? There is opportunity and what is the real chance to produce new business models in Industry 4.0 contexts (Casadesus-Masanell and Ricart 2010; Zott et al. 2011; Gatautis 2017)? What is the impact of new emerging technologies or implementations for Industry 4.0—is there any opportunity for 5.0, 6.0, etc., waves, based on artificial intelligence evolution?

These trends, recommended to be addressed in future studies, are shown here to complete a picture in what Industry 4.0 can be considered in modern times.

All these events and complex, dynamic scenario, produce the precise context where we propose digital maturity concept, stated in its analytical dimensions, followed by an explanatory research, testing its formulation, to become a set of parameters of analysis on how one organization, sector or even a region can become a candidate for Industry 4.0 implementation or improvement, in the case it is partially or consistently adopted. With digital maturity notion, one can understand how digital transformation, an associated concept for Industry 4.0, has chances to be perceived as a concrete solution for organizations.

1.3 From Digital Transformation to Digital Maturity

The emergence and consequent adoption of new digital technologies presented a new way of organizational transformation: digital (Hanna 2016). Digital transformation (DT) became a quite popular expression in the last years, also among managers, consultants and related professionals.

This immense interest, however, did not produce a consensual conceptual definition for DT. If, for some practitioners, the simple implementation of a digital technology implement, like an Enterprise Resource Planning (ERP) system can be regarded as an initiative of digital transformation, for others DT must be considered an evolutionary process, that takes time to be developed and implemented (Wang et al. 2016). While researchers relate digital transformation to new business models and corporative strategy, others consider it as an organizational process to be fully planned and implemented (Berman and Marshall 2014).

This way, there are various definitions of digital transformation in the literature, as the consensus is still in sight, not completely reached. Some authors define it as a development of information technology (IT) strategy, as it unfolds to all areas of one company (Mithas et al. 2013). Others consider it as the insertion of new digital technologies aiming, for the main goal, to improve performance of organizational processes, even reaching the level of a radical innovative proposition (Liu et al. 2011; Fitzgerald 2013; Piccinini et al. 2015). Analyzing in a broader view, other authors believe that DT is related to the inclusion of new technologies with the objective to change or create new business models (Lucas et al. 2013; Henriette et al. 2015; Schuchmann and Seufert 2015; Hess et al. 2016), redefining products and associated product lines and services, processes, relationships with external suppliers and customer engagement.

For Kane et al. (2015, 2017), the idea of digital transformation must be criticized. For these authors, business companies are following a process of digital maturity (DM), understood like the ability of one organization to react consistently to changes in business environments, progressively more influenced by the application of technologies. These reactions are usually learned, not becoming an instinctive, impulsive event, according to the same research work.

Still considering Kane et al. (2015), some factors compose the digital maturity level of one organization:

- It is a gradual process, which must be shared to all organization context, along a significant period. Therefore, one organization does not become digitally mature from one day to another. There are different and progressive stages of DM and this process is never complete.
- Analog to unpredictability related to development processes, like "children becoming adults", organizations do not have a complete idea on what they will be as they evolve through digital maturity levels. One organization can have a clearer picture as it moves forward, developing through subsequent evolutive phases of maturity.
- Digital maturity is not an automatic or spontaneous process. It must be regarded as a process of continuous learning, when the organization learns to coherently react to new competitive digital environment. Managers shall develop practical knowledge about digital trends, enabling them to adapt their companies in a planned, adequate way.

One aspect about digital transformation planning is to consider that an expressive need for its development occurs outside organizational boundaries, not strictly under its control, leading to a reaction or adaptation to customers, business partners, employees and competitive pressure demands, which somehow impose DT implementations. Beyond that, installing a new digital technology is just a small part of the change intended for a more complex project. Other aspects, such as strategic factors, plans and goals, human resources management, organizational structure, competitiveness and leadership are important, compared to technologic development.

Digital maturity (DM) defines an evolutionary process, which looks more reasonable for companies which need to face changes along some time, as they were not created as digital ones. Moreover, managers should modify a temporary business view focused on immediate change, considered radical, to evaluate digital maturity, regarding it as a transformation which is endless, taking expressive time to show perceivable results, being a constant adaptation to a new competitive scenario, increasingly digital.

1.4 Digital Maturity Model Dimensions

A maturity model consists in dimensions and associated criteria that defines the way to improve, to achieve the desired maturity level. Maturity models are tools that allow to check the actual level in which the examinee—an organization, a process, etc—is at this moment and how it must develop its evolution, defined by indicators, parameters and planning guidance (Becker et al. 2009). More than ten digital maturity models were developed in the last decade, by academic and science people and institutions and consulting companies. Each one of these models were proposed and adapted for one specific sector or company type (Berghaus and Back 2016). Some focus technological factors, others overlook the organizational composition, considering the complete corporative context, and, finally, there are some which concentrate in one internal aspect or competence or eventually just evaluate intended answers from digital transformation results, as the process is considered ended.

It is challenging to choose one model as the best option, from all these propositions, considering it the completely adapted or suitable to implement for one specific company, as a process model (Mullaly 2014). Proposing a new model, on the other hand, is likely a repetitive definition, eventually not reaching the desirable level, intended for practical applications. Taking these options into consideration, after a general evaluation, this work main objective was defined: to identify the main dimensions to be tackled inside organizations, allowing them to pursue digital maturity.

According to this comparative analysis, ten dimensions are conceived to propose the digital maturity model, to be exercised in a research that will be presented in the following, producing an opportune level to understand digitalization for a consistent digital transformation in Industry 4.0 applications:

1. *Digital Strategy*: Definition of wide plan, encompassing organizational characteristics, issues and specific digital technology-based goals. This is considered a fundamental factor for digital maturity evolution. Various authors cite strategy, and not technology, as the main source of digital transformation (Matt et al. 2015; Kane et al. 2015). Organizations which show lower levels of digital maturity usually do not have digital strategy definitions and tend to concentrate in few or impulsive technology isolated initiatives, frequently towards operational results. Otherwise, those which present a higher level of maturity planned digital strategies aligned with their main business transformation.
2. *Digital Technologies*: Even not regarded as the main components for digital maturity, new technologies are considered as basic elements for this process. Internet of things (IoT), Artificial intelligence, Big Data, mobile devices and its connection resources, 3D printers, intelligent sensors, augmented reality, cloud computing are among the set of emerging technologies, commercially offered in the last years, potentially producing big transformations in business models of several companies. However, organizational value is not strictly related to technology, otherwise to the way it is proposed and applied to produce value for the organization and its *stakeholders*. This technology factor is related to

organizational attitude on adopting these new technologies and its ability on exploring it as innovative resources.

3. *Analytical and predictive abilities*: adding to the former item, it is important for digital transformation that a company develops a consistent predictive ability. Digital transformation richness is not on the huge amount of data generated and put available, otherwise in the ability that will be applied to treat this data, information production from it, allowing managers to improve their decision-making choices. In a previous technical report, done by author Tadeu argued about the needed conditions to innovate, calling attention the fact that organizations should develop their abilities to manage their datasets, capturing and producing the correspondent information in a managed way, for effective and faster decisions, beyond other conditions to produce innovations. One organization, considered digitally mature, must have the ability to predict and adapt rapidly to new market situations.
4. *Customer relationships*: This dimension encompass all aspects that refer to customer relationships—from pre-sales to pos-sales. New digital technologies advances provoked additional relevance to this theme, for organizational competitiveness. New technologies allow increasingly perception about customers′ needs and position customized offers, with result in richer experiences and closer market relationship. Agility on problem solutions and better-quality levels of interactions, achieved and improved by digital means, are determinant for digital transformation. Customer centricity became a challenge for all corporative sectors, demanding a higher level of digital maturity.
5. *Network relationship*: A digitally mature organization needs to understand itself as an element of a network, which involves suppliers, startups, governments, universities, investors, among others. New digital technologies allow better interaction and integration among these players, with faster and more transparent and precise communication. Global productive and value-aggregated chains can be developed with increased degree of responsiveness and efficiency, becoming a real aspect of digital maturity. Network connectivity, data access and analytical methods application—for the entire network—allow risk reduction and performance gains.
6. *Organizational structure and digital processes*: this dimension seek to evaluate the organizational structure, from an internal point of view and its related processes. These processes consist in a series of related and connected activities, which demand personnel and departmental interactions. Going further from the merely automation for these and other processes, organizations digitally mature share internal and external resources, aiming to consistently reduce their functional areas isolation (sometimes considered "silos"). Processes automation, associated with resources such as artificial intelligence, turn decision-making faster, agile and assertive. Internal processes with flexible structures can increase the focus on customers relations, improving the value perception. Frequently, job positions, teams' compositions and key process indicators, along with benefits and rewards policies must be also reviewed, adopting this faster decision-making managerial context.

7. *People and culture*: According to a research by Capgemini (2017), culture is the biggest difficulty one organization faces when attempting a digital transformation. Culture refers to the importance attributed to attitudes and behaviors for organizational performance. Corporations with higher level of digital maturity tend to develop a more collaborative culture, opened to innovations and customer focused, also using data analysis to improve decisions (referred as data-driven decision-making). People formation is essential for digital maturity. Organizations that matured over time present an increased probability, by a factor of four times, to proportionate to employees those needed digital abilities than others with lowest maturity levels (Kane et al. 2015). Moreover, the capacity on evaluate how digital technology can impact business is an expertise not dominated by several corporations when they are in the early stages of digital maturity. Thus, digital fluency does not demand technological domain, otherwise requires the ability on articulate digital technologies for organizational future (Kane et al. 2015). Leadership role is vital in the creation and strengthening for a digital culture, as to promote the focus and digital mindset of all employees and related agents.
8. *Risks and investments*: investments bring risks and risk taking is part of mature organizations culture. Leaders need to face the associated risk level when considering technology introduction and advances, which means to accept and understand failures and development of innovation-friendly teams. Traditionally, organizations focus their policies on investments based on regular, classical financial models, based on metrics like return on investment, net present value and internal return ratio. For a digital transformation approach, is recommended an estimate of value referring to final offers, being products, services or new business, considering its "valuation" indexes. In general, this value can be estimated comparing the present and future value of cash flow. Organizations which adopt disruptive technologies and manage to build entry barriers to be faced by new competitors, may present a superior level for future value and an increased predisposition for investments in new projects.
9. *Legal and ethical aspects*: Ethical and legal aspects must be considered and faced when thinking about new technologies application, becoming a significant worry for a digitally mature enterprise. Anticipated identification of legal issues which are impacted by new digital technologies can be an important corporative source for competitive advantage. Contract models must be modified, frequently with risk and/or reward sharing with other business partners. Another important question is the customer and business agents' data protection, against digital hacking and other criminal actions. It is mandatory, for one digital mature organization, to effectively address these risks in a competition increasingly digital.
10. *New digital business models*: Supported by almost all previous events, digitally mature organizations can create new competitive business models. Mature

enterprises tend to expand their offer of products and services through the application of digital integrated platforms. Organizations that negotiate digital products and services can offer complete solutions for their customers, based on an integrated digital ecosystem.

1.5 A Brazilian Picture for Digitalization: Research with Potential Industry 4.0 Agents

From this approach, it is possible to verify the need of understanding about the difference from traditional business culture to the new emerging technologies. Figure 1.1 present these dimensions and the relationship among them. It is considered that digital maturity is the result of the process related to those ten dimensions. Defining a digital strategy, setting specific long-term goals to achieve, can be the first step and relevant driver for these model dimensions. Customer, networks, organizational structure, processes, culture, people, financial risks, legal and ethical aspects must be taken into consideration, with variable degree of significance, depending on the business which the organization competes or develop its actions. New business models, frequently, result from

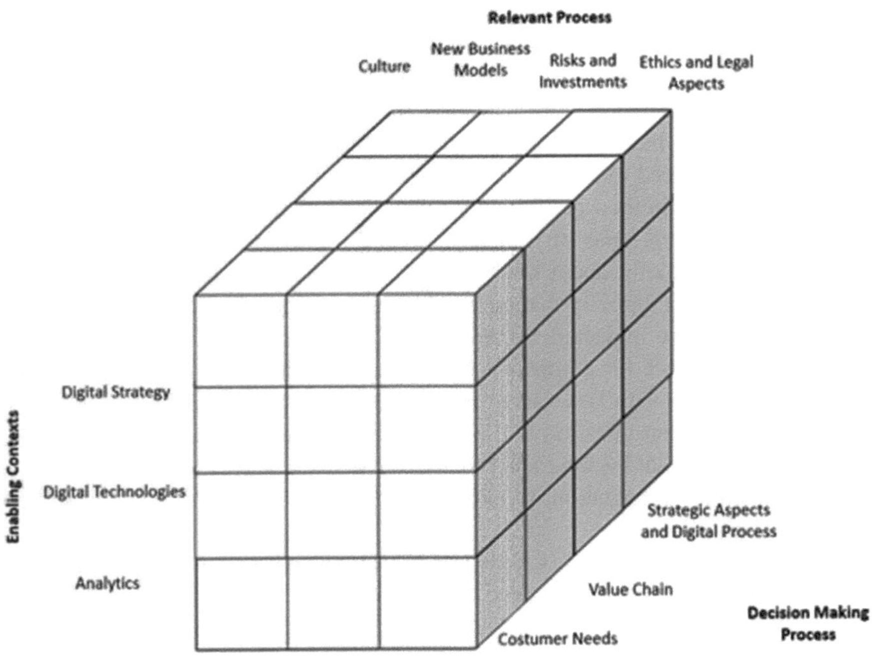

Fig. 1.1 Structure of dimensions for digital maturity analysis FDC (2015)

combined plans from some or several of those dimensions, presented in the previous description. By its turn, technology and analytical abilities dimensions are strong facilitators and tools for a process of digital maturity. Therefore, digital maturity must be evaluated from organizational processes maturity and not exclusively as a simple isolated result.

From this approach, it is possible to verify the need of understanding about the difference from traditional business culture to the new emerging technologies. Figure 1.1 presents these dimensions and the relationship among them. It is considered that digital maturity is the result of the process related to those ten dimensions. Defining a digital strategy, setting specific long-term goals to achieve, can be the first step and relevant driver for these model dimensions. Customer, networks, organizational structure, processes, culture, people, financial risks, legal and ethical aspects must be taken into consideration, with variable degree of significance, depending on the business which the organization competes or develop its actions. New business models, frequently, result from combined plans from some or several of those dimensions, presented in the previous description. By its turn, technology and analytical abilities dimensions are strong facilitators and tools for a process of digital maturity. Therefore, digital maturity must be evaluated from organizational processes maturity and not exclusively as a simple isolated result.

A risk for organizational long-term survival emerges from the application of disruptive technologies, posing problems for management practices. As seen in the previous discussion, to propose new digitally supported business models, it is necessary to emulate an open, autonomous culture, monitor risks and tolerate failures, learning also from mistakes. At the same time, it is important to create more flexible processes and try to engage people towards the essential organizational goals. Resources must not only be strictly allocated, otherwise aligned to answer the major strategic propositions, aiming relevance and the predicted view for the future. Finally, new digital business models must focus on innovation, leadership and talents management, true digital assets (Westerman et al. 2014).

A robust multiple case analysis about digital transformation was surveyed in Brazil, consolidating several studies conducted by the Innovation and Entrepreneurship research group of Fundação Dom Cabral, a major executive educational and preparation institute in Latin America. One of these studies, "Digitalização no Brasil" (Digitalization in Brazil) was updated, in a partnership with a German research institute. The main goal of this study was to evaluate the digital transformation level in several companies, in a planned, arbitrary sample, approaching questions like industrial automation, data management, innovation and investments. One questionnaire, composed from literature sources and examined by a group (or "college") of experts was composed and applied, through Internet web support, for a set of 246 managers who act in the top-level management—directive, board, committees, etc.—in these big Brazilian cases.

1.5.1 Research Results

This study revealed that, for the most of those interviewed companies, digitalization is considered just the change from analog, manual or interventive processes to versions where they are supported by digital technologies. These findings were followed by themes such as automation, data management, processes efficiency, among others.

In the following, the research results are shown and briefly discussed:

The graph shown above, Fig. 1.2, shows the predominant relation of digitalization to a limited perception of merely changing something from analog to digital infrastructure. Processes and service offering automation and data management were among the subsequent choices made by interviewees (Fig. 1.3).

More than 90% of the interviewees considered interface connection as the most representative conceptualization for digitalization. This connection refers to the communication of machines, productive chain, and supply management systems, etc. This option had almost the same relevance as resources optimization—when digitalization was regarded as the optimal replacement of physical resources by digital machines and devices—and automated manufacturing. See Fig. 1.4.

When asked about the existence of a digital strategy in his/her organization, most of the interviewees answered positively. This result can indicate that new knowledge regarding DT is being searched, to access cutting-edge technologies which allow to offer innovations to the market. See Fig. 1.5.

Answers concentrated on information technology and production departments or units, show the evidence of the search for technologic knowledge support for processes and operational alignment to strategy. See Fig. 1.6.

These results show the search for systems control, automation, process digitalization and data management are the most representatives for its organizational strategies. An important evidence that presents organizational interests on

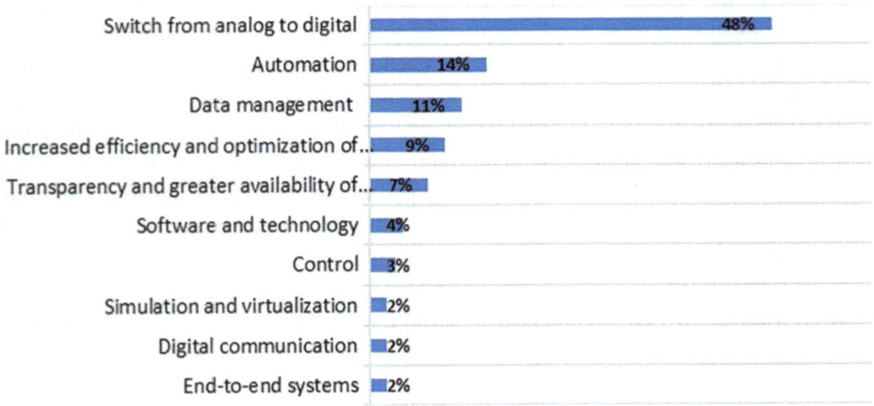

Fig. 1.2 What do you understand by digitalization? FDC (2015)

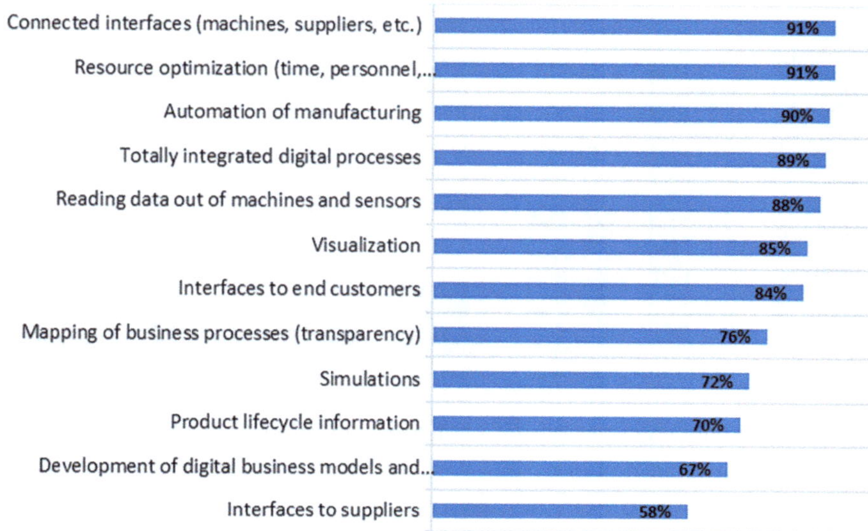

Fig. 1.3 From the following expressions, what is the most representative for digitalization concept? FDC (2015)

Fig. 1.4 Is there a digital strategy in your organization? FDC (2015)

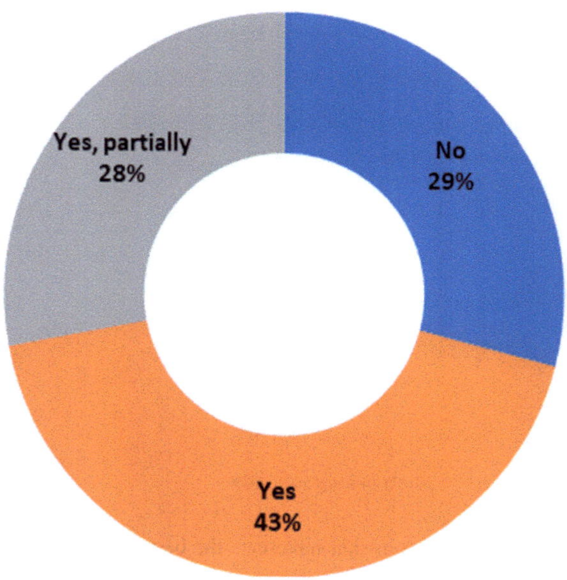

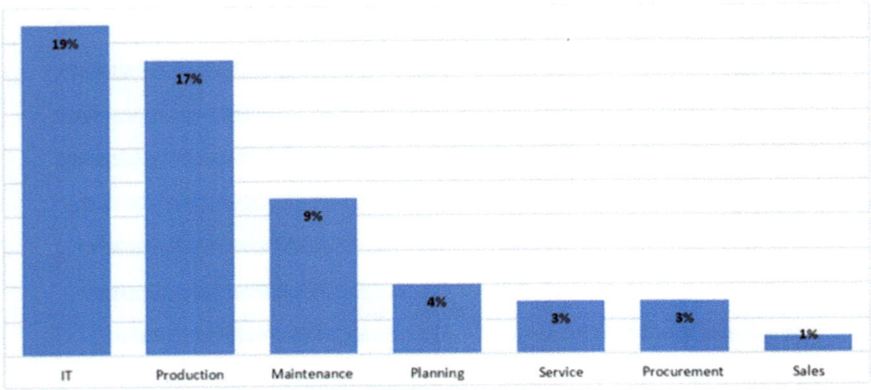

Fig. 1.5 What are the organizational areas which have a strategy to develop digitalization? FDC (2015)

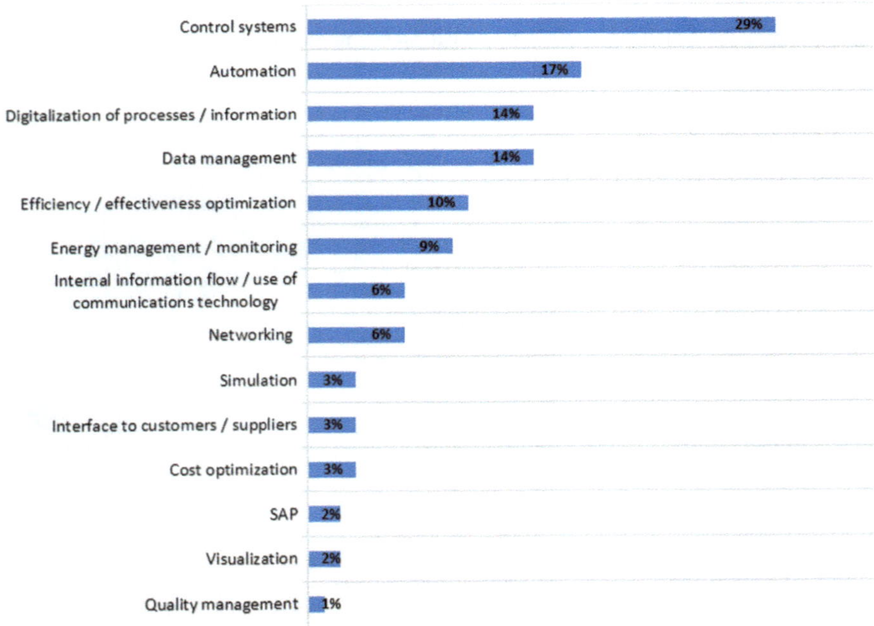

Fig. 1.6 What expression represents the best description for your digitalization strategy? FDC (2015)

productivity gains for their processes. However, it can also point a delay on automation investment, a signal which can be related to an undeveloped Economy situation, differently from others, at a higher level of frontier technologies adoption. See Fig. 1.7.

Fig. 1.7 Has a process analysis or a feasibility study been already performed? FDC (2015)

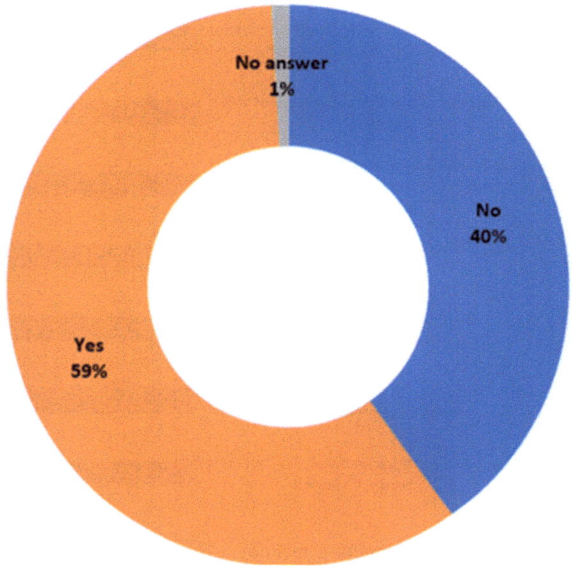

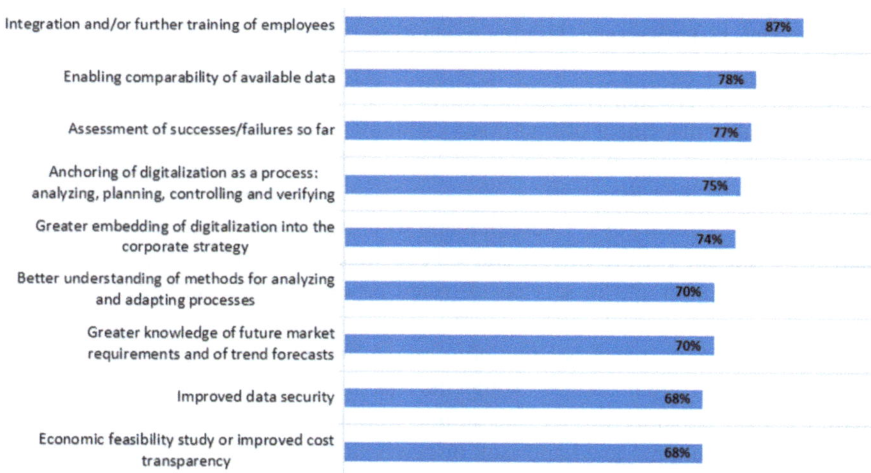

Fig. 1.8 What could be done, or is needed to advance digitalization in your organization? FDC (2015)

There is an evidence about the sustainability of DT in the companies' sample, supported by feasibility studies or process analysis. These two themes, so, define a fair situation for digital transformation in those companies. See Fig. 1.8.

Actions perceived as necessary to allow digitalization continuous improvement in organizations, towards better personnel integration and preparation were prioritized by interviewees, succeeded by standardization of data comparison (increase

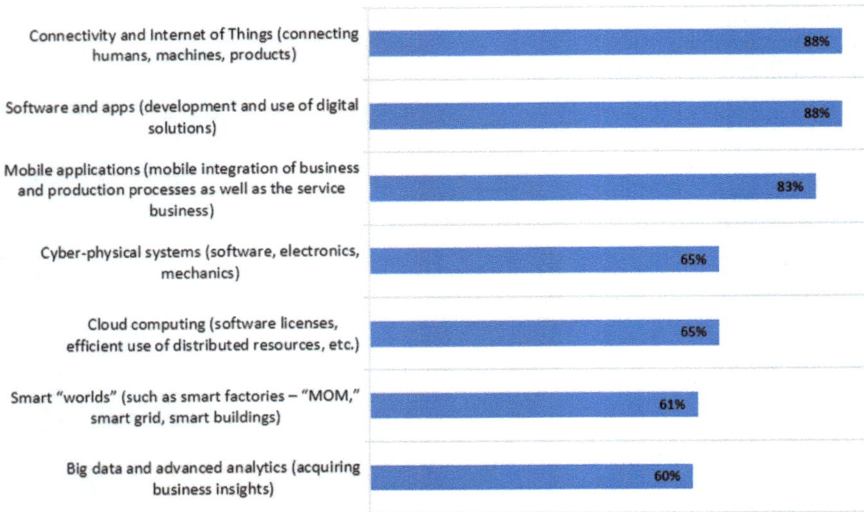

Fig. 1.9 Choice, from these technological trends, the most important that could be applied in your business environment FDC (2015)

on analytical expertise) and failures evaluation, probably linked to risk management. See Fig. 1.9.

Machine connections, systems application, mobile infrastructure and cybernetic (technological replacement of traditional) systems were the most cited evidences. It is a signal that intelligent systems must be considered to manage digital transformation practices. See Fig. 1.10.

Referring to benefits of DT adoption and its related impacts, the answers suggest that improved application of resources, better decision-making, improvements in energy consumption and in services-oriented processes were considered the most appreciable. See Fig. 1.11.

At the same time when benefits are noticed by interviewees, regarding digital transformation in organizations, some internal barriers (difficulties) are verified. Organizational structure (lack of flexibility), culture factors, high costs of investment in the operational level and lack of perception of its returns were signaled in the research as the main initial internal problems in DT adoption process. See Fig. 1.12.

For external barriers to improve the application of digital technologies and processes in organizations, the research shown that data safety procedures (note: this research was held when the NSA/Snowden news were a hot topic, in the headlines of the international media), no fiscal advantages for these expressive investments, lack of standards for technical and regulatory matters were mainly considered by interviewees. See Fig. 1.13.

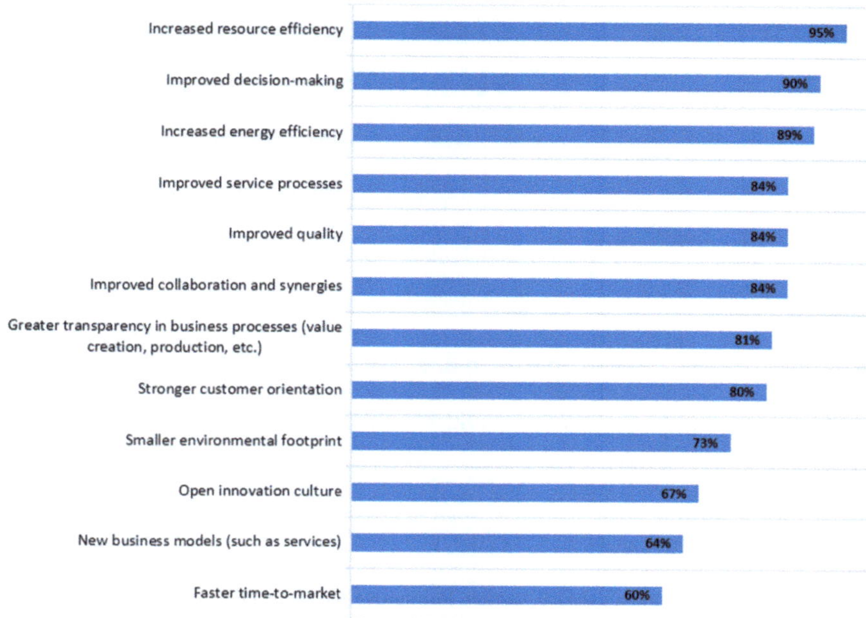

Fig. 1.10 What are the main digital transformation benefits for your organization? FDC (2015)

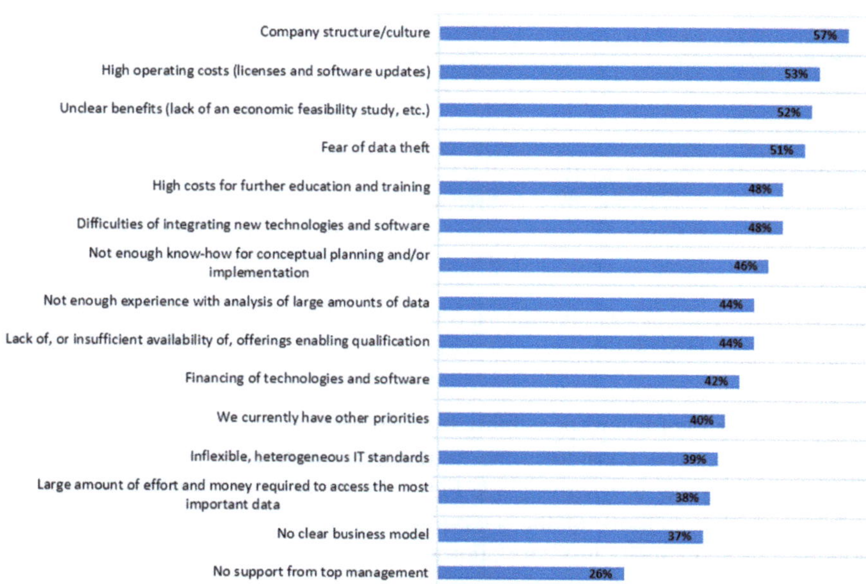

Fig. 1.11 What are the internal barriers for increased usage of digital technologies and processes in your organization? FDC (2015)

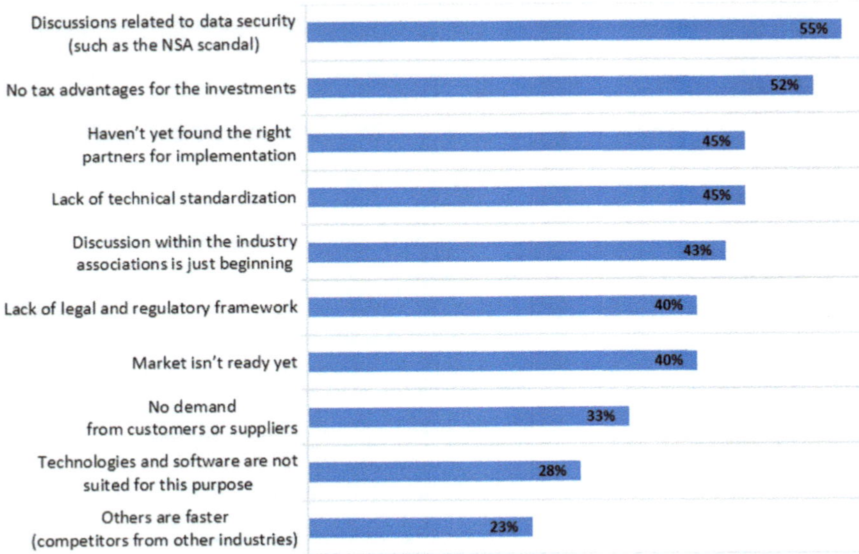

Fig. 1.12 What are the main external barriers to improve the application of digital technologies and digital processes in your organization? FDC (2015)

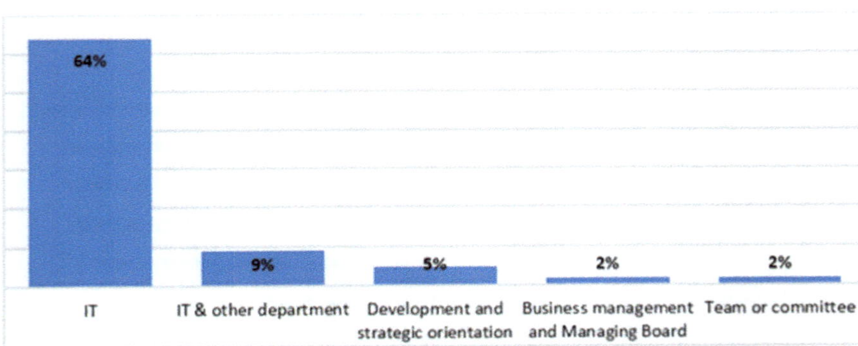

Fig. 1.13 What sector, in your organization, is the main responsible for topics related to digital strategy decisions? FDC (2015)

Approaching the main responsibility over digital strategy decisions, information technology sector was considered the most important, with an expressive level of perception over other alternatives. See Fig. 1.14.

An expressive majority of the interviewees answered positively to a possible participation in an "alpha test" or "pilot project" about digitalization technologies application. See Fig. 1.15.

Concluding this research, a question about the initiative on creating projects of digitalization was proposed to the interviewees, with results that indicate a

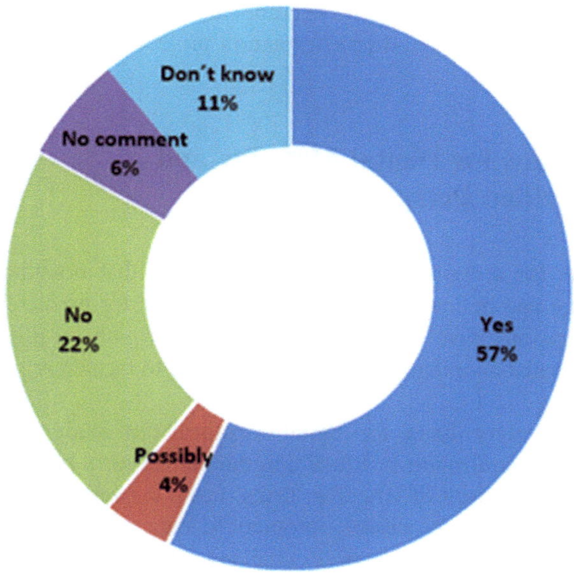

Fig. 1.14 Are you interested on participate in an "alpha test" (pilot project) of digitalization? FDC (2015)

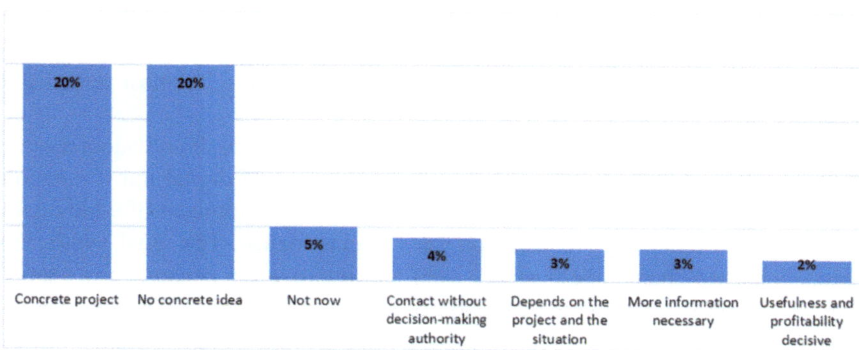

Fig. 1.15 Are you interested on creating pilot projects about digitalization? FDC (2015)

concentration on "concrete" (precise, well defined and proposed) projects, oriented for business application. There is a potential contradiction between the results exposed in Figs. 1.14 and 1.15, possibly related with the test project leadership.

These results, for the research entitled "Digitalization in Brazil" shown a great commitment of interviewed executives and organizations towards digitalization. However, digital transformation in these organizations still face several challenges, like organizational culture, operational structure, resources availability, and regulatory aspects. It is concluded that there is also difficulties on executing testing, or pilot, initiatives, although some of these companies suggest they already have units or sectors responsible for digital transformation projects.

1.6 Dimensions for Digital Maturity and the Brazilian Perspective

The proposition for a digital maturity model enables that organizations evaluate their actual stage and define its way to develop further, for predicted goals. In this paper, ten dimensions were suggested for a consistent path for digital transformation. It is evident that an integration must be searched to face emerging technologies advance.

As a complementary theme for digital transformation analysis, it was held the research called "Digitalization in Brasil", in the pursue for a better comprehension for organizational maturity of Brazilian firms. From the results, it was possible to learn that focus was for processes automation, data management and machine connection. Digital transformation poses challenges for Brazilian enterprises, which present a model too centralized for strategic decisions, mainly in information technology and production areas. It is also an evidence for training and integration needs for employees, partners and agents, to improve the outcomes of digital transformation advances.

Concluding, there is a perception about the impact of new digital technologies, with increased efficiency search, better decision-making processes, quality improvement, and cultural change for a new paradigm being the most remarkable unfolding.

From these results, it was noticed that Brazilian companies should adopt a series of actions, towards its growth and better comprehension of digital transformation. As a starting point, a fundamental point would be the development of wide digital strategy, not exclusively focused on new technologies. The goal, with this approach, should change the business itself and improve the perception of market changes. From this strategy, the adoption of new digital Technologies will ignite a new process for organizational growth, from the abilities on usage of available data.

Finally, the capacity of customer demands answer, customizing services and implementing digital networks, could enable a rapid advance towards digital transformation. But it will be possible only with the processes redesign and legal risks mitigation for new business models.

1.7 Conclusion

Important reflections from the "Digitalização no Brasil" research can be obtained, analyzing it together with the proposition of those ten dimensions for digital maturity. Although held through different methodological approaches, both studies produce convergent results.

The connection between main theme—digital maturity—and the elaboration of digital strategy is one of the main findings of these research work. Approaching Brazilian companies considered in the sample, it is possible to identify an evidence of greater data usage and its centralized management and control by information technology and production areas. This result is not aligned to the best practices found in compatible global players. Working these diverse evidences, Brazilian companies are trying to get consistency with their digital strategies, tackling other signals that were found also in the maturity model research—the ten dimensions—presented in this paper.

Another conclusion from this study is the need, by these Brazilian companies, to create a new culture for digital transformation. Considering the maturity model, here developed, it is urgent for these enterprises to better comprehend that a digital culture is a consequence of new organizational structure, risk-taking postures, new sources of investments and design of new digital business models.

In this sense, the search for digital transformation has been one alternative to surpass the topic's lack of meaning for several players and related risks for the business. Therefore, it is basic to compose a wider business agenda that combines new processes, enabling technologic conditions and decision levels committed with a new business environment.

As a limitation of these studies, only indirect, secondary data was used for analysis, because of methodological decisions. This way, "Digitalização no Brasil" research reflects the attitude of big Brazilian companies, making it impossible to generalize this research context and results for small and medium enterprises.

Further works, a deeper approach for case studies, supported by a longer and more detailed questionnaire, composed from a detailed literature review and with a wider sample selection can produce more evidences about digital transformation expectations, perspectives, and plans.

References

Andal-Ancion A, Cartwright PA, Yip GS (2003) The digital transformation of traditional businesses. MIT Sloan Manag Rev 44(4):34–41

Ansoff I (1990) Implanting Strategic Management, 2nd edn. Prentice Hall, New York

Becker J, Knackstedt R, Pöppelbuß J (2009) Developing Maturity Models for IT, Management. Bus Inf Syst Eng 1(3):213–222

Berghaus S, Back A (2016) Stages in digital business transformation: results of an empirical maturity study. In: 10th Mediterranean conference on information systems (MCIS 2016)

Berman SJ, Marshall A (2014) The next digital transformation: from an individual-centered to an everyone-to-everyone economy. Strategy Leadersh 42(5):9–17

Bishop B (2018) Production machining. March 2018, pp. 28–30

Bleicher J, Stanley H (2016) Digitization as a catalyst for business model innovation a three-step approach to facilitating economic success. J Bus Manag 12:62–71

Boyd P (2017) Factory of the future: 3D printers, process automation and specialists needed. Appliance design October 2017, pp 20–25

Capgemini Consulting (2017) The digital culture challenge: closing the employee-leadership gap. Available at: https://www.capgemini.com/consulting-de/wp-content/uploads/sites/32/2017/06/digital-culture-report-2017.pdf. Retrieved 31/03/2017

Casadesus-Masanell R, Ricart JE (2010) From strategy to business models and to tactics. Long Range Plan 43(2–3):195–215. https://doi.org/10.1016/j.lrp.2010.01.004

Celaschi F (2017) Advanced design-driven approaches for an Industry 4.0 framework: the human-centered dimension of the digital industrial revolution. Strateg Des Res J 10(12):97–104

Chui M, Loffler M, Roberts R (2010) The internet of things. McKinsey Quaterly. Available at https://www.mckinseyquarterly.com/ghost.aspx?ID=/The_Internet_of_T. Retrieved on May, 2016

Colquhoun L (2017) IoT: Utopia or Anarchy? Available at https://www.telecomasia.net/content/iot-utopia-or-anarchy. Retrieved on April, 2018

FDC (2015) Dom Cabral Foundation—digitization in Brazil. Available at www.fdc.org.br/inovacao. Retrieved on April, 2018 [In Portuguese]

Ford M (2018) Industry 4.0: If only I had known. SMT Magazine. Available at: http://smt.iconnect007.com/index.php/article/109693/industry-40-technologies-if-only-i-had-known/109696/?skin=smt. Retrieved on May 06, 2018

Fitzgerald M (2013) How Starbucks has gone digital. MIT Sloan Management Review. Available at https://sloanreview.mit.edu/article/how-starbucks-has-gone-digital/. Retrieved on May 06, 2018

Gatautis R (2017) The rise of platforms: business models innovation perspectives. Inzinerine Ekon-Eng Econ 28(5):585–591. https://doi.org/10.5755/j01.ee.28.5.19579

Hanna NK (2016) Mastering digital transformation: towards a smarter society, economy, city and nation. Emerald Group Publishing, Bingley

Henriette E, Feki M, Boughzala I (2015) The shape of digital transformation: a systematic literature review. In: MCIS 2015 Proceedings, Paper 10, pp 1–19

Hess et al. (2016) Options for formulating a digital transformation strategy. MIS Q Executive 15 (2):123–139

Kane GC, Palmer D, Phillips AN, Kiron D, Buckley N (2015) Strategy, not technology, drives digital transformation. MIT Sloan Manag Rev Deloitte Univ Press, 14

Kane GC, Palmer D, Phillips AN, Kiron D, Buckley N (2017) Achieving digital maturity. MIT Sloan Manag Rev Deloitte Univ Press, July

Karimi J, Walter Z (2015) The role of dynamic capabilities in responding to digital disruption: a factor-based study of the newspaper industry. J Manag Inf Syst 32(1):39–81

Kocsi B, Oláh J (2017) Potential connections of unique manufacturing and industry. LogForum 13 (4):389–400. http://dx.doi.org/10.17270/J.LOG.2017.4.1

Liu LY, Chen SW, Chou TC (2011) Resource fit in digital transformation—lessons learned from The CBC Bank global e-banking project. Manag Decis 49(10):1728–1742

Lucas HC, Agarwal R, Clemons RK, El Sawy OA, Weber B (2013) Impact research on transformational information technology: an opportunity to inform new audiences. MIS Q 37 (2):371–382

Maier D (2017) Industry 4.0: a radical shift in mindset and investments. Manufacturing Today, September–October 10–13

Matt C, Hess T, Benlian A (2015) Digital transformation strategies. Bus Inf Syst Eng 57(5):339–343

Meyer K, Han A (2017) Bossard AG: enabling industry 4.0 logistics worldwide. Available at https://www.thecasecentre.org/main/products/view?id=146968. Accessed in May, 2018
Mintzberg H, Alhstrand B, Lampel J (2008) Strategy Safari. Pearson Education, New York, USA
Mithas S, Tafti A, Mitchell W (2013) How a firm's competitive environment and digital strategic posture influence digital business strategy. MIS Q 37(2):511–536
Mullaly M (2014) If maturity is the answer, then exactly what was the question? Int J Managing Projects Bus 7(2):169–185
Piccinini E, Hanelt A, Gregory R, Kolbe L (2015) Transforming industrial business: the impact of digital transformation on automotive organizations. In: 36th International conference on information systems, Fort Worth
Porter M (2008) On Competition. Harvard Business School Press, Cambridge, USA
Roblek V, Mesko M, Krapez A (2017). A complex view of industry 4.0. Available at SAGE Open Services at http://journals.sagepub.com/doi/abs/10.1177/2158244016653987. Accessed on April, 2018
Rose J, Berndtsson M, Mathiason G (2017) The advanced analytics jumpstart: definition, process model and best practices. J Inf Syst Technol Manag—Jistem USP 14(3):339–360. https://doi.org/10.4301/s1807-17752017000300003
Saturno M, Pertel VM, Deschamps F, Loures E (2018) Proposal of an Automation Solutions Architecture for Industry 4.0. LogForum 14(2):185–195, http://dx.doi.org/10.17270/J.LOG.2018.266
Schuchmann D, Seufert S (2015) Corporate learning in times of digital transformation: a conceptual framework and service portfolio for the learning function in banking organizations. Int J Adv Corp Learn 8(1):31–39
Schwab K (2017) The fourth industrial revolution. Crown Business, London
Serban RA (2018) The impact of big data, sustainability and digitalization in company performance. Stud Bus Econ 12(3):181–189
Utterback JM (1971) The process of technological innovation of the firm. Acad Manag J 14(1):75–88
Wanga Y, Kung AL, Byrd TA (2016) Big data analytics: understanding its capabilities and potential benefits to Healthcare Organizations. Technol Forecast Soc Change 126(January):3–13. https://doi.org/10.1016/j.techfore.2015.12.019
WEF—World Economic Forum (2016) Digital transformation in industries. Available at http://reports.weforum.org/digital-transformation/wp-content/blogs.dir/94/mp/files/pages/files/digital-enterprise-narrative-final-January-2016.pdf. Retrieved on February, 2018
West A (2017) Industry 3.0 versus 4.0: Vision of the new manufacturing world. HIS Markit world. Available at https://www.coilwindingexpo.com/berlin/__media/Speaker-presentations-2017/Day-3-Alex-West,-Industry-3-0-v-4-0-compressed.pdf. Retrieved on April, 2017
Westerman G, Bonnet D, McAfee A (2014) The nine elements of digital transformation. MIT Sloan Manag Rev, Opin Anal, January 07, 2014. Retrieved on June 2017
Zott C, Amit R, Massa L (2011) The business model: recent developments and future research. J Manag 37(4):1019–1042. https://doi.org/10.1177/0149206311406265

Chapter 2
Modelling and Analysis of the Apples Export Supply Chain Business Processes: Experiences from Chile

Mónica López-Campos, Rosa Guadalupe González-Ramírez, Lorena Bearzotti and Salvatore Cannella

Abstract In this chapter, we present an analysis of the fresh apple export supply chain in Chile. Fresh fruits are one of the main exports in Chile, after the mining products. This supply chain is particularly complex due to its perishable nature and the fact that the fruit export season occurs during the summer period, overlapping with the arrival of cruises. For this reason, maritime ports present high levels of utilization and congestion. We follow the Business Process Analysis (BPA) methodology proposed by United Nations to model the current situation of the shipment business processes, from the warehouse or packing facility of the shipper to the port of departure where it will be loaded to the corresponding vessel. With this analysis, we identify the main challenges and opportunities and propose a redesign of the current processes. The proposed redesigned business processes consider the implementation of electronic data interchange procedures instead of the paper-based ones currently employed.

Keywords Business process management · Trade facilitation · Supply chain management

---

M. López-Campos
Industrial Engineering Department, Universidad Tecnológica Federico Santa María, Valparaíso, Chile
e-mail: mlopez@usm.cl

R. G. González-Ramírez (✉)
Faculty of Engineering and Applied Sciences, Universidad de Los Andes, Chile, Santiago, Chile
e-mail: rgonzalez@uandes.cl

L. Bearzotti
School of Transport Engineering, Pontificia Universidad Católica de Valparaíso, Valparaíso, Chile

S. Cannella
Università Degli Studi Di Catania, Catania, Italy

J. L. García Alcaraz et al. (eds.), *Best Practices in Manufacturing Processes*,
https://doi.org/10.1007/978-3-319-99190-0_2

2.1 Introduction

Foreign trade has an important impact on the economic growth of a country and its benefits have been extensively documented in the literature (Helpman 2014; Laursen 2015; Neary 2016). Most of the manufactured products are not only produced to satisfy domestic demand, but also international markets. For this reason, the efficiency on the operations of foreign trade has a significant impact on the final costs of products payed by the final consumers, impacting on the competitiveness of manufacturing systems.

The competitiveness of foreign trade is highly dependent on the efficiency and effectiveness of the export and import logistics chains, where a significant number of private and public actors interact among them. Current levels of globalization demand for more integrated supply chains and given that maritime transport accounts for around 90% of worldwide foreign trade, seaports play an active role in the performance of global supply chains (Rodrigue and Notteboom 2009). Because of that, every change to have a better port logistic chain has a positive impact in the foreign trade and the country competitiveness.

Latin American and Caribbean ports accounted for 88 percent of the net container growth of the Americas during the interval 2000–2010. This growth is putting pressure on freight distribution systems, which need to develop better logistical capabilities such as acquisitions and development of port technology as well as a better hinterland coordination (Rodrigue 2012). In this sense, there is a need to better understanding the logistics business processes in relation with the main products in import and export operations.

Chile shows a high dynamism of foreign trade. The country is characterized as an open economy with several free trade agreements worldwide. More than 90% of the traffic is transported by sea. The ports of San Antonio and Valparaíso are the most important in terms of TEUs transferred (Jean et al. 2014), where the port of San Antonio occupied the 12th place and the Port of Valparaíso the 17th in terms of containerized port throughputs at Latin America and Caribbean according to the 2016 Ranking of container throughput in the region elaborated by ECLAC.[1] The main products exported by Chile are minerals, being the copper the most important. This type of products is produced in the North-Centre Region. Other important types of exports correspond to fruits. Grapes and apples are the main ones in terms of volume and produced in the Centre-South Region. Chile is the first exporter of fruits in South America and the fresh fruit is the third sector more important in the Chilean economy. At present, it has a vital importance in the Chilean growing economic. In the recent years, raspberries, blueberries and cherries are other products being exported with high value due to its shorter shelf life.

During the fruit season (around November until April), a lot of congestion occurs, generating significant inefficiencies in the process of transporting the fruit from the hinterland to the corresponding port where it will be loaded into a

[1]https://www.cepal.org/en/infographics/ports-ranking-top-20-latin-america-and-caribbean-2016.

ship. Because of that, inland transport represents an important challenge and it is currently affecting total landed costs. Hence, better coordination and integration of the ports with their hinterland is demanded at Chilean ports (Monios and Wilmsmeier 2013; Ascencio et al. 2014; Merk and Notteboon 2015; Wilmsmeier et al. 2015; Wilmsmeier and Sanchez 2017). With the aim to improve competitiveness, reduce logistical costs and address the afore-mentioned aspects, the Chilean government is promoting a series of measures and reforms (Port Community Systems, Logistic Observatory, etc.) where one of its final objectives is the Implementation of Single Window for foreign trade operations. These initiatives are based on the need to carry out studies and surveys of the logistics processes associated with import and export processes; for which it is necessary to have tools and methodologies that facilitate the communication between the technicians and the actors of the chains under study.

The purpose of this chapter is to illustrate how Business Process Modelling, in combination with UML and in the context of UN-ESCAP model, is a timely and appropriate technique to carry out a structured analysis of the supply chain process, as a previous and necessary step before optimization and/or redesign, in other words realizing a Business Process Analysis (BPA). Specifically, this paper describes and models, using UML standard and BPMN (Business Process Modelling Notation), the physical and documentary procedures carried out by the actors involved in part of the export logistics chain of apples. The aim of the modelling is to provide recommendations for the current business processes involved that may reduce delays and non-value operations that influence logistics costs.

This modelling provides and initial analysis of requirements to eliminate the paper-based documental procedures in the port logistics chain, contributing to the ongoing projects of the National Single Window SICEX and a potential Port Community System for the San Antonio's Port. The employed methodology takes as a reference the United Nations UN-ESCAP guide (UN-ESCAP 2012), which states that in order to improve processes and information flows of international supply chains, it is necessary that the "as-is" conditions of relevant business processes may be well understood prior to the selection of trade facilitation measures. The business process analysis permits to simplify the trade procedures. Likewise, this paper fills the gap found in literature, due to the shortage of articles modelling, through Business Process Modelling (BPM) and UML language such as the documental procedures related to foreign trade.

This chapter is structured as follows. The literature review is presented in the Sect. 2, summarizing the literature related to the use of BPM for the analysis of agri-food supply chains, and describing some studies case of redesign and facilitation of foreign trade processes in several ports around the world, through the implementation of BPM and the UN-ESCAP guide. In the Sect. 3, the methodology based in the integration of BPM with UML in the context of the UN-ESCAP with BPA guide is presented. The case study of the apples exports supply chain in Chile is analyzed in Sect. 4, from its economical and processes point of view. Particularly, we consider the operations in of the Port of San Antonio. In this case study the integration is of the different approaches and languages is presented. Finally, in Sect. 5 the conclusions and recommendations are detailed.

2.2 Literature Review

The fresh-produce supply chain is a research area with a growing attention in the last decades where the goal is satisfying the new requirements of final customs related with quantities, timing and qualities of the products. However, the agricultural supply chain is complex systems with many actors involved, where the complexity is increasing because of the globalization and the final client demands. In front of this situation, the practitioners require new information systems and methodologies to have sustainable and resilient agricultural supply chains with a trade facilitation and using the new advances of information and communication technologies. In this scenario is a key factor to have the ability to share information among the actors in the supply chain and, in this way, to have better making decision processes in the different levels.

The agricultural supply chain considering the export process is a central element in the analysis and design of solution related with the increase of competitive indicators. The first step in this process to be more competitive knows the different supply chains related, identifying the actors, their roles and the activities doing for them, the different kind of documents, the information systems used, etc. Modelling permits to understand complex problems by different interrelated diagrams to visualize concepts and elements (Dunugbo et al. 2013).

BPMN and UML are employed for modelling specific processes. UML is a modelling language used by different disciplines. BPMN emerged in the last years as the notation to represent business process with an easy integration to business architectures and information systems (García-Holgado et al. 2015; Gromoff et al. 2017). BPMN can be used to represent the different business process in the organizations (Dumas and Pfahl 2016) and provide more transparency and compliance to the external and internal stakeholders (Betke et al. 2013); or for monitoring and handling (Stavirou et al. 2014). Furthermore, BPMN can be used as a base to redesign the business processes aiming to foster supply chain integration (Palma-Mendoza and Neaily 2015). The UML diagrams and BPMN can be integrated to represent different aspects of the problem under study.

Section 2.1 provides a review of the use of BPMN and UML in agri-food supply chains. Section 2.2 provides a review of some case studies in which the aim is to redesign foreign trade business process.

2.2.1 BPMN and UML in Agri-Food Supply Chains

The agricultural supply chain, as any other supply chain, is a network of organizations working together in different processes and activities to bring products and services to the market, with the purpose of satisfying the demand of its customers (Ahumada and Villalobos 2009). Management of this type of supply chain is more complex due to the perishable nature of products, high fluctuations in demand and prices, increasing consumer concerns for food safety (Van der Vorst and Beulens 2002), and dependence on climate conditions.

Thus, agricultural supply chain, are facing nowadays several complications due to some food specific characteristics such as long production lead times, seasonable production, quality variations between producers and between lots, short required delivery time (Verdouw et al. 2010). In fact, growers of perishable agricultural products, such as fresh fruits and vegetables, very often face complex planning problems such as deciding the level of technology to use, how much of a particular crop to plant, the timing of planting and harvesting (Ahumada and Villalobos 2009). Hence, agricultural and food supply chains are often more complex and more difficult to manage because the food products are perishable and have a short shelf life (Aung and Chang 2014).

To improve the performance of the production and delivery process of perishable products, the identification of deficiencies and related improvement strategies is crucial. In this context, the adoption of suitable tools, such as Business Process Analysis (BPA) and BPM, for the management of agri-food supply chain, is required. Reference process models can be a valuable means to support the challenges in the design and implementation of fruit supply chains (Verdouw et al. 2010).

A business process is the set of activities a business performs to create value for customers, where the main elements are the inputs required, the activities to do and the outputs to generate. There are various notations for the creation and analysis of business process models. They all graphically represent the activities, events, decisions control and actors involved. A well-known standard is the UML (Verdow et al. 2010). According to (López-Campos and Cannella 2011), only few studies have explored the benefits of BPM in supply chain. Most of those studies mainly focus on the general functioning of the production-distribution multi-echelon system.

In the last decade, the common subjects of research are related with the analysis of the benefits and implications of using BPM and information technology in the supply chain area. Other topic is the proposals of BPM reference frame-works or BPM meta-models for supply chain simulation and analysis; following by the development of supply chain model classifications; and several applications of business process modelling to implement SCOR framework (Murthy et al. 2009).

In this context, to the best of the authors' knowledge, the work of (Verdouwet et al. 2010) is the first and currently unique example of a reference process model for fruit supply chains. In their pioneer work, they show how reference process models can be used to model the demand-driven fruit supply chains and to support the implementation of supporting information systems. They suggest that future research should focus on the extension of the BPM to other processes, such as strategic planning, product development of fruit. In this chapter, we contribute to this advocated effort adopted for the improvement of the foreign trade process of apples using BPMN with UML in the representation of the different processes and their relationships in two situations, AS-IS models and TO-BE models.

2.2.2 Case Studies of Foreign Trade Business Process Redesign

The study of supply chains related with the foreign trade is an active research area because there is a growing interest into solve the different problems related with the complexity of this systems (Lan and Tian 2013; Nakandela et al. 2015; Mehmann and Teuteberg 2016; Bae 2016; Nedic et al. 2016; Dunugbo et al. 2013; Soto-Silva et al. 2016).

Several countries have been interested in trade promotion through the simplification of trade processes and procedures, using the UN-ESCAP recommendations for this purpose. Bangladesh, China, India, Nepal, Sri Lanka and Thailand are some of the countries that have applied the UN-ESCAP methodology and analysis to improve their import/export processes. For this, each country selected as pilot one import and export supply chain that represents important commodities in order to identify the main challenges and associated costs. The objective of those projects was to design a set of policy recommendations to simplify general trade operations (Saifuddin 2010; Ramasany 2010; de Prabir 2011; Rajkarnikar 2010; de Mel et al. 2011; Keretho and Nakkada 2011).

Bangladesh started modelling four selected products (two for export: woven garments and shrimps, and two for import: cotton fabrics and sugar) to identify the steps, actors, documents, time and cost involved in the overall transaction procedure. Thereafter, in several steps, the processes were automatized. The results of this project were measured in terms of reduction in the number of required documents, time and cost of import/export of the mentioned goods (Saifuddin 2010).

China executed a similar project, choosing as supply chains of analysis the exports of garments to Japan, the export of electronic products to Thailand and the import of fabric/accessories and automobile components from Japan. One of the conclusions of this project indicate that to facilitate trade operations, control, and inspections, as well as decreasing errors, the number of documents involved has to be reduced (Ramasamy 2010).

To perform its corresponding trade facilitation project, the cotton supply chain and agricultural supply chain in India was selected for export operations, and the tyres and tubes supply chain for imports. The analysis showed that bottlenecks are mainly in transportation, customs clearance and getting payment, especially referred to exports (De 2011). Nepal analyzed the hydrogenated vegetable oil export supply chain, known as the vegetable ghee; and the import supply chain textile, as it is one of the major import items of Nepal, especially from India. The main obstacle identified by the analyzed export processes is the problem of obtaining export quota and certification of quality and origin (Rajkarnikar 2010).

A similar effort has been made by Sri Lanka, documenting the export of rubber tyres and tea, and the import of textile and motor vehicles. Likewise, several situations and non-optimal procedures were identified as source of delays. That diagnosis constitutes a first step to improving trade processes (de Mel et al. 2011).

Thailand also implemented this methodology and analyzed four strategic export products to improve its logistics supply chain: frozen shrimp, frozen chicken, durian fruit and automobile; and four strategic selected import products to analyze were tuna, beef, grapes and auto parts. Most of the identified problems during the studied processes would be eliminated or reduced via implementation of a National Single Windows (Kertho and Naklada 2011).

Finally, this methodology was applied to analyze the export business process of rice and mango and import of palm oil in Myanmar. This study has been conducted has part of initial diagnosis studies for trade facilitation with the goal to identify constraints and recommend measures with regard to export or import process of selected products. After the analysis they detected some redundant activities in the different process and the needs to have better infrastructure, information technology, reviewing the banking processes and the licensing regime (Ksoll et al. 2013).

Accordingly, we can identify a trend to carry out studies of the most important import and export chains for the countries. The analysis of the supply chains is the first step in the process of improving the foreign trade operations tending to have better logistics indicators and consequently the competitiveness of the country. This result in a better positioning of products shipped abroad. The methodology used in the cases presented is based on UN-ESCAP. This methodology does not detail how to carry out the modelling and representation of processes relevant for analysis and improvement.

Due to this, the proposal presented in this chapter consists of the integration of the UN-ESCAP methodology with UML and BPMN in order to facilitate communication in the work team and with the stakeholders involved, taking advantage of the potentialities of both modelling languages and the possibility of its integration with business architectures and the development of information systems. The proposal is applied to the case of the apples export supply chain in Chile. These topics are presented in the following section.

2.2.3 *Integrating UN-ESCAP and BPA with UML and BPMN*

2.2.3.1 The UN-ESCAP Reference Model

To improve the efficiency and the effectiveness of process and information flows along the international supply chain, it is required to model the relevant business processes involved (UN 2012). According to UN-ESCAP reference model (UN 2012), an international supply chain, such as the fruit export, consists of three macro-processes: Buy, Ship and Pay, which they are represented in Fig. 2.1.

The key players of the international supply chain are the authorities, intermediaries, suppliers, and customers. The Buy macro-process is focused in the sales contract and how the suppliers are related with the customer. The Pay macro-process describes the payment of goods.

In this chapter, the focus of the modelling of the apple export process is the Ship macro-process, which is associated with the distribution process from the hinterland to the corresponding port. The case study corresponds to the Port of San Antonio. Six processes, composing the Ship macro-process, have been identified and they are presented in Table 2.1, followed by a brief description of each of them.

The name in parentheses of each process in Table 2.1 corresponds to the identification of each process in the UML Diagrams; these will be presented in next section. These macro-processes are composed by processes and activities, then is necessary to have a methodology to described them and presenting the relations among them.

2.2.3.2 Business Process Analysis (BPA)

The Business Process Analysis methodology provides a reference framework to evaluating the business processes and procedures, their rationale, the time required to complete them and the associated costs (Ksoll et al. 2013).

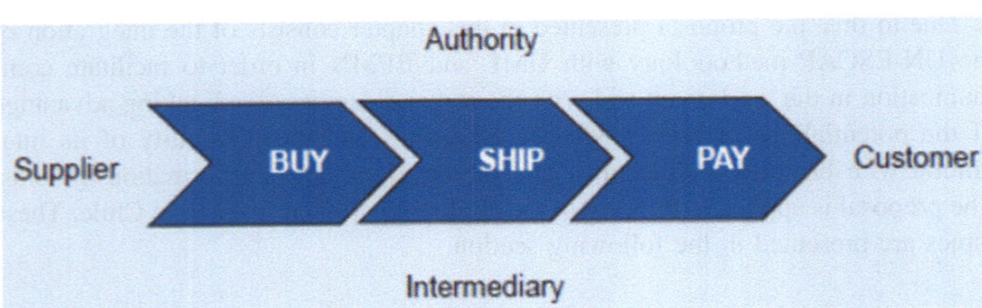

Fig. 2.1 Macro-processes of the international supply chain. *Source* "Business Process Analysis Guide to Simplify Trade procedures", 2009

Table 2.1 Macro process ship and it processes of the apple export logistic chain

Processes
1. BUY
2. SHIP
2.1 Request booking cargo with the shipping line, and organize inland transport (Arrange Transport)
2.2 Prepare export documentation and export permissions (Prepare documentation)
2.3 Stuff cargo at a container and comply clearance procedures at exporter facilities. Transfer the container to the port of departure (Consolidate Cargo)
2.4 Deliver cargo at the port terminal and stack it in the yard (Transport Cargo to Port)
2.5 Handle cargo and load it to the ship (Transfer cargo to ship)
2.6 Prepare documents required by the importer (Issue documents)
3. PAY

The idea behind BPA is to determine how the process works and how it can be improved. The main outputs of BPA are: the AS-IS Model (a description of the current process), the TO-BE Model (a proposal redesign of the process with the required adjustments to improve it in time, cost and use of resources), Customer and Supplier definition, Process Ownership and Governance, Roles and Responsibilities, Process Impact, Organization Impact, System Impact, Risks, Impact Type, Impact Level and Expected Outcomes (Stavirou et al. 2014).

The cycle of BPA starts with the mapping of processes (the current procedure and practice), following with the monitor (this is a verification of actual situation) and the analyses (to identify area to improve). The next step corresponds to the research and development to fill knowledge gaps and finishes with the implementation (changes in the procedures and practices).

An important part of BPA is the representation of the different processes to understand their behaviour and the interactions among them. This is especially important when the object in study is supply chains. In this case, the entire supply chain is mapped; analyzed and key problems will be identified. The language used to map the supply chain is BPMN, this is a universal language which enables all involved parties to communicate the processes clearly and completely (Stavirou et al. 2014). In the next section, BPMN is detailed.

2.2.3.3 Business Process Modelling Notation UML

A business process is a collection of activities logically related that takes one or more inputs and creates one or more outputs with value to the customer (Succi and Pedronzani 2000). Considering the context of trade facilitation, a business process can be defined as a chain of logically connected activities to move goods and information across borders from buyer to seller and providing related services.

It should be noted that delay or lack of integrity when processing information through business process is a factor that slows the movement of cargo. On average, each day of delay reduce trade volume by at least 1% to about 7% if the products are perishable. Modelling of business process, understood as the use of methods, techniques and software to design, enact, control and analyze operational processes involving humans, organizations, applications, documents and other sources of information (Van der Aalst 2003), is an important technique to improve operations and to prevent losses. Related to the foreign trade operations, business process modelling can serve as a starting point for the implementation of trade facilitation measures such as

- Simplification of business processes (including commercial transport, regulatory and financial procedures).
- Simplification of documentary requirements and harmonization of them according international standards requirements.
- Automation of international business transactions and electronic documents, in order to converge to a single window that does not require papers.

Process modelling not only describe processes, but in addition, it represents a preparatory stage for the improvement of business processes, process reengineering, technological transference and process standardization. BPM serves as a tool that facilitates:

- The analysis of activities, documents and information flow in international trade procedures.
- The identification and prioritization of problem areas that cause delays in the movement of goods from seller to buyer.
- The design of plans to improve these problem areas (for example, simplification of processes and data and removing of redundancies).

To have a better representation including other dimensions not only the process the Unified Modelling Language (UML) can be used. UML provides a standard set of graphical notations for modelling business process and it is internationally accepted and widely used, not only among practitioners in the business communities, but also in information technology and software development. Since the aim of this chapter is to enhance the automation of trading procedures, interacting through e-commerce exchangeable documents, the use of a common standard graphical notation for BPM is vital, especially because a standard notation allows business experts to communicate procedural and documentary requirements.

Regarding software-modelling tools, there are several platforms. Some of them are non-proprietary. The selection of the most appropriate tool depends on the particular modelling requirements and the project scope.

2.2.3.4 Modelling Framework Based on the Integration of UN-ESCAP and BPA with BPMN and UML

The qualitative research method is commonly employed for data collection. According to (Goulding 2005) a qualitative research approach is useful for gathering and analyzing exploratory data. Min et al. (2005) point out that qualitative research helps to collect data from organizations providing valuable insights that are not possible to gain in a case study or in quantitative style research. Islam and Olsen (2014), highlight that qualitative research allows the flexibility to ask why, how or when, and to use probing questions to encourage participants to talk spontaneously about the area of interest.

For this project, in-depth interviews with the main stakeholders involved in the export logistic chain of apple were performed. Respondent selection was done under the scope of the Logistics Committee of the Port-Logistics Community of San Antonio. In total six interviews were performed including: custom agent, exporters, container terminal operator, customs, and transport carrier. Semi-structured and unstructured questions were employed to obtain insights of the current export processes.

According to the UN-ESCAP reference model, in general the phases applied in the business process analysis of the export logistic chain, are the following:

- Phase 1: Defining the scope and project plan. For this particular project, the analysis is focused to the export shipment of Chilean apple through San Antonio Port. Specifically, this paper studies the first process related to shipment (*Arrange Transport* in the Fig. 2.1).
- Phase 2: Acquire background information, conduct interviews and document captured data. This is the collection of data and process documentation (AS-IS). As a diagnosis is the first step to improve, documentation of processes using diagrams showing the current situation is a prerequisite for the further improvement. The project documents, through UML and BPMN. For example, the *Arrange Transport process* is presented in this paper.
- Phase 3: analyze the AS-IS process, identify bottlenecks and develop some recommendations and changes. This means doing process analysis and development of recommendations (TO-BE). In this stage is where changes and improvements in the process, activities and documentation are made.
- Phase 4: implementation of recommendations. This is the most substantial step, where new processes are physically implemented through staff training, implementation of information systems, new documents, etc.

In the Phases 2 and 3 is necessary the representation of the different processes, in both phases the strategy is using BPMN to represent the processes with the different activities and actors related in their different roles, in combination with UML to represent more information about the supply chain under studies for example using Use Case Diagram and Interaction Diagram.

Modelling is a classic approach to understand complex systems that can be achieved diagrammatically to visualize concepts and mathematically to analyze attributes and concepts (Dunugbo et at. 2013). The diagrams make easier to understand organizational requirements, capturing activities and interactions among them. In this sense BPMN and UML make it easier to communicate processes and in a compact way across the different stakeholders. The advantage to use BPMN and UML together is that both languages are almost standard in the representation of process and complex systems. In the next section the case of Chilean apples export logistic chain is presented showing how the proposal can be used.

2.3 Case Study: Export Apples Supply Chain in Chile

In this section, the case of study is developed showing how the proposal is developed. The different parts are related with the phases defined previously. First the scope is defined following with the AS-IS modelling and TO-BE modelling ending with some ideas about the implementation.

2.3.1 *Phase 1: Scope Definition*

2.3.1.1 Statistical Description of Apples Production and Exports

The apples are originated from Central Asia. There are almost 7500 different types of apples. According to data from 2014, the apples exports have generated $7500 million. Worldwide production of apples accounts for around 4.7 million-acres and 75 million tonnes in 2011. Worldwide foreign trade of apples in 2012 accounted for 8.2 million tones, which represented 10% of increment compared to 2008. The main importer country is Russia with 14.8% followed by Germany (8.2%) and then United Kingdom (6.6%). China is the main exporter of apples in terms of volume (12.8%), followed by Poland (12.4%), Italy (12.3%), United States (11.4%) and in fifth place, Chile (10%) (Odepa 2013). In terms of the value of the export (FOB), Chile occupies the fourth place in the world.

Fruit industry in Chile is a relevant sector due to the geographical conditions and natural resources at the nation. During the last decade, the fruit industry has been consolidated as one of the main export products at the country and around 80% of fruit production at Chile is exported. According the Chilean Association of Exporters (ASOEX), fruits exports accounted for 2,625,718 tonnes during the season of September 2012–August 2013. In 2013, fresh fruit exports increased about 5.4% with respect to 2012. Apples (red and green) accounted for 30% of the total exports of fruit in Chile with a 2.8% increment with respect to previous period. The main destination countries of the Chilean apples exports are United States (18%), Holland (10%), Colombia (8.1%), Saudi Arabia (7.1%) and Taiwan (6.9%).

Fig. 2.2 Types of apples produced in Chile. *Source* www.chileanfreshfruit.com

Although Chile is according with the changes in the global production with focus in the new varieties that they are obtained better prices compared with the traditional varieties.

According to ASOEX, fresh fruit production is concentrated among the III and VII regions of Chile. The Region of Libertador Bernardo O′Higgins (VI) is the main production zone of the country, with 34% of the total exports, followed by the Region of Maule with the 24% and the Region of Valparaíso with 13% of exports. Particularly, apples are produced at the cold regions of the Central and South Regions of Chile: V, VI and VII, where the VI and VII Regions account for more than 70% of the production. More than 20 different types of apples are produced, which are illustrated in Fig. 2.2. The Royal Gala apples are the main type of apples produced and exported.

Motivated by the importance that apple's exports represent for the country; the export apples supply chain was selected as case study in this chapter. Given that a significant proportion of the fruit's exports transferred by the Port of San Antonio, the next sub-section provides a brief overview.

2.3.1.2 Description of San Antonio Port Logistic Chain

The Port of San Antonio is one of the main ports in Chile, located in the central region. The hinterland of the port considers the metropolitan area of Santiago, the Chilean central regions and the region of Cuyo at Argentina. The port has connection with the railway and roadways. In 2016, the Port of San Antonio transferred 18 million tonnes that corresponds to 1.5 million TEUs, 200,000 Ro-Ro cargo and 2.5 million bulk cargo. During the 2017 the movement of TEUs is estimated in 3.0 million.

The Port consists of four terminals, which operate under a concession. Two of them are for bulk cargo and the other two terminals are mainly dedicated to container operations, San Antonio International Terminal (STI by its acronym in Spanish) and Central Port. The latter has been recently given in concession and it is currently performing the expansion investments in infrastructure, reason for which the focus of the analysis presented herein is STI.

Logistics operation model of the Port of San Antonio considers a direct flow of trucks to the terminal of the port, which differs from the operations at the Port of Valparaíso, in which all truck flows are directed to the pre-terminal ZEAL, where they wait until they are allowed to enter the corresponding terminal.

STI terminal has three berths with a depth alongside of 13.5 m. Total terminal area considers 31 ha with a quay of 800 m and a depth of 15 m. The terminal has 2000 reefer container units and it has a static storage capacity of 20.252 TEUs. Equipment includes 6 Gantry Cranes, 41 internal trucks, 9 forklifts, 2 Rubber Tyred Gantry cranes and reach-stackers.

2.3.2 *Phase 2: Background and AS-IS Modelling Overview of the Apple Export Logistics Processes*

2.3.2.1 Overview of the Apple Export Supply Chain Business Processes

Foreign trade transactions encompass a wide range of activities, related to the establishment of commercial contracts, the arrangement of inland and cross-border transportation of goods, the export and import formalities to meet regulatory requirements and the payment of the purchased goods. It requires the cooperation of many actors (traders, government, agencies, government authorities, logistics operators, transport operators, etc.). Furthermore, preparation of transport and regulatory documents demands a lot of effort, especially for those cases in which no adequate Information System are implemented and hence several of the procedures are still paper-based (UN-ESCAP 2012).

A huge number of parties are involved in the port logistics chain, which needs to exchange information on a large scale, often frequently and urgently. Unfortunately, lack of coordination among stakeholders is commonly observed.

Particularly at the port of San Antonio, fruits are exported only by reefer containers. Only the Port of Valparaíso and Coquimbo export bulk fruit cargo. Hence, the focus of the analysis herein considers only containerized cargo. Export Containers are received at the port during the stacking time window, which is established about 5 days prior the estimated time arrival of the corresponding ship. Only containers that have been assigned to a ship are received. Export containers must comply all the documentary requirements of the government agencies (i.e. Customs) and the Port, prior arriving to the gate. Those containers that have fulfilled the requirements are considered as approved or cleared and can be received at the port.

Based on the statistics of STI Terminal in San Antonio Port, around the 85% of the container arriving to the terminal under this condition and they are allowed to enter the terminal, the rest are rejected and directed to a parking area that has recently been established, where containers should wait until the corresponding procedures are cleaned. There is not Booking or Appointment System for export operations at the port, and hence, trucks arrive randomly during the stacking time window. During fruit season, there exists a lot of uncertainty and most of the fruit containers arrive at the last day of the stacking time window. A lot of congestion is observed at this period, with long queues of containers at the gate, representing an opportunity area for improvements.

As a solution of the mentioned lack of coordination, Single Windows and Port Community Systems (PCS) projects have been implemented worldwide to facilitate data interchange among the parties involved in foreign trade processes, as part of trade facilitation measures. Nevertheless, currently Chilean ports have not implemented a PCS yet and most of the procedures are paper-based. At 2015, there is in Chile an ongoing project for the National Single Window (SICEX) and the Port of Valparaiso is currently developing a PCS named SILOGPORT. However, none of them has been implemented even on a pilot. Currently foreign trade procedures continue to be paper-based at several of the echelons of the port supply chain, which produces high costs and inefficiencies.

All the information was recollected throw èxtensive interviews with key stakeholders (Government agencies, exporters, traders, logistics service supplier, manager in the port, etc.) although some key site have been visited to better understand the trade, for example the STI installations and a packing site. The objective with the field visits and the interviews was obtaining information on business processes and dealing of time has been spent on explaining this assignment. To validate findings, the study team had several rounds of interviews with the various actors in the logistics chain. However, in this background recompilation some limitations occurred, for example the data availability and consistency.

2.3.2.2 Modelling the AS-IS Apples Export Process

Following a top-down approach, the top value chain process of the export logistic chain (or level 0) is constituted by the already mentioned three macro-processes: *Buy, Ship* and *Pay* (Fig. 2.1). Subsequently, each macro-process is compounded by processes (level 1) and each process can be subdivided in sub-process (level 2), in activities (level 3) and so on. The scope of this work is focused on the modelling of the *Ship* macro-process (Table 2.1).

The previously identified six *Ship* processes relate to actors in the supply chain, these actors execute the involved activities. This relation between actors and processes is presented in Fig. 2.3. For a simplification, actors are classified into three types: *Exporter, Intermediaries* and *Authorities.* In the group of *Intermediaries,* we can find the Shipping Agency, Land Transportation, Customs Agent, Container Depot, Cross-docking Plants Packing, Port Terminal, Maritime Transportation. The *Authorities* group includes the National Customs Service and the Chamber of Commerce.

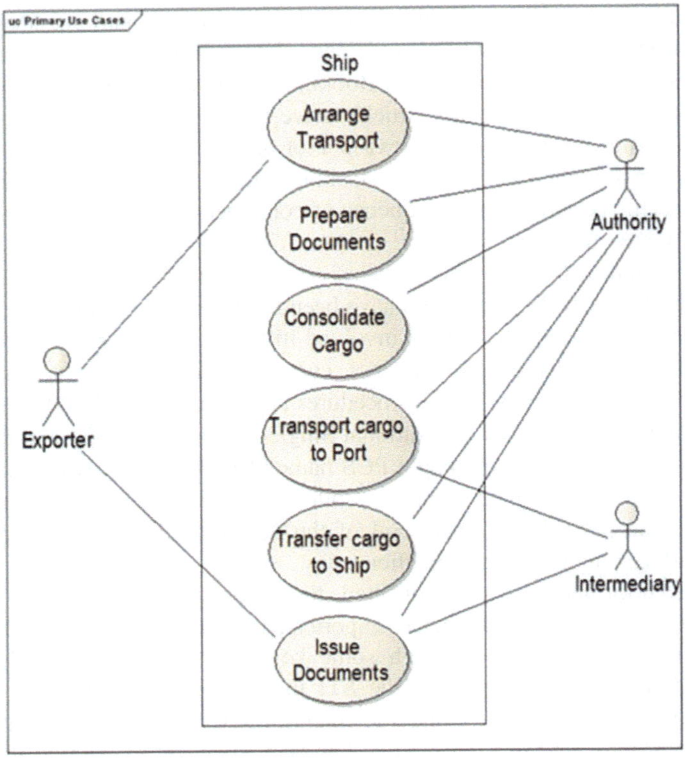

Fig. 2.3 UML uses case diagram for ship processes and actors

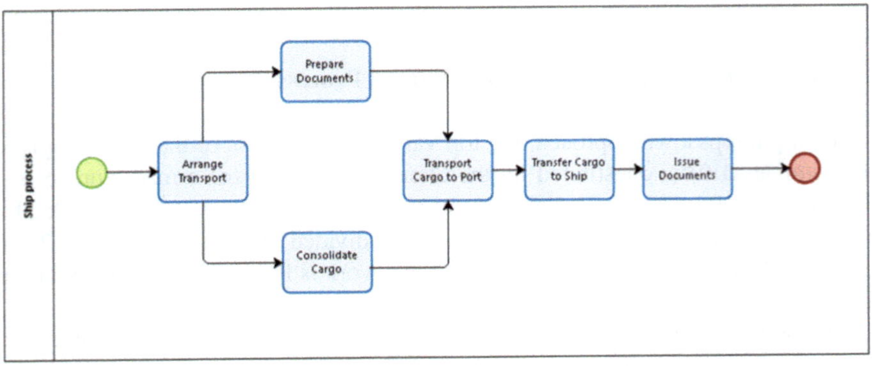

Fig. 2.4 BPMN diagram for ship process

Moreover, in Fig. 2.4 is shown the *Ship* macro-process describing the interrelationship between its processes. This diagram reveals that the processes are sequential, with the exception of the *Prepare Documents* and *Consolidate Cargo*

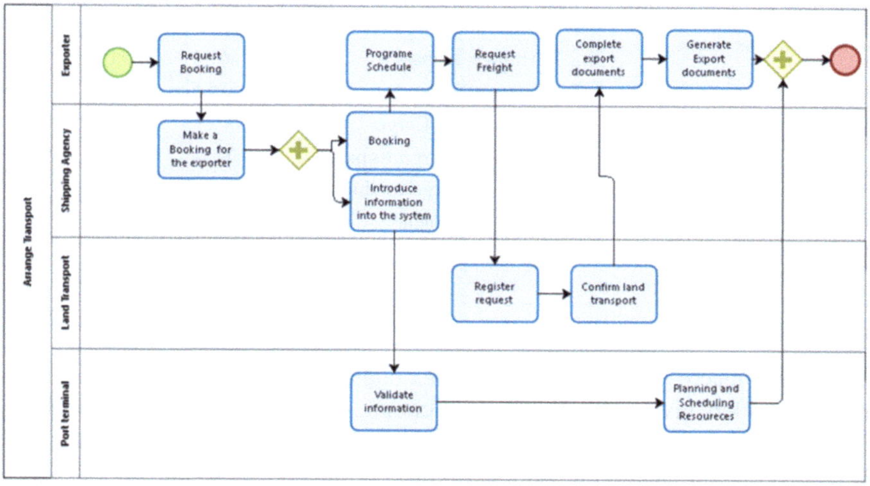

Fig. 2.5 BPMN diagram AS-IS arrange transport process

processes that are performed in parallel, requiring both to be finished before continuing with other process operation.

Finally, Fig. 2.5 shows the BPMN diagram for the AS-IS *Arrange Transport* process. The purpose of this process is to book a space on the ship and arrange land transportation from the hinterland to the port terminal. The main actors involved are Shipping Agency, Exporter, Land Transport and Port Terminal.

Once the exporter has a fruit-purchasing contract, he requests the shipping space to the shipping agency. The Shipping Company generates a Booking document to confirm and provide information related to the arrival of the ship to the port. Then, the exporter must schedule the collection of the container from an empty container depot and organize land transportation. Thereafter, the exporter concentrates all the export information, for example, the instructional performs for different agents (land transport, packing and customs agent) which is sent along with the Proforma invoice to the appropriate parties.

2.3.2.3 Analysis and Diagnosis from the AS-IS Apples Export Process Models

Analyzing the AS-IS diagrams and doing time control analysis (Table 2.2) during a deep interview with a custom agent, some factors that are the source of most inefficiencies were identified. These factors are low training of employees responsible for land transportation of cargo (especially about knowledge of documents and perishable cargo handling), redundancy of information in formats, non-coordinated actors, and lack of standardization in documents and in activities.

Table 2.2 Times involved in the ship process

2. Ship	Average time (h)
2.1 Arrange Transport	6
2.2 Prepare Documents	15.5
2.3 Consolidate Cargo	
2.4 Transport cargo to port	0.5
2.5 Transfer cargo to ship	48
2.6 Issue Documents	60
Total	130
	5.4 days

The Ship macro-process is very complex due to the high number of documents and stakeholders involved in this logistic circuit, considering also the perishability of the product and the dynamism of the international markets. Consequently, any delay affects significantly the competitiveness of the logistic chain.

Different recommendations can be provided to improve the current situation. One of the most important activities that may affect different process is related to the design and implementation of a paperless system, where the current documental procedures can be done electronically. As a first step, standardization of documents is a mandatory step that may influence all the processes. For instance, a paperless system may reduce current delays when information is exchanged, and delays at the gate that are partially due to trucks arriving at the gate without completing authorization procedures, then these trucks are rejected generating delays for the rest of the trucks at the queue.

Based on the AS-IS diagrams presented herein, and on the information provided at the interviews, Table 2.3 presents a summary of the main findings and current challenges for each analyzed process.

Different recommendations can be provided to improve previous challenges, observed in the current situation. One of the most important activities that may impact different process is related to the design and implementation of a paperless system, where the current documental procedures can be done electronically. As a first step, standardization of documents is a mandatory step that may impact all the processes. For instance, a paperless system may reduce current delays when information is exchanged and delays at the gate that partially are due to trucks arriving at the gate without completing authorization procedures, then these trucks are rejected generating delays for the rest of trucks at the queue.

2.3.3 Modelling the TO-BE Apples Export Process

Based on the preceding analysis, this section aims to detail the proposed redesign of the *Arrange Transport* macro-process, describing standards that are used, the improvement proposal and the benefits it brings. The current situation of increased

Table 2.3 Opportunity areas per processes of the apple export logistic chain

Processes	Challenges per process	Common challenges
2.1 Arrange Transport	Lack of training of truck drivers	Re-digitation of information Lack of coordination of parties involved in the export logistic chain Lack of standards for export documents and activities
2.2 Prepare documents	Lack of standards for export documents Delays of documents and information exchanges with custom agents. Delays when registering data at the Terminal Operator System	
2.3 Consolidate cargo	Shipping companies require Gate-Out payment before the assignment of an empty container Insufficient availability of empty containers Insufficient inspectors of the agricultural and livestock services for inspections at packings and cross-docking	
2.4 Transport cargo to port	Delays and long queues at gate of the port terminal Insufficient reach-stackers equipment at the port terminal	
2.5 Transfer cargo to ship	Delays for the stowage planning confection	
2.6 Issue documents	Inflexibility for modifying export documents	

international trade and traffic related data has meant that traditional systems cannot meet the demands of processing. Besides, the supply chain actors work with numerous documents and forms with repeated and redundant data, for example in the case of the apple supply chain the name of the exporter appears 17 times in different documents. Another aspect is that documents are not governed by standards, so completing them becomes a complex process.

Excessive physical documentation flow causes: longer times in the process, increased use of resources and consequently an increase in transaction costs. The experiences in both private and public areas suggest that electronic documentation and use of standards in both documents and processes favour conducting business transactions; this is also the first step towards the presentation of electronic documents and customs automation.

The proposal presented in this section is related with the introduction of standardized documents and the creation of a unique system where different actors in the chain can interact to address the two issues identified as major problems: data redundancy and lack of standardization.

The benefits of information systems for submitting business documentation may include: increased speed in processing documentation, tax revenue growth and faster payments, a compilation of trade statistics more accurate and faster, lower costs per transaction, fewer documents, easier to complete documents, etc.

The proposed system allows prepare, transmit and store digital documents, with benefit to the actors of the chain under study. For example, for industry actors, this system brings the benefit of standardization documents, decreasing time and cost of operations and information integration. The objective of this initiative is increasing the competitiveness of the apple industry against its international peers, facilitating export of fresh apples to different foreign markets at a competitive price.

So, the proposal is twofold, the identification and standardization of the documents considered key and the development of an information system that allows the creation, storage and sharing of documents the different actors in the chain require originate. Figures 2.6 and 2.7 show the BPMN diagram for Arrange Transport, indicating the documents related to the activities carried out by different actors and which of these activities will be carried out in the system.

The impact of this proposed redesign can be displayed on the standardization of documents, reduction of time/operating costs, integration of information and standardization of procedures. Note that in the first stage of implementation, the proposed redesigned processes and considers only those most accessible and documents critical to the actors, the advantage of this stage is to better coordinate the actors and progressively integrating documents. With this initiative, the work of customs agents and exporter, through standardization and digitization of documents is easier.

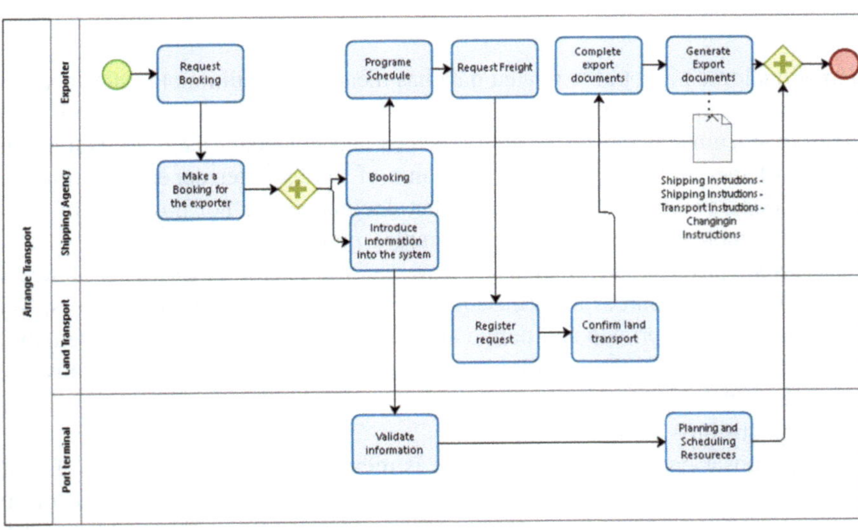

Fig. 2.6 BPMN diagram arrange transport process indicating the different documents

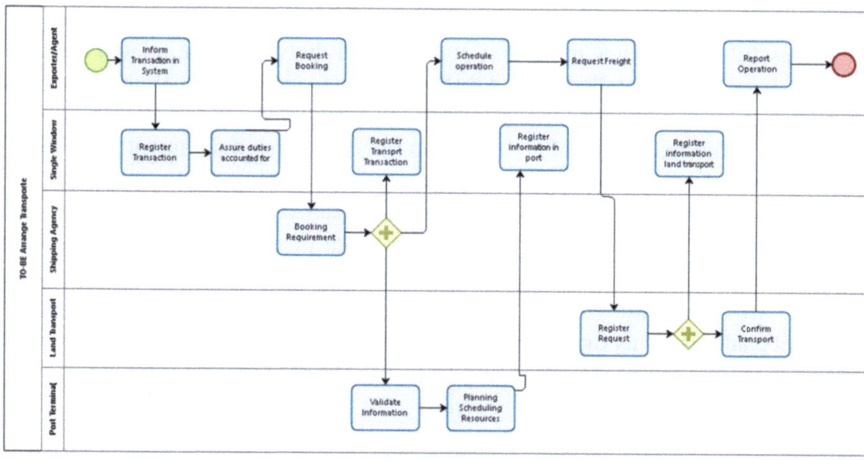

Fig. 2.7 BPMN TO-BE diagram arrange transport process

2.4 Discussion and Conclusions

The successful implementation of trade facilitation measures not only requires the political and government support in terms of policy guidelines, human and financial resources, but also an accurate diagnosis of the existing business processes.

According to the UN-ESCAP, the transition to a Single Window system is done step by step. Before undertaking any other trade facilitation activity, the current processes are modelled to understand current situation and identify opportunity areas. Then, it is important to simplify, harmonize and automate trade procedures and documents.

In this paper, the apples export supply chain business processes have been modelled considering the case study of Chile. For the country, fresh fruits represent an important export product, and particularly grapes and apples are one of the main fresh fruits exported by the country in terms of volume. Fresh fruit export business processes are complex, given the large number of documents and actors involved in the logistics circuit and the speed of operations required to carry the product to target, within conditions and agreed dates. In comparison to other type of products, fresh fruit export business is more dynamic and is affected by the product perishability. Consequently, it is prone to errors and mistakes, as well as subject to delays and high levels of congestion during the fruit export season. Delays at some point in the chain can provoke millions in losses to the exporter. For these reasons, modelling and subsequent automation of shipping operations is a major issue.

Regarding the analysis of the current situation of shipping activities, six processes have been determined. Each of them is vital when exporting apples and take place chronologically in the order: 1. *Arrange Transport,* 2. *Prepare Documents,* 3. *Consolidate cargo,* 4. *Transport cargo to port,* 5. *Transfer cargo to ship* and 6. *Issue documents.* Within each process, several stakeholders participate with their own documents and procedures. According to the UN-ESCAPE recommendations and successful case studies in several ports around the world, the related business processes have been modelled using UML standard. Generated AS-IS diagrams and in-depth interviews with personnel allowed to identify opportunity areas. In general, we notice that the main obstacle to speed up the documented shipment activities is the lack of standardized information and the lack of training.

Modelling of the *Buy* and *Pay* macro-process can create an integrated system that allows the creation, transfer, endorse and store of each document. This integrated system can serve to coordinate in a better way the actors and processes involved. An impact study of the current paper-based procedures on the logistics costs can be addressed as further research, with the aim to quantify the benefits of a standard paperless-based operation and to justify the required investments.

References

Ahumada O, Villalobos JR (2009) Application of planning models in the agri-food supply chain: a review. Eur J Oper Res 196(1):1–20

Ascencio L, Gonzalez-Ramirez RG, Bearzotti LA, Smith NR, Camacho-Vallejos JF (2014) A collaborative supply chain management system for a maritime port logistics chain. J Appl Res Technol 12(3):444–458

Aung MM, Chang YS (2014) Temperature management for the quality, assurance of a perishable food supply chain. Food Control 40(1):198–207

Bae H (2016) The moderating effect of logistics information systems on inter-organizational collaboration and performance of Korean Shipping and Logistics Firms. Int J e-Navigation Marit Econ 5:085–096

Betke V, Kittel K, Sackmann S (2013) Modelling controls for compliance—an analysis of business process modelling languages. Advanced Information Networking and Applications Workshops (WAINA), 2013 27th international conference on advanced information networking and applications workshops

De Mel P, Jayarathe S, Premarathe P (2011) Improving import—export procedures and processes in Sri Lanka. Asia-Pacific research and training network on trade. Working paper series, no. 91, January 2011

De P (2011) Trade facilitation in India: an analysis of trade process and procedures. Asia-Pacific research and training network on trade. Working paper series, no. 95, February 2011

Dumas M, Pfahl D (2016) Modelling software process using BPMN: when and when not? In: Managing software process evolution. Traditional Agile and beyond—how to handle process change

Dumugo C, Tiwari A, Alcock J (2013) Modelling information flow for organisations: a review of approaches and future challenges. Int J Inf Manage 33:597–610

García-Holgado A, Hernandez-García A, Llorens-Largo F (2015) Analysis and improvement of knowledge management processes in organizations using the business process model notations. New Inf Commun Technol Knowl Manag Organ 222:93–101

Goulding C (2005) Grounded theory, ethnography and phenomenology: a comparative analysis of three qualitative strategies for marketing research. Eur J Mark 39(3):294–308

Gromoff A, Bilinkis Y, Kazantseu N (2017) Business architecture flexibility as a result of knowledge-intensive process management. Glob J Flex Syst Manag 18(1):73–86

Helpman E (2014) Foreign trade and investment: firm-level perspectives. Economica 81:1–14

Islam S, Olsen T (2014) Truck-sharing challenges for hinterland trucking companies: a case of the empty container truck trips problem. Bus Process Manag J 20(2):290–334
Jean S, Mulder N, Ramos M (2014) A general equilibrium ex-post evaluation of the EU-Chile Free Trade Agreement. Econ Model 41:33–45
Keretho S, Naklada S (2011) Analysis of export and import processes of selected products in Thailand. Asia-Pacific research and training network on trade working paper series, no. 103, June 2011
Ksoll C, Myint A, Gi T (2013) Business process analysis export of rice and mango, and import of palm oil in Myanmar. Final report in SatNet Asia, December 2013
Lan H, Tian Y (2013) Analysis of the demand status and forecast of food cold chain in Beinjing. J Ind Eng Manag 6(1):346–355
Laursen K (2015) Revealed comparative advantage and the alternatives as measures of international specialization. Eurasian Bus Rev 5:99–115
López-Campos M, Cannella S (2011) On business process modelling in collaborative supply chains. In: ICOVACS 2011, 14–16 2011, Leuven, Belgium
Manzini R, Accorsi R (2013) The new conceptual framework for food supply chain assessment. J Food Eng 115(2):251–263
Mehmann J, Teuteberg F (2016) Process Re-engineering by using 4PL approach: a case study on transportation processing in the agricultural bulk logistics sector. Bus Process Manag J 22 (4):879–902
Merk O, Notteboon T (2015) Port hinterland connectivity. International Forum Discussions. Paper no 2015–13
Min S, Roath AS, Daugherty PJ, Genchev SE, Chen H, Arndt AD, Richey RG (2005) Supply chain collaboration: what's happening? Int J Logistics Manag 16(2):237–256
Monios J, Wilmsmeier G (2013) The role of intermodal transport in port regionalisation. Transp Policy 30:161–172
Murthy DS, Sudha M, Gajanana TM (2009) Marketing and post-harvest losses in fruits: its implications on availability and economy. Indian J Agric Econ 64(2):259–275
Nakandela D, Samaranayake P, Lau H (2015) A representative model for information sharing needs in fresh food Supply Chain. In: Proceedings of the 13th ANZAM operations, supply chain and services management symposium, 31 May–2 June, 2015. Melbourne Victoria
Neary P (2016) International trade in general oligopolistic equilibrium. Rev Int Econ 24(4):669–698
Nedic V, Despotovic D, Cvetanovic S, Despotovic M, Eric M (2016) Innovation of IT metasystems by means of event-driven paradigm using QDMS. Enterp Inf Syst 10(8):893–910
Odepa (2013) Apples: a season of high value of exports. Agricultural Studies and Policies Office, Ministry of Agriculture, Chile (ODEPA). Website: www.odepa.gob.cl
Palma-Mendoza J, Neaily K (2015) A business process re-design methodology to support Supply Chain integration: an application in an Airline MRO SC. Int J Inf Manag 35:620–631
Rajkarnikar P (2010) Improving regional trade procedures and processes: a business process analysis of export of vegetable ghee from Nepal to Indica and China and Import of Textile from Indiato Nepal. Asia-Pacific research and training network on trade working paper series, no. 89, December 2010
Ramasany B (2010) An analysis of import-export procedures and process in China. Asia-Pacific research and training network on trade working paper series, no. 88, December 2010
Rodrigue JP, Notteboon T (2009) The terminalization of supply chains: reassessing the role of terminals in Ports/Hinterland logistical relationships. Marit Policy Manag 36:165–183
Rodrigue JP (2012) The benefits of logistics investments: opportunities for Latin America and the Caribbean. Inter-American Development Bank. Technical Notes, IDB-TN-395
Saifuddin H (2010) Simplification of trade presentation on processes and procedures in Bangladesh: results from Business Process Analysis (BPA) of export and import procedure for selected Commodities. CPD-ARTNeT Dialogue on Trade Promotion Through Trade Facilitation Dhaka

Soto-Silva W, Nadal-Roig E, Gonzalez-Araya M, Pla-Aragonés L (2016) Operational research models applied to the fresh fruit Supply Chain. Eur J Oper Res 251:345–355

Stavirou V, Kandias M, Karoulas G, Gritzalis D (2014) Business process modeling for insider threat monitoring and handling. In: Trust, privacy and security in digital business, vol 8647 of the series lecture notes in computer science, pp 119–131

Succi G, Predonzani P (2000) Business process modelling with objects, costs and human resources. In: Systems modeling for business process improvement. Artech House, London, 2000, pp 47–60

United Nations ESCAP (Economic and Social Commission for Asia and the Pacific) (2012) Business process analysis guide to simplify trade procedures

Van der Vorst JGAJ, Beulens AJM (2002) Identifiying sources of uncertainty to generate supply chain redesign strategies. Int J Phys Distrib Logistics Manag 32(6):409–430

Van der Aalst W (2003) Don't go with the flow: web services composition standards exposed. IEEE Inteligent Syst 18(1):1–7

Verdouw CN, Beulens AJM, Trienekens JH, Wolfert J (2010) Process modelling in demand-driven supply chains: a reference model for the fruit industry. Comput Electron Agric 73(2):147–187

Wilmsmeier G, Monios J, Rodrigue JP (2015) Drivers for outside in port hinterland integration in Latin America: the case of Veracruz, México. Res Transp Bus Manag 14:34–45

Wilmsmeier G, Sanchez R (2017) Evolution of national port governance and interport competition in Chile. Res Transp Bus Manag. Available online 17 January 2017

Chapter 3
Considerations for the Latin American and Caribbean Region in Light of the Global Move Towards Low Carbon Shipping

Sukhjit Singh and Vivian Rambarath-Parasram

Abstract The developed world has made firm steps towards mitigating the impact of shipping to climate change through both national and regional actions. The European Union has been leading the charge, making billions of dollars available for climate action and supporting initiatives within various international organizations, within the framework of the United Nations, such as the International Maritime Organization, United Nations Development Programme and the United National Economic Commission for Latin America and the Caribbean. While climate change mitigation, adaption, and resilience are complex issues affecting countries and regions in different ways, this chapter will highlight some realities and developments on climate action in the international maritime industry that will impact the economies of the Latin American and Caribbean region. The region must pay attention to these developments in order to capitalize on opportunities for capacity building, knowledge transfer, to successfully overcome well documented barriers to the implementation and adoption of maritime technological advances related to mitigating climate change.

Keywords Climate change mitigation · Maritime transport · Energy efficiency

3.1 Introduction

International shipping provides a vital link in the world economy with 80% of international trade dependent on ships (UNCTAD 2017), shipping lines are critical to the global economic growth. The need for shipping is in direct response to demand of transport by international commodity trade. Being an economic, reliable and safe mode of transport, ships providing sea mode of transportation are key

S. Singh (✉) · V. Rambarath-Parasram
Centre for Maritime and Ocean Studies, The University of Trinidad and Tobago, 2nd Avenue North Western Main Road, Chaguaramas, Trinidad and Tobago
e-mail: sukhjit.singh@utt.edu.tt

J. L. García Alcaraz et al. (eds.), *Best Practices in Manufacturing Processes*,
https://doi.org/10.1007/978-3-319-99190-0_3

enablers of globalization (Rehmatulla and Smith 2015). Maritime transport accounts for the largest share of tonne-kilometres. In the baseline scenario of ITF study 2017 (OECD/ITF 2017) growth in international shipping is predicted to grow from 71% in 2015 to 75% by 2050. Maritime shipping remains the most carbon efficient and cost-effective medium of transport for long-distance trade, accounting for around 80% in volume and over 70% in value of global trade.

Decarbonization of shipping is a major challenge to the global shipping industry. The third GHG study led by International Maritime Organization, updated the future of GHG emissions in shipping based on various scenarios taking into consideration several factors and in particular the growth in world trade (IMO 2015). Greenhouse gas emissions from shipping presented around 3% of total global emissions. This study also established the clear link between trade evolution and air emissions. With maritime transport flows expected to grow, GHG emissions from shipping are projected to increase by 50–250% by 2050. According to some other studies, the result could be that the share of global shipping emissions increases to 17%. These predictions of increased trade have also led to rising concerns on the associated GHG emissions caused by the combustion of fuel on-board ships. The Paris Climate Agreement requires that contracting parties undertake measures to keep the global temperature rise below 2 °C above pre-industrial levels. Shipping however, due to its international and complex nature is not included in the scope of these measures.

This paper refers to the Latin American and Caribbean region as those countries falling within the purview of the Economic Commission for Latin America and the Caribbean (ECLAC). The 33 countries of Latin America and the Caribbean (LAC) along with several Asian, European and North American nations that have historical, economic and cultural ties with the region, comprise the 46 member states of ECLAC. Fourteen non-independent territories in the Caribbean are Associate members of the Commission (ECLAC 2018) (Table 3.1).

In this regard, the paper will provide an overview of the regional approaches to climate action with specific focus on the maritime sector. It also brings into focus the challenges, opportunities and realities facing the regional maritime industry in response to emerging international standards driving towards low carbon shipping.

3.2 The Global Challenge of Climate Action in the Context of the Latin American and Caribbean Region

Contemporary society is faced with the challenge of understanding the economic, social and environmental benefits of achieving the sustainable development goals (SDGs) and preventing climate change. As a matter of urgency, a comprehensive understanding is required of the interrelationships between the human development, economy growth, technology advancement and environmental impacts. This understanding then must be translated into effective policies and plans. In Latin

Table 3.1 Member states and associate members of ECLAC (ECLAC 2018)

Member states		
Antigua and Barbuda	France	Peru
Argentina	Germany	Portugal
Bahamas	Grenada	Republic of Korea
Barbados	Guatemala	Saint Kitts and Nevis
Belize	Guyana	Saint Lucia
Bolivia (Plurinational State of)	Haiti	Saint Vincent and the Grenadines
Brazil	Honduras	Spain
Canada	Italy	Suriname
Chile	Jamaica	Trinidad and Tobago
Colombia	Japan	Turkey
Costa Rica	Mexico	United Kingdom of Great Britain and Northern Ireland
Cuba	Netherlands	United States of America
Dominica	Nicaragua	Uruguay
Dominical Republic	Norway	Venezuela (Bolivarian Republic of)
Ecuador	Panama	
El Salvador	Paraguay	
Associate members		
Anguilla	Curaçao	Puerto Rico
Aruba	French Guiana	Saint Maarten
Bermuda	Guadeloupe	Turks and Caicos Islands
British Virgin Islands	Martinique	United States Virgin Islands
Cayman Islands	Montserrat	

America and the Caribbean, the much-needed shift towards a more sustainable development pattern has specific characteristics. In recent years, ECLAC has analysed extensively the constraints on the region's economic, institutional and social development, and the urgent need for a leap towards sustainability.

The region has been slow in taking decisive action on climate change mitigation and there are economies and societies within the region that are highly vulnerable to the adverse impacts of climate change with production structures and consumption patterns that are carbon intensive. In a 2015 report, ECLAC also stated that a more egalitarian and socially cohesive society that is on a sustainable development path will be less vulnerable to climate related shocks and will enable a better capacity to meet mitigation targets. It further noted that the region is in an asymmetrical position in relation to climate change as countries in the region has historically had a small contribution to climate change, yet it is highly vulnerable to its effects. Based on a 2011 study the region contributes 9% of world emissions. The economic costs of climate change to the region are estimated to be 1.5–5% of the region's Gross Domestic Product (GDP) (ECLAC 2015).

ECLAC also acknowledges that the current development pattern in the region is unsustainable. The region is sustained by a boom in exports and prices of renewable and non-renewable natural resourced which has helped to reduce poverty and improve social conditions but with considerable impact to atmospheric pollution and the climate. Adapting to climate change requires that the region effectively embrace global efforts underway in instituting mitigating measures needed to meet climate targets. However, it also asserts that climate change can only be met by building consensus founded on the acceptance of common but historically differentiated responsibilities. Climate change, the reports states, manifests and intensifies the economic, social and environmental consequences and pressures associated with the current development lifestyle and these challenges can be dealt with by transitioning to a sustainable form of development (ECLAC 2015).

Over 80% of the population of region lives in urban areas and cities. While the cities in the region do not account for a significant percentage of the world's GHG emissions, the small islands States of the region are particularly vulnerable to sea level rise and extreme weather events such as hurricanes and droughts. Cities are often cited as the main drivers for economic growth and perhaps this is why the aforementioned report concludes that cities have a wide array of tools at their disposal for implementation and mitigation measures. The report states that they have the power to zone their territories, plan land use, invest in eco-efficient socially inclusive infrastructure, set up disaster prevention and early warning systems and develop economic and regulatory instruments to promote reductions on GHG emissions (ECLAC 2015). These submissions present a very naïve perspective on the autonomy and the funding available to cities. Most cities in the region are highly dependent on subventions from their respective central Governments for recurrent and capital expenditure. Consequently, a city's ability to aggressively impact a climate agenda in the absence of a national agenda on sustainable development and climate action is naively ambitious.

ECLAC has undertaken a comprehensive review of the region on the progress of the sustainable development agenda 2030. The report comes out of the Forum of the Countries of Latin America and the Caribbean on Sustainable Development which was held in Mexico City in 2017. At this inaugural meeting, six pillars were defined for action and cooperation around the 2030 Agenda namely (ECLAC 2018):

- The creation of an inter-institutional and inter-sectoral architecture at the highest level in each country;
- The incorporation of SDG's into national budgets and development plans;
- The strengthening of national statistical capacities;
- The need to prioritize the means of implementation (financing, technology, trade and accountability);
- The strengthening of regional architecture; and
- The promotion of dialogue between government, private sector, and citizens, fostering coordination with the united nations system, regional organizations and development banks.

This report highlights the importance of global, regional and national coordination to facilitate realization of the goals of the 2030 Agenda. It also underscores the need for the region to strengthen intraregional trade, promote a trade facilitation agenda; nurture industrialization and innovation by increasing local and regional contents in export and work together to build global and regional fiscal governance. While the need for managing maritime interests and infrastructure is implicit in these undertakings, its importance for the success of these measures cannot be overstated. While the region has not mapped the impact of emissions from shipping and maritime infrastructure such as ports, studies out of Europe bear sufficient evidence to make maritime interventions for climate action and well-being of coastal communities an imperative. One such study has highlighted that emissions from international ship traffic are responsible for external costs related to impacts on human health of €58 billion per year (Brandt 2011).

Sustainable development should be ensured not only from an environmental perspective but also from the port development and extensions. It should also not be limited to the existing ports but extended to the hinterland transport operations. Therefore, for any port project environmental sustainability is as vital as economic viability. This demands a very systematic approach where impacts from ports and associated development are considered within wider environmental policies. Hence it is imperative to highlight the policy effectiveness, strategic planning and implementation using specific environmental management tools to local governance and regulatory settings. Osthorst and Manz (2012) highlighted that lack of environmental policies and tools in ports can lead to imbalance between benefits and costs associated with such ports, especially for the local community. The imbalances can form the cause of socioeconomic confrontations related to port operations and developments.

In the early 1900s, to accommodate and promote the harbour commerce, navigation and fisheries, the State conveyed the Port tidelands to Los Angeles and Long Beach, as trustees for the people of the State of California. The economic benefits of the Ports are felt throughout the State; however, the environmental impacts of trade are more locally concentrated. It was evident in the report released by South Coast Air Quality Management District (SCAQMD) in year 2000. It raised concerns about the impact of emissions from ships, trucks and trains in the vicinity of the Ports and major transportation corridors. As a mitigation plan ports developed the Clean Air Action Plan (CAAP). This plan was the first of its kind in the country, linking the emissions reduction efforts and visions of the two largest ports in the United States with similar efforts and goals of the regulatory agencies in charge of ensuring compliance with air quality standards. CAAP included the preliminary cost estimates for selected strategies, economic and workforce considerations and potential emission reductions. The CAAP identified strategies to reduce pollution from every source—ships, trucks, trains, harbour craft (such as tugs and workboats), and cargo-handling equipment (such as cranes and yard tractors). Since 2005, these strategies have resulted in emission reductions exceeding 85% for particulate matter, 50% for nitrogen oxides, and 95% for Sulphur oxides (POLA 2014). The CAAP 2017 update provides new strategies and emission reduction targets to cut emissions from sources operating in and around the Ports, setting the Ports firmly on the path toward zero-emissions goods movement. See Fig. 3.1.

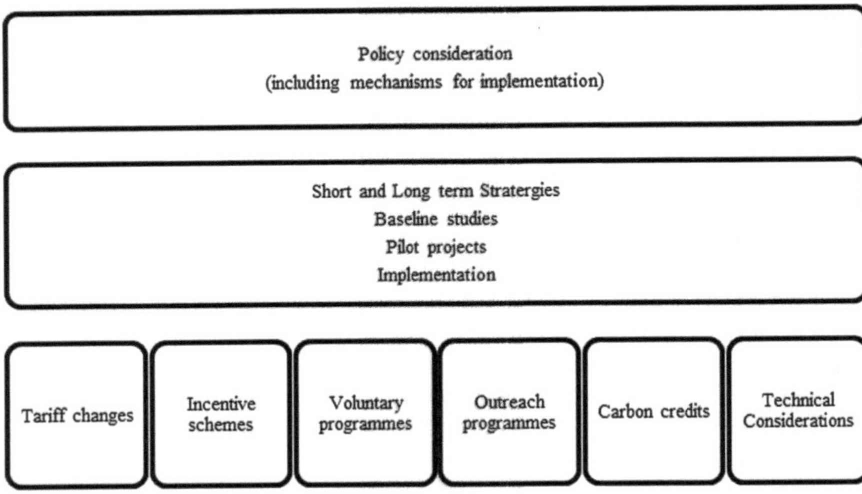

Fig. 3.1 Approach towards reducing port generated GHG emissions

3.3 Impact of Climate Change on Coastlines, Maritime Infrastructure and Shipping

The reports of the Intergovernmental Panel on Climate Change (IPCC) have confirmed that the coastal areas of Latin America and the Caribbean are being impacted by climate change. These changes adversely impact ports and maritime infrastructure. Within the region, seal level appears to have climbed between 2 and 7 mm per year between 1950 and 2008 (ECLAC 2015). Adverse impacts to port operations and other maritime infrastructure will directly impact the regions' economies as existing infrastructure need to be reassessed and redesigned to facilitate low carbon emissions in alignment with emerging global standards and now entrenched shipping standards regulating emission from ships.

3.4 Existing Challenges and Barriers to Energy Efficiency in the Maritime Shipping Sector

The energy efficiency gap is an inconsistency between actual and optimal use of energy and is often explained due the existence of barriers. This existing gap between current and optimal use was defined by Jaffe and Stavins (1994) as the energy efficiency gap. The energy efficiency gap concept has been debated between economists and technologists. In general, where economists consider non-adoption of energy saving measures is due to their economic inefficiency. Technologists point it to the failure of cost effectiveness. But in the maritime domain there are operational measures involving human interface in addition to these technologies, which can inhibit or persuade energy-efficient measures. The IPCC divided the barriers to energy efficiency into four major categories—limited capital access, lack of skilled personnel, lack of relevant information and other barriers (IPCC 2013). However, given that the causes of barriers are often interlinked, it is difficult to find the exact reason for the lack of action.

Even in the presence of economic and environmental benefits associated with controlling GHG emissions and uptake of energy efficiency, there are known barriers to the adoption of new technologies. These barriers, especially in the realm of institutional and technological capacities of the region, can be defined as the mechanisms inhibiting the diffusion of energy-efficient measures. As highlighted by Jafarzadeh and Utne (2014) there exists an intra and inter organizational barriers in addition to economic and technological barriers, split incentives, crew responsibility, commercial contracts, maintenance and measurement of equipment. The non-market failures in the context of shipping are also highlighted in the literature (Rehmatulla and Smith 2015). Even when organization behaves rationally for existence, the obstacles contributing to the slow adoption and diffusion of energy efficiency measures are called non-market failures. A summary of obstacles to energy efficiency in relation to shipping is highlighted below.

3.4.1 Heterogeneity in Ships

As highlighted by Wang et al. (2010) the cost incentives and cost of reduction in emissions for each option varies widely as a function of ship type, age and size. Therefore, the technology or operational options available do not offer the same level of cost savings and level of efficiency for various ship owners and operators. For example, the reduction in fuel consumption and CO_2 emissions is much greater on container ships (fast moving) compared to bulkers and tankers (relatively slower vessels). Thus, such options can be an overstatement for the opportunity available for a particular sector when compared to the others given the heterogeneity of the factors.

3.4.2 Risk

As highlighted by Faber et al. (2009) in order to cover the investment risk, early adopters of technology are skeptical on the prospects of technology, especially when there is a demand on investment. For example, a ship with innovative design may depreciate faster than expected compared to a conventional design. Such an assessment may not favour the most fuel-efficient ship. Further Sorrell et al. (2004) divide risk into three dimensions.

3.4.2.1 External Risk

This is highly representative of challenges to be tackled by shipping companies and includes policy, regulation, fuel price and overall economic trends. The uncertainties of forthcoming mandatory regulations have to be factored in design which could be potentially implemented in the ships. As fuel costs are dominant in shipping operations and any investment in energy efficiency is shaped on its expectations.

3.4.2.2 Business Risk

Financing cost of the ship and its repayment is a major focus for ship owners. Certain shipping markets stands risk from intertemporal choices such as use of LNG as fuel or development of a new Emission Control Area (ECA). Thus, financial and sectoral trends pose business risk.

3.4.2.3 Technical Risk

This refers to the unreliability of the energy efficiency measure and its technical performance. This will apply equally to the retrofit technologies for existing ships and new build design. Therefore, the risk aversion is seen as a major barrier to energy efficiency.

3.4.3 Hidden Costs

There are substantial hidden costs associated with energy efficiency investments. These may not be cited by analysts performing techno-economic modelling. As highlighted by Koomey and Sanstad (1994) these hidden costs result in overestimation of the energy efficiency potential. Life cycle costs are direct costs incurred in search of optimal energy-efficient measure. Further, project appraisal cost includes the cost of evaluating relevant energy-efficient measure with respect to the profiles of the fleet. Commissioning, maintenance, and decommissioning of any energy efficiency method also contributes to life cycle costs.

The inability to spread the cost over a large number of ships is a challenge for smaller ship owners. Retrofitting of abatement methods such as a waste heat recovery system requires vessel to be out of service and thus are costly to implement unless synchronized with major survey or dry docking. This in turn causes time-lag in the implementation of energy-efficient methods.

Reliability, service, extra maintenance reduces the benefits associated with relevant energy-efficient options. For example, waste heat recovery abatement method recovers exhaust gas heat from boilers and utilizes it for producing steam and electricity. However, if power requirement for the system to be effective is more than the charterers requirement, it will be costly and any saving from less fuel consumption may not be worth.

3.4.4 Access to Capital

Capital rationing is used as the means by firms while allocating capital for investments. This results in realization of investment which will yield an expected return (Bhattacharyya 2011). The competition between projects within a company can place energy efficiency low on priority especially if there exists more cost-effective opportunities (Faber et al. 2009).

3.4.5 Informational Problem

Lack of information, accuracy of information and cost of obtaining information in different forms are the major sources of market failures accounting for an energy efficiency gap. Faber et al. (2011) also suggested that lack of trusted data on relevant energy-efficient measures from an independent third party is a major barrier in implementation. Further adoption is dependent on the perceived credibility and trust by the decision maker in the information provider.

3.4.6 Split Incentives

Split incentives occur when benefits and costs of energy efficiency accrue to different agents involved in the operation of vessels. It arises when two parties involved in a contract have different goals and different levels of information. Due to the complexity of various charter parties existing among ship owners and charterers, split incentives are most likely to occur due to divided responsibility for fuel costs.

EEDI with other indicators of fuel efficiency may not qualify the owner operating energy-efficient ship to command higher charter rates. Failure of owners to obtain premiums of energy efficiency in time charter market presents split incentives for charter and ship owner. Therefore, neither charter rates nor second hand price of a ship, fails to reflect any positive economic benefit of vessels fuel efficiency (Faber et al. 2009).

3.4.7 Bounded Rationality

In addition to economic barriers Sorrell et al. (2004) suggest behavioural and organizational barriers stemming from behavioural science contributing to energy efficiency gap. Bounded rationality tends to make satisfactory decisions instead of optimum decisions. Decisions are more based on rule of thumb than being based on perfect information. It highlights 'satisficing' which is combination of 'Satisfy and Suffice'. Therefore, decisions do not realize the optimal value but seek just 'Good enough'. Bounded rationality speaks of the decision maker's importance to arrive at the acceptable outcome. Being a behavioural and physiological concept, it has a direct impact on decisions (operational and commercial) within shipping operations in general and energy efficiency, in particular. Further as highlighted by Sorrell et al. (2011) orthodox economics relies on unrealistic and mathematical models when assuming human decision-making, which is in contrast to 'Behavioral economics' that decision-making is not only 'Bounded rational' but can be systematically erroneous and biased.

Figure 3.2 captures the above energy efficiency barriers from existing research based on organizational theory, behavioural perspectives, economies, and theoretical concepts (Sorrell et al. 2004, 2011).

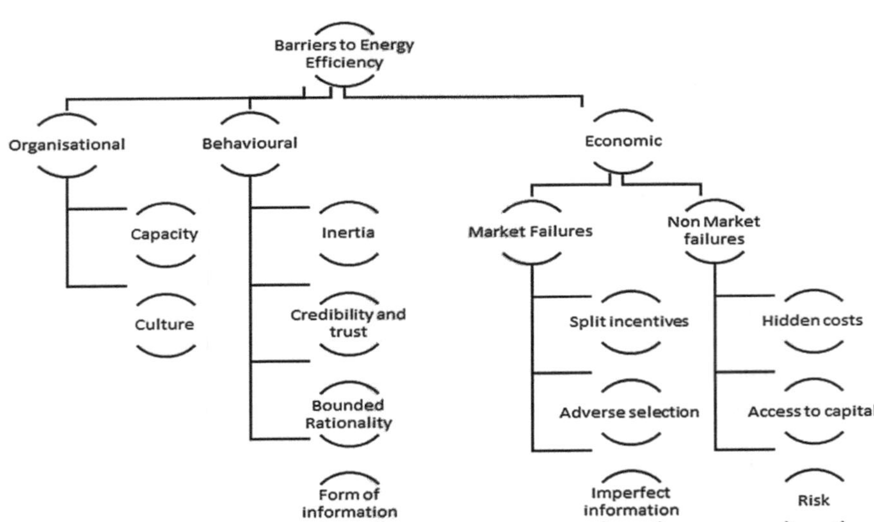

Fig. 3.2 Classification of barriers to energy efficiency (Singh 2017)

3.4.8 *Challenges in Admiralty Jurisdiction*

Another significant advantage of Europe and the developed world is the state of their admiralty law and jurisprudence. Archaic maritime and admiralty laws are the dominant reality of the regional maritime regulatory framework. 75% of Caribbean territories lack the enabling legislation to enforce Annex VI of MARPOL (Rambarath-Parasram et al. 2018). The statistics are perhaps not much different for the Latin American region. As a consequence, many states rely on the principles of 'No More Favorable Treatment' (NMFT) to facilitate PSC inspections. NMFT means that IMO conventions, on entry into force, cover all ships, regardless of the flag they fly as ships of non-convention states are subject to 'no more favorable treatment'. In the absence of enabling legislation in the port states they are in fact limited as to the international conventions they can inspect vessels for compliance. Moreover, if coastal states were to assert jurisdiction using principles of nuisance and common law doctrines to support environmental transgressions they can be challenged by flag states.

3.5 Key Commercial Considerations and Energy Efficiency in Shipping Operations

To provide a distinctive view of the challenges within LAC region it is imperative to understand that cargoes range from dry commodities, liquid energy and various consumer goods including semi-manufactured goods. The freight rates can change overnight and considered volatile based on supply and demand. Demand for shipping can shift suddenly but usually it takes anything between 6 months and 2 years to build a new vessel, supply is slow to respond to such changes. Charterers requiring transportation are on the demand side and shipping companies providing the ship forms the supply side. The commercial conditions settled through freight markets are written under charter parties. These charter parties vary in terms of costs, risk distribution and duration. The cost is usually further subdivided into capital costs, operating costs and voyage costs.

- Capital cost—(investment in ship)
- Operating cost—(maintenance, crew wages, other supplies and marine insurance)
- Voyage costs—(port dues, canal dues, fuel costs)

Based on the distribution of these costs three major types of charter parties exist. Voyage charters also known as spot charters and are where all costs (and risks) are assumed by ship owner and charter pays based on the quantity of cargo carried and rate per unit cargo. Time charters are however beyond the duration of a voyage and can have durations from months up to several years. Under time charter, charters pay based on a daily hire rate for the duration of contract. Capital costs and operating costs are paid by ship owner and charter in addition to daily rate pays voyage costs including fuel costs. Under Bareboat charter, charter pays daily hire rate for the duration of charter. All costs and operational decisions except capital costs are also paid by charterers. Table 3.2 summarizes the type of charter.

Operational aspects of the vessel have a high impact on the energy efficiency. Therefore, how charter party terms are stipulated effect the scope and possibility of energy consumption optimization given guarantees on bunker and speed.

Given the complex nature of commercial operations of ships, it is clear that to benefit most from the energy-efficient measures, each of the parties must have flexibility and incentives to join in such efforts. It is more important that they should not have incentives while contributing to inefficient behaviour. The 'Split incentives' needs to be overcome. They exist especially in a situation where benefits from energy efficiency are reaped directly by the party not paying for it.

There is a dearth of research and literature on the microeconomics of freight rates explicitly impacting energy efficiency on the pricing of contracts. As highlighted above, the vessels can be chartered mainly either via voyage charters or time charters. The differentiation between these two main types of contracts is closely related to the issue of split incentives. In this case, split incentives occur when the costs and benefits of energy efficiency accrue to different agents. In the context of a time charters, the cost of building or upgrading an energy-efficient vessel is borne by the ship owner, while the benefits in terms of fuel cost savings accrue to the charterer. There are no clear indications by prospective charterers for any preferences for vessels utilizing alternative fuels. Therefore, any such investment is left as the last resort until there is a regulatory demand for such investments. Further, lack of any incentives for going beyond the minimum requirements does not promote voluntary use of alternative fuel technologies. However, sharing of the savings in fuel costs can reduce split incentives. Agnolucci et al. (2014) estimate a microeconomic model for TC rates in the Panamax dry bulk market and find that approximately 40% of the energy efficiency savings are shared with the ship owner for the period 2008–2012. Adland et al. (2017) expand the analysis to several dry bulk vessel sizes over a 15-year period and find that only 14–27% of the fuel cost savings are shared with ship owners by way of higher time charter rates, but only during normal market conditions. But this situation is differing when compared to strong market conditions. In strong markets, revenue-enhancing characteristics of a vessel such as speed and capacity outperform the value of fuel savings and in a way award inefficient tonnage. Therefore, presently any pricing of energy efficiency in time

Table 3.2 Summary of type of ship charter

Charter type	Duration	Capital costs	Operating costs	Voyage costs
Voyage charter	Duration of voyage usually few weeks	Ship owner	Ship owner	Ship owner
Time charter	Months to years	Ship owner	Ship owner	Charter
Bareboat charter	Months to several years	Ship owner	Charter	Charter

charters, is market dependent. And thus, the ship owners of energy-efficient tonnage could even be penalized in very strong markets.

While financial considerations and regulatory compliance are noted drivers for energy efficiency, the inadequacy of the enforcement of international standards in the region creates an environment where ship operators do not perceive the need to comply. Government intervention is likely to play a key role in facilitating low carbon technology transfer by developing a suitable policy (Shujing 2012). The business case for the uptake of energy efficiency can only be made when operating costs are reduced or capital costs reduced when compared to existing operational conditions.

National and international policy interventions are critical in achieving low carbon technology transfer. At the national level, domestic policies that provide incentives for the use of low carbon can play a strong role in overcoming cost barriers and developing markets for new low carbon technologies. National level efforts are also required in developing national systems of innovation, actively engaging with international collaborative initiatives, and ensuring appropriate infrastructure is in place to foster technological development and change.

3.6 Policy Interventions Targeting Climate Change Mitigation in the Regional Maritime Domain

3.6.1 United National Environment—Caribbean Environment Program

Within the Latin American and Caribbean region, there a few other initiatives that could be leveraged to complement the regional agenda for climate action. The United Nations Environment championed the creation of the first regional framework for ocean governance in the wider Caribbean region through the founding of the Caribbean Environment Programme in 1981. In 1983 the Convention for the Protection and Development of the Marine Environment in the Wider Caribbean Region (Cartagena Convention) was adopted and entered into force in 1986. The convention covers the marine environment of the Gulf of Mexico, the Caribbean Sea, and the areas of the Atlantic, ocean adjacent thereto, south of 30 north latitude and with 200 nautical miles of Atlantic Coasts of States (CEP 2018). Having reviewed the major project and focus of the CEP, it can be concluded that climate action and climate action with particular emphasis on shipping is not an area of priority. While Article 5 of the Cartegena convention mandates that the region take action to combat vessel source pollution, there is no project or initiative of emphasis focussing on air emissions from ships. Cartegena has a member complement of 25 States compared to ECLAC's 46 States and associated members. The Cartegena convention therefore represents a sub-region of the Latin American and Caribbean region.

The United National Environment Programme was established in 1972 to be the primary driver for the global environmental agenda. In addition to the CEP, UNEP has also established the regional office for Latin America and the Caribbean (ROLAC) in Panama City. ROLAC works closely with 33-member states of the region all of which are also members of ECLAC. ROLAC, like the CEP and ECLAC pledges to work with member States and government for enhance national and regional capacity for managing environmental issues. One notable initiative is the Regional Gateway for Technology Transfer and Climate Change Action in Latin America and the Caribbean (REGATTA). REGATTA ironically does not appear to have an emphasis on maritime climate action.

Perhaps the primary reason most of the regional organizations appear to focus on the terrestrial side of climate action is because the Kyoto protocol puts the onus on the IMO to address the impact of air emissions from ships. This view was supported by Nast (2013) that the Kyoto protocol passed the ball to the IMO to regulate ocean shipping emissions through Annex 1 of the protocol which states: The Parties included in Annex 1 shall purse limitation or reduction of emissions of greenhouse gases not controlled by the Montreal protocol from aviation and marine bunker fuels, working through the International Civil Aviation Organization and the International Maritime Organization respectively. The protocol excludes emissions from ships due to the global nature of shipping and in recognition of the complex corporate and regulatory regime that impact international shipping.

3.6.2 International Maritime Organization

The International Maritime Organization was formed through an international conference in 1948. It was previously referred to as the Inter-Governmental Maritime Consultative Organization until 1982. The primary purpose of the IMO was to create the mechanism for cooperation among Governments to promote safer international shipping. However, with the growth in size and volume of ships, the IMO was entrusted with the international mandate to regulate vessel source pollution. With the initial emphasis on oil pollution and then to chemical and hazardous and noxious substances and then to all aspects of operational waste emanating from vessels, the IMO's mandate to managing vessel source pollution continues to emerge both in scope and complexity. Through years of concerted effort, the IMO was able to facilitate the adoption of the most comprehensive regime for tackling vessel source pollution, the International Convention for the Prevention of Pollution from Ships 1973. It is a core convention addressing accidental and operational pollution. The initial convention did not enjoy significant success until it was amended by the Protocol of 1978. MARPOL 73/78 therefore covers not only accidental and operational oil pollution but pollution by chemicals, goods in packaged form, sewage, garbage and air pollution. This paper focusses on its management of air pollution from ships.

IMO conducted various studies from 2000 till 2015. These studies established the link between ship-based GHG emissions and growth in trade. Highlighting various operational and technological measures to reduce emissions these studies aimed at energy efficiency and emission controls through its MARPOL (Marine Pollution) convention.

Air pollution from ships is managed through Annex VI of MARPOL 73/78/. This includes requirements applicable to the manufacture, certification, and operation of vessels and engines, as well as fuel quality used on-board vessels. It provides and legally binding regime for flag states registering ships in relation to energy efficiency and emission control. Sulphur content from I January 2020 onwards for any fuel to be used on board a ship must be limited to 0.5%. However inside a dedicated Environmental Control Area (ECA), the SOx limits were reduced to 0.10%, effective from 1 January 2015. Therefore, investment in technology and other operational management procedures is also required to comply with mandatory requirements affecting ships under MARPOL Annex VI. International Air Pollution Prevention Certificate (IAPP) is required by all ships of 400 gross tonnage and above (MARPOL Annex VI/6). An International Energy Efficiency Certificate is issued once the vessel clears the survey under the provisions of the Annex VI regulation 5.4. Chapter 4 of annex VI regulations target design of new ships (EEDI) and energy efficiency management plan (SEEMP) for existing ships for improving the energy efficiency of a ship whilst operating at sea and in port.

This commitment of IMO is targeted at reduction of greenhouse gases emitted by ships globally. Under the new Regulation 22A on Collection and reporting of ship fuel oil consumption data, ships of 5000 gross tonnage and above are required to collect consumption data for each type of fuel oil they use, as well as other, additional, specified data including proxies for transport work. These ships account for approximately 85% of CO_2 emissions from international shipping. The initial strategy is expected to include, inter alia, a list of candidates short-, mid-, and long-term further measures, with possible timelines, to be revised as appropriate as additional information becomes available. The data collected under the mandatory reporting system will help inform the Marine Environmental Protection Committee (MEPC) when it comes to adopting a revised strategy in 2023 (IMO 2017).

In response to increasing concerns about the IMO's effort to mitigate climate change from the impact of shipping, the IMO, through its Marine Environment Division has undertaken bold initiatives: Global Maritime Energy Efficiency Partnerships (GLOMEEP) a GEF-UNDP-IMO project aimed at supporting the uptake and implementation of energy-efficient measures in shipping and the Global MTCC Network Project (GMN).

3.6.2.1 GLOMEEP

The GLOMEEP project supported 10 lead pilot countries for the purposes of championing:

- Legal, policy and institutional reforms
- Awareness raising and capacity building activities
- Establishment of public private partnerships to encourage technology transfer.

The Lead pilot countries are Argentina, China, Georgia, India, Jamaica, Malaysia, Morocco, Panama, Philippines and South Africa.

While the GLOMEEP project has offered considerable assistance to the global community in the technical requirements for energy efficiency in shipping, the much-anticipated toolkit/technical report for Legal, policy and institutional reforms remain unavailable at the time of writing. A component and document with such a title is certainly of interest to the maritime industry and national maritime administrations.

3.6.2.2 Global MTCC Network

The GMN project represents another bold move to build capacity across the developing world in undertaking climate action to mitigate the impacts of international shipping. The project is formally titled 'Capacity Building for Climate Mitigation in the Maritime Shipping Industry'. The project is funded by the European Union and implemented by the IMO through contractual partnerships with five centres of excellence across the developing world. The MTCCs act as regional focal points for a wide range of activities including (IMO 2017):

- Building compliance with existing and future international energy efficiency regulations
- Help participating countries develop national energy efficiency policies and measures for their maritime sectors
- Promote uptake of low carbon technologies and operations in maritime transport
- Establish voluntary pilot-data collection and reporting systems to feedback into the global regulatory process.

While the GMN project has an initial 4-year horizon, having started in 2016 with the wrap up of the project in December 2019, the MTCC themselves are expected to champion long-term capacity building for climate action in the maritime sector. Table 3.3 highlights the regions each MTCC is responsible for building capacity.

The individual MTCCs are positioned to have considerable impact in their respective regions through pilot projects, capacity building and stakeholder engagement with all level of industry including policy makers and operational personnel within governmental bodies. Many of the MTCCs have adopted the

Table 3.3 MTCC's regional membership

MTCC	Host Institution	# of Member States
MTCC Caribbean	The University of Trinidad and Tobago	16 Island States of Caribbean
MTCC Latin America	International Maritime University of Panama	17 Countries
MTCC Africa	Jomo Kenyatta University of Agriculture and Technology	42 Countries
MTCC Asia	Shanghai Maritime University	32 Countries
MTCC Pacific	Pacific Community and SPREP	13 Island States

approach of engagement with academia, government, and private sector industry partnerships.

The Latin American and Caribbean region is impacted by the operations of two MTCCs which provides a significant opportunity for the region to make strides in addressing the complex issues of climate action in the maritime industry.

3.7 Considerations in Light of Global Agenda for Moving Shipping into a Low Carbon Industry

3.7.1 Increased Costs of Shipping

The approach for curbing emissions from ships, thus far, has been very technical, with emphasis on technology uptake and capacity building. Given the role of shipping in world trade, the socioeconomic considerations cannot be ignored. The increased cost of operations either due to use of alternative fuels (including low sulphur fuel oil) or abatement technologies such as scrubbers, will ultimately be passed on to consumers. Therefore, the wider question about the impact of the increased costs within logistics chain on trade still needs to be answered, especially in the Caribbean, as the countries are heavily import-dependent.

3.7.2 Need for Monitoring, Reporting and Verification

Monitoring, reporting, and verification systems are essential for the region to facilitate successful implementation of Annex VI of MARPOL 73/78. Effective technology transfer is therefore necessary to achieve energy efficiency and reduction of emissions. Given that many maritime administrations in the region lack sufficient resources and access to appropriate technology, the innovative use of existing institutional infrastructure and especially the market-based measures will be an imperative. Creating the enabling environment and capacity building through a tailored approach will require incentives not only for compliance but towards a behaviour promoting efficient use of energy and a focus on energy abuse related to ship emissions. The aim should be not only to enable technology transfer but sustain and absorb the associated benefits. Therefore, as highlighted through EU proposal 525/2013 a strong and accurate 'monitoring, reporting and verification' (MRV) system must be created as it will provide a robust foundation for market-based measures. Further research into cultural and socioeconomic factors may be required in this sector to support policy decisions (Rambarath-Parasram et al. 2018).

Maritime administrations will need to enhance existing flag and port state control measures. The absence of any existing reporting system related to the use of fuel and emissions regionally, no designated focal points for collecting and analysing such data, limits development of regional MBM. MRV constitutes the first stage for the development of MBM. The MTCC's global network is a step closer to the formation of that baseline and providing platform for such technology cooperation. Mapping the existing technology used on-board vessels by major energy consumers and reporting on fuel type and consumption data, MTCC Caribbean and MTCC Latin America, will provide the much-needed baseline to the regional administrations for facilitating the improved policy frameworks.

3.7.3 Institutional Partnerships

The introduction of stricter environmental regulations, rising fuel costs and market-based measures in the near future will require investments by all stakeholders, including maritime administrations. Even while investments in capacity building and uptake of technology continue, many barriers still exist which can only be cleared with active support and participation from industry stakeholders. The research and data collection, particularly in developing economies is at an embryonic stage and voluntary in nature. Therefore, the support and participation from private operators will help in more informed and shared decisions.

The decision-making and planning process often lacks sufficient coordination and creates the potential for conflicts. Policy coherence can be enhanced by using common platforms to take advantage of interrelationships. This can be further enhanced by conducting comprehensive policy analysis, defining institutional mechanisms for monitoring and collecting data and proposing strategies that will take advantage of identified interrelationships.

3.7.4 *Opportunities to Facilitate Compliance and Enforcement*

The Latin American and Caribbean region is also home to two Port State Control Regimes governed by regional agreements. Port state control is a mechanism through which foreign flagged vessels, that have voluntarily entered a national port, can be surveyed for compliance with generally accepted and relevant international standards. The Caribbean Memorandum on Port State Control (CMOU) as well as the Latin American Agreement on Port State control represents existing avenues that can be leveraged for monitoring, reporting and verification of data. These agreements have given rise to regional databases of ships, and vessel compliance that could be leveraged as an avenue for the assisting in creating of the baseline of the state of technology adoption in the maritime sector.

3.8 Conclusion

The challenge and issues of reducing GHG emissions and achieving energy efficiency in shipping industry has been able to receive considerable attention as industry is highly energy intensive and fuel costs can constitute up to 65% of the voyage operating costs. With attention on emission control and energy conservation, keeping abreast of these international standards will demand the optimal use of energy. While global shipping emissions are regulated by the IMO, emission targets are yet being established. It is expected that IMO's Marine Environmental Protection Committee (MEPC) will decide on Initial GHG Strategy for shipping in April 2018 and which most likely will include targets and measures. However due to the complexity on maritime operations, achieving the optimal energy efficiency results are affected by various factors, which range from various commercial, regulatory and operational agreements.

The change to more stringent fuel specifications will have knock-on effects throughout the global energy system. The cost of these regulatory changes on the shipping industry is unknown, but it is expected to be significant. It will impact shipping lines, refiners, crude producers, bunker suppliers, and emissions and air quality. On the positive side, the reductions in emissions will positively impact human health and well-being, particularly for coastal communities and port cities.

Barriers to emission control, energy efficiency and the effectiveness of existing regulatory frame work are the focus of many researchers. Wider research supports technology as the key to achieve climate change mitigation. With limited research and development into maritime technology being undertaken in the region, technology transfer from developed States to the Latin America and Caribbean is a key requirement for facilitating compliance with the applicable international standards. But the enabling environment required for technology transfer involves the complex interface of human factors, institutional knowledge and institutional capacity.

However, technology transfer is not an end in itself. States of the region must be vested with the appropriate legal and institutional framework, to create the enabling environment to promote technological uptake, and drive compliance and enforcement. The inadequacy of the regional and national institutional framework is often difficult to overcome due to other legislative priorities. Greater political awareness and measures may also unlock the full potential of existing resources as the region shifts towards low carbon shipping.

Even while investments in capacity building and uptake of technology continue, many barriers still exist which can only be cleared with active support and participation from industry stakeholders. The research and data collection, particularly in developing economies is at an embryonic stage and voluntary in nature. It can be considered that despite a body of knowledge, the adoption of best practices, lessons learnt, and modern technologies continues to remain a challenge as part of main stream shipping practices. Countries will need to combine a wide range of policies, from support to technology, research, and development to behavioural measures, adequately supported by the required legal and institutional framework.

References

Adland R, Alger H, Banyte J, Jia H (2017) Does fuel efficiency pay? Empirical evidence from the drybulk time-charter market revisited. Transp Res A Policy Pract 95:1–12

Agnolucci P, Smith T, Rehmatulla N (2014) Energy efficiency and time charter rates: energy efficiency savings recovered by ship owners in the Panamax market. Transp Res Part A 66:173–184

Bhattacharyya S (2011) Energy economics. Springer, London

Brandt J (2011) Assessment of health-cost externalities of air pollution at the national level using the EVA model system, CEEH scientific report no. 3. Centre for Energy, Environment and Health Report Series. Aarhs University, National Environmental Research Institute, Roskilde, Denmark

CEP (2018) UN Environment—the Caribbean Environment Programme. Available from: www.cep.unep.org/cartegena-convention

ECLAC (2015) The economics of climate change in Latin America and the Caribbean. Paradoxes and challenges of sustainable development. United Nations, Santiago, Chile

ECLAC (2017) Retrieved from Economic Commission for Latin America and the Caribbean: https://www.cepal.org/en/estados-miembros

ECLAC (2018) Economic Commission for Latin America and the Caribbean, second annual report on regional progress and challenges in relation to the 2030 agenda for sustainable development in Latin America and the Caribbean (LC/FDS.2/3). United Nations, Santiago, Chile

Faber J, Markowska A, Nelissen D, Davidson M, Eyring V, Cionni I, Selstad E, Kågeson P, Lee D, Buhaug O, Lindtsad H (2009) Technical support for European action to reducing green house gas emissions from international maritime transport. Available from: https://ec.europa.eu/clima/policies/transport/shipping/docs/ghg_ships_report_en.pdf

Faber J, Behrends B, Nelissen D (2011) Analysis of GHG marginal abatement cost curves. Available from: https://www.cedelft.eu/publicatie/analysis_of_ghg_%20marginal_abatement_cost_curves/1155

IMO (2015) Third IMO greenhouse gas study 2014. Available from: http://www.imo.org/en/OurWork/Environment/PollutionPrevention/AirPollution/Documents/Third%20Greenhouse%20Gas%20Study/GHG3%20Executive%20Summary%20and%20Report.pdf
IMO (2017) Global MTCC network. Available from: https://gmn.imo.org/about-gmn/
IPCC (2013) Climate change 2013—the physical science basis, contribution of working group I to the fifth assessment report of the intergovernmental panel on climate change. Cambridge University Press, Cambridge
Jafarzadeh S, Utne I (2014) A framework to bridge the energy efficiency gap in shipping. Energy 69:603–612
Jaffe A, Stavins R (1994) The energy-efficiency gap—what does it mean. Energy Policy 22 (10):804–810
Koomey J, Sanstad A (1994) Technical evidence for assessing the performance of markets affecting energy efficiency. Energy Policy 22(10):826–832
Nast T (2013) The response of the international shipping industry to global climate change. J Marit Law Commer 44(1):29–45
OECD/ITF (2017) ITF transport outlook 2017. OECD Publishing, Paris
Osthorst W, Manz C (2012) Types of cluster adaptation to climate change. Lessons from the port and logistics sector of Northwest Germany. Marit Policy Manag 39(2):227–248
POLA (2014) Actions to reduce greenhouse gas emissions by 2050. Available from: https://www.portoflosangeles.org/DOC/PV_FINAL_POLA_GHG_Assessment_Sept_2014.pdf
Rambarath-Parasram V, Singh S, Aiken D (2018) Overcoming the challenges to maritime energy efficiency in the Caribbean. In: Ölçer AK (ed) Trends and challenges in maritime energy management, vol 6, 1st edn. Springer, Malmo
Rehmatulla N, Smith T (2015) Barriers to energy efficiency in shipping: a triangulated approach to investigate the principal agent problem. Energy Policy 84:44–57
Shujing Q (2012) The analysis on barriers of low carbon technology transfer. Energy Procedia 14:1398–1403. In: 2011 2nd international conference on advances in energy engineering (ICAEE)
Singh S (2017) Determining the role of human factors influencing energy efficiency within a commercial and organisational framework of a shipping multinational company. MS international management (oil and gas). University of Liverpool, Liverpool
Sorrell S, O'Malley E, Schleich J, Scott S (2004) The economics of energy efficiency: barriers to cost-effective investment. Edward Elgar Publishing, Cheltenham
Sorrell S, Mallett A, Nye S (2011) Barriers to industrial energy efficiency: a literature review. Working Paper No. 10/2011. United Nations Industrial Development Organization
UNCTAD (2017) United Nations Conference on Trade and Development review of maritime transport 2017. New York. Available from: http://unctad.org/en/PublicationsLibrary/rmt2017_en.pdf
Wang H, Faber J, Nelissen D, Russel B, St Amand D (2010) Marginal abatement costs and cost-effectiveness of energy efficiency measures. IMarEST, London

Chapter 4
Risk-based Strategic Inventory Supply Model for New Products

María-del-Rosario Sánchez-Vega, Santiago-Omar Caballero-Morales, Diana Sánchez-Partida and José-Luís Martínez-Flores

Abstract Companies must maintain a high competitive level to stay in the global market. This objective is even more complex when considering the introduction of new products (NPs) in the market because these present greater uncertainty and sudden changes in demand. Because of this, an agile Supply Chain (SC) is required to respond quickly to the dynamic behavior of the NP. Thus, the present work contributes with a Risk-based Strategic Inventory Supply Model for New Products (RSISNP) that integrates diverse inventory management tools to address the introduction of NPs within the SC. For this purpose, a comprehensive study was performed with 36 managers of related companies to determine the most important factors that may affect the supply of these products. Then, these factors were integrated as cost elements within an adapted periodic-continuous review strategy which is the basis for the RSISNP model. When tested on a case study it was found that the proposed model can reduce excess-of-inventory and obsolescence costs for NPs for different type of demands and life cycles. Thus, for professionals in the logistics and production area, this model can represent a tool to improve the response of the SC to these factors throughout the development process and launch of NPs.

Keywords Supply chain risk · RSISNP · Inventory · New products

4.1 Introduction

Currently, having the means to innovate products and services has become one of the main priorities for companies (Liao and Marsillac 2015). This is due to the pressure of globalization, the rapid development of technologies and the continuous

M.-del.-R. Sánchez-Vega (✉) · S.-O. Caballero-Morales
D. Sánchez-Partida · J.-L. Martínez-Flores
Universidad Popular Autónoma del Estado de Puebla A.C.,
17 Sur 901 Barrio de Santiago, C.P. 72410 Puebla, Mexico
e-mail: mariadelrosario.sanchez03@upaep.edu.mx

J. L. García Alcaraz et al. (eds.), *Best Practices in Manufacturing Processes*,
https://doi.org/10.1007/978-3-319-99190-0_4

change in the demands from the markets. Because of this, companies compete to be the first to present new products (NPs) within the global and regional markets with the purpose of obtaining a competitive advantage and improve their position.

The great competition that exists, the demanding market, the new distribution platforms, and the accelerated advance of technology, require that the development process of NPs be readjusted to comply with increasingly shorter delivery times. However, the development of NPs requires new structures of tactical knowledge to generate or keep ideas and concepts that may be useful to achieve the desired state for the NPs (Park et al. 2015).

Particularly, the development process of NPs depends of inter-organizational collaboration (Hashiba and Paiva 2016). Due to this, the individual internal processes must be efficient to obtain the best global collaborative result that ensures the successful launch of the NP and the permanence of its innovation (Chaudhuri and Boer 2016).

The supply chain (SC) is one of the fundamental areas in the development process of NPs and its efficient execution can ensure the permanence of the product in the market. This is why research in the management of innovations has emphasized the importance of flexibility in the SC network for the success of innovation in companies (Liao and Marsillac 2015).

The ability to respond quickly to customer requirements requires the rapid execution of the development of NPs and services. This, combined with the efficient movement of the product through the SC, represents a powerful competitive advantage (Khan et al. 2016). Likewise, the SC is involved in the entire development process of NPs, covering risk assessment operations, sustainability, material selection, and approach design, among others (Parlings and Klingebiel 2017).

In this context, the contribution of the present work consists of the development of a Risk-based Strategic Inventory Supply Model for New Products (RSISNP) to make more efficient the SC for NPs. To do this, a comprehensive study was done to determine the risk factors and elements of the SC which are present in the development of NPs. This study provided the theoretical and practical support for the development of the RSISNP model. The advances of the present work are presented as follows: Sect. 4.2 presents the information regarding the importance of the SC within the context of production of NPs. Section 4.3 presents the details of the study performed to identify the most important risk factors within the context of NPs. Later in Sect. 4.4, the analysis of these factors and the development of the RSISNP model are presented. Results of this model are presented and discussed in Sect. 4.5 with a case instance. Finally, Sect. 4.6 presents the conclusions and future work.

4.2 Considerations in the Supply Chain for New Products

The SC is constituted as the fundamental base for the launching of NPs because it contributes with: (a) the establishment of inventory levels, (b) the determination of production lots and the purchase of materials, and (c) the determination of routes for distribution. The efficient integration of the SC in the development of NPs contributes to their permanence through: (a) the adequate selection of suppliers, (b) the establishment of long-term relationships with suppliers, (c) the development of 3PL (third party logistics), (d) the establishment of alternative modes of transport, (e) the integration of suppliers in the process of product development, and (f) the standardization of production (Khan et al. 2016). Therefore, the integration of product design in SC provides companies with strategies to establish a sustainable competitive advantage in a volatile and globally competitive market (Khan et al. 2016).

However, the high competition to have new and innovative products in the market as soon as possible tends to put pressure on innovation project managers (Schilling 2017). This impacts the SC, forcing it to cut time in the analysis processes, reducing normal delivery times, and/or adjusting production times. As shown in Fig. 4.1, this involves *opportunity forces* and challenges given the changes required by the SC to respond to the competitive objective.

As presented in Fig. 4.1, the development process of NPs involves challenges for the SC. Thus, robust processes are required to support the rapid and efficient execution of SC activities and avoid the following risks: effects on the level of service and shortage, variations in the final cost of the product, production delays, excess of inventories, obsolescence, delays in launch dates, and inefficient distribution of the product. Due to the above, it is necessary that the SC has processes and tools to support the management of NPs by reducing risks and collaborating to generate more efficient results in shorter times.

A key challenge regarding the integration of the SC with the development process of NPs is that the functions within it are regularly executed independently of other processes (Pero et al. 2010). Therefore, the tools proposed for the present

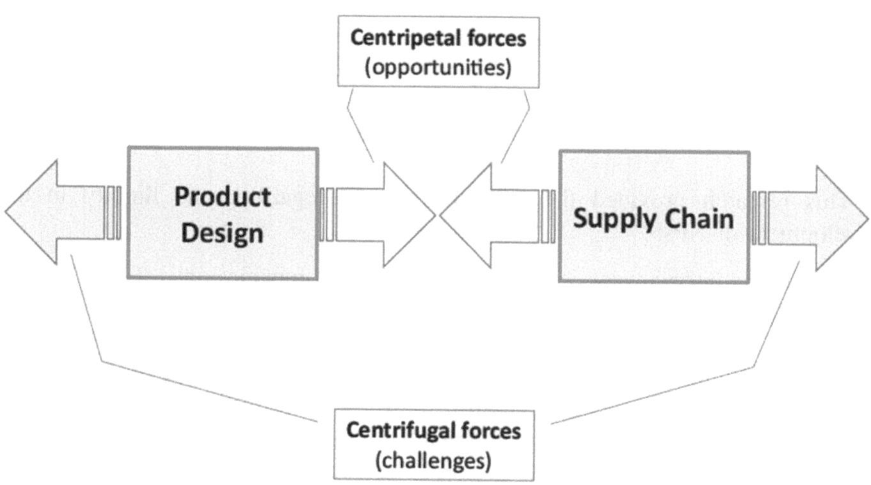

Fig. 4.1 Opportunity forces and challenges in the integration of NPs and the SC (Khan et al. 2016)

context should increase communication between all areas and functions within the SC (Felekoglu et al. 2013; Morales et al. 2013).

In the specialized literature there are works that have addressed the importance of having a process within the SC for NPs while few works have proposed the modification of inventory and production tools as the mean to address the needs of NPs. The most recent work proposed a stochastic inventory policy of continuous review with uncertain demand and temporal dependence through an autoregressive moving average (ARMA) model with explicative variables for NPs (Rojas 2017). Although the model presented in (Rojas 2017) addressed the uncertainty of demand, which is a characteristic of NPs, the additional risks factors that the management of NPs entails were not considered.

To address this situation a Risk-based Strategic Inventory Supply Model for New Products (RSISNP) is proposed. This model consists of a group of tools aimed to mitigate and control the risks and impacts on the SC caused by the need to speed up and accelerate the process of development of NPs. This model can lead to the following benefits: reduction of production costs, reduced time to market needs, and improved customer service.

4.3 Field Research: Identification of Risk Factors

The initial stage of the present research consisted of determining a reference on the affectations and implications that the SC can bring about the lack of interaction and the incorrect administration in the development process of NPs (Hoo-Gon and Jungon 2010). For this, a field research was performed with 36 managers and heads of different industries in the State of Puebla in Mexico. Among the different industries in the State of Puebla, those with the higher presence in the development of NPs were considered. These industries were: Manufacturing (10%), Food (40%), Automotive (35%), Textile (5%) and Services (10%). Finally, to avoid a bias in the results, only people exclusively involved in the process of developing new businesses (40%) and people with secondary roles in the processes (60%) were considered.

This research provided the following results regarding the fluidity in the development of NPs:

- 45% of the interviewees considered that in their companies only 30–60% of the NPs launched in the year were developed under a harmonious, clear and smooth process.
- 30% of the interviewees considered that in their companies only 20% of the launches in the year were successful.
- 25% of the interviewees considered that more than 70% of NPs in the year were designed under a well-defined and organized process.

Regarding the most frequent problems that have been found in the development of NPs, the following were reported: loss of communication between all the involved areas (18%), changes in the original idea of the product and/or service (14%), changes in the release or launch date (14%), confusing or unclear communication (14%), changes in sales volumes (9%), lack of visibility of risks (10%), unilateral decision-making (10%), changes in costs (6%) and others (5%).

About the risks that cause a greater impact on the inventory cost of NPs, the most frequent are presented in Table 4.1. For the objective of inventory management, the costs C_{EI} and C_{OB} have a direct impact on the cost of the NP.

In addition to these risks, others were identified that, even though they do not have a direct impact on the cost of the product, have general financial implications in the company's medium-term SC. These risks are the following: low employee morale, administrative reworks, low quality of the product or materials, conflicts within the organization, and unnecessary investments.

4.4 Strategic Inventory Supply Model

Derived from the literary review and the findings presented in Sects. 4.2 and 4.3, we corroborated the need to generate a model to:

- Support the SC to integrate it into the development process of NPs.
- Reduce the risks associated with the integration process.
- Support the company's strategy to position itself with a competitive advantage with innovations designed, built and sustained by the SC.

To achieve these objectives, we propose a Risk-based Strategic Inventory Supply Model for New Products (termed as RSISNP) which consists of an adaptation of diverse logistic inventory tools specifically designed for the demand and risk characteristics of NPs. Hence, the tools considered within the RSISNP model

Table 4.1 Contribution of each individual risk to the inventory cost of NPs based on their unit cost

Risk cost	Contribution to cost increase (%)
Impact on the service level (C_{SL})	36
Product shortage (C_{SH})	31
Variation in the cost of the product/service (C_{VP})	27
Negative impact on production (C_{IP})	26
Excess of inventory (C_{EI})	25
Obsolete materials (C_{OB})	21
Delayed release/launch date (C_{DR})	20
Early discontinuation (C_{ED})	19
Inefficient distribution of the product (C_{ID})	17
Additional freight (C_{AF})	14

are tools for control and administration of production and inventories, which have been modified and adapted to the needs of the dynamic SC to interact with the development of NPs. Specifically, these tools are aimed at reducing the risks identified in Table 4.1, where the most important are the following:

- *Impact on the service level*: By having a direct impact on the service level, the risks can affect both the NP and all the base products (including raw materials). As consequence, having a service level below the expectation of the client can lead to penalties, loss of sales, expedition costs, and even the potential loss of a client. Therefore, this risk is considered as the one that mostly can affect the cost of NPs.
- *Product shortage*: The risk of not having the quantity of finished NPs at the required time can be present in the launch phase or in the penetration phases after their release. This risk also refers to having enough, but poorly distributed, finished NPs through the different distribution centers. Among the causes of this risk the following can be mentioned: poor production planning, incorrect estimation of safety stock, and unknown variability in NP demand.
- *Variation in the cost of the product/service*: Changes in the optimal quantities of production and in the levels of inventory can directly affect the cost of the product. These changes can generate a risk of unfavorable variation between the established cost and the estimated cost of a NP.
- *Negative impact on production*: This risk is linked to the previous ones and it has its origin in the change of the optimal quantities to be produced and the production cost. Therefore, this risk can negatively affect the NPs as well as the entire company's base product line. As causes of this the following can be mentioned: unbalanced production lines, the increase of line adjustments, readjustments of shifts.
- *Excess of inventory*: This risk, like the previous ones, is also related to the impacts on inventories and production. This risk is present when, due to an incorrect establishment of planning parameters in production lots, purchase lots, inventory safety levels, reorder points or variations in the predicted demand, an over-inventory is generated in both, finished products and raw materials. This causes storage and handling costs of additional inventory over the budgeted one. This situation can also cause a risk of obsolescence.

These risks are directly associated with the inventory, its costs and production. Therefore, these risks are a fundamental part of the supply chain and they are always latent. However, if the SC is integrated efficiently from the initial phases of the development process of NPs, these risks can be reduced through their different phases of development.

Given the above, the RSISNP model consists of the strategic integration of inventory control tools. These tools must have specific characteristics to allow the integration of the SC to the business development process from its initial phases with flexibility. In the same way, these tools must be adapted to the characteristics of the NPs, since the traditional tools of inventory control and production are designed for products with assumptions of stability regarding demand patterns and risks throughout the planning horizon (which does not happen with NPs).

Two main factors which are not certain when considering NPs are: (a) demand and (b) risks over a planning horizon. Although this information is generated through analysis and forecasting tools based on the market studies, this information will always be less certain than that of established base products that already have historical sales data (which improves the forecast of their demand). Since NPs can have different demand patterns at each stage of the NP's life cycle, it is necessary to develop and/or adapt the most appropriate planning tools considering these characteristics.

Table 4.2 describes the inventory control tools that were adapted for each type of demand at each stage of the NP's life cycle. Based on the field research described in Sect. 4.3 it was determined that the annual life cycle of a NP can be segmented into four quarters (four periods of three consecutive months). This is because the uncertainty of the NP's demand can change based on the stage of its life cycle (e.g., the uncertainty may be greater in the first months of the NP's life cycle). Table 4.3 presents the impacts of the associated risks on the NP's inventory costs at each stage of its life cycle and type of demand.

As presented in Tables 4.2 and 4.3, the integration of tools must be specific to the stage of the NP's life cycle and type of demand. As the product gets established in the market (last stages of the NP's life cycle) the uncertainty and risks may decrease, and standard tools can be applicable. In this way, the uncertainty of the forecast for a NP can be managed more efficiently and thus the risks associated with inventories and production can be reduced.

Since these benefits have a direct effect on the planning of inventories, they will be present throughout the different links of the SC. As presented in Table 4.2, the RSISNP model is considered for the early stages of a NP's life cycle. Specifically, this model is integrated by two main non-deterministic models for inventory replenishment: The Periodic Review (P) and the Continuous Review (Q, R) models.

Table 4.2 Relationship of life cycle and type of demand with the adapted tool for NP planning

Type of demand/life cycle	0–3 months	3–6 months	6–9 months	9–12 months
Demand with Seasonality	RSISNP model P + (Q, R)	RSISNP model P + (Q, R)	Standard model used for base products	Standard model used for base products
One-Shot Demand	Newsvendor model	Newsvendor model	Newsvendor model	Newsvendor model
Constant Demand	RSISNP model P + (Q, R)	RSISNP model P + (Q, R)	Standard model used for base products	Standard model used for base products
Demand with Trend	RSISNP model P + (Q, R)	RSISNP model P + (Q, R)	RSISNP model P + (Q, R)	RSISNP model P + (Q, R)

Table 4.3 Percentage of impact of the risks (%Risk) on NP inventory costs in relation to life cycle and type of demand

Type of demand/ life cycle	0–3 months	3–6 months	6–9 months	9–12 months
Demand with Seasonality	100%	75%	Depends of seasonality	Depends of seasonality
One-Shot Demand	100%	100%	100%	100%
Constant Demand	100%	75%	Depends of seasonality	Depends of seasonality
Demand with Trend	100%	75%	50%	25%

These were considered to balance the planning horizon in which the uncertainty of the demand cannot cause significant risks.

The P model can contribute to a volume benefit which lies in the unit cost of the supplies or finished products when the option for larger lots is available with the supplier. This periodic planning can also support the supplier to plan its purchases.

The integration of the P model with the (Q, R) model can reduce the risk of an excess or lack of inventory during the planning horizon defined by the time interval between orders of the P model. Therefore, this integration can result in a planning that interacts with the supplier and the company, where the requirements of the company and its demand pattern for the NP are known quickly. This can reduce the risks due to obsolescence and lack of inventory in all the links of the SC.

Thus, by integrating the P and (Q, R) models, risks and their impacts can be transferred to lower levels where their costs can be less (for example, before raw materials have been transformed into finished products, or before raw materials begin the transformation process). In the same way, this integration can balance the risks and optimal inventory levels of NPs within the company for different periods of time. By optimizing and balancing production and inventories with costs and risks, a less stressed SC can be obtained. This can improve the service levels by allowing the products to flow through the SC in a smoother way (from production and storage to distribution) to reach the right place, with the right quantity at the right time.

Particularly, for the development of NPs, this integrated planning can favor their optimal availability in the launching and growth stages. This is due to the problems inherent in this type of products. Just the delay in a release date due to lack of availability of the NP can lead to several additional costs for the company. A NP requires a prior investment in marketing, advertising, rental of exhibition spaces, promotions, etc., and a delay in the launching date can have significant costs associated with reprogramming and penalties by customers.

This planning can ensure compliance with the release date with the adequate inventory to cover all locations and launch centers, which is vital for the sustainability of the NPs' innovation.

4.4.1 Technical Background

Due to the need for inventory control and production with the uncertainty of demand and planning horizon, combined with additional and specific costs associated to the risks of a NP, the integrated RSISNP model is proposed considering the base models of Periodic Review (P) and Continuous Review (Q, R) for inventory supply with uncertain demand (Adamu 2017; Lieberman and Hillier 2000). The proposed tool can favor the quick adjustment of the supply strategy to changes in demand over time that cannot be adjusted during normal planning cycles. The visualization of the strategies of the base models is presented in Fig. 4.2.

In the Periodic Review (P) model the inventory is planned through a review of the same which is performed every period of time T (time interval between reviews which can be two weeks, a month, etc.) which is fixed. At the time of the review, the estimate of the required inventory (lot size Q) is determined by Eq. (4.1):

$$Q = d \times (T + \mathrm{LT}) + z\sigma\sqrt{T + \mathrm{LT}} - I \tag{4.1}$$

In Eq. (4.1) d is the average demand during the smallest unit of time within T (for example, daily, weekly, etc.), LT is the lead time (it must be defined according to the unit of time in which d is measured), σ is the standard deviation of the demand during the lead time, z is the number of standard deviations considered for a required service level, and I is the inventory level at the time of the review. Because the inventory level at the time of the review can be variable given the non-deterministic nature of the demand, the required inventory Q can be different for all periods.

In the Continuous Review (Q, R) model, the review is performed in short intervals and the lot request (order) is performed when the inventory reaches the level determined by the Reorder Point (R). Therefore, in this model the time between reviews is variable. The lot size Q, which is fixed, is determined by the following iterative process (Nagi 2001):

- Step 1: Start with iteration $i = 0$ and estimate the lot size Q with the following equation:

$$Q_i = \sqrt{\frac{2D[C_\mathrm{o} + pn(R_i)]}{C_\mathrm{h}}} \tag{4.2}$$

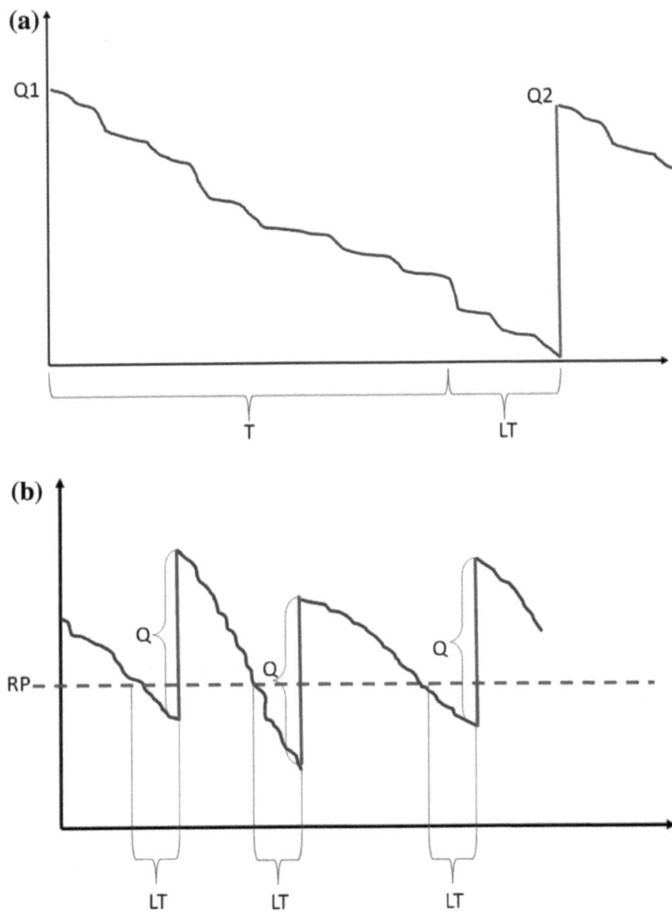

Fig. 4.2 Independent base models of **a** Periodic Review (P) and **b** Continuous Review (Q, R)

For the initial iteration ($i = 0$) the component $pn(R_0) = 0$.

- Step 2: Estimate the reorder point R with the following equation:

$$R_{i+1} = \mu_{\mathrm{LT}} + z\sigma_{\mathrm{LT}}, \tag{4.3}$$

where $z = \Phi^{-1}(F(R_{i+1}))$ and the element $F(R_{i+1})$ is estimated by the equation $F(R_{i+1}) = 1 - (Q_i C_{\mathrm{h}})/(pD)$.

- Step 3: Estimate the number of products not supplied with the following equation:

$$n(R_{i+1}) = \sigma_{\mathrm{LT}} L(z) \tag{4.4}$$

Once this element is estimated, the iteration number is increased ($i = i + 1$) and the process is repeated from Step 1. The process is repeated until there are no significant changes in the estimation of R.

In this model, D is the total demand to be covered or served during the considered planning horizon (for example, semi-annual, annual, etc.), Q_i and R_i are the i-th estimate of the lot size and the reorder point respectively, C_{h} is the cost of maintaining a unit of product in inventory (holding cost), C_{o} is the cost of issuing a lot order (order cost), p is the unit cost of a product not supplied, $n(R_{i+1})$ is the estimated number of products not supplied considering the reorder point, μ_{LT} and σ_{LT} are the average and the standard deviation of the demand of the product during the lead time LT, z is the number of deviations considered for a required service level, and $L(z)$ is the associated loss function.

Another non-deterministic model, known as the newsvendor (newsboy, single-period, or perishable) model, is also used for uncertain demand. In contrast to the P and (Q, R) models, the newsvendor model assumes a single inventory cycle. Thus, this model can be used for perishable products and one-shot demands (a single and limited launch of a product for marketing and promotion purposes) (Stevenson 2017).

Assuming normality, the inventory level with the newsvendor model is estimated by the following equation:

$$Q = \mu + z\sigma, \tag{4.5}$$

where $z = \Phi^{-1}((v - C)/v)$, μ and σ are the mean and standard deviation of the demand, and v and C are the sale price and purchase cost of the product.

4.4.2 *RSISNP Model: Integrated Periodic Review (P) Model with Continuous Review (Q, R)*

Once that the foundations of the base models were obtained, we proceeded to develop the integrated model. This was performed by the following steps:

- Initially it is necessary to define a planning horizon for the NP and the associated net requirement. Therefore, this horizon must be a manageable period of time in accordance to the NP's unique life cycle. In general, for NPs this time should range between one and six months for the early stages of their life cycle (see Table 4.2) and be increased as they reach the maturity stages (e.g., the periods established by each company for the rest of their products). For the integrated model, the planning horizon is determined based on the T + LT interval of the P model. Based on the same model, the lot size Q_P is determined as shown in Eq. (4.1). This lot represents the net requirement to cover during the T + LT period and can support the availability of a manageable base quantity and to delimit the impact of the risks during the different life cycles of the product.

- Once the planning horizon is defined, a continuous review scheme is assumed to cover the Q_P requirement. Therefore, for this general Q_P to be covered throughout the period T + LT, sub-lots are estimated by the (Q, R) model. In Eq. (4.2), this is equivalent to define $Q_P = D$.
- Note that this integration favors the flexibility that the NP's process requires. This is because, in order to define the sub-lots, the variation of the real demand that is found during the NP's life cycle is continuously monitored. As it can be assumed, these adjustments in the early stages of the product will be more frequent and, as the product reaches maturity, these adjustments should be reduced until it can be planned in common with the base products.
- As with any tool, the best way to evaluate it is by means of the economic implications of its implementation. In this case, a total cost equation can provide a tangible parameter to objectively assess the outcomes of the RSISNP model. Thus, the specific associated costs of a NP have to be added into the inventory planning and production processes. This is so that the balance between the sub-lots and the risks can be as real as possible and the estimated lot size can minimize the associated costs-risks. Among the costs associated to NPs which can be integrated into the equation the following can be considered: administration costs, storage costs, obsolescence risk, loss of sale cost, cost of loss of innovation opportunity and competitive advantage.

Thus, the development of the integrated inventory planning model and the associated total cost equation (CT) is presented as follows:

- By considering a planning horizon defined by T + LT, the total requirement to be served in that period is determined by the P model:

$$D = Q_P = d \times (T + \mathrm{LT}) + z\sigma\sqrt{T + \mathrm{LT}} - I, \tag{4.6}$$

where d = average demand of the NP and T = fixed time between periods which depends of the product's stage and it could vary from 3 to 6 months.

- To make the integration of the models, Eq. (4.6) is integrated into Eq. (4.2) which gives the following iterative equation:

$$Q_{\mathrm{NP}i}=\sqrt{\frac{2\left(d\times(T+\mathrm{LT})+z\sigma\sqrt{T+\mathrm{LT}}-I\right)\times[C_{\mathrm{oNP}}+pn(R_i)]}{C_{\mathrm{hNP}}}},\tag{4.7}$$

where Q_{NP} = lot size of the NP, C_{oNP} and C_{hNP} are the order and holding costs associated to the NP respectively. These costs, based on Table 4.1 and the risk percentages presented in Table 4.3, are computed as:

$$C_{\mathrm{oNP}}=\text{standard } C_{\mathrm{o}}+[\text{increase due to initial run}]\times\%\mathrm{Risk}\tag{4.8}$$

$$C_{\mathrm{hNP}}=\text{standard } C_{\mathrm{h}}+[C_{\mathrm{EI}}+C_{\mathrm{OB}}]\times\%\mathrm{Risk}\tag{4.9}$$

Finally, p = is the loss associated to a non-supplied product. For this model, the reorder point $R_{\mathrm{NP}i+1}$ is estimated according to Eq. (4.3).

- The total cost ($\mathrm{CT_{NP}}$) of the integrated model for inventory planning is defined as:

$$\begin{aligned}\mathrm{CT_{NP}}=&\text{ Total Order Cost}+\text{Total Holding Cost of Cycle Inventory}\\&+\text{Total Holding Cost of Safety Stock}+\text{Total Shortage Cost},\end{aligned}\tag{4.10}$$

where

$$\text{Total Order Cost} = (C_{\mathrm{oNP}}\times D)/Q_{\mathrm{NP}}\tag{4.11}$$

$$\text{Total Holding Cost on Cycle Inventory}=C_{\mathrm{hNP}}\times(Q_{\mathrm{NP}}/2)\tag{4.12}$$

$$\text{Total Holding Cost of Safety Stock} = C_{\mathrm{hNP}}\times[R_{\mathrm{NP}}-\mu_{\mathrm{LT}}+\sigma_{\mathrm{LT}}L(z)]\tag{4.13}$$

$$\text{Total Shortage Cost} = pn(R)\times(D/Q_{\mathrm{NP}})\tag{4.14}$$

4.5 Application and Results

In order to exemplify the benefits of using an adequate inventory planning model for NPs, such as the RSISNP model, an application was performed on a case instance. This instance consists of data from a confectionery company which has invested a large amount of money in the development of NPs within the last year. Its most profitable product family is lollipops with a baseline of four products which corresponds to four flavors: strawberry, grape, blackberry and watermelon.

For this company, the market studies and prototypes have shown that a multi-flavor lollipop would be a great success in its market. Although this NP implies a change of technology and processes (as well as the consumption of new

Table 4.4 Estimated monthly demand for the NP (millions of multi-flavor lollipops)

Months	Lollipops	Months	Lollipops	Months	Lollipops
January	1.40	May	1.68	September	2.24
February	1.68	June	1.40	October	5.04
March	1.68	July	1.40	November	4.50
April	2.80	August	1.68	December	5.32

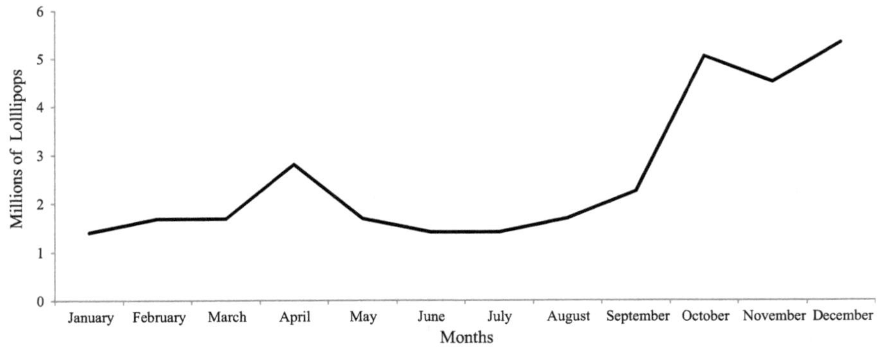

Fig. 4.3 Volume and trend of the estimated monthly demand for the NP (multi-flavor lollipops)

and unique materials from those used for the base products) the company has finished its definition and design stages and it is ready to start with its supply and launch plan.

Based on the market studies, the company estimates a volume of sales for the lollipop of 30.82 million within its first year. This volume was estimated from the market tests that were run in past months and the results that were obtained, with great acceptance among the people in the sample. If accomplished, this volume would represent an increase of 9% in the company's annual sales.

Table 4.4 presents the estimated demand for each month within the first year of the NP. Figure 4.3 presents the visualization of the monthly demand. It is observed that, particularly for the last semester of the year, the demand presents some degree of seasonality. This is corroborated by the Halloween and Christmas seasons.

4.5.1 Parameters for the RSISNP Model

The company has an operational calendar of 52 weeks. Due to the seasonality and variability of the demand, the company considered planning horizons of 3 months ($T = 12$ weeks). The unit cost-value of the NP was established at a price higher than the base lollipops (almost 15% more expensive). This was established due to the new raw material which is significantly more expensive than the raw materials of the base lollipops. The unit sale price during its release was set to \$4.37 ($v = 4.37$ for the newsvendor model) with a manufacturing cost of C = \$3.06.

Due to the complex production process of the NP the company's plant has established unique production runs for the NP. Hence, the company established a lead time LT = 2 week for the NP.

Regarding the costs associated to inventory, the company has defined a base (standard) order cost C_o = \$136.00 after the first 3 months of the launch of the NP. During the first 3 months, C_o is considered to be 30% higher due to the required additional managerial tasks. It is considered that C_o would decrease once that the NP gets established in the marked and becomes a base product. The standard holding cost C_h was established at 5% of its unit cost C.

The agreements with the suppliers have led to establish that the material in-transit is not part of the company's inventory. On the part of its customers, the agreement has led to establish that the company is responsible for the management and administration of the inventory within the plant. The costs associated to this responsibility represent 7% of the cost of the product C.

The company has a service level (SL) of 97%. This is because the contracts with customers have established that SL must not be smaller than 95% in order to avoid fines or contract recessions. Due to these potential penalties, the cost of a non-supplied unit (loss) was set to p = \$5.46.

In the development of NPs there are latent risks that could generate a negative monetary impact. As presented in Table 4.1, this impact can directly or indirectly affect the inventory cost of the product. Thus, from Table 4.1, Table 4.5 presents the estimated risk costs for the NP based on its unit cost C.

4.5.2 Results of the RSISNP Model

From Table 4.5 the updated inventory costs were estimated as follows:

- C_{oNP} = \$136 + (0.30 × \$136) × %Risk
- C_{hNP} = (0.05 × \$3.06) + [\$0.76 + \$0.64] × %Risk

Then, as the year consists of 12 months, we proceeded to estimate the average weekly demand for each month (considering 4 weeks per month). This was performed to consider the seasonal demand patterns of each period of 3 months

Table 4.5 Contribution of each individual risk to the inventory costs of the NP based on its unit cost (multi-flavor lollipops)

Risk	Unit cost (\$)	Risk	Unit cost (\$)
C_{SL}	1.10	C_{OB}	0.64
C_{SH}	0.95	C_{DR}	0.61
C_{VP}	0.83	C_{ED}	0.58
C_{IP}	0.80	C_{ID}	0.52
C_{EI}	0.76	C_{AF}	0.43

(quarter) through the year. This led to the estimates (millions of units) presented in Table 4.6. Note that the total sum of all weekly demands is equal to 30.82 million.

By estimating weekly data, the weekly average demand and standard deviation could be estimated for each quarter. This was performed for risk assessment purposes based on the correlation matrices presented in Tables 4.2 and 4.3. The weekly data for each quarter is presented in Table 4.7.

By considering the variability within each quarter, the estimated demand to be served per quarter is computed as defined by Eq. (4.6) of the RSISNP model. This data is presented in Table 4.8 where T = 12 weeks, LT = 2 weeks, and I = 0. For comparison purposes, the constant demand per quarter (from data of Table 4.4) is also presented.

Table 4.6 Estimated weekly demand for each month (millions of multi-flavor lollipops)

Quarter	Month	Week 1	Week 2	Week 3	Week 4
1st	Jan	0.350	0.350	0.350	0.350
	Feb	0.420	0.420	0.420	0.420
	Mar	0.420	0.420	0.420	0.420
2nd	Apr	0.700	0.700	0.700	0.700
	May	0.420	0.420	0.420	0.420
	Jun	0.350	0.350	0.350	0.350
3rd	Jul	0.350	0.350	0.350	0.350
	Aug	0.420	0.420	0.420	0.420
	Sep	0.560	0.560	0.560	0.560
4th	Oct	1.260	1.260	1.260	1.260
	Nov	1.125	1.125	1.125	1.125
	Dec	1.330	1.330	1.330	1.330

Table 4.7 Estimated weekly average demand and standard deviation for each quarter (millions of multi-flavor lollipops)

Quarter	d	σ	$CV = \sigma/d$
1st	0.397	0.034	0.09
2nd	0.490	0.158	0.32
3rd	0.443	0.091	0.21
4th	1.238	0.089	0.07

Table 4.8 Estimated cumulative demand per quarter according to the RSISNP model (millions of multi-flavor lollipops)

Quarter	D—Constant	D—Eq. (4.6) periodic review
1st	4.76	5.80
2nd	5.88	7.97
3rd	5.32	6.85
4th	14.86	17.96
Total	30.82	38.58

As observed, the demand per quarter is slightly higher when considering the variability. This is further explained by the coefficient of variability (CV) which in some cases is higher than 20% (i.e., 0.32 and 0.21 for the second and third quarters, see Table 4.7). Nevertheless, it is important to mention that the available inventory at the end of each quarter is assumed to be zero ($I = 0$). Thus, in practice, the actual lot size to be ordered (and the cumulative demand) may be smaller. By using the RSISNP model, the periodic review approach can be performed within quarters, and this could minimize the over and under stock risks, thus allowing the tracking of the available inventory I to adjust the quarterly demand to be served (e.g., periodic Q for the next period).

By using the P model to determine D for each quarter, we proceeded to apply the (Q, R) model to define the inventory supply strategy for each quarter. In accordance to the correlation matrix presented in Table 4.3 (NP's cycle time versus type of demand), the supply strategies were obtained considering different risk percentages. Table 4.9 presents the (Q, R) strategy for each risk percentage. Implementation was performed with the optimization software Lingo (Lindo Systems Inc. 2003).

For a better understanding of the implications of the RSISNP model, the visualization of the results reported in Table 4.9 is presented in Figs. 4.4, 4.5 and 4.6. For Demand with Seasonality (see Fig. 4.4), Q_{NP} is very small when compared to R_{NP} during the first quarter of the NP's life cycle. This means that Q_{NP} is ordered very frequently due to the high risks in C_{oNP} and C_{hNP}. As the life cycle progresses

Table 4.9 Inventory supply strategies for each quarter based on the %Risk associated to the NP's cycle time and type of demand

Type of demand/life cycle	1st quarter 0–3 months	2nd quarter 3–6 months	3rd quarter 6–9 months	4th quarter 9–12 months
Demand with Seasonality	%Risk = 100% (RSISNP) $Q_{NP} = 36{,}327$ $R_{NP} = 888{,}545$ $CT_{NP} = 613{,}127$	%Risk = 50% (RSISNP) $Q_{NP} = 50{,}418$ $R_{NP} = 1{,}459{,}989$ $CT_{NP} = 1{,}542{,}049$	%Risk = 0% (standard (Q, R)) $Q_{NP} = 105{,}044$ $R_{NP} = 1{,}206{,}298$ $CT_{NP} = 154{,}108$	%Risk = 0% (standard (Q, R)) $Q_{NP} = 175{,}559$ $R_{NP} = 2{,}810{,}426$ $CT_{NP} = 160{,}897$
One-Shot Demand	Newsvendor (weekly orders) $Q_{NP} = 461{,}490$	Newsvendor (weekly orders) $Q_{NP} = 787{,}055$	Newsvendor (weekly orders) $Q_{NP} = 614{,}839$	Newsvendor (weekly orders) $Q_{NP} = 1{,}405{,}470$
Constant Demand	%Risk = 100% (RSISNP) $Q_{NP} = 36{,}327$ $R_{NP} = 888{,}545$ $CT_{NP} = 613{,}127$	%Risk = 75% (RSISNP) $Q_{NP} = 42{,}455$ $R_{NP} = 1{,}444{,}744$ $CT_{NP} = 2{,}174{,}170$	%Risk = 25% (RSISNP) $Q_{NP} = 60{,}855$ $R_{NP} = 1{,}175{,}764$ $CT_{NP} = 519{,}894$	%Risk = 0% (standard (Q, R)) $Q_{NP} = 175{,}559$ $R_{NP} = 2{,}810{,}426$ $CT_{NP} = 160{,}897$
Demand with Trend	%Risk = 100% (RSISNP) $Q_{NP} = 36{,}327$ $R_{NP} = 888{,}545$ $CT_{NP} = 613{,}127$	%Risk = 75% (RSISNP) $Q_{NP} = 42{,}455$ $R_{NP} = 1{,}444{,}744$ $CT_{NP} = 2{,}174{,}170$	%Risk = 50% (RSISNP) $Q_{NP} = 46{,}731$ $R_{NP} = 1{,}162{,}563$ $CT_{NP} = 874{,}051$	%Risk = 25% (RSISNP) $Q_{NP} = 98{,}555$ $R_{NP} = 2{,}827{,}481$ $CT_{NP} = 324{,}027$

%Risk applied as defined in Eqs. (4.8) and (4.9)

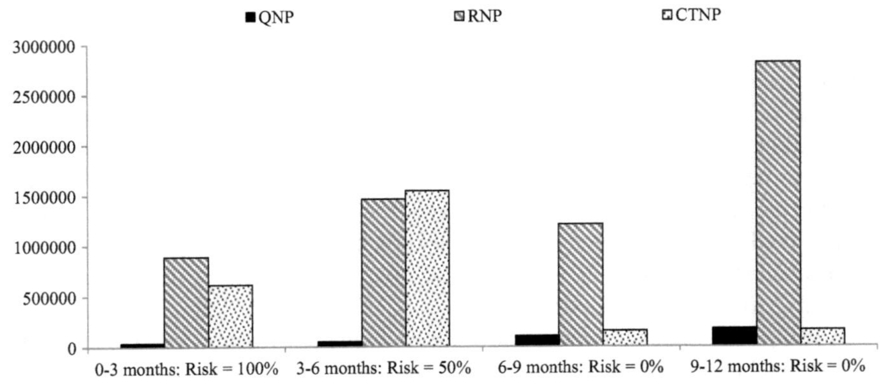

Fig. 4.4 Visualization of results: Demand with Seasonality (multi-flavor lollipops)

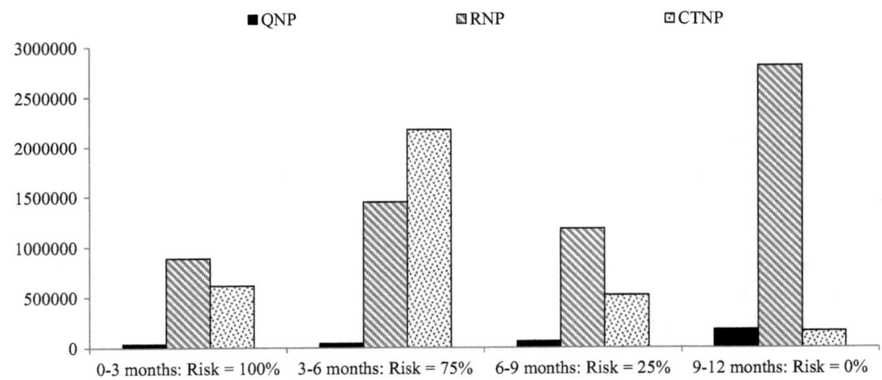

Fig. 4.5 Visualization of results: Constant Demand (multi-flavor lollipops)

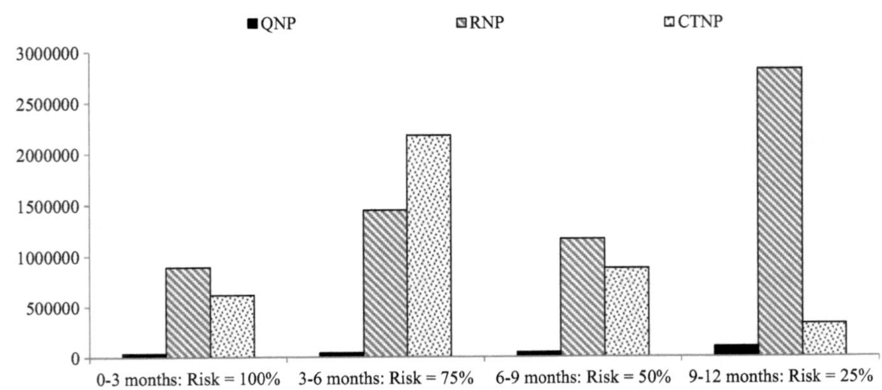

Fig. 4.6 Visualization of results: Demand with Trend (multi-flavor lollipops)

to the second and third quarters, Q_{NP} and R_{NP} increase. By the fourth quarter, R_{NP} significantly increases due to the season of high demand. It is important to observe that the total inventory cost CT_{NP} is significantly high during the first and second quarters, and it decreases as risks become zero during the third and fourth quarters.

For *Constant Demand* and *Demand with Trend* (Figs. 4.5 and 4.6) a similar pattern is observed. During the initial first and second quarters of the NP's life cycle, CT_{NP} increases. However, as the NP becomes more stable in the third and fourth quarters, the CT_{NP} decreases independently of the season's demand.

4.6 Conclusions

In this work an integrated inventory supply model based on risks for new products (NPs) was presented. These risks were determined from an extensive field research performed with the participation of managers and personnel involved in the development of NPs. Then, costs associated to these risks were defined in order to integrate them into an inventory model which was termed as RSISNP. This model is based on the periodic and continuous review models for inventory supply. Thus, the application of this model led to minimize inventory costs considering the inherent risks associated to NPs.

When the RSISNP model was tested, it was observed that the lot size Q_{NP} can be adjusted in accordance to the risks existent at the different stages of the NP's life cycle. Thus, at the time of launch (0–3 months) small batches which are frequently ordered (and/or produced) are considered. As the NP's life cycle progresses, Q_{NP} increases as risks decrease and the balance of Q_{NP} versus order frequency gets adjusted so CT_{NP} can be minimized to near non-risk levels.

Particularly for the RSISNP model, the costs associated to excess of inventory (C_{EI}) and inventory obsolescence (C_{OB}) were integrated into the holding cost (C_h). However, depending of the nature of the NP and the company's policies, other costs identified by our field research (i.e., C_{SL}, C_{ED}) can be added within C_h (or C_o).

Hence, if the production of the NP consists of just a line extension, the risks' costs that can be added to C_h can be only the basic costs C_{OB} and C_{EI}. However, if the NP requires a technology change and/or it is characterized by complete new and innovative features, its associated risks may be higher. In such case, almost all risks' costs defined in Table 4.1 could be added to C_h and C_o.

Depending of the priorities of the company regarding specific metrics of performance, some costs may be more recommended to be integrated within C_h than others. For example, if the service level is a priority, the costs C_{SL} and C_{SH} should be added to C_h. If the company is focused on inventory policies, or if it is focused on the cash flow, then all the costs related to inventories should be added. On the other hand, if the company is focused on the innovation and competitive features of the NP, the costs C_{VP}, C_{DR}, and C_{ED} should be added to C_h.

In future research the model can be extended to cover the risks that affect the distribution process of the NPs. This is because having the correct amounts in the right places within all locations in a distribution network is essential for the success of NPs. Also, it is important to extend on the methodology to describe in more detail the risks associated to NPs. This can be addressed by means of stochastic modeling, so a probability can be associated to a risk. This would lead to a more detailed risk percentage which can improve the assessment of the scenarios presented in Table 4.9. Finally, the use of integer-linear programming (ILP) can be considered as a tool to extend the RSISNP to consider different NPs and the variables associated to their distribution network.

References

Adamu I (2017) Reorder quantities for (Q, R) inventory models. Int Math Forum 12(11):505–514. https://doi.org/10.12988/imf.2017.610142

Chaudhuri A, Boer H (2016) The impact of product-process complexity and new product development order winners on new product development performance: the mediating role of collaborative competence. J Eng Technol Manag 42:65–80. https://doi.org/10.1016/j.jengtecman.2016.10.002

Felekoglu B, Maier A, Moultrie J (2013) Interactions in new product development: how the nature of the NPD process influences interaction between teams and management. J Eng Technol Manag 30(4):384–401. https://doi.org/10.1016/j.jengtecman.2013.08.004

Hashiba L, Paiva E (2016) Incorporating sustainability in the new product development process: an analysis based on the resource-based view. BASE Rev Adm Contab Unisinos 13(3):188–199. https://doi.org/10.4013/base.2016.133.01

Hoo-Gon C, Jungon A (2010) Risk analysis models and risk degree determination in new product development: a case study. J Eng Technol Manag 27(1–2):110–124. https://doi.org/10.1016/j.jengtecman.2010.03.006

Khan O, Stolte T, Creazza A, Hansen Z (2016) Integrating product design into the supply chain. Cogent Eng 3(1):1–24. https://doi.org/10.1080/23311916.2016.1210478

Liao Y, Marsillac E (2015) External knowledge acquisition and innovation: the role of supply chain network-oriented flexibility and organisational awareness. Int J Prod Res 53(18):5437–5455. https://doi.org/10.1080/00207543.2015.1008106

Lieberman G, Hillier F (2000) Introduction to operations research, 7th edn. McGraw-Hill

Lindo Systems Inc. (2003) Inventory, production, and supply chain management. In: L. S. Inc. Optimization modeling with LINGO, 5th edn. Lindo Systems Inc, Chicago, Illinois, United States of America, pp 469–506

Morales M, Riaga C, Cante M (2013) Determinants of innovation processes: a look at the situation in Latin America. Rev EAN 72:148–163. https://doi.org/10.21158/01208160.n72.2012.573 (in Spanish)

Nagi R (2001) Notes of IE 505 production planning and control: inventory control (uncertain demand). University at Buffalo, The State University of New York, New York

Park H, Chang H, Park Y (2015) Firm's knowledge creation structure for new product development. Cogent Bus Manag 2(1):1–19. https://doi.org/10.1080/23311975.2015.1023507

Parlings M, Klingebiel K (2017) A framework for aligning the supply chain throughout a radical product innovation life cycle. Int J Manuf Technol Manag 31(1/2/3):37–61. https://doi.org/10.1504/IJMTM.2017.082011

Pero M, Abdelkafi N, Sianesi A, Blecker T (2010) A framework for the alignment of new product development and supply chains. Supply Chain Manag Int J 15(2):115–128. https://doi.org/10.1108/13598541011028723

Rojas F (2017) A methodology for stochastic inventory modelling with ARMA triangular distribution for new products. Cogent Bus Manag 4:1–13. https://doi.org/10.1080/23311975.2016.1270706
Schilling M (2017) What's your best innovation bet? Harv Bus Rev 87–93
Stevenson W (2017) Operations management, 13th edn. McGraw-Hill Education

Chapter 5
Fuzzy QFD and TOPSIS for Dispatching Prioritization in Maritime Transportation Considering Operational Risk

Juan Carlos Osorio-Gómez and Diego Fernando Manotas-Duque

Abstract Supply chain risk management is an important activity in current supply chain management. Operational risk is one of the most important risks in supply chains. The participation of Third-Party Logistics providers (3PL) in supply chains has been increasing, and it is important to consider how their presence affects risk management. Maritime transportation is a fundamental activity in global supply chain and it is essential for the commercial trade. A maritime transportation company operates as a 3PL and is responsible for both the management of the supply chain in general and the risk management for the company and their customers in particular. Dispatching prioritization is an important activity to the companies, especially to 3PL because they need to satisfy the demand with the best efficiency of the resources. We propose a multicriteria approach to prioritize dispatching according to the availability of resources and considering in prioritization criteria, the risks associated with this decision. The multicriteria model uses Fuzzy QFD and TOPSIS for the prioritization. The approach is applied in maritime transportation to prioritize container shipments according to the availability of resources (ships) and considering in prioritization criteria, the risks associated with this decision in an international company with a branch office located in Buena Ventura Port in Colombia.

Keywords Operational risk management · Dispatching prioritization
Maritime transportation · Fuzzy QFD · TOPSIS

J. C. Osorio-Gómez (✉) · D. F. Manotas-Duque
Grupo de Investigación en Logística y Producción, Universidad del Valle,
Colombia, Street 13 No 100-00, Cali, Valle, Colombia
e-mail: juan.osorio@correounivalle.edu.co

J. L. García Alcaraz et al. (eds.), *Best Practices in Manufacturing Processes*,
https://doi.org/10.1007/978-3-319-99190-0_5

5.1 Introduction

According to Bravo Bastidas et al. (2009) due to the exigencies of new commercial trades, enterprises have seen forced to increment their competitive levels to become more efficient in the management of their limited resources, and they continued telling that in this situation. It is natural that the organizations consider the transportation in general as an important field to study.

A considerable amount of liquid and solid bulk cargoes such as grain, coal, iron ore, crude oil, petrochemical products, etc., are carried by ships around the world. In this context, efficient port operations are necessary to ensure continuous trade and to avoid congestion at berth (Celik and Akyuz 2018).

The increasing need of the companies to focus on the core business object has generated a trend oriented to outsource different activities. In this context, supply chain activities have evolved from a first stage where we have companies that are responsible for their logistics processes up to the current trend with companies who have delegated all their logistics activities to specialized agents (Manotas et al. 2016).

According with Sener and Ozturk (2015) maritime transportation is one of the most important industries with its immense share in the global trade. In addition, the same authors establish that it is a cost-effective method that allows companies to transfer their goods of an international way between two ports. In conditions of global cargo loaded, their share increased from 60% in 2012 to 61% in 2013. In the meantime, their import command, as considered by the volume of freight unloaded, increased from 58 to 60% (Kherbash and Mocan 2015).

Finally, the management of the risk in the supply chains as been configured as a fundamental activity inside the management of the modern supply chains, which makes necessary to link risks in the business decision-making.

Considering the three previous aspects and taking into account that among the activities associated with maritime transport is the prioritization of shipments, that is, the order in which the containers of the client companies will be shipped so that the shipped goods arrive at their ports destination in the time and conditions expected by international customers, this activity becomes a central issue for both the shipping company, because they must manage their resources in the best way possible by providing the best service to their customers and achieving profitability of their operation, as for customers who expect to minimize the risks associated with the transport of their merchandise and succeed in their international trade activities.

In this sense, prioritization can clearly be seen as a problem of multicriteria decision-making, for which the company needs to have adequate tools that allow it to make the best decision when making this prioritization.

Considering the importance of maritime companies in the supply chain management in general and in supply chain risk management in particular, we developed a proposal for decision-making related to risk in maritime transportation. A proposal is then presented to prioritize the shipments of a shipping company considering inside the multiple criteria, the risks incurred by the company and the supply chain.

This proposal uses a combination of two tools known in the bibliography but that have not been used for this type of activities, these tools are the fuzzy quality function deployment (Fuzzy QFD) and the technique for ordering preferences based on the solution ideal (TOPSIS). The proposal was applied in a company dedicated to maritime transport with headquarters in the city of Buenaventura in Colombia.

This chapter is organized of the following way: first the literature review is presented, then the methodological proposal is developed, this proposal is applied in a case of study. Finally, the discussion and conclusions of the application and the proposal are presented.

5.2 Literature Review

5.2.1 Risk in the Supply Chains

According with Heckmann et al. (2015) the risk of the supply chain is the potential loss of a supply chain in terms of its efficiency and effectiveness objective values due to the uncertain evolution of the characteristics of the chain whose changes were caused by the occurrence of detonating events.

To achieve successful supply chains requires successful logistics activities, is for it that the role of 3PL suppliers has changed from some simple tasks to a total outsourcing, becoming not only a provider of logistics services but also a strategic supplier of the chain, working simultaneously with multiple partners in the supply chain (Kumar and Singh 2012).

The importance of logistics outsourcing has been continuously dominating the global supply chain due to the growth of companies and their expansion in the world (Singh et al. 2010). This makes companies seek to expand their business globally (Kumar and Singh 2012).

Pfohl et al. (2011) add that both national and international outsourcing activities result in more geographically dispersed supply chains, which are exposed to all kinds of risks.

Yazdani et al. (2017) work in a decision support system to deal with a sustainable agriculture supply chain with respect to risk and uncertainty.

5.2.1.1 Maritime Transportation

With the development of the production regionalization and economic globalization, the international trade has been booming in the past few decades. The import and export freights in the international trade are mainly distributed by maritime transportation (Yu et al. 2018). Shipping is one main way for mass freight transportation in international trading (Bao et al. 2017).

The importance of sea transportation in global good trade in unique, mainly in conditions of weight as it handles about 90% of the global trade (Kherbash and Mocan 2015).

Maritime transportation has critical importance in terms of economic development to countries since almost 80 percent of the world trade is carried by sea (Celik and Akyuz 2018). According to Yu et al. (2018) the performance of global supply chain depends crucially on the efficiency of the ocean container transportation system. And nearly all intercontinental transport of goods takes place by sea, and an increasing share of this transport is containerized. Containerized ocean transport has become the lifeline of almost any global supply chain (Fransoo and Lee 2013).

Flows of goods have increasingly growing and development of international intermodal transport networks. With the arrival of container, handling is standardized and transit time from one mode of transport to another has significantly decreased (Mabrouki et al. 2014).

The passage through the port terminal is still the weakest link in the intermodal transport chain, for this reason the need to optimize port management in order to accelerate and reduce the cost of moving the goods through the port (Mabrouki et al. 2014).

The proposal presented in this chapter looking to improve these processes.

5.2.1.2 Risk in Maritime Transportation

Maritime traffic poses various risks in terms of fatalities, environmental pollution, and loss of property (Montewka et al. 2014; Goerlandt and Montewka 2015).

According to Akyuz and Celik (2014) safety is the one of the key aspects of sustainable maritime transportation. It directly deals with the management and operation of ships. The achievements of the International Maritime Organization (IMO) related to maritime safety and marine environmental protection are marvelous. They continued telling that studies show that maritime safety is playing a critical role on shipboard managements and operations. They proposed a hybrid-decision model to monitor the implementation performance of SMS via monitoring the KPIs data. The model is based on AHP and TOPSIS technique to prioritize and use KPIs data under unique frame on which safety performance is measured.

Risk analysis methods for maritime transportation have received a growing interest in recent years, even to the extent that international organizations have provided recommendations on the use of specific risk analysis and management tools (Goerlandt and Montewka 2015).

Mabrouki et al. (2014) propose a decision support methodology for risk management within a port terminal. The methodology is based on three steps. The first step is to define and identify the risk factors and they used the brainstorming approach for this. The second step to describe risks quantitatively and finally, the development of criteria and their weighting, the most probable risks are assessed under the AHP method and the result is to establish preventive measures.

Their work is more interested by the identification, analysis and the assessment of risks in the management operating system within the port terminal, specifically vehicle traffic activity roll-on/roll-off cargo (RO–RO).

5.2.1.3 Fuzzy Quality Function Deployment (FQFD)

In recent years, fuzzy QFD has become a widely used quality tool developed to satisfy customer need in product design and development. Fuzzy QFD provides a mean of translating customer requirements into appropriate technical requirement for each stage of product development and production (Dat et al. 2015) but in recent years its scope has been expanded towards multicriteria decision-making.

The basis of QFD is to obtain and translate customer needs into engineering characteristics, and subsequently into part characteristics, process plans and production requirements (Sener and Ozturk 2015).

Although there are many papers related with QFD and FQFD applications in maritime transportation (Lam and Bai 2015; Lam 2015; Razik et al. 2015; Sener and Ozturk 2015; Liang et al. 2012; Ding 2009; Celik et al. 2009), none of them applied FQFD to operational risk assessment.

Besides, there are some papers related to QFD applications in risk management such as Gento et al. (2001) and Costantino et al. (2012) still there is a gap in risk assessment research and Fuzzy QFD.

QFD approach to enhance maritime supply chain resilience taking both customer requirements and maritime risks into consideration is presented in Lam and Bai (2015) however still is necessary to include risk prioritization which is the aim of this research.

Table 5.1 shows the scope of some papers related to QFD and FQFD applications in supply chain, 3PL and 4PL, risk and maritime transportation.

Table 5.1 Papers related to QFD and FQFD applications

Applications	Papers
QFD and risk management	Costantino et al. (2012), Gento et al. (2001)
FQFD in supply chain management	Bevilacqua et al. (2006), Bottani and Rizzi (2006), Sohn and Choi (2001), Zarei et al. (2011), Hassanzadeh and Razmi (2009), Kazançoğlu and Aksoy (2011)
FQFD in 3PL (4PL) applications	Leina et al. (2010), Wang et al. (2007)
QFD in maritime transportation	Lam (2015), Razik et al. (2015), Sener and Ozturk (2015)
FQFD in maritime transportation	Liang et al. (2012), Ding (2009), Celik et al. (2009)
QFD in maritime transportation and risk	Lam and Bai (2015)
QFD in supply chain decisions (reverse logistics)	Li et al. (2018)

The quality function deployment (QFD) has been linked traditionally with the decisions of design and development of product, but in the last years its field of action has been extended to multicriteria decision-making, often complemented with the fuzzy logic (Fuzzy QFD). In QFD, the importance of the customer requirements, the strength relationships between the customer requirements and the design requirements, and the merits of the design schemes regarding to the design requirements, are difficult to be expressed by crisp data because the available information is usually subjective and imprecise and fuzzy logic is used to solve this (Wu and Liao 2018).

Although the table presents applications related to risk management, applications in supply chains, applications for 3PL companies, and developments in maritime transport, none of them presented an approach similar to developed in this chapter, becoming this proposal in a novel subject from the point of view of the application of the QFD in multicriteria decision-making, and in the management of shipments for the maritime transport of goods.

5.2.1.4 TOPSIS

TOPSIS (technique for order preference by similarity to ideal solution) method was firstly proposed by Hwang and Yoon. The basic concept of this method is that the chosen alternative (appropriate alternative) should have the shortest distance from the positive ideal solution and the farthest distance from negative ideal solution. The underlying principle of TOPSIS is a bipolar comparison of each alternative under consideration with both the positive ideal (PIS) and the negative ideal (NIS) solutions (Wachowicz 2011). Positive ideal solution is a solution that maximizes the benefit criteria and minimizes adverse criteria, whereas the negative ideal solution minimizes the benefit criteria and maximizes the adverse criteria (Shemshadi et al. 2011; Wachowicz 2011).

With respect to applications of TOPSIS in maritime transportation is important to mention Kandakoglu et al. (2009) in their paper proposes a structured multi-methodological approach to shipping registry selection under multiple criteria in the maritime transportation industry. They proposed a combination of the AHP and the TOPSIS methodologies with the SWOT analysis is determined to be used as the fundamental of the proposed approach. Another paper is Wu et al. (2016) they proposed a hybrid group decision-making approach to facilitate NUC (not under control) ship safety control by incorporating fuzzy logic, consistency-based linear programming and Technique for Order Preference by Similarity to an Ideal Solution (TOPSIS).

To conduct TOPSIS analysis we assume that the decision-making problem is presented in the form of a matrix as show in Fig. 5.1 (Wachowicz 2011).

Where A_j describes the alternative j under consideration ($j = 1, \ldots, m$), C_k describes the criterion k for measuring the alternatives' performance ($k = 1, \ldots, n$) and x_{jk} is the resolution level (performance) of alternative A_j with respect to criterion C_k. Furthermore, the criteria importance is specified in the form of a vector

Fig. 5.1 TOPSIS matrix (Wachowicz 2011)

	C_1	C_2	…	C_n
A_1	x_{11}	x_{12}	…	x_{1n}
A_2	x_{21}	x_{22}	…	x_{2n}
…	…	…	…	…
A_m	x_{m1}	x_{m2}	…	x_{mn}

of weights $w = (w_1, w_2, \ldots, w_n)$, where $\sum_{k=1}^{n} Wk = 1$. In this case, the weights were obtained using FQFD.

5.2.1.5 QFD and TOPSIS

Considering the combination of these two techniques, there are also important works in the literature, which are presented in Table 5.2. However, it is important to emphasize that none of these applications has solved the problem of prioritizing shipments for maritime transport. In addition, the combination of these techniques is not done in the same way as presented in this proposal and will be explained in the next section.

However is important to mention the most relevant papers that using QFD and TOPSIS, they are Yazdani et al. (2017), they propose a decision support system for selecting logistics providers based on the QFD and the technique for order preference by the similarity to ideal solution (TOPSIS) for agricultural supply chain in France. Initially, weights of the decision criteria (e.g., weights of technical supply chain requirements) are determined using fuzzy QFD. The second task is to generate the ranking of alternative logistic providers using fuzzy TOPSIS.

Although Lima Junior and Carpinetti (2016) do not use QFD with TOPSIS, they combine the fuzzy QFD technique for weighting the criteria with a procedure for assessing the difficulty to obtain information to evaluate the suppliers on each criterion. The fuzzy QFD technique is used for weighting the requirement and criteria since it combines the modeling of linguistic judgments from fuzzy set theory with a technique for prioritization of criteria based on a systematic

Table 5.2 Main applications of the QFD - TOPSIS

Applications	Papers
Evaluate and select better e-CRM approach	Zandi and Tavana (2011)
Prioritize characteristics of Telecommunications Engineering	Khademi-Zare et al. (2010)
Evaluate and select market segments	Dat et al. (2015)
Suppliers selection	Kumaraswamy et al. (2011)
Product design	Chang and Tseng (2008), Karimi et al. (2012)

deployment. The final choice of the criteria is based on a classification procedure that considers both the weight of the criteria and their associated degree of difficulty of data collection. The degree of difficulty of data collection is evaluated based on linguistic judgments related to information availability, human resources and time required and additional resources.

In the same way Wu and Liao (2018) presented a multi-expert multicriteria decision-making method to solve the innovative product design selection problem by developing an QFD method combined with the complicated fuzzy linguistic representation model, the probabilistic linguistic term set, and the ranking method, ORESTE_(organísation, rangement et Synthèse de données relarionnelles, in French).

Haldar et al. (2012) presented a multi-dimensional approach on the selection of a supplier considering multicriteria by integrating QFD tool with TOPSIS. A number of decision determinants for supply chain resilience have been chosen as technical requirements and customer criteria`s. AHP has also been embedded to deal with linguistic judgments expressing relationships and correlations required in QFD. TOPSIS is used to develop the ranking of a number of suppliers considering different multi-dimensional parameters in a uniform system.

Hsu et al. (2017) in order to select important elements in the wide range of sustainability indicators suggested in the literature and launch performance factors for improving the sustainability of manufacturing small and medium-sized enterprises, this research utilizes QFD approach as the basic structure, in which it combines with fuzzy Delphi method, modified fuzzy extent analytic hierarchy process and TOPSIS method to prioritize the performance factors. The methodology of modified fuzzy extend AHP is used in deciding the importance weights of the related sustainability criteria, and the integrated QFD and TOPSIS technique is utilized to determine the performance priority. The weights obtained from the modified fuzzy extend AHP method are used as input in the QFD and TOPSIS technique, and the identified performance indicators are ranked to obtain the priorities in terms of the implementation of sustainability of SMEs.

Khademi-Zare et al. (2010) developed two prioritization models based on QFD to rank the strategic actions or of Iran mobile cellular telecommunication. On the other hand, taking into consideration the gap between the current state and the positive ideal state for each customer attributes, TOPSIS is used to rank these.

5.2.1.6 Multicriteria in Maritime Transportation

With respect to the multicriteria applications in maritime transportation related to the decision involves in this chapter it is important to mention Celik and Akyuz (2018), they present a multicriteria decision-making (MCDM) method extended by interval type-2 fuzzy sets (IT2FSs) for selecting appropriate ship loader type in maritime transportation. In their model they used AHP and TOPSIS. According with they, the studies have tendency to focus on more specific case rather than systematic approach. However, most of ship owners and port managers are seeking systematic approaches which enable to enhance efficiency of cargo handling process. They paper aims to introduce a methodological approach utilizing AHP and TOPSIS under IT2FSs environment which is able to deal with uncertainty of experts' judgement and expression in decision-making. The proposed approach is capable of analyzing performance of dry bulk cargo ship loaders from the perspective of their operational and technical indicators.

Another of the most recent papers is Soner et al. (2017), they proposed hybrid approach integrates AHP and VIKOR under interval type-2 fuzzy (IT2F) environment to the selection of proper hatch cover in the equipment for bulk carriers.

Gagatsi et al. (2017) discusses on the applicability of a combined multicriteria method developed to facilitate policy making in the area of maritime transport. The methodology involves two MCDA techniques (PROMETHEE and AHP) in a multi-actors evaluation environment. This combination seeks to improve both methods' applicability and decrease their potential deficiencies and limitations.

Nikolova et al. (2012) analyses several alternative plans for the movement of the port of Varna outside the city limits and development of a recreation center at its place (The Port of Varna is the largest maritime port in Bulgaria). They used an original procedure, called REPOMP (Randomized Expert Panel Opinion Marginalizing Procedure).

Sharghi et al. (2016) consider a hierarchical multiattribute decision problem of an optimal port choice under Z-number-based information. The solution of the problem is based on the use of Z-number-valued weighted average aggregation operator.

Bao et al. (2017) proposed a MADM method on the basis of intuitionist fuzzy set (IFS), evidential reasoning approach (ERA) and prospect theory for performance evaluation of shipping companies. Through the evaluation results, they obtained the ranking order of all alternative enterprises and the strength and weakness of each alternative.

Methodological Approach This proposal includes two phases that could be defined as long-term or strategic (Phase 1) and short-term or operational (Phase 2). Figure 5.2 shows the proposed methodological scheme in which, by means of the application of the fuzzy QFD, the weighting of the criteria that the organization defines for the prioritization of its shipments is obtained (long term), and in phase 2, through the application of the TOPSIS, the decision of which requests to attend first, that is, the priority of the shipments is defined (short term). It is important to clarify that a team responsible for the decision-making must be established and they will be responsible for the successful development of the methodology. Next, each of the phases is detailed.

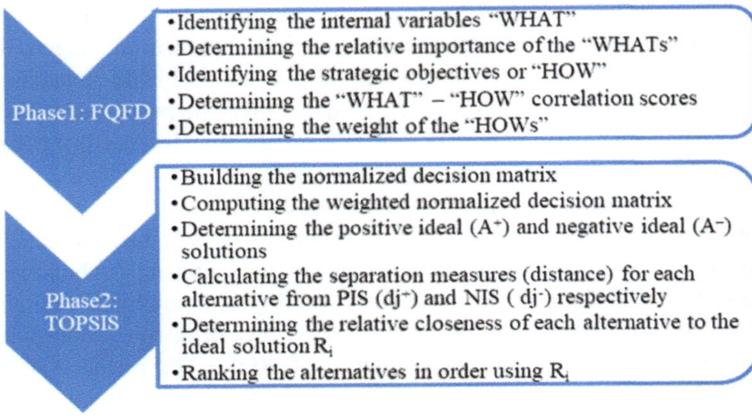

Fig. 5.2 Methodological approach

5.2.2 *Phase 1: Weighting of the Criteria Through the Application of FQFD*

As mentioned, this proposal is based in Bevilacqua et al. (2006), only steps 1, 2, 3, 4, and 6 will be used. The expected result is the weight (weighting) of the criteria that the company will use to define the priority of its shipments. The sequence of the steps is:

- *Stage 1. Identify the What or internal variables*: In this stage the decision-making team establishes the basic attributes it expects to be considered when priorities of the dispatches are defined.
- *Stage 2. Determine the importance of the What*: using the linguistic scale presented in Table 5.3, the team defines the level of importance of each of the What identified in the previous point.

Through the application of fuzzy mathematics, a triangular number is obtained that represents this level of importance. Equation (5.1) shows this calculation.

Table 5.3 Linguistic scale for the FQFD

Linguistic variable	Triangular fuzzy number
Very low (VL)	(0, 1, 2)
Low (L)	(2, 3, 4)
Medium (M)	(4, 5, 6)
High (H)	(6, 7, 8)
Very high (VH)	(8, 9, 10)

Source Bevilacqua et al. (2006)

$$\begin{aligned}&\text{Importance of What} = \{w_i,\ \text{where},\ i = 1,\ldots,q\},\\ &w_i = \frac{1}{n} \otimes (w_{i1} \oplus w_{i2} \oplus \cdots \oplus w_{in}),\end{aligned} \tag{5.1}$$

where $\boldsymbol{q}$ is the number of what's, $\boldsymbol{n}$ is the number of the decision makers and $\boldsymbol{w_i}$ is a triangular fuzzy number.

- *Stage 3. Identify the relevant variables for the evaluation of the alternatives—How's*: How's are the criteria that the company considered when defining the priority of their shipments.
- *Stage 4. Determining the "WHAT"—"HOW" correlation scores*: Through the application of the same linguistic scale presented in Table 5.3, the decision-making team will establish the relationships between the What's and the How's. That is, it will establish in which level each one of the defined What's relates to the criteria (how's) considered in the stage 3. Equation (5.2) shows this calculation.

$$\begin{aligned}&\text{Relationships} = \left\{\begin{matrix} r_{ij}, & \text{where}\, i = 1,\ldots,q\,\text{and} \\ & j = 1,\ldots,c \end{matrix}\right\}\\ &r_{ij} = \frac{1}{n} \otimes \left(r_{ij1} \oplus r_{ij2} \oplus \cdots \oplus r_{ijn}\right),\end{aligned} \tag{5.2}$$

where c is the number of the how's, q and n were mentioned above. Every element r_{ij} represents the relationship between the ith-what with the jth-how and is a triangular fuzzy number.

- *Stage 5. Define the importance of the How*: Applying fuzzy mathematics and from the relationships established in the previous point the weight of the How is calculated. This is the result of multiplying the average of the evaluations carried out by experts in each WHAT-HOW relationship by the weights of the WHATs obtained in stage 2. The result is a fuzzy triangular number. Equation (5.3) shows this calculation.

$$\begin{aligned}&\text{Importance of How} = \left\{H_j,\ \text{where},j = 1,\ldots,c\right\},\\ &H_j = \frac{1}{q} \otimes \left[\left(r_{j1} \otimes w_1\right) \oplus \cdots \oplus \left(r_{jq} \otimes w_q\right)\right],\end{aligned} \tag{5.3}$$

where c and q were defined above.

We need to "defuzzified" this triangular fuzzy number to obtain the ponderation of each criteria. Once these weights are established, we continued with phase 2, which consists of the traditional application of TOPSIS.

5.2.3 Phase 2: Application of TOPSIS

According with (Wachowicz 2011) the TOPSIS algorithm consists of six subsequent steps, but in this chapter we include one more, because before the step 1 presented for (Wachowicz 2011), is important to build the decision matrix. So the approach in this case is:

- *Stage 1. Establish the decision matrix*: this is built with the information of pending requests when establishing the priorities of the dispatches. This matrix is similar to presented in Fig. 5.1.
- *Stage 2. Building the normalized decision matrix*: for the normalization of the values, Eq. (5.4) that is traditionally applied in the development of this technique will be used. Figure 5.3 shown the normalized matrix.

Where $\hat{x}_{ij}$ is presented in Eq. (5.4).

$$\hat{x}_{ij} = \frac{x_{ij}}{\sqrt{\sum (x_{ij})^2}} \tag{5.4}$$

- *Stage 3. Computing the weighted normalized decision matrix*: the matrix is normalized and each of the values is weighted. This weighting is done with the weights obtained in the FQFD of phase 1. Figure 5.4 shown this stage.

$$N = \begin{bmatrix} \hat{x}_{11} & \hat{x}_{12} & \cdots & \hat{x}_{1n} \\ \hat{x}_{21} & \hat{x}_{22} & \cdots & \hat{x}_{2n} \\ \cdots & \cdots & \cdots & \cdots \\ \hat{x}_{m1} & \hat{x}_{m2} & \cdots & \hat{x}_{mn} \end{bmatrix}$$

Fig. 5.3 Normalized decision matrix (Wachowicz 2011)

$$V = \begin{bmatrix} w_1 \hat{x}_{11} & w_2 \hat{x}_{12} & \cdots & w_n \hat{x}_{1n} \\ w_1 \hat{x}_{21} & w_2 \hat{x}_{22} & \cdots & w_n \hat{x}_{2n} \\ \cdots & \cdots & \cdots & \cdots \\ w_1 \hat{x}_{m1} & w_2 \hat{x}_{m2} & \cdots & w_n \hat{x}_{mn} \end{bmatrix} = \begin{bmatrix} v_{11} & v_{12} & \cdots & v_{1n} \\ v_{21} & v_{22} & \cdots & v_{2n} \\ \cdots & \cdots & \cdots & \cdots \\ v_{m1} & v_{m2} & \cdots & v_{mn} \end{bmatrix}.$$

Fig. 5.4 Weighted normalized decision matrix (Wachowicz 2011)

- *Stage 4. Determining the positive ideal (A+) and negative ideal (A–) solutions*: For each of the criteria, the positive ideal solution and negative ideal solution are calculated considering whether there are cost or benefit criteria. For cost criteria, the positive ideal solution is the minimum value and for the benefit criteria it is the maximum value as shown in Eq. (5.5). For the negative ideal solution is considered the opposite as shown in Eq. (5.6).

$$A^{+} = \left(v_1^{+}, v_2^{+}, \ldots, v_n^{+}\right) \text{ where } v_k^{+} = \max_j\left(x_{jk}\right), \quad \text{for } k = 1, 2, \ldots, n \tag{5.5}$$

$$A^{-} = \left(v_1^{-}, v_2^{-}, \ldots, v_n^{-}\right) \text{ where } v_k^{-} = \min_j\left(x_{jk}\right), \quad \text{for } k = 1, 2, \ldots, n \tag{5.6}$$

- *Stage 5. Calculating the distances for each alternative from positive ideal solution (d_j+) and negative ideal solution (d_j−) respectively*: calculate the Euclidean distance of each one of the alternatives to the ideal positive solution and to the ideal negative solution determined in the previous stage according with Eqs. (5.7) and (5.8)

$$d_j^{+} = \sqrt{\sum \left(v_{jk} - v_k +\right)^2}, \quad \text{for } j = 1, 2, \ldots, m \tag{5.7}$$

$$d_j^{-} = \sqrt{\sum \left(v_{jk} - v_k -\right)^2}, \quad \text{for } j = 1, 2, \ldots, m \tag{5.8}$$

- *Stage 6. Determining the relative closeness of each alternative to the ideal solution*: this index is calculated according to Eq. (5.9).

$$R_j = d_j^{-} / \left(d_j^{+} + d_j^{-}\right), \tag{5.9}$$

where $0 \leq R_j \leq 1$. The closer the alternative A_j to positive ideal solution is, the larger the value of R_j.

- *Stage 7. Ranking the alternatives in descending order using R_j*: the alternatives (dispatch requests) are organized according to the value of the index from highest to lowest, where the highest values imply higher priority.

Once the methodology is applied, the company will have the priority of its shipments. The loading of the ship is then carried out in the order of priority established. However, at the moment when the space of the ship is being depleted if the next shipment in the order of priority cannot be assembled in its totality, the shipment mentioned is skipped and the next one that can be completely dispatched is continued.

In the next section a case of study is presented to show how the methodology was applied in the real context.

5.3 Case of Study

The methodology was applied in a 3PL company of maritime transport. The company is based in Colombia, specifically in the port of Buena Ventura in the Valle del Cauca. The company is one of the largest and most recognized worldwide in the transport of cargo in containers. The company operates 480 offices in 150 countries and it is owner of 480 ships and operates 200 routes around the world.

5.3.1 *Phase 1: Weighting of the Criteria by Applying the FQFD*

The first step was to establish the decision-making team, in this case, it is integrated by the operations manager (OM), the commercial manager (CM), the export manager (EM) and the branch manager (BM).

- *Stage 1. Identify the What or internal variables*: This team established the Whats presented below:
  - Maximum utilization of resources—MUR
  - Provide very good service to the customer—PSC
  - Greater economic benefit—GEB
  - Positioning of the company—POC
  - Increase market share—IMS
- *Stage 2. Determine the importance of the What*: Using the linguistic scale presented in Table 5.4 they defined the importance's shown in Table 5.5 and

Table 5.4 Qualification of the What's

What's	OM	CM	EM	BM
Maximum utilization of resources	H	VH	VH	H
Provide very good service to the customer	VH	H	H	H
Greater economic benefit	VH	VH	VH	VH
Positioning of the company	H	M	H	M
Increase market share	M	H	H	VH

Table 5.5 Importance of the What's

What's	Importance of the Whats (FTN)		
Maximum use of resources	7	8	9
Provide very good customer service	6.5	7.5	8.5
Greater economic benefit	8	9	10
Positioning of the firm	5	6	7
Increase market share	6	7	8

using fuzzy mathematics and Eq. (5.1) the importance of the Whats was defined in fuzzy triangular numbers (FTN—see Table 5.5).

It is important to note that the judgments of the members of the decision-making team were very similar respect to the importance of the Whats.

- *Stage 3. Identify the relevant variables for the evaluation of the alternatives—How's*: Continuing with the methodology, the criteria (How's) that are briefly explained:
  - *C1, Quantity of containers x shipment*: refers to the number of containers that the customer has in his shipment inside the terminal.
  - *C2, Type of client (trust - business relationship)*: it refers to the trust that exists with the client given the commercial relationship, antiquity, volume that they handle, etc. This criterion *measures the risk associated with a non-compliance* and the probability that the client will continue in case that said non-compliance occurs. It has been qualitatively defined as a very low risk, low risk, medium risk, high risk and very high risk.
  - *C3, Days in transit*: is the number of days that the ship will be at sea before arriving at the port of destination. To more days in transit, higher rates and greater economic benefit for the company.
  - *C4, Profit of the shipment*: economic benefit of the shipment in US dollars.
  - *C5, Number of shipments/year*: it refers to the volume of containers that the customer moves per year, this reflects the size of the client and its importance for the shipping line.
  - *C6, Potential value per claim*: this is another *criterion related to risk* and refers to the value of the claim for each customer considering the possibility that their shipment is not loaded on this Ship. The values are presented in US dollars.

- *Stage 4. Determining the "WHAT"—"HOW" correlation scores*: Once the criteria were defined, the relationship between the What's and the How's was established, as shown in Fig. 5.5.
- *Stage 5. Define the importance of the How*: and as a result of this activity, the weighting value of each of the criteria shown in Table 5.6. With these defined values, continue with phase 2.

Whats	Hows																							
	C1				C2				C3				C4				C5				C6			
	OM	CM	EM	BM	OM	CM	EM	BM	OM	CM	EM	BM	OM	CM	EM	BM	OM	CM	EM	BM	OM	CM	EM	BM
MUR	H	H	H	VH	VH	H	H	H	VH	M	H	VH	M	VH	H	VH	H	VH	VH	VH	VH	M	H	M
PSC	H	H	VH	L	H	L	M	L	VH	L	M	VL	H	M	M	M	H	M	H	M	VH	VH	VH	VH
GEB	VH	H	VH	VH	VH	M	M	VH	M	M	M	M	M	M	M	H	VH	VH	VH	VH	M	L	L	L
POC	M	M	H	H	M	L	L	L	VH	H	H	H	VH	L	VL	L	VH	M	M	M	VH	L	L	M
IMS	M	H	H	H	M	H	H	H	M	H	H	H	H	M	H	L	M	L	M	L	M	VH	VH	VH

Fig. 5.5 Whats-Hows correlation matrix

Table 5.6 Weighting of the criteria (How's)

	*C*1			*C*2			*C*3			*C*4			*C*5			*C*6		
Triangular fuzzy number	40	53	69	32	45	59	33	46	61	33	45	60	38	52	67	35	47	62
Crisp number	54			45			46			46			52			48		
Weight of the criteria (%)	18.5			15.5			15.9			15.6			17.9			16.4		

5.3.2 *Phase 2: Application of TOPSIS*

Table 5.7 presents a list of pending shipments to prioritize. Table 5.8 shows the weighted normalized matrix and the ideal positive and negative ideal solutions calculated for this case.

Finally, Table 5.9 presents the order of priority of dispatches according to the calculated value of the relative proximity index.

As shown in Table 5.9, the first shipment must be *E*2, followed by *E*6. As already mentioned, shipments are made in the established order until the capacity of

Table 5.7 Data for the case study

	Number of containers x boarding *C*1	Relationship history *C*2	Transit days *C*3	Utility USD/cnt *C*4	Number of containers/year *C*5	Potential value per claim *C*6
*E*1	8	3	22	1800	120	$1.600
*E*2	20	5	3	300	800	$4.000
*E*3	12	2	15	1500	60	$2.400
*E*4	4	1	44	4800	20	$800
*E*5	6	4	24	1800	200	$1.200
*E*6	2	1	60	6500	10	$400
*E*7	10	4	9	700	180	$2.000
*E*8	5	2	8	500	30	$1.000

Table 5.8 Weighted normalized matrix and ideal positive and negative solutions

	*C*1	*C*2	*C*3	*C*4	*C*5	*C*6
*E*1	0.0527	0.0466	0.0420	0.0326	0.0251	0.0469
*E*2	0.1318	0.0155	0.0057	0.0054	0.1675	0.1172
*E*3	0.0791	0.0621	0.0286	0.0271	0.0126	0.0703
*E*4	0.0264	0.0776	0.0839	0.0869	0.0042	0.0234
*E*5	0.0395	0.0310	0.0458	0.0326	0.0419	0.0352
*E*6	0.0132	0.0776	0.1145	0.1176	0.0021	0.0117
*E*7	0.0659	0.0310	0.0172	0.0127	0.0377	0.0586
*E*8	0.0330	0.0621	0.0153	0.0090	0.0063	0.0293
Positive ideal solution	0.1318	0.0155	0.1145	0.1176	0.1675	0.0117
Ideal negative solution	0.0132	0.0776	0.0057	0.0054	0.0021	0.1172

Table 5.9 Priority order result of the applied methodology

*E*2	*E*6	*E*4	*E*5	*E*1	*E*7	*E*3	*E*8
0.5302	0.4698	0.4136	0.3767	0.3306	0.3207	0.2886	0.2735

the ship has been covered. In the cases in which the shipment cannot be fully dispatched, it is passed to the next in the order.

This order of priority has considered both quantitative and qualitative criteria and the risk associated with the prioritization, which has already mentioned is included in the *C*2 and *C*6 criteria.

The proposed methodology allows a wide variety of criteria to be included. Therefore, the company will consider all the important aspects in the prioritization of the shipment.

5.4 Conclusions

Considering the great importance of maritime transportation in the global market and the effectiveness of the modern supply chain management as well as the current trend focus in mitigate or eliminate the operational risks in supply chains, our proposal seeks to contribute in both of them.

A novel model has been presented to prioritize the shipping of a shipping company, considering multiple criteria, including operational risk related to this decision.

There is not similar proposal in the literature for this activity, therefore the proposal is innovative and is aimed at improving the performance not only of the 3PL company but of the whole chain or related supply chains.

Although in the literature there are combined applications of the FQFD and TOPSIS, the application is not similar to the one presented in this paper, any of them has been designed to prioritize dispatches in maritime transport.

It is important to continue exploring the usefulness of the tools presented in this chapter for other types of decisions both in the field of supply chains, companies providing logistics services and aimed at strengthening risk management.

References

Akyuz E, Celik M (2014) A hybrid decision-making approach to measure effectiveness of safety management system implementations on-board ships. Saf Sci 68:169–179. https://doi.org/10.1016/j.ssci.2014.04.003

Bao TT, Xie XL, Long PY (2017) Shipping enterprise performance evaluation under uncertainty base on multiple-criteria evidential reasoning approach. Transp Res Procedia 25:2761–2772. https://doi.org/10.1016/j.trpro.2017.05.220

Bevilacqua M, Ciarapica FE, Giacchetta G (2006) A fuzzy-QFD approach to supplier selection. J Purch Supply Manag 12:14–27. https://doi.org/10.1016/j.pursup.2006.02.001

Bottani E, Rizzi A (2006) Strategic management of logistics service: a fuzzy QFD approach. Int J Prod Econ 103:585–599. https://doi.org/10.1016/j.ijpe.2005.11.006

Bravo Bastidas JJ, Osorio Gómez JC, Orejuela Cabrera JP (2009) Dynamic priorization model of dispatching vehicles using analytic hierarchy process. Rev Fac Ing, 201–215

Celik E, Akyuz E (2018) An interval type-2 fuzzy AHP and TOPSIS methods for decision-making problems in maritime transportation engineering: the case of ship loader. Ocean Eng 155:371–381. https://doi.org/10.1016/j.oceaneng.2018.01.039

Celik M, Cebi S, Kahraman C, Er ID (2009) An integrated fuzzy QFD model proposal on routing of shipping investment decisions in crude oil tanker market. Expert Syst Appl 36:6227–6235. https://doi.org/10.1016/j.eswa.2008.07.031

Chang SH, Tseng HE (2008) Fuzzy topsis decision method for configuration management. Int J Ind Eng Theory Appl Pract 15:304–313

Costantino F, De Minicis M, González-Prida V, Crespo A (2012) On the use of quality function deployment (QFD) for the identification of risks associated to warranty programs. In: ESREL Conference, Helsinki, pp 4440–4449

Dat LQ, Phuong TT, Kao HP et al (2015) A new integrated fuzzy QFD approach for market segments evaluation and selection. Appl Math Model 39:3653–3665. https://doi.org/10.1016/j.apm.2014.11.051

Ding JF (2009) Applying fuzzy quality function deployment (QFD) to identify solutions of service delivery system for port of Kaohsiung. Qual Quant 43:553–570. https://doi.org/10.1007/s11135-007-9138-7

Fransoo JC, Lee CY (2013) The critical role of ocean container transport in global supply chain performance. Prod Oper Manag 22:253–268. https://doi.org/10.1111/j.1937-5956.2011.01310.x

Gagatsi E, Giannopoulos G, Aifantopoulou G, Charalampous G (2017) Stakeholders-based multi-criteria policy analysis in maritime transport: From theory to practice. Transp Res Procedia 22:655–664. https://doi.org/10.1016/j.trpro.2017.03.062

Gento AM, Minambres MD, Redondo A, Perez ME (2001) QFD application in a service environment: a new approach in risk management in an university. Oper Res 1:115–132. https://doi.org/10.1007/BF02936289

Goerlandt F, Montewka J (2015) Maritime transportation risk analysis: Review and analysis in light of some foundational issues. Reliab Eng Syst Saf 138:115–134. https://doi.org/10.1016/j.ress.2015.01.025

Haldar A, Banerjee D, Ray A, Ghosh S (2012) An integrated approach for supplier selection. Procedia Eng 38:2087–2102. https://doi.org/10.1016/j.proeng.2012.06.251

Hassanzadeh S, Razmi J (2009) An integrated fuzzy model for supplier management: a case study of ISP selection and evaluation. Expert Syst Appl 36:8639–8648. https://doi.org/10.1016/j.eswa.2008.10.012

Heckmann I, Comes T, Nickel S (2015) A critical review on supply chain risk—definition, measure and modeling. Omega 52:119–132. https://doi.org/10.1016/j.omega.2014.10.004

Hsu CH, Chang AY, Luo W (2017) Identifying key performance factors for sustainability development of SMEs—integrating QFD and fuzzy MADM methods. J Clean Prod 161:629–645. https://doi.org/10.1016/j.jclepro.2017.05.063

Kandakoglu A, Celik M, Akgun I (2009) A multi-methodological approach for shipping registry selection in maritime transportation industry. Math Comput Model 49:586–597. https://doi.org/10.1016/j.mcm.2008.09.001

Karimi BH, Mozafari MM, Asli MN (2012) Applying a hybrid QFD-TOPSIS method to design product in the industry (case study in sum service company). Res J Appl Sci Eng Technol 4:3283–3288

Kazançoğlu Y, Aksoy M (2011) A fuzzy logic-based quality function deployment for selection of e-learning provider. Turk Online J Educ Technol 10:39–45

Khademi-Zare H, Zarei M, Sadeghieh A, Saleh Owlia M (2010) Ranking the strategic actions of Iran mobile cellular telecommunication using two models of fuzzy QFD. Telecomm Policy 34:747–759. https://doi.org/10.1016/j.telpol.2010.10.001

Kherbash O, Mocan ML (2015) A review of logistics and transport sector as a factor of globalization. Procedia Econ Financ 27:42–47. https://doi.org/10.1016/S2212-5671(15)00969-7

Kumar P, Singh RK (2012) A fuzzy AHP and TOPSIS methodology to evaluate 3PL in a supply chain. J Model Manag 7:287–303. https://doi.org/10.1108/17465661211283287

Kumaraswamy AH, Bhattacharya A, Kumar V, Brady M (2011) An integrated QFD-TOPSIS methodology for supplier selection in SMEs. In: Proceedings of CIMSim 2011 3rd international conference on computational intelligence, modelling & simulation, pp 271–276. https://doi.org/10.1109/cimsim.2011.55

Lam JSL (2015) Designing a sustainable maritime supply chain: a hybrid QFD-ANP approach. Transp Res Part E Logist Transp Rev 78:70–81. https://doi.org/10.1016/j.tre.2014.10.003

Lam JSL, Bai X (2015) A quality function deployment approach to improve maritime supply chain resilience. Transp Res Part E Logist Transp Rev. https://doi.org/10.1016/j.tre.2016.01.012

Leina Z, Tiejun P, Guoqing Y (2010) Process control and simulation for the fourth party logistics using value chain analysis. Int Conf Electr Control Eng 2010:1503–1506. https://doi.org/10.1109/iCECE.2010.371

Li Y, Kannan D, Garg K et al (2018) Business orientation policy and process analysis evaluation for establishing third party providers of reverse logistics services. J Clean Prod 182:1033–1047. https://doi.org/10.1016/j.jclepro.2017.12.241

Liang GS, Ding JF, Wang CK (2012) Applying fuzzy quality function deployment to prioritize solutions of knowledge management for an international port in Taiwan. Knowl Based Syst 33:83–91. https://doi.org/10.1016/j.knosys.2012.03.012

Lima Junior FR, Carpinetti LCR (2016) A multicriteria approach based on Fuzzy-QFD for choosing criteria for supplier selection. Comput Ind Eng 101:269–285. https://doi.org/10.1016/j.cie.2016.09.014

Mabrouki C, Bentaleb F, Mousrij A (2014) A decision support methodology for risk management within a port terminal. Saf Sci 63:124–132. https://doi.org/10.1016/j.ssci.2013.09.015

Manotas DF, Osorio JC, Rivera L (2016) Operational risk management in third party logistics (3PL). In: Alor-Hernández G, Sánchez-Ramírez C, García-Alcaraz JL (eds) Handbook of research on managerial strategies for achieving optimal performance in industrial processes

Montewka J, Ehlers S, Goerlandt F et al (2014) A framework for risk assessment for maritime transportation systems—a case study for open sea collisions involving RoPax vessels. Reliab Eng Syst Saf 124:142–157. https://doi.org/10.1016/j.ress.2013.11.014

Nikolova ND, Armenski IS, Tchikova KT, Tenekedjiev KI (2012) Expert assessment of alternatives to change the location of the port of Varna. IFAC Proc Vol, pp 138–143. https://doi.org/10.3182/20120912-3-bg-2031.00027

Pfohl H, Gallus P, Thomas D (2011) Interpretive structural modeling of supply chain risks. Int J Phys Distrib Logist Manag 41:839–859. https://doi.org/10.1108/09600031111175816

Razik MA, Tahar RM, Hasrulnizzam W et al (2015) Integrated quality function deployment (QFD) model for dry bulk terminal improvements (DBTI) in Malaysian ports. J Econ Bus Manag 3:4–7. https://doi.org/10.7763/JOEBM.2015.V3.219

Sener Z, Ozturk E (2015) A QFD-based decision model for ship selection in maritime transportation. Int J Innov Manag Technol 6:202–205. https://doi.org/10.7763/IJIMT.2015.V6.602

Sharghi P, Jabbarova K, Aliyeva K (2016) Decision making on an optimal port choice under Z-information. Procedia Comput Sci 102:378–384. https://doi.org/10.1016/j.procs.2016.09.415

Shemshadi A, Toreihi M, Shirazi H, Tarokh MJ (2011) Supplier selection based on supplier risk: An ANP and fuzzy TOPSIS approach. Science (80-)1:111–121

Singh RK, Sharma HO, Garg SK (2010) Interpretive structural modelling for selection of best supply chain practices. Int J Bus Perform Supply Chain Model 2(3/4):237–57. https://doi.org/10.1504/IJBPSCM.2010.036201

Sohn SY, Choi IS (2001) Fuzzy QFD for supply chain management with reliability consideration. Reliab Eng Syst Saf 72:327–334. https://doi.org/10.1016/S0951-8320(01)00022-9

Soner O, Celik E, Akyuz E (2017) Application of AHP and VIKOR methods under interval type 2 fuzzy environment in maritime transportation. Ocean Eng 129:107–116. https://doi.org/10.1016/j.oceaneng.2016.11.010

Wachowicz T (2011) Application of topsis methodology to the scoring of negotiation issues. In: Trzaskalik T, Wachowicz T (eds) Multiple criteria decision making '10-11. Publisher of The University of Economics in Katowice, Katowice, p 300

Wang F, Li XH, Rui WN, Zhang Y (2007) A fuzzy QFD-based method for customizing positioning of logistics Service products of 3PLS. Int Conf Wirel Commun Netw Mob Comput WiCOM 2007:3326–3329. https://doi.org/10.1109/WICOM.2007.825

Wu B, Yan X, Wang Y, Guedes Soares C (2016) Selection of maritime safety control options for NUC ships using a hybrid group decision-making approach. Saf Sci 88:108–122. https://doi.org/10.1016/j.ssci.2016.04.026

Wu X, Liao H (2018) An approach to quality function deployment based on probabilistic linguistic term sets and ORESTE method for multi-expert multi-criteria decision making. Inf Fusion 43:13–26. https://doi.org/10.1016/j.inffus.2017.11.008

Yazdani M, Zarate P, Coulibaly A, Zavadskas EK (2017) A group decision making support system in logistics and supply chain management. Expert Syst Appl 88:376–392. https://doi.org/10.1016/j.eswa.2017.07.014

Yu M, Fransoo JC, Lee CY (2018) Detention decisions for empty containers in the hinterland transportation system. Transp Res Part B Methodol 110:188–208. https://doi.org/10.1016/j.trb.2018.02.007

Zandi F, Tavana M (2011) A fuzzy group quality function deployment model for e-CRM framework assessment in agile manufacturing. Comput Ind Eng 61:1–19. https://doi.org/10.1016/j.cie.2011.02.004

Zarei M, Fakhrzad MB, Jamali Paghaleh M (2011) Food supply chain leanness using a developed QFD model. J Food Eng 102:25–33. https://doi.org/10.1016/j.jfoodeng.2010.07.026

Chapter 6
Logistics Planning for the Synchronization of Key Functional Areas of a Latin American Bottling Company

Diana Sánchez-Partida, Dulce Arroyo-García, José-Luis Martínez-Flores and Emmanuel Juárez-García

Abstract Currently, organizations of all kinds are forced to compete providing value to their customers effectively. Within any organization, the supply chain plays a strategic role that determines the efficiency and competitiveness of the same through the optimization of its processes in order to maximize the profitability of the organization and meet the requirements of its customers. Generally, companies base their operations on a sales forecast whose level of variation directly impacts the performance of production planning, material planning, and the management of their inventories. Faced with a state of uncertainty in demand, the control of inventory of their products becomes complex when it comes to materials with a limited lifetime for the food industry. Therefore, the objective of this case study focuses on defining the most appropriate forecasting method for a finished product from a beverage bottling company settled in Mexico that smooths the variability of the demand. The company experiences a forecast variability of minimum 135% with respect to its actual sales based on naive systems. In addition, it is also considered convenient to synchronize the demand of the finished product with the optimal inventory level of its raw materials through the analysis of different stochastic inventory models like a periodic and continuous review in order to avoid MXN 14'805,188.00 in spills due to the expiration date. As a result of this proposal, we obtained better forecasting reducing the variability up to 95%, and the developer of new inventory policies can ensure the availability of materials for production lines at 98%.

Keywords Planning synchronization · Time series forecasting
Smoothed demand · Stochastic inventory models · Beverage industry
Service level

D. Sánchez-Partida (✉) · D. Arroyo-García ·
J.-L. Martínez-Flores · E. Juárez-García
Department of Logistics and Supply Chain Management, Universidad Popular Autónoma del Estado de Puebla, 17 Sur 901, Barrio de Santiago, CP 72410 Puebla, Puebla, Mexico
e-mail: diana.sanchez@upaep.mx

J. L. García Alcaraz et al. (eds.), *Best Practices in Manufacturing Processes*,
https://doi.org/10.1007/978-3-319-99190-0_6

6.1 Introduction

The term Logistics as a military discipline related to the supply of materials for the army and the control of supplying for the fulfillment of a mission has evolved in recent decades to become an ally of business management and customer satisfaction (Ballou 2004). This new environment leads companies to be more efficient and productive in their supply chain processes to be able to compete in a globalized environment. Therefore, today's Logistics is a competitive advantage for companies; is the integral combination of demand management, inventories, transportation, storage, handling of materials, and packaging activities within a supply chain (Bowersox et al. 2013).

Many decisions are involved in the isolated way causing misalignment in the objectives of each area. Is important to point out that alignment in the objectives can represent competitive advantages such a saving costs into the company, particularly in a competitive global economy (Sachan and Datta 2005; Gunasekaran and Kobu 2007; Arzu and Erman 2010). Some researchers have successfully documented the savings associated with integrated decision-making for logistical planning like (Rodríguez-Méndez et al. 2015) that synchronized the Demand, Materials, Manufacturing Process, and Warehousing areas to have a positive impact on manufacturing and inventory level performance. Because of involving many management levels enhance the collaborative decision-making process in the company.

Ellinger (2000) that show how the internal behaviors may positively affect collaborative marketing/logistics integration by examining relationships between the organization's evaluation and reward system, cross-functional collaboration, effective marketing/logistics interdepartmental integration, and distribution service performance.

Other authors like Sandberg (2007) present theory results from a survey that investigates the situation in real-world supply chains concerning logistics collaboration. Based on a questionnaire that covered important topics from the literature, the questionnaire was sent to the logistics manager at Swedish manufacturing companies, and the information was analyzed by statistics techniques. As a result, he obtained a clear relationship between the intensity of the collaboration and the positive effects experienced from the collaboration. For this reason, is important to have integrated decision-making within any company.

In accordance with Comhaire and Papier (2015) planning is the fundamental part of efficient decision-making, because developing a planning tool determines optimal production volumes by considering the various levels of uncertainty. But a planning synchronization can improve not only the production area but also all the interdepartmental areas.

The main objective of this research is to elaborate a Logistics Planning for the synchronization of key functional areas. First, we will define the most appropriate time series forecasting method that can smooth the variability of the demand for the finished products for a beverage bottling company. Then, this result will be used for obtaining the optimal inventory level of raw materials, evaluating it through different stochastic models of inventory. Trying to obtain the availability of materials for production lines, diminish the value of the raw materials, prevent them from expiring, and improve the customer service level.

6.2 Description of the Real-World Instance

The company produces beverages and has more than four thousand employees and more than 30 production plants in the United States of America, Canada, United Kingdom, and Mexico. In general, the production plants have three packaging lines such as can, cold filling (PET), and hot filling for carbonated, sports, alcoholic, teas, and energy drinks.

In Mexico the bottling company currently serves three major sales channels (a) co-packing market for the manufacture of international brands in the market; (b) supermarket that includes the own brands of the main commercial chains; and (c) the traditional market where the brand products are sold to different retailers.

The scope of this study covers the improvement the synchronization of six areas that make up the company's functionality:

- *Business Intelligent Area* prepares the sales forecast and develops the Annual Operational Planning (AOP) and the Sales and Operations Planning (S&OP). Demand Management area receives the sales forecast for the next 3 months, which should show the actual sale of the first month, a projection with 90% reliability in the second month and a projection with 80% reliability in the third month. Quarter by quarter the forecasted demand is adjusted to the real conditions and for this the Demand Management area directly deals with the placed orders of the customers. These areas are aware of the demand and maintain contact with the customers.
- *Production Planning Area* disaggregates the demand and plans the production through the administration of a Master Production Schedule (MPS). Its main input is the information contained in the S&OP and the orders received by Demand Management area.
- *Material Planning Area* guarantees the existence of raw material in storage through the Material Requirement Planning (MRP) for executing correctly the MPS. The Purchasing area is the commercial part that maintains contact with the suppliers and determines the supply agreements such as price, delivery times, delivery and payment terms.
- *Warehouse Area* receives storage and supervise the correct movement of raw materials. For the output of material, the company follows the First In First Out method (FIFO) under the premise of respecting the expiration date.

When the AOP Forecast of the finished product elaborated by Business Intelligent area cannot cover the demand; the supply of raw material, the warehousing, and production process start to stress. Trying to relieve this situation the Material Planning area start to request raw material at the moment in order to supply the production line and in consequence the customer demand can be satisfied incurring high dispatching cost. These arrivals of raw material also cause problems in Warehouse area because it begins to lack locations for the accommodation of the raw material losing control of the FIFO system and thus spilling raw material due to the expiration date.

6.3 Methodology and Methods

Planning requires the estimation of variables of a probably uncertain environment and constant changes (Rodríguez-Méndez et al. 2015). Even with this, the need to forecast becomes relevant in a competitive context and in this way, forecasting provides the company the opportunity to improve the decision-making process (Hanke and Wichern 2006; Rodríguez-Méndez et al. 2015).

According to Ballou (2004) and Chopra and Meindl (2008), the forecast of demand levels is vital for the company since it provides the input data for the planning and control of all functional areas including marketing, production, and finance. Some elements that are necessary to know are the historical demand, the delivery time of the product, the marketing campaigns, economy, the discounts of planned prices, and the actions that competitors have taken. Also, Chase et al. (2009) comment that demand forecasts are based on patterns such as temporality, the degree of variability and its randomness. Thus, it is important because identifying the different patterns of behavior of the data will discover the adequate forecasting method.

So, the forecast is a corporative planning method that will help to manage future demand. This research follows the systematic way that proposes Render et al. (2012) which happens independently of the forecasting method selected; (a) determine what the forecast will be used for; (b) Select the elements or quantities that are going to be forecast; (c) determine the forecast time horizon (short, medium or long-term); (d) Select the model or forecast models; (e) collect the data or information necessary to carry out the forecast; (f) validate the forecast model; (g) carry out the forecast; and (h) implement the results.

There are different methods or techniques of forecasting, which can be qualitative or quantitative (Chapman 2006). The technique used in this research is time series that belongs to quantitative techniques which trying to predict the future based on past information. The time series models that will be discussed are Simple Moving Average, Weighted Moving Average, Simple Exponential Smoothing or Brown Method, Exponential Smoothing with tendency also known as Holt Method, and Exponential Smoothing with tendency and seasonality also known as Winters Method.

6.3.1 Methods and Errors of the Forecasting

For these subsections only, we will introduce a brave description of the forecasting methods include the mathematical model for the more used method in this research, and the kind of forecasting errors.

- *Simple Moving Average* can be useful when the historical data does not have seasonal characteristics (Chase et al. 2009).
- *Weighted Moving Average* allows assigning percentage importance to each item to be forecast as long as the sum of all the weights is equal to one. According to (Chase et al. 2009), the simplest ways to determine the weights are experience and tests. As a general rule, the most recent past is the most important indicator of what is expected in the future and, therefore, should have a higher weight.
- *Simple Exponential Smoothing or Brown Method* is a suitable model for series with a linear trend and without seasonality. In this method, only three elements are needed to forecast the future, (1) the most recent forecast, (2) the real demand that occurred during the forecast period, and an alpha (α) uniformity constant. The smoothing constant determines the level of uniformity and the speed of reaction to the differences between forecasts and real occurrences. The value of a constant is determined both by the nature of the product and by the sense of an expert (usually a manager). The reason why it is called exponential smoothing is that each increase in the past is reduced ($1 - \alpha$) (Chase et al. 2009).
- *Exponential Smoothing with Tendency* also known as *Holt Method*, with this method the forecasts are smoothed exponentially and can be corrected by adjusting the trend. To correct this trend, two constants are required (α) and (δ). Where α the same function as the previous method and δ has reduced the impact of the error that occurs between the real demand and the predicted value. Without the smoothing constants, the trend reacts unfavorably to errors. Equation (6.1) is used to calculate the future projections while Eq. (6.2) helps to calculate the level of uniformity and calculates the trend Eq. (6.3) (Chase et al. 2009).

$$\mathrm{FIT} = Ft + T_t \tag{6.1}$$

$$Ft = \mathrm{FIT}_{t-1} + \alpha(A_{t-1} - \mathrm{FIT}_{t-1}) \tag{6.2}$$

$$T_t = T_{t-1} + \delta(Ft - \mathrm{FIT}_{t-1}), \tag{6.3}$$

where:

Ft the forecast smoothed exponentially for period t.
T_t the trend smoothed exponentially for period t.
FIT_t the forecast including the trend for period t.
FIT_{t-1} the forecast including the trend made for the previous period.
A_{t-1} actual demand for the previous period.

α smoothing constant.
δ smoothing constant.

- *Exponential Smoothing with Tendency and Seasonality Also Known as Winters Method*, with this method the forecast is smoothed exponentially and can be corrected by adjusting the trend and seasonality. To correct these elements three smoothing constants are required (Chase et al. 2009).

A perfect forecast is not possible. The forecasts are subject to an error. An error is understood as the difference between the value forecasted and the real demand which are known as statistically speaking how residuals. The smaller the error among the forecasting method tells us the more accurate the forecast will be and vice versa. The literature mentions that the terms most used to define the validity and confidence of the forecasting method is the degree of error. The usual forecasting errors are the Mean Square Error (MSE), the Mean Absolute Deviation (MAD), and the Mean Absolute Percentage Error (MAPE) (Ballou 2004; Rodríguez-Méndez et al. 2015; Chase et al. 2009).

- *Mean Square Error (MSE)* is the most common measurement to make comparisons and is obtained from the average of the square of the errors according to the following Eq. (6.4).

$$\mathrm{MSE} = \frac{\sum (\mathrm{error})^2}{n} \tag{6.4}$$

- *Mean Absolute Deviation (MAD)* is the average error of the forecast by using absolute values, and in the same way as the standard deviation, it measures the dispersion of an observed value in relation to an expected value, Eq. (6.5).

$$\mathrm{MAD} = \frac{\sum |\text{forecasting error}|}{n} \tag{6.5}$$

- *Mean Absolute Percentage Error (MAPE)* is the average of the absolute values of the errors expressed as percentages of the real demand as indicated by the following Eq. (6.6).

$$\mathrm{MAPE} = \frac{\sum \left|\frac{\text{forecasting error}}{\text{real demand}}\right|}{n} 100\% \tag{6.6}$$

Also, the forecasting error serves to determine a Safety Stock (SS) (Handley 2004). In a complementary manner, Chase et al. (2009) also mention that the MAD is useful to establish security inventory levels in the case of inventory control.

6.3.2 Importance of Inventory Levels

It is important to establish the adequate levels and control of inventory because if there is an inaccurate inventory, negative effects will occur. For example, the existence of unnecessary inventory that would increase the cost and reduce the cash flow or a shortage of material that can cause lost sales (Sánchez-Partida et al. 2017). In other words, an increase in the inventory of raw materials may indicate an exceeded supply of material, the existence of an inactive or low turnover inventory, a reduction in the production cycle or low speed of distribution. However, an increase in inventory can also be justified by an increase in the physical production of the company. The good management of inventories involves the determination of a level that ensures both the requirements of the production process and market demand while achieving a compliance target indicator (Burja and Burja 2010).

The SS level is designed both the requirements of the production process and market demand based on the variation of the demand and the level of service desired. Even if a company wants to provide an excellent service level, the logistics cost can increase exponentially (Dooley 2005; Aguilar 2012).

For that reasons, inventory topic is considered one of the most important assets of an organization (Ruankaew and Williams 2013). Render et al. (2012) define it as a stored resource that serves to satisfy current or future market requirements. Other authors consider that inventories are a reserve to support the variations between supply and demand (Torres Navarro and Córdova Neira 2014), and also represent an expenditure of money until they become a finished product and are sold generating a profit (Burja and Burja 2010). This expenditure is expressed in terms of raw materials, products in the process or finished products. The administration of this resource is part of the activities of the supply chain, and its management has always been a challenge.

Undoubtedly, the administration of inventories is in function on demand for finished products (Dooley 2005). Any change in demand has an impact on the strategic planning of a company, its inventories, and the level of service to its customers, among others (Aguilar 2012). One of the goals of the inventory optimization is to minimize the total logistics costs.

The previous literature shows how demand forecasting should be generated in order to use it for the inventory optimization.

Before evaluating the selection of an adequate inventory model, it is necessary to have knowledge of the stored products and segment them according to their frequency of historical consumption or even their cost to define a classification system (Torres Navarro and Córdova Neira 2014).

6.3.3 ABC Classification

A classification system allows companies to design decision models for each product, and the most used is an ABC classification. It is based on the Pareto principle well known as law 80-20. In 1906, the Italian Vilfredo Pareto established that 20% of the products constitutes 80% of the monetary value of a warehouse, the suggest inventory model for each classification is shown in Table 6.1 (López et al. 2013).

6.3.4 Inventory Models

As we already know, the variability presented by the finished products in this study is quite high, so stochastic inventory models are proposed because these models allow the creation of SS.

To specify the characteristics of the variation in consumption are suggested the used as a criterion the Variability Coefficient (VC) (Sánchez-Partida et al. 2017). This measures the stability of the demand for a product.

Once, the uncertainty is confirmed, and the classification is obtained the stochastic inventory models can be selected. In this research, the *continuous review (q, R) and periodic review (S, T) and (s, S) models* will be taken into account.

- Continuous Review Model (q, R):

$$S_d' = \sqrt{L * s_d^2 + d^2 s_L^2} \tag{6.7}$$

Table 6.1 A classic and effective approach to inventory management

Kind of product	General rules
A	Continuous review policy with frequent calculations of the quantity to be ordered, the reorder point and the safety stock Frequent cyclic counts Constant review of the forecast and the forecasting method Detailed and accurate records including product losses Great efforts to reduce inventory and delivery time
B	Less frequent review Good records to detect significant changes in consumption
C	Periodic review system The calculation of the EOQ model is not required Simple counting Order large quantities and use a safety inventory to avoid lack of orderly inventory with low frequency Low priority for planning

Source López et al. (2013)

$$\mathrm{R} = \mathrm{d} * \mathrm{L} + \mathrm{Z}_{\mathrm{CSL}} * S'_d \tag{6.8}$$

$$\mathrm{q} = \sqrt{(2\mathrm{kd}/\mathrm{h})} \tag{6.9}$$

$$\mu(\mathrm{q}) = \mathrm{kd}/\mathrm{q} + \mathrm{h}(\mathrm{q}/2 + \mathrm{Z}_{\mathrm{CSL}} * S'_d) + \mathrm{cd} + \mathrm{d}/\mathrm{q}(\mathrm{u} * S'_d * \mathrm{Ez}) \tag{6.10}$$

$$\mathrm{Ez} = \mathrm{z} * [\mathrm{Fs}(\mathrm{z}) - 1] + \mathrm{fs}(\mathrm{z}) \tag{6.11}$$

$$\mathrm{FR} = 1 - [(\mathrm{d}/\mathrm{q}) * \mathrm{S'd} * \mathrm{Ez}]/\mathrm{d} = 1 - (\mathrm{s'd} * \mathrm{Ez})/\mathrm{q}, \tag{6.12}$$

where:

Z_{CSL} value of z for the standard normal distribution with a cyclic service level (CSL).
s′d standard deviation of the demand in the period of delivery.
u stock-out cost.
Ez stock-outs proportion for a given z.
Fs(z) cumulative distribution function for the normal standard distribution.
fs(z) probability density function for the standard distribution.

In this model, when the inventory decline below the reorder point (R) a new order of size (q) is requested.

- Periodic Review Models (S, T) and (s, S):

We review the inventory at certain intervals of time (*T*), and we request an order enough to elevate the inventory to a given level (*S*). In the (*S*, *T*) model the size of the order (*q*) is the difference between the maximum level (*S*) and the current inventory level at the moment of the review. The (*s*, *S*) model is associated with a min-max system, where orders are requested with less frequency.

$$T = \sqrt{\frac{2k}{dh}} \tag{6.13}$$

$$\mathrm{S} = \mathrm{d} * (\mathrm{T} + \mathrm{L}) + \mathrm{Z}_{\mathrm{CSL}} * S'_d \tag{6.14}$$

$$S'_d = \sqrt{s_d^2(T + L) + s_L^2 d^2} \tag{6.15}$$

$$\mathrm{AIL} = \frac{dT}{2} + \mathrm{Z}_{\mathrm{CSL}} * \mathrm{S'd}, \tag{6.16}$$

where

T time between reviews.
S maximum inventory level.
s′d standard deviation of the demand in the period of delivery.
Z_{CSL} value of z for the normal distribution with a cyclic service level (CSL).

AIL Average Inventory Level.
s minimum inventory level.

It is worth indicating the two types of service that were used in this work. Cycle service level (CSL), is the proportion of order cycles where there is no stock-out. And the item Fill rate (FR) is defined as the fraction of demand satisfied from stock, which is a widely used measure of customer service for inventory systems (Douglas 2005).

6.4 Results

The results presented in this section arise from comparing the current situation with the proposal one. We explain the selection criteria of the forecasting model taken into account for each finished product and why the use of a smoothed demand in the finished products. Additionally, we compare the current purchasing cost versus the results obtained from stochastic inventory models for raw materials.

6.4.1 Results of Monthly and Smoothed Forecasts

First, we applied all the methods above-mentioned from Simple Moving Average Method to Winters Method by each of the 67 finished products considering a monthly demand. The methods were applied in accordance with the demand characteristics and optimized smoothing constants to decrease the forecasting error. Using monthly demand, the minimum forecasting error reported was 135%, and the error maximum was 520%, this because the data have a great variability period by period, do not have tendency or seasonality, and its pattern is totally irregular.

Therefore, the smoothing of the demand quarterly was made for the 67 finished products; this smoothing let to reduce the forecasting error and thus to have a more accurate forecast, Table 6.2 shows the forecasting method defined by the finished products in accordance to obtaining the minimum error. The most used method is Holt, so the smoothing demand let to reduce the variability and mark more the trend of the data. In fact, with the smoothing demand was not necessarily used the Winters Method.

Table 6.2 Forecasting method selected for quarterly smoothing demand of the finished product

Method	Frequency
Simple Moving Average	1
Weighted Moving Average	9
Brown	11
Holt	46
Total, finished products	67

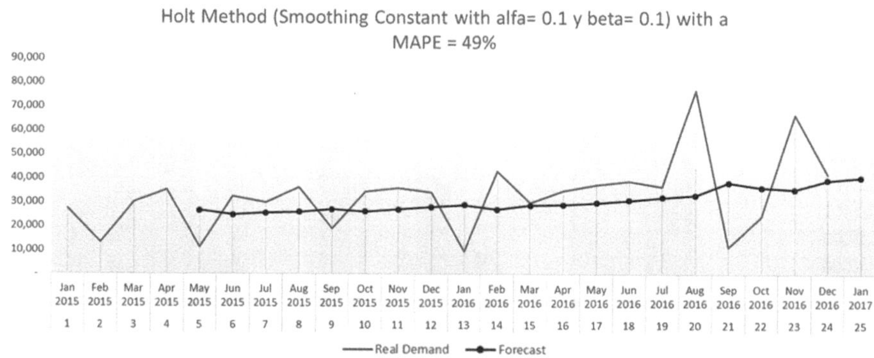

Fig. 6.1 Monthly forecast chart with MAPE of 49%

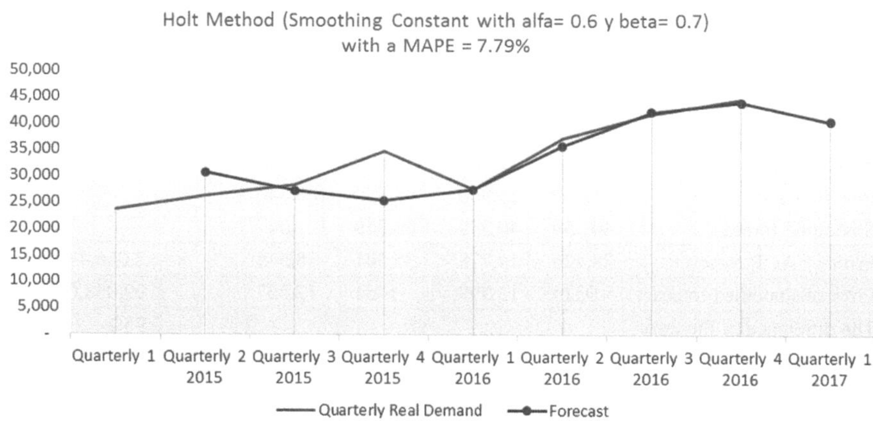

Fig. 6.2 Quarterly forecast chart with MAPE of 7.79%

For having a punctual comparison reference, we show the process and results obtained for product 9. We compare the best MAPE obtained from a monthly forecast with a quarterly smoothed forecast. It was possible to see that using a smoothing demand, substantially decreases the forecasting error from 49 to 7.79%. Figure 6.1 shows the forecast of a monthly demand with a resulting MAPE of 49% for the month of January 2017, and Fig. 6.2 shows the effect of the smoothing demand with a resulting MAPE of 7.79% for the quarterly one 2017 (January, February, and March) in both analysis the Holt Method was selected using their own smoothing constants that minimized the error.

By May 2017, there was already access to the real demand for the first quarter of the year, which allowed establishing a point of comparison between the AOP forecast and the proposed smoothed forecasts. In Fig. 6.3 is compared the AOP forecast made for the company, real or historical demand, and the estimated smoothed forecast for the first 3 months of 2017.

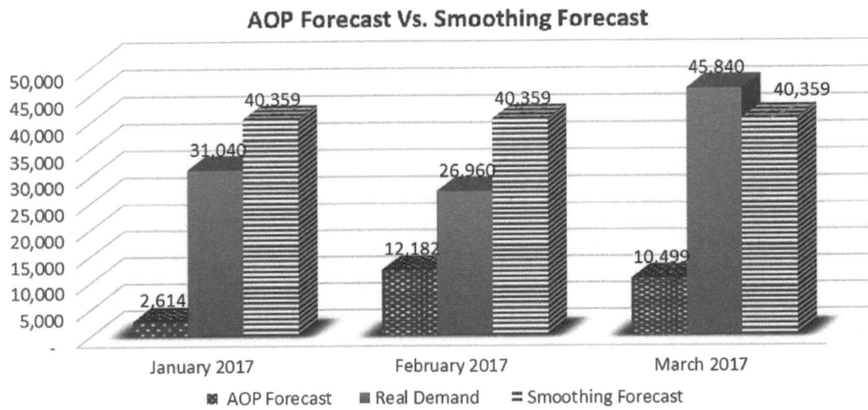

Fig. 6.3 Comparison between the forecast in AOP, the real demand and the smoothed forecast

Table 6.3 Comparison of error between the AOP forecast and the smoothed forecast

	January	February	March	First quarter 2017	MSE
AOP forecast	2614	12,182	10,499	25,295	
Real demand	31,040	26,960	45,840	103,840	
Smoothed forecast	40,359	40,359	40,359	121,077	
Error of AOP forecast	28,426	14,778	35,341	78,545	2,056,439,008
Error of smoothed forecast	9319	13,399	5481	17,237	99,038,056
The difference of the error					95%

In quantifiable terms, the difference of the error between the AOP forecast estimated by Business Intelligence area for product 9, and the smoothed forecast estimated through the Holt method is 95%, it is shown in Table 6.3. Clearly, AOP Forecast did not cover 75.6% of the demand while the smoothed forecast completely covered the demand and remained as safety stock 16.5%.

Through the adoption of the forecasting methods proposed in Table 6.2, can improve the level of assertiveness substantially, also can improve the planning the supply of raw materials and production batches.

Considering all the company's products in the first quarter of 2017, the difference average error rate between the AOP forecast and real demand was 241.31%; while the average error rate between the proposed smoothed forecast and real demand was of 194.94%. Although the percentage of the smoothed forecast is still considerable, the improvement is about 46.36%.

Analyzing the same features by each of the three sales channels (supermarket, co-packing market, and traditional market) see Table 6.4.

Specifying in the sixth column of Table 6.4, it is possible to notice a reduction in the error of the forecasts per sales channel with the exception of the co-packing

Table 6.4 Difference between the real demands, AOP forecast, and smoothed forecast by sales channel

Period	Real demand	AOP forecast	Smoothed forecast	Difference between real demand versus AOP forecast	Difference between real demand versus smoothed forecast
Supermarket					
2015	152,636	N/A	N/A	N/A	N/A
2016	336,494	462,177	N/A	37.4%	N/A
January 2017	44,916	25,434	31,386	43.4%	30.1%
February 2017	30,710	27,613	31,386	10.1%	2.2%
March 2017	33,618	35,643	31,386	6.0%	6.6%
Co-packing market					
2015	5,242,288	N/A	N/A	N/A	N/A
2016	6,695,360	3,393,408	N/A	49.3%	N/A
January 2017	157,670	233,285	385,330	48.0%	144.4%
February 2017	89,959	281,270	385,330	212.7%	328.3%
March 2017	62,059	318,131	385,330	412.6%	520.9%
Traditional market					
2015	763,259	N/A	N/A	N/A	N/A
2016	916,652	815,344	N/A	11.1%	N/A
January 2017	67,331	58,336	78,935	13.4%	17.2%
February 2017	50,849	118,013	78,935	132.1%	55.2%
March 2017	72,991	133,308	78,935	82.6%	8.1%

market division. This happened because it was forecasted based on historical data, obtaining a low forecast error, but when the customer's orders were received, it could be observed that the production request decreased. Concluding that this channel is the most difficult to predict since the order depends on large companies that require the bottling service, because at that time they did not have the production capacity, requesting the service from the bottling company.

6.4.2 Results of Inventory Models

Once we obtained the smoothed forecast of the finished products, we proceeded to exploit them into its components or raw material. The total of raw materials was 397, which were classified in accordance with the utility. The ABC classification of the 397 materials is summarized in the following Table 6.5. Certainly, all the products in accordance to the last information have a certain degree of uncertainty resulting greater than 0.20 in each one of the raw material, which indicates that the data have a great variability and thus we need to use stochastic models for each raw material in order to fix inventory levels. In addition, this table shows the different models assigned depending on the type of classification. The "A classification" contains a group of expensive raw materials, thus the model (q, R) is the best in this context due to its operation. The (q, R) is a model of continuous review, which avoids having inventory storage and ask for it when the raw material can be needed. The "B Classification" and the "C Classification" are treated with periodic review models (S, T) and (s, S) respectively, these types of model operate very similar.

The unique difference is the frequency of orders, for example, when is fixed a period of time for reviewing the inventory level, in the (S, T) model is necessary to ask in each period the missing raw material in order to rise to the maximum point fixed, whereas in the (s, S) model if the inventory level do not pass the minimum level fixed in each period there is no need to ask some order. In accordance to the characteristics the "B Classification" is materials less expensive than the "A Classification", which implies that these materials until certain optimized level can be storage. And for last, the "C Classification" are the group of raw materials of low cost, and the model selected can help fixing an adequate inventory level considering that is cheaper to have these materials in storage than pay continuously for transport.

The models used in this research are the continuous review model (q, R) and the periodic review models (S, T) and (s, S). The following results are based on three scenarios in order to compare them:

- *Scenario* 1: Considering the AOP forecast for the first quarter of 2017 and the current purchasing policies, where the amount of raw material required only cover the S&OP for the following 3 months. Generally, the AOP information is the same that the S&OP for the first 3 months of the year.

Table 6.5 Summary of ABC classification

Classification	Amount of raw materials	Rate of materials (%)	Kind of demand	Inventory model applied
A	31	7.80	Stochastic	q, R
B	51	12.85	Stochastic	S, T
C	315	79.35	Stochastic	s, S
Total	397	100		

- *Scenario* 2: Considering the smoothed forecast obtained and applying the stochastic inventory model according to the ABC classification.

6.4.2.1 Results of the Scenario 1

Under *Scenario* 1, the following considerations should be taken into account:

- Since it is not possible to obtain the available inventory for a certain historical period, it was assumed that the inventory was zero.
- The purchasing orders are assumed that were placed in period zero (January 2017) for the first 3 months.
- The expiration date of some raw materials is 90 days which means that, if they are not consumed as predicted, they will be obsolete shortly after the end of the first quarter.

On the one hand, considering the information of the AOP forecast and the current purchasing policies, a total of MXN 3’701,297.00 of obsolete raw material was estimated for the first quarter of 2017. This situation represents a loss for the company which it is not contemplating material destruction, storage, and other opportunity costs.

On the other hand, the stock-out cost was also estimated, which directly affects the lack of availability to produce and comply with the delivery of orders. The lack of capacity to respond to emergencies and to support the inaccuracies of the demand is due to the fact that an SS is not handled by the company. The stock-out cost for the first quarter of 2017 was estimated at MXN 13’195,867.00.

In summary, during January, February, and March 2017 the bottling company assumed the loss of MXN 16’897,164.00.

6.4.2.2 Results of the Scenario 2

Under *Scenario* 2, the forecasting method that best smoothed the demand for each product was used and, based on this finished products were exploded as raw materials to covering demand of the first 3 months of 2017.

- It is considered the total sum of all those raw materials that are used in more than one finished product.
- For each raw material, the corresponding inventory policy was applied according to the ABC classification.
- Since it is not possible to obtain the available inventory for a certain historical period, it was assumed that the inventory was zero.
- We manage a Fill Rate (FR) of 90%.

- It is assumed that the minimum purchasing amount agreed between the company and the suppliers can be renegotiated. Then, the current minimum purchasing quantities would be invalidated.
- It is assumed that the delivery time of all raw materials is constant.
- It is desired to include within the SS calculation of the raw material the MAD error of the finished product instead of the standard deviation of the real demand. However, it is not possible since a raw material can be used in more than one finished product and each finished product has a different MAD error. Therefore, the standard deviation of the real demand will be used.
- The forecasts are not linked to any macroeconomic variable in the country that could affect their projections.

Following the continuous review model (q, R) for the 31 raw materials class "A" results in the following purchasing policy:

> Each time where the inventory reaches the reorder point (R), an order of size (Q).

Under this policy, a total quarterly inventory cost is MXN 33.03 million, which MXN 32.91 million is the cost of purchasing. The level of compliance guaranteed in deliveries of the finished product is 98% on average.

According to the periodic review model (S, T), for the 51 raw material class "B," the following purchasing policy is suggested:

> Every 30 days (T units of time), an order is made with a size Q equal to the desired or maximum inventory level S minus the current inventory level.

Under this policy, a total quarterly inventory cost is estimated at MXN 5.3 million, which MXN 5.25 million are the cost of purchasing. We selected a time T equal 30 day for placing the orders to unify the deliveries of the suppliers, avoiding to request small orders that mean an additional cost due to a greater frequency of delivery.

According to this policy, a level of delivery compliance is between 95 and 100%. Finally, by applying the periodic review model (s, S) for the raw materials class C, the following purchasing policy is suggested:

> Whenever available inventory falls to the reorder point or minimum point s or below it, an order is placed for a quantity Q to increase the inventory level to a maximum level S.

In terms of this policy, a total quarterly inventory cost of MXN 1.86 million is estimated, which MXN 1.503 is the purchase cost.

Although levels of compliance oscillate between 70 and 80% are appreciated, it is no longer possible to increase the inventory level greater than 30 days because the expirations date do not allow it.

Table 6.6 shows a proportion monthly of the inventory cost and the purchasing cost for the first month of 2017. This separation of these costs is made in order to compare it with the current scheme of costs that the company manages. The difference between the inventory cost and the purchasing cost corresponds to the maintaining and ordering cost.

Table 6.6 Total monthly inventory cost in January 2017 according to the proposed inventory models

Inventory model applied	Inventory cost (January 2017)	Purchasing cost (January 2017)	Service level estimated
q, R	$11,012,038.00	$10,977,870.00	98%
S, T	$1,767,313.00	$1,751,186.00	99%
s, S	$621,975.00	$501,262.00	98%
	Total: $13,401,327.00	Total: $13,230,318.00	Average: 98%

Table 6.7 Current inventory cost in January 2017

Inventory model	Real purchasing cost (January 2017)	Target inventory value	Real customer service level
None	$12,924,922.00	$10,000,000.00	70–80%

In contrast, the following Table 6.7 shows the current inventory cost that the company reported during the month of January 2017 which considers only the purchasing cost versus the target inventory value that is controlled in the Balanced Scorecard. It is important to comment that the target inventory value established by the company is not met 24% in this month, but the customer service level increases up to 28%.

In quarterly terms, Table 6.8 presents the proposed costs and compares them with current costs. The reference is made during the first quarter of 2017, the current purchasing cost calculated by the company was MXN 40’064,419.00 compared to MXN 39’690,954.00 of the proposal, which helps to have a customer service level of 98%. This difference represents an improvement of 1%. This cost is the only compared because the company does not put focusing and do not have records of the maintaining and order costs but considering that the company request every time that need raw material the reduction in order frequency likely can reduce the ordering cost. Also, is important mention that the annual spills can reduce MXN 14’805,188.00 annually and the customer service level can increase up to 28%.

Table 6.8 Comparison of the purchasing cost of raw materials proposed versus current

Proposal				Current
Classification	Proposed total cost	Maintaining and order cost	Purchasing cost	Purchasing cost
A	$33,036,114.00	$102,505.00	$32,933,610.00	Not ungrouped
B	$5,301,940.00	$48,381.00	$5,253,559.00	
C	$1,865,927.00	$362,141.00	$1,503,785.00	
Quarterly cost	$40,203,981.00	$513,027.00	$39,690,954.00	$40,064,419.00
Monthly cost	$13,401,327.00	$171,009.00	$13,230,318.00	$ 13,354,806.00

Table 6.9 Final comparison of results

First quarterly 2017 (real)		First quarterly 2017 (proposal)		
ABC classification	Unidentified	A	B	C
Purchasing cost	$40,064,419.00	$32,933,610.000	$5,253,559.00	$1,503,785.00
Total cost of inventory	Unidentified	$33,036,114.00	$5,301,940.00	$1,865,927.00
Inventory model	N/A	Continuous Review (Q, R)	Periodic Review (S, T)	Periodic Review (s, S)
Placement of orders	Every time it is required	When falling to the reorder point	Each 30 days	Each 30 days
Customer service level	70–80%	98%	99%	98%

Table 6.9 summarizes the current results obtained during the first quarter of 2017 versus the scenario proposed.

6.5 Concluding Remarks

It is important to know that the function of forecast based on intuitions, assumptions, judgment, and personal experience must complement with quantitative techniques and technological advances in data management for transforming the making decisions of companies.

This study achieved to collaborate with the Business Intelligence department to increase the assertiveness of the sales forecast in the S&OP by smoothing demand quarterly. Also, it was possible to adapt each adequate inventory model to the type of raw material optimizing the inventory level that does not generate destruction of raw material due to the expiration date.

The smoothing demand has advantages, on the one hand, a constant production can be obtained avoiding stress in the production lines due to the reduction in the variability of the demand. On the other hand, considering that the expiration date of the raw material is 3 months and the expiration date as a finished product is 10 months. It is preferable that the raw material can be transformed and stored quarterly as a finished product rather than risk not using it and disposing of it according to the expiration date.

In addition, the purchasing parameters were determined and changed in the Systems Applications and Products System or Systeme, Anwendungen und Produkte for its acronym in German (SAP System), such the lot size, reorder point and safety stock in order to of fix the amount and day of purchasing.

At the same time, the proposal minimized the inventory cost of raw material, the target of the purchasing cost was changed and the customer service level increased.

In conclusion, a good estimate of forecasting makes the planning and production processes more efficient, guarantees the supply of materials, optimizes the inventories, and increases the level of satisfaction of its customers maintaining a supply chain synchronized and affected significantly the company's profitability.

References

Aguilar PA (2012) An inventory classification model to increase the level of customer service and the profitability of the company. Pensamiento Gestión 32:142–164 [in Spanish]

Arzu G, Erman T (2010) Supply chain performance measurement: a literature review. Int J Prod Res 48(17):5137–5155

Ballou RH (2004) Logistics. Supply chain management, 5th edn. Pearson Education, Mexico 1:1–32 [in Spanish]

Bowersox D, Closs D, Cooper MB, Bowersox J (2013) Supply chain logistics management, 4th edn. McGraw Hill International Edition. ISBN: 9780078024054

Burja C, Burja V (2010) Analysis model for inventory management. Ann Univ Petroşani Econ 10 (1):43–50

Chapman S (2006) Planning and production control, 1st edn. Pearson Education, Mexico [in Spanish]

Chase RB, Jacobs FR, Aquilano NJ (2009) Operations management: production and supply chain, 12th edn. McGraw-Hill, Mexico [in Spanish]

Chopra S, Meindl P (2008) Supply chain management: strategy, planning and operation, 3rd edn. Pearson Education, Mexico [in Spanish]

Comhaire P, Papier F (2015) Syngenta uses a cover optimizer to determine production volumes for its European seed supply chain. Interfaces 45(6):501–513

Dooley F (2005) Logistics inventory control and supply chain management. Choices 20(4): 287–291

Douglas J (2005) Measuring item fill-rate performance in a finite horizon. Manuf Serv Oper Manag 7(1):74–80

Ellinger AE (2000) Improving marketing/logistics cross-functional collaboration in the supply chain. Ind Mark Manage 29(1):85–96

Gunasekaran A, Kobu B (2007) Performance measures and metrics in logistics and supply chain management: a review of recent literature (1995–2004) for research and applications. Int J Prod Res 45(12):2819–2840

Handley SW (2004) A modern view of inventory. How do you decide what level of stock is right for you? Strateg Financ 1:30

Hanke J, Wichern D (2006) Business forecasts, 8th edn. Pearson Education, Mexico [in Spanish]

López JA, Mendoza A, Masini J (2013) A classic and effective approach to inventory management. Int J Ind Eng 20:372–386

Render B, Stair R, Hanna M (2012) Quantitative methods for business, 11th edn. Pearson Education, Mexico [in Spanish]

Rodríguez-Méndez R, Sánchez-Partida D, Martínez-Flores JL, Arvizu-Barrón E (eds) (2015) A case study: SMED & JIT methodologies to develop continuous flow of stamped parts into AC disconnect assembly line in Schneider Electric Tlaxcala Plant. IFAC-PapersOnLine 48 (3):1399–1404. ISSN 2405-8963. https://doi.org/10.1016/j.ifacol.2015.06.282

Ruankaew T, Williams P (2013) The impact of inventory inaccuracy in the food manufacturing industry: a case of study. Bus Manag Dyn 2(10):28–34

Sachan A, Datta S (2005) Review of supply chain management and logistics research. Int J Phys Distrib Logist Manag 35(9):664–705

Sánchez-Partida D, Martínez-Flores JL, Cruz-Hernández J, Chancey E (2017) The impact of planning tools on production and inventory activities. In: Proceedings of the 2017 international symposium on industrial engineering and operations management (IEOM) Bristol, UK, July, pp 185–195

Sandberg E (2007) Logistics collaboration in supply chains: practice vs. theory. Int J Logist Manag 18(2):274–293

Torres Navarro CA, Córdova Neira JA (2014) Expert system design for material purchase decision making. Sci Electron Libr Online 52:20–30 [in Spanish]

Chapter 7
Demand Management to Optimize the Supply Chain of an Agribusiness Company

Alicia Pérez-Pérez, José Claudio Villamonte-Cornejo, César López-Rosas, Roberto Pérez-Franco and Mario Gustavo Chong Chong

Abstract The growing consumption in the world of super-foods, such as quinoa (*Chenopodium quinoa*), presents new opportunities for a small business that is active on the transformation and marketing, which allow them to make the most out of the positive conditions of the market in a rapid and flexible way. The current research proposes the optimization of the supply chain for BAMSA Company, using strategic diagnosis methodologies and operational analysis to carry out an analysis in multiple dimensions that allows decisions to be made in the short, medium, and long term. As a result, it is determined that the strategic flexibility of an efficient supply chain gives priority to operational activities which allow growth objectives to be achieved. For that matter, an adjustment in the supply chain's sourcing strategies of a pull–push type, based on forecast demand, should allow an increase in sales of USD $239,475 dollars annually.

Keywords Supply chain · Flexibility · Optimization · Forecast

---

A. Pérez-Pérez · J. C. Villamonte-Cornejo · C. López-Rosas · M. G. Chong Chong (✉)
Graduate School, Universidad del Pacífico, Av. Gral. Salaverry 2020,
Jesús María 15072, Lima, Peru
e-mail: m.chong@up.edu.pe

A. Pérez-Pérez
e-mail: ae.perezp@alum.up.edu.pe

R. Pérez-Franco
Centre for Supply Chain and Logistics, Deakin University, 221 Burwood Hwy,
Burwood Vic 3125, Melbourne, Australia

J. L. García Alcaraz et al. (eds.), *Best Practices in Manufacturing Processes*,
https://doi.org/10.1007/978-3-319-99190-0_7

7.1 Introduction

Currently, the global market of super-foods exceeds USD $39 billion dollars. Peru is part of this market, with a broad portfolio of exportable supply, which is based on Andean grains such as quinoa, cocoa beans, and its derivatives, among others. The present document is framed under the opportunities currently facing this market and how BAMSA Company can benefit from them, optimizing their supply chain.

In this regard, it has been identified that the biggest obstacles the company faces, and its business model are the limited visibility of demand, the seasonality of productive supply of raw materials and restricted planning, which do not allow a continuous supply of raw material and therefore put at risk the timely care of customer demands. In the past 3 years, the company has experienced a rapid business growth due to its varied portfolio of products that meet customers' needs, which places the company in situations of continuous launch and/or discontinuity in their product presentations, among others. In that sense, defining the role of demand forecast is key to determine technical, operational, and human resources for company operations (Sunil and Meindl 2013). The limited visibility of demand does not allow managing purchasing needs and available resources appropriately. Accordingly, there is a quest for recruiting the largest quality of sales, reducing internal operational constraints which have generated loss of sales for a value of approximately USD $316,750 annually, taking the last two years as a reference.

The aim of redrafting a supply chain strategy is to improve it to provide better support to business strategies. It intends to produce a new systematic statement describing a supply chain strategy that is more valuable for the company than what is currently available (Pérez-Franco 2016). The specific objectives that are expected from this optimization are the following:

- To reduce the percentage of lost sales due to the lack of raw material, buying quality, quantity and much-needed opportunity to meet the customer's value proposition.
- To generate financially sound levels of safety stocks, to avoid lost sales and improve customer service time.

We estimate that a supply chain optimization along the lines proposed in the present article will lead to an increase in sales by USD $239,475 thousand dollars, equivalent to obtaining the NPV of PEN 167.602 and the IRR of 57% in an optimistic scenario with a year of recovery.

7.2 Literature Review

7.2.1 About the Supply Chain

To minimize the risk during market share, from a commercial multi-product portfolio and production strategies pull type, is not necessarily enough, but there is an adequate demand management that can lead to the establishment of strategies and to define the operational and financial needs. In that sense, recent studies such as Pereira (2015) have emphasized the importance of identifying risk sources which have impeded supply networks to create effective actions to mitigate them.

Thus, changes to the strategy, context or environment of a business unit may necessitate a revision of its supply chain strategy. As we attempt to conceptualize the supply chain strategy of a company, it must be acknowledged as an entity that is interconnected to other functional strategies. From this originates the CSAR methodology (Pérez-Franco et al. 2016), which seeks to capture the strategy's current situation and to reformulate it, identifying principal activities that are performed in relation to supply operations, as well as suggest improvements, which are hierarchical by using multi-criteria tools. The application of methodology CSAR has multiple levels (Sakihara et al. 2017) as Pedregal's case.

Mentzer et al. (2001, p. 3–5), defines supply chain, as a group of entities directly involved in product flows, services, finances, and information provided by customers. It attempts to conceptualize only one functional strategy, it could hardly succeed on its own, since functional strategies are not isolated entities. A company's multiple operational objectives often require multiple functional strategies, to be achieved, simultaneously through a coordinated effort, so they are consistent and congruent with each other (Hines 2009, p. 39). Furthermore, Pérez-Franco et al. (2016) define supply chain strategy as a compilation of general and specific objectives, policies and choices made in a supply chain to align their operations with the company's general strategy.

According to Ferreira et al. (2015), considering that almost every supply chain faces interruptions, given the type and several difficulties of supply (Wieland 2013; Gogeci and Ponomarov 2013, Treiblmainer 2014), Christopher and Peck (2004), classified these interruptions into: internal, external, and environmental. Being prepared for any disruptive future event allows companies generate effective responses and are therefore less vulnerable to disturbances (Ponomarov and Holcomb 2009; Pettit et al. 2010; Azevedo et al. 2013; Scholten et al. 2014).

On the other hand, a method for collecting data, related to necessary actions to take material through the value chain (material flow), based on the use of value stream map—VSM methodology, in which each product has a different value stream; for instance, for a finished product it will be from the supplier to the customer and for a sub-product, from the supplier until becoming an input for new material (Rother 2008).

Moreno (2016), quotes Fawcett and Magnan (2002), who have carried out an extensive research to optimize company's operations in the fields of production planning, inventory, storage, among others, and that have been addressed as independent problems for several researchers. According to Ganeshan and Harrison (1995), classify the supply chain's functions in four categories: location, production, inventory, and transport.

To take decisions at the financial and managerial level, the understanding of the field where the company develops and the behavior of product demand, is necessary. In that regard, the process and information analysis are a key to the extraction, purification, and homogenization of data (Sunil and Meindl 2013).

7.2.2 *Case Study*

BAMSA is a company was created in 2010 that produces and sells products from three lines of business such as Andean grains, cocoa and its derivatives and dried fruits, has the presence in more than 40 countries, and has more than 80 workers. Currently its sales exceed 15 million dollars. The complex portfolio of products it sells requires a quick supply chain management to reach the destination markets on time.

In that sense, the methodologies used, seek to separate supply chain from the company in a series of activities to identify the key bottlenecks who face handling purchasing management of the company, in its strategic role with the supply of different raw materials, material, and services. For it, they are divided in two parts: the methodology used in the diagnostic phase and in the optimization phase or proposal for improvements, organized in the following way:

1. Situational analysis
2. Operational analysis
3. Strategic analysis.

In that sense, it seeks to answer the following questions:

- The current strategy of the supply chain of the company is aligned with the objectives of fulfilling the promise of value towards the client?
- The operational activities of the company generate value towards the company?
- Can short, medium, and long-term decisions be made in a practical way that allows for positive financial results?

7.3 Methodology

Figure 7.1 shows the methods and procedures used to generate the situational, operational and strategic diagnosis of the company are presented to pro-optimize the supply chain of the company.

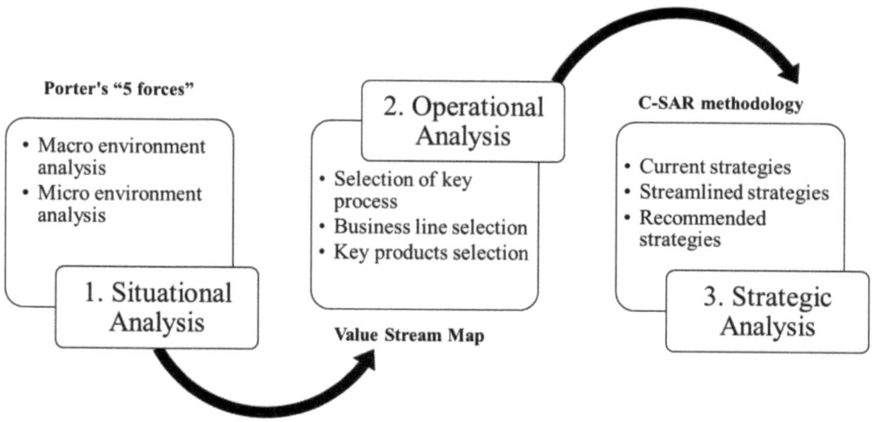

Fig. 7.1 Summary of the methodological process

7.3.1 Diagnosis Phase

7.3.1.1 Situational Analysis

Understanding competitive forces, and their underlying causes, reveal the origins of the current profitability of a sector and provides a framework to anticipate competition and flow in it (and profitability) in the long term. A healthy structure of its sector; it should be as important to a strategist as the position of your company. Understand the structure of a sector is also key to an effective strategic positioning (Porter 2008).

We will explain about the context in which the company competes, breaking down the macroenvironment into its political, economic, social, technological, infrastructure and environmental components. To later, describe the microenvironment of the company to determine the business model behavior in the three business lines for the company.

7.3.1.2 Operational Analysis

For the operational analysis, the Value stream map (VSM) has been used as a visual tool to analyze the current situation of the productive system and channel it towards a true implantation of a lean flow devoid of waste. The complete flow is considered, from the supply to the client, defining a starting and future situation called lean objective. An important aspect refers to the volume of material waiting to be processed between two operations, as well as the time in which the material will be between them. The times of operation on the product and those that are waiting to constitute a stock between operations are reflected also the value stream map, which will allow to calculate the lead time between supplier and client, going through all the operations of the process (Cuatrecasas 2011).

7.3.1.3 Strategic Analysis

For the strategic analysis, the CSAR methodology was used since it allows us to reflect in a graphic and systemic way the strategy of the supply chain of a company that is often implicit in a greater strategy.

With regard to the implementation of CSAR methodology, to identify current activities and how to propose improvement opportunities of supply chain's strategy through its analysis in multiple dimensions and as a logical bridge between the company's strategy and the activities carried out by the supply chain, in two moments: (i) the initial capture, which reflects current activities; and (ii) the final capture, which identifies activities as opportunities for improvement that serve to update and/or rethink the supply chain strategy, defining the pillars, specific objectives and operational objectives of the strategy.

For the identification of the strategic objective or objectives proposed with its purpose or pillar, it has, or they have greater impact within the succession of supply chain's strategy. To understand how describing supply chain strategy in terms of principles and imperatives can help logically connect the business strategy and the operations in the field which are represented in a segmentation tree and have definite characteristics that are presented in Fig. 7.2.

In other words, the organization can change at will its overall and supply chain strategies and practices, and—with time—its assets, culture, and capabilities All the internal elements fall inside the dotted circumference because they are within our control. Beyond our control are external elements. The expectations of the parent

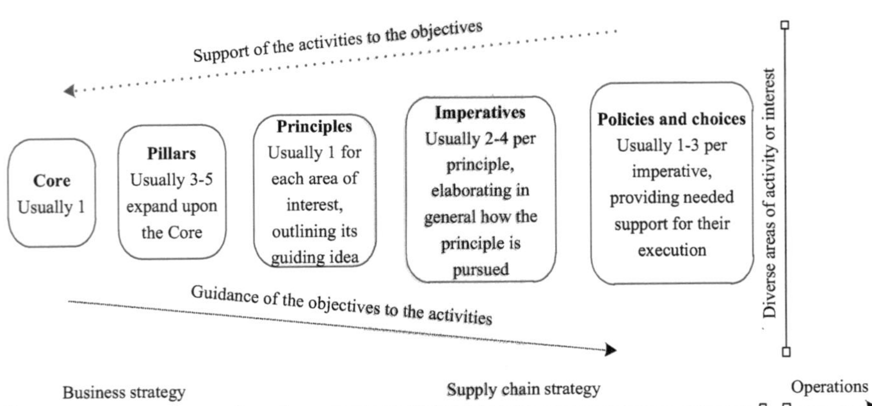

Fig. 7.2 Working framework of the supply chain strategy of a business unit (Pérez-Franco et al. 2018)

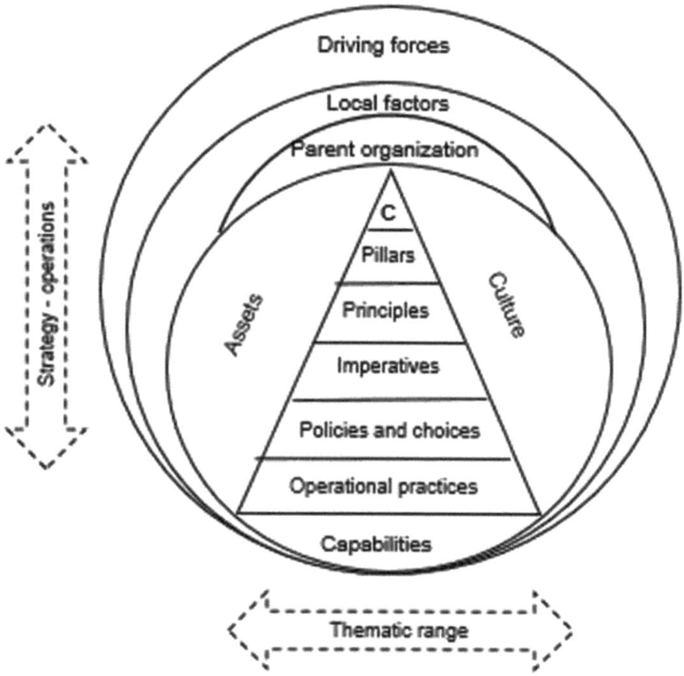

Fig. 7.3 The internal and external elements what the organization controls. Model for rethinking supply chain strategy

organization, the local factors and the driving forces. Thinking about supply chain strategies requires us to consider multiple dimensions. The first and most obvious dimension run from supplier to consumer. The overall task of a supply chain strategy along this dimension is to successfully match demand and supply. A second dimension cut across all the supply chain relevant functions of the business. A third dimension we must consider runs from the top down, from lofty statements of the business strategy all the way down to the activities and operations of the supply chain, along what we call the strategy operations continuum (Pérez Franco 2016).

The dotted circumference in Fig. 7.3 indicates what falls within the control of the organization.

A hierarchical structure has been used in CSAR methodology, which consists of the development of four matrices: (i) Compatibility between strategic pillars matrix: Nst = X, X by X, (ii) synergies matrix between strategic pillars and main objectives: Nst = X, Nft = Y, (iii) synergies matrix between main objectives: Nst = Y, Y by Y, (iv) synergies matrix between operational practices and main objectives: Nft = Y, Not = Z, Y by Z (parent/child approach); of which use two-way approach scales and full spectrum, thus, it seeks to measure the synergies and compatibilities between strategic pillars, main objectives, operational objectives of supply chain's strategy.

Table 7.1 Compatibility between strategic pillars (Matrix 1)

Bidirectional approach scale	Assessment
Yes, they are fully compatible	+2
They are somewhat compatible	+1
They are in something incompatible	−1
No, they are completely incompatible	−2
I'm not sure	0

Table 7.2 Synergies between strategic pillars and main objectives (Matrix 2)

Bidirectional approach scale	Assessment
Yes, it is necessary	+2
I could help something	+1
It does not make any difference	0
It could be harmful to something	−1
No, it's harmful	−2

Table 7.3 Synergies matrix between main objectives (Matrix 3)

Full spectrum scale	Assessment
Yes, it is crucial	+4
Yes, it helps significantly	+2
I could help something	+1
Does not have any effect (unrelated concepts)	0
It could hurt something	−1
No, it significantly impairs	−2
No, completely harmful	−4

Table 7.4 Synergies matrix between operational practices and main objectives (Matrix 4)

Father-Son approach	Assessment
If required	+2
Help, but is not necessary	+1
It does not make any difference	0
Could cause harm	−1
No, it is clearly harmful	−2

The techniques-tools matrix is significant as a pioneering effort to operationally define and represent a supply chain strategy (Pérez-Franco et al. 2018). Tables 7.1, 7.2, 7.3 and 7.4 are the scales used for each of the qualitative evaluation matrices used, which are applied to assess in-depth interviews with the work teams involved in the company's supply chain.

7.3.2 Optimization Phase

- *Product of business portfolio selection:* As of sales analysis according to its commercial portfolio for each business line.
- *Forecast development:* Commercial data with the use of tools such as Tableau® and Qlick® are analyzed and a moving averages method was used for forecast demand. According to Sunil and Meindl (2013), moving average of several periods is evaluated as follows, Eq. (7.1):

$$\mathrm{Lt} = (\mathrm{Dt} + \mathrm{Dt} - 1 + \cdots + \mathrm{Dt} - N + 1)/N \tag{7.1}$$

- *Safety stock estimate:* The suggested formula by Sunil and Meindl (2013) will be used, Eq. (7.2):

$$\mathrm{SS} = \mathrm{ROP}\ - \mathrm{D} * \mathrm{L} \tag{7.2}$$

- *Definition of operational needs and economic and financial assessment*: The need for a safety stock redefines storage, as well as operational and financial resources needs for its operation. The analysis of NPV and IRR, seeks to measure the results of the optimization's implementation.

7.3.3 Numerical Setting

- *Business line selection and business portfolio product*: In 2016, the business line of Andean grains was the one with the greatest growth with a 52% of annual turnover, quinoa represents 80% of participation in this business line.
- *Identification of strategies*: The methodology is applied in a current, enhanced and post-implementation scenario.
- *Identification of key processes*: Quality protocol for product's acceptance and rejection fulfills a key role since it directly affects defined times of customers' order fulfillment.
- *For the definition of forecast*: Sales product is analyzed, due to the lack of information on real demand, under the supposition that the calculation of safety stock will cover the satisfied demand and it will define the basis of the creation of information that helps determine real demand. To reduce the level of imprecision that every forecast has, the prognostic error will be measured, which is established lower than 6% defining monthly adjustments and subsequently quarterly adjustments.
- The time range analyzed corresponds to the analysis of quarterly intervals, the annual behavior and the waiting time for the replenishment of the raw material.

- The level of aggregation will be constituted by the actual sales of the final customer, in this case the orders of the intermediaries and operators in order to have the smallest standard deviation of the error in relation to the average. As well as monitoring the behavior of the main SKUs that have higher production and more are profitable.

7.4 Results

7.4.1 Situational Analysis

As of macro and micro-setting, it is determined that external conditions are favorable to the agro-industrial sector investments, oriented towards the production of super-foods. Nevertheless, the business line of Andean grains, particularly quinoa, is the least attractive; this is because of the pressure exerted by combined strengths in which buyers, substitute products, and the industry's attraction, possess a high negotiating power in different times of the production cycle.

Situations of raw material shortage directly affected the fulfillment of the company's value proposition to their customers and loss of sales in 2016 by USD $316.750 annually. In this sense, the proposal for improvements aims to attend the most responsive process line and product to market fluctuations and avoid lost sales. Figure 7.4 shows the results of the three business lines evaluated.

7.4.2 Operational Analysis

The value stream map (VSM)—raises information about the actions (that add and do not add value) required to carry a material through the value chain (material flow). Each product has a different value stream; for example, for a finished product

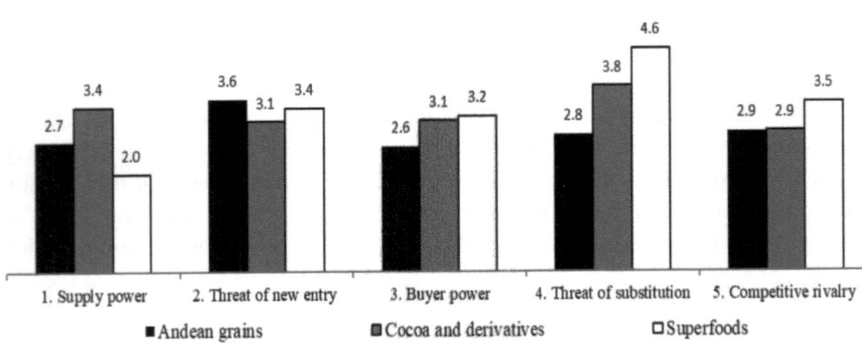

Fig. 7.4 Five Porter's forces

it will be from the supplier to the customer, from a by-product to the supplier until it becomes an input for a new material (Rother 2008).

Currently, the company has a pull process to start production operations, orders can be shipments of a single product or shipments consisting of several products, which are transferred by air or sea by different logistics operators. The operations of the company are organized in different processes that can be structured in directional, core or key processes and support.

For this present investigation, VSM has been used to identify the bottlenecks during the entry of the raw material and its distribution to the final customer as a finished product. The main impacts are:

- The days elapsed for the approval of the raw materials have an impact on the warehouse because it can become saturated with this material; in the process of approval and/or waiting to be pick up due to non-compliance with quality parameters.
- The warehouse current situation has a direct impact on the company's finances. As mentioned above, the mechanism for leveraging operations is via factoring, a delay in customer service has a high financial cost.
- On the other hand, not having the suitable raw material in terms of quality, quantity and/or opportunity affects the rate of operation in the plant, which generates a direct impact on the unit cost of the products, reducing the profitability of the operation.

The process to approve and/or refuse supply, specific for raw material, could take up to 30 days. The waiting time related to the use of raw material's approval cause an impact on warehouse's capacity, which can lead to overload of suitable raw material, during the approval process and/or product in idle expected to be picked up by uncompleted quality parameters. On the other hand, not having enough suitable raw materials in quality, quantity and in timely manner terms, affect the plant operational rhythm; as a consequence, a direct impact on the product's unit cost reduces the operational profitability.

The operations of the company are organized in different processes that can be structured in directional, core or key processes and support.

- *Directional processes*: They are integrated by the processes related to (i) the corporate policies and guidelines, which make up the value proposition to the client and which are established in the corporate vision and mission; (ii) human resources, include processes related to the management of human capital; (iii) the management of the sustainable production program, related to the technical training of the farmers, the supervision and audit of the production processes exclusively for the production of aguaymanto.
- *Key processes*: Conformed by the processes related to (i) customer management, (ii) the logistics of entry, (iii) operations and (iv) logistics of exit.
- *Support processes*: They are linked to (i) project management, which contribute to the continuous improvement of the company, (ii) industrial maintenance,

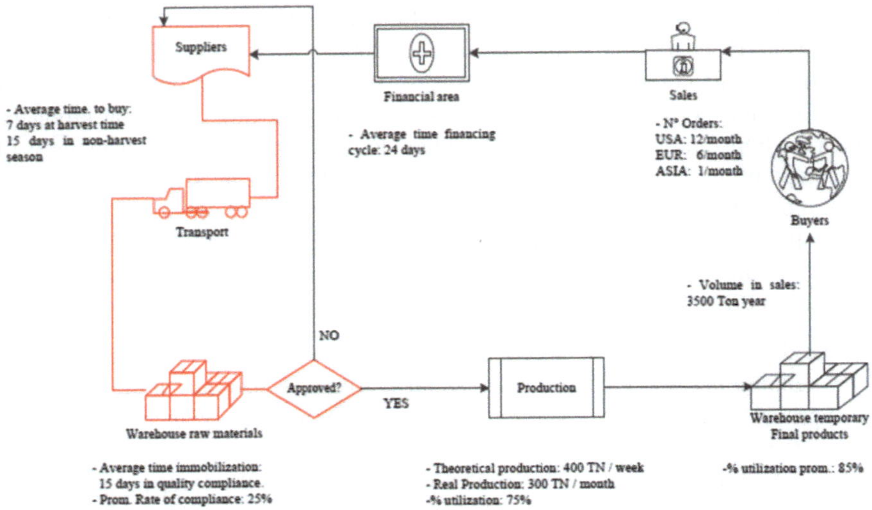

Fig. 7.5 Value Stream Map: map of the processes of the supply chain of the company

(iii) inventory management, (iv) quality management and (v) financial and accounting management.

In Fig. 7.5, the main activities involved in the transformation of raw materials and the main bottlenecks facing the current supply chain can be observed. In this regard, it is important to see that the critical stage is between the time it takes to obtain the raw material and the time it takes to meet the quality protocols.

7.4.3 Strategic Analysis

Based on the business model core and as of the implementation of CSAR methodology, current strategies of supply chain are identified, they are short-term oriented and are about three defined purposes as pillars for purchasing management: (i) utilize financial, operational and human resources efficiently, (ii) contribute with the quality parameters required, (iii) keep production process going at the lowest cost.

There is strong dependence between the supply operations and the plant operations in the spirit of not increase the unit cost. In turn there are 17 main objectives of multifunctional nature, and even duplicities.

The analysis of the matrices allows us to see that, in order to efficiently use the resources financial, operational and human resources (A1), it is necessary to improve the payment terms of suppliers (B2) to ensure the sourcing program (B8) and maintain a production continues at the lowest cost (A3). In turn, promote production flexibility (B9) as alternative to breakage of stock of raw materials does

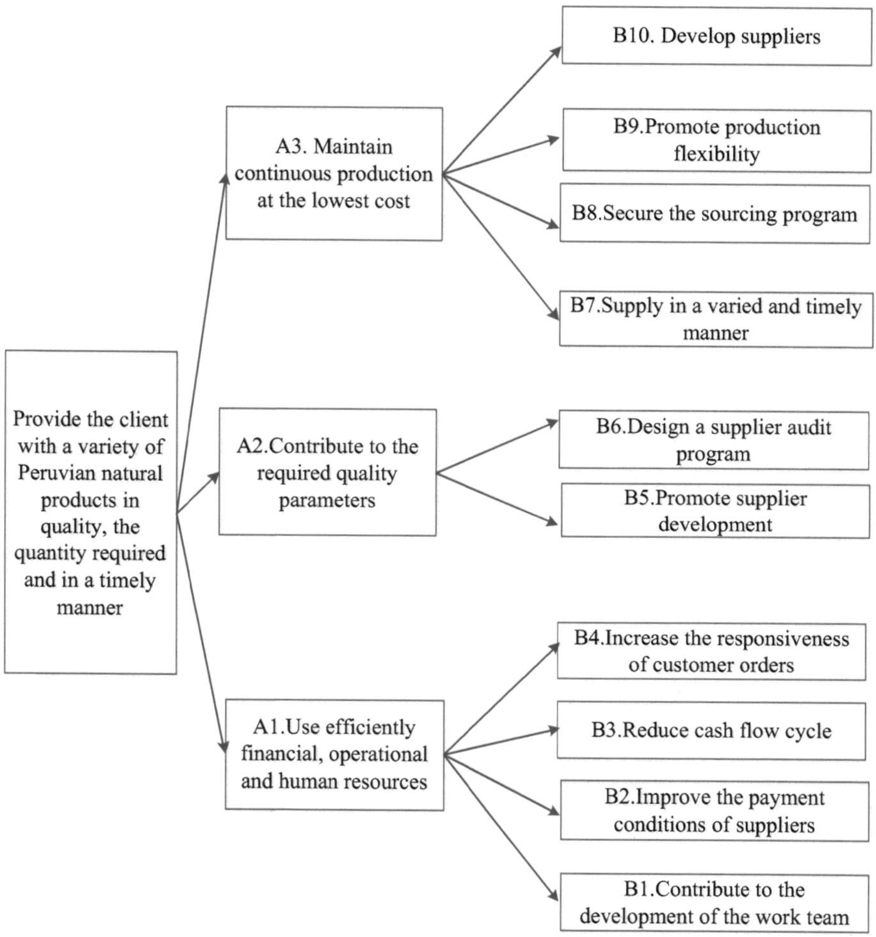

Fig. 7.6 The initial capture

not necessarily contribute to maintain a continuous production and even does not allow the achievement of the strategic pillars of the operation. See Fig. 7.6.

The optimization proposal seeks to strengthen the three identified pillars incorporating one more pillar, which is defined as supply in quality, quantity and in a timely manner. See Fig. 7.7. In this sense, the incorporation of key actions as planning and forecasting fulfill, are proposed, in order to serve customers in a timely manner. Performance measurements are suggested as indicators: (i) Purchase gap fulfillment and (ii) Forecast error measurement.

In this regard, buy without planning that generates complications at the cycle level of cash flow, which affects the payment of the suppliers that arrived at the date or out of date, but that meets quality parameters or even with higher working capital needs for reduce the replacement or replenishment times, which generates cost overruns.

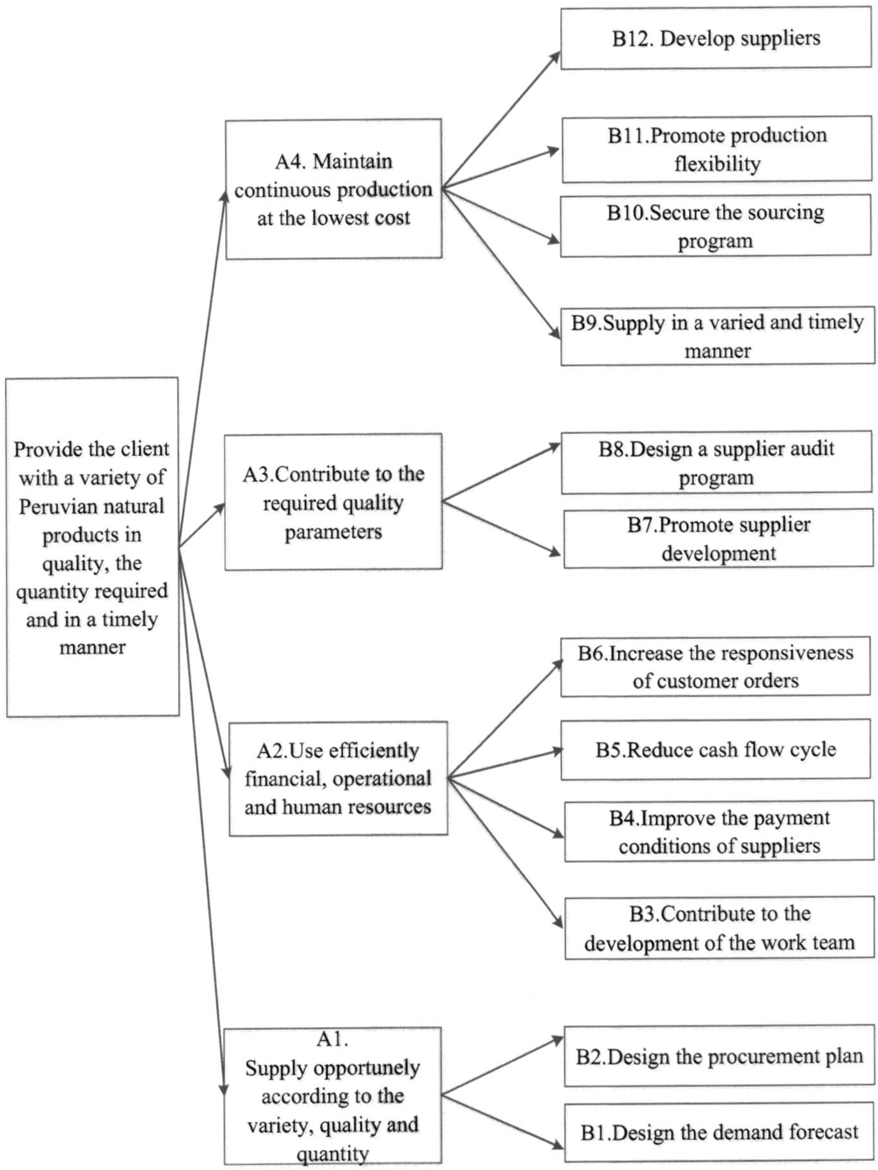

Fig. 7.7 Optimized strategy

From the results obtained, it is clear that to supply product in quality, quantity and in a timely (A1) is completely compatible with the compliance of quality parameters (A3) and maintain continuous production at the lowest cost (A4); In addition, it is compatible with efficient use of financial, operational and human resources (A2). Also, the crossing of prioritization between the pillars and main

Matrix 3		Functional strategy							
		B1	B2	B3	B4	B5	B6	B7	B8
Functional strategy	B1								
	B2	+			+				
	B3		+		+			+	
	B4								
	B5								
	B6					+			
	B7	+	+		+		+		
	B8	+	+		+		+	+	

Fig. 7.8 Evaluation of coherence of functional strategies

objectives show that the timely supply According to the variety, quality and quantity is the most important, since it has more than 50% of importance with respect to the total, while the rest of the pillars maintain an average relevance.

In this regard, in the matrix between strategic pillars and functional strategies it is observed that all the strategic pillars receive significant support from at least one functional strategy. The most significant are those related to designing the demand forecast (B1), designing the procurement planning (B2) and control management indicators (B4).

From this analysis, we can conclude that the implementation of activities that lead to demand forecasting design, as well as in the planning of acquisitions, will have a greater impact on the achievement of the nuclear goal of the area, as shown in Fig. 7.8.

CSAR methodology is participatory, and it incorporates good practices related to collaboration between operative unities inside of an organization. Once the objective is accomplished for which it was designed, it allows its adjustment, identifying new actions that allow the achievement of the business value proposition. After a successful implementation of the optimization proposal, a new scenario recommends attention to the pillars of efficient use of resources to meet the growth.

The application of CSAR methodology has allowed the identification of the actions to where the procurement management's efforts must be oriented in short and medium term. In this sense, to efficiently serve out grains of quinoa according to type, quality and quantity expected, it proposes the design of the demands forecast and planning.

7.4.3.1 Forecast Demand

Due to the lack of information about real demand, commercial data is analyzed under the assumption of safety stock's calculation covering the unsatisfied demand and defining the creation's base of information that helps to determine the real demand, it means, a forecast based on history and the expected growths is made: if sales are higher than expected, then the safety stock could attend the lost previous sale.

For its calculation, an analysis of commercial data has been made, using tools such as Tableau® and Qlick® based on the following criteria: (i) The annual total quinoa sales behavior making a difference on product's quality, such as the result obtained, where organic quinoa presents a growing trend; (ii) The annual total organic quinoa sales behavior incorporating the time value, its results present a growing trend in the quarterly assessment; (iii) Analysis of organic quality sales behavior of quarterly manner incorporated to color variable, results that organic white quinoa presents a growing trend.

In that sense, the white organic quinoa's forecast of sales for 2018 was calculated, analyzing the information from 2015 to May 2017, using the moving average method. For this purpose, it proceeded to eliminate the seasonal factor, determining if the business line is in a growing or decreasing state as shown in Fig. 7.9. In that sense, the behavior of sales of organic quinoa was elected, which presents a positive growing trend defined by the following formula: $Y = 6171.6x + 177354$

At this level, the procurement's analysis made by the top ten clients in quinoa was incorporated, resulting complex because it was observed that those clients do not present a monthly consumption. It is changing per month and volume, that is the reason why the forecast by global volume, considering if a client does not consume, then another will. As a result, the overall total shows a growth of 27%. It seeks for a balance between objective and subjective factors that can have a great impact on the forecast, considering the human aspect, the economics state and

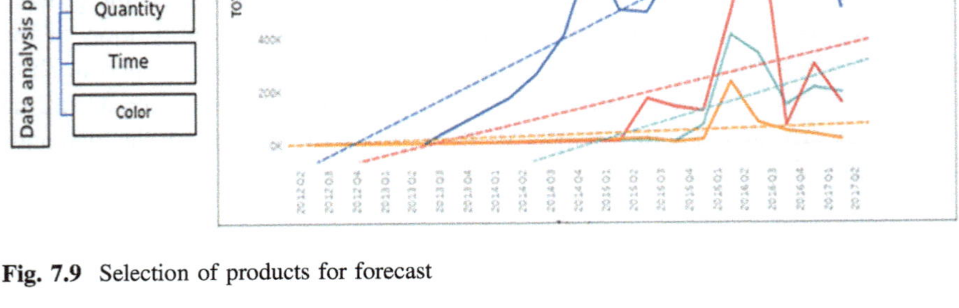

Fig. 7.9 Selection of products for forecast

Table 7.5 Forecast

Forecast	Conservative	Probable	Optimist
Quinoa	3,966,851	4,297,725	4,725,725
2018	5,037,900	5,458,111	6,001,671
2019	6,398,133	6,931,801	7,622,122
2020	8,125,629	8,803,387	9,680,095

competition actions such as a dynamic process and in parallel. For it, three scenarios of forecast have been designed: optimal, probable, and conservative. See Table 7.5.

Knowing sales volumes for the next 3 days should allow the operative and financial needs to scale up adequately, as Moreno (2016) manifested, the same that has carried out an extensive research, in order to optimize production planning, inventory, warehouse location and vehicle's routing, which have been boarded as independent issues by several researches (Fawcett and Magnan 2002).

Not knowing the size of the lost sale affects the forecast, since it is not known, it cannot be assumed. In that sense, as an alternative we will assume that the safety stock will cover it, that is, a forecast based on the history and expected growths is made; if it is sold more than expected, it is probable that the previous lost sale will be taken care of with the safety stock. In summary, with the movement of the safety stock (SS) it will be seen that the lost sale is being "recovered" and as the time progresses, less SS will be used and over time it can be reduced. In this sense, based on the demand and the number of containers that are served in the following month, the optimal SS has been determined to avoid lost sales and to properly size the purchase of raw material and the capacity of the warehouse.

The SS is determined based on two containers, also the cycle inventory (Q/2) is of one container. In that sense, of the estimated average inventory, and the result of the sum of the safety inventory and cycle inventory result in three containers. The coverage rate average is 1.09 month or equivalent to 33 days. This methodology has allowed to estimate the capacity of the warehouse which is based on the demand unattended monthly average that is equivalent to four containers or 44 tons. The results are shown in Table 7.6.

Table 7.6 Calculation of the safety stock

Criteria	Value
Average demand not met—D (containers)	4
Standard deviation—σD	0.86
Waiting time for replenishment—L (month)	0.50
Reorder point—ROP (containers)	4
Average lot size—Q (containers)	2
Safety stock (containers)	2

As long as the variation of the forecast error remains within the expected (a new challenge would be represented in decreasing the expectation of error, and continuously improve), the company you should continue to use the proposed forecasting method.

Meanwhile the company would have of more information related to sales, which would allow you to perform correlation analysis to validate if the proposed model is still valid or corresponds to modify another of greater correlation coefficient of the opposite the forecast model is no longer adequate or there are strong variations in demand, so it is recommended to evaluate the use of a new model forecast as the exponential smoothing, Holt model or Winter model, which improvement is adapted to the new market situation. Additionally, consider that the point of reorder would vary in the new situation.

7.5 Future Work

The present investigation considers that after a successful implementation of the improvement proposal the permanent challenges that define the market and the flexibility of the business model of the company would pose a new scenario where the strategy can be adjusted. In this sense, the CSAR methodology can be reapplied.

Making a rapid run of the methodology the next challenge that the company faces would be "efficiently use financial, operational and human resources", defining a new segmentation tree. It can be seen in Fig. 7.10.

7.6 Conclusions and Industrial Implications

- Situational analysis determines the need to concrete actions about the business line of Andean grains, resulting with lowest score in relation to the three business lines in which the Company operates with 2.89 points. The deficient purchase in terms of quantity and opportunity, affect profitability directly.
- Sales of quinoa's behavior are not aligned to a sales and purchase strategy. The implementation of a supply mix strategy as pull–push should eliminate the breach between the demand and the raw material's supply. The purchase of a safety stock of 120 tons is required, in order to secure sales and as of its results, build real information simultaneously about the demands' behavior.
- Using CSAR methodology, has allowed redesign the supply chain strategy, defining four strategic pillars where it can supply in a timely manner, in quality, the same that should reduce the lead time of clients' attention. Highlight, that the method has allowed training to three challenges: (i) Identify if the current strategy works; (ii) Anticipate the future needs of the supply chain; (iii) Create an improved scenario oriented to growth objective achievements.

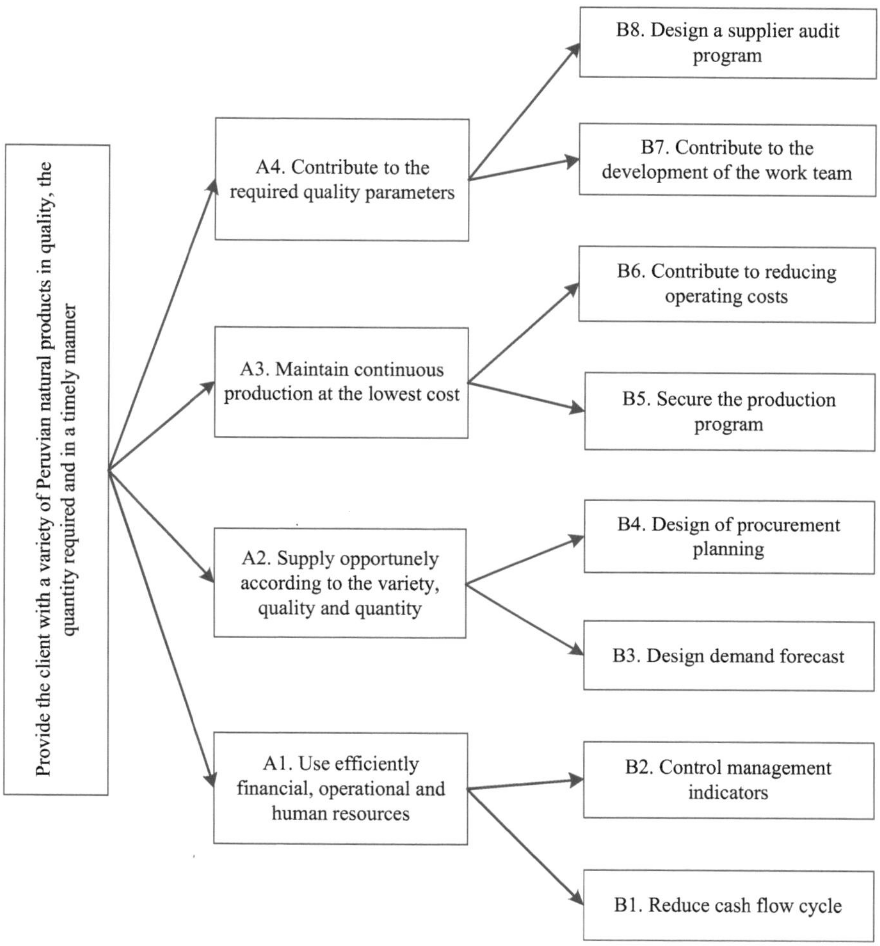

Fig. 7.10 Recommended future strategy

- Using forecast and planning management of supply chain tools, should improve substantially positives synergies between quality processes, productive processes, economic efficiency, and should improve indexes in the ranking of stronger relationship between principal objectives and strategic pillars.
- The implementation of the pull–push strategy requires an investment of PEN 134,898.00 derived from the warehouse improvement, and the implementation planning unit.
- The measurement calculation of the error between the real demand and the forecast is of 2.73% for a period from January to May 2017. This value allows to define an error percentage about the forecast method of moving average lower than 6% as a measure indicator of the forecast accurate degree.

- The optimization is technically viable to obtain in a probable scenario, NPV of PEN 36.839 and an IRR of 57%. It would stop losing sales by a value of USD $239 thousand dollars per year.

References

Azevedo S, Govindan K, Carvalho H, Cruz-Machado V (2013) Ecosilient Index to assess the greenness and resilience of the upstream automotive supply chain. J Clean Prod 56:131–146
Christopher M, Peck H (2004) Building the resilient supply chain. Int J Logistics Manag 15(2): 1–14
Cuatrecasas Arbos L (2011) Organization of production and management of operations: current systems of efficient and competitive management. Diaz de Santos Editorial, Barcelona, Spain [In Spanish]
Fawcett SE, Magnan GM (2002) The rhetoric and reality of supply chain integration. Int J Phys Distrib Logistics Manag 32(5):339–361
Ferreira A, Lopes M, Olivo F (2015) Characterization of cross-functional integration level: a multi case study in Agribusiness Organizations. J Oper Supply Chain Manag 8(2):36–51
Ganeshan R, Harrison TP (1995) An introduction to supply chain management. Retrieved from http://lcm.csa.iisc.ernet. in/scm/supply_chain_intro html
Golgeci I, Ponomarov SY (2013) Does firm innovativeness enable effective responses to supply chain disruptions? An empirical study. Supply Chain Manag Int J 18(6):604–617
Hines T (2009) Supply chain strategies: customer driven and customer focussed, 2nd edn. Routledge. https://doi.org/10.4324/9780203631669
Mentzer J, DeWitt W, Keebler J, Min S, Nix N, Smith C, Zacharia Z (2001) Defining supply chain management. Journal Bus Logist 22(2):1–25
Mike-Rother JS (2008) Learning to see, VSM to create. Value and eliminate muda. The Lean Enterprise Institute, Inc., Cambridge, MA
Moreno E (2016) Integrating supply chain and production chain: a genesis in the ethanol industry. http://dx.doi/10.12660/joscmv9n1p129–146
Pereira C (2015) Key organizational factors for building supply chain resilience: a multiple case study of buyers and suppliers. J Oper Supply Chain Manag 8(2):77–95
Pérez-Franco R, Phadnis S, Caplice C, Sheffi Y (2016) Rethinking supply chain strategy as a conceptual system. Int J Prod Econ 182:384–396
Pérez-Franco R (2016) Rethinking your supply chain strategy. A brief guide. Massachusetts Institute of Technology, Cambrigde
Pettit TJ, Fiksel J, Croxton KL (2010) Ensuring supply chain resilience: development of a conceptual framework. J Bus Logistics 31(1):1–21
Ponomarov SY, Holcomb MC (2009) Understanding the concept of supply chain resilience. Int J Logistics Manag 20(1):124–143
Porter ME (2008) The five competitive forces that shape strategy. Harvard Bus Rev 86(1):57–71
Sakihara L, Rios F, Ochoa J (2017) Supply and replacement management improvement in an agroindustrial company Universidad del Pacífico, Lima [In Spanish]
Scholten K, Scott PS, Fynes B (2014) Mitigation processes—antecedents for building supply chain resilience. Supply Chain Manag Int J 19(2):211–228

Sunil C, Mendl P (2013) Administration of the supply chain. Strategy, planning and operation. 3a edn. Pearson Education, México DF

Treiblmaier H (2014) The next step in supply chain resilience research: from an assessment tool toward theoretical integration, 21th Annual IPSERA Conference. Palermo, Italy

Wieland A (2013) Selecting the right supply chain based on risks. J Manuf Technol Manag 24 (5):652–668

Chapter 8
Courier Logistics Strategy to Create Commercial Impact in Peru. Application of Correlation and Regression Analysis, Lean Manufacturing

Daniel Núñez-Pauca, Gisela Landa-Gálvez and Mario Gustavo Chong Chong

Abstract International supply chains of different companies use a logistical network, such as Express or Courier to facilitate their business operations. In this study, Peruvian Express services' analysis has been made, presenting its direct relation with an economic growth of the country measured in commercial terms. Correlation and regression studies have been made as well, and these have shown a strong relation between the external trade of revenues and imports express shipment, this last one being, partially, the performance of the first one. A direct and positive relation between both variables has been evidenced, estimating an increase of USD 161.4 million of commercial exchange for each USD 1.00 million of shipping express increase. The supply chain, which represents the international Express service in Peru has been analyzed and it shows strategic improvement opportunities, which allow driving the sector and encouraging its use. As a consequence, it provides an economic growth in the country.

Keywords E-commerce · Courier logistics · Foreign trade · Economic growth Customs control

---

D. Núñez-Pauca (✉) · G. Landa-Gálvez (✉)
Escuela de Post Grado, Universidad del Pacífico, Jr. Sánchez Cerro 2121, Jesús María, Lima, Peru
e-mail: fd.nunezp@alum.up.edu

G. Landa-Gálvez
e-mail: Gjlandag@alum.up.edu.pe

M. G. Chong Chong (✉)
Facultad de Ingeniería, Universidad del Pacífico, Av. Gral. Salaverry 2020, Jesús María, Lima, Peru
e-mail: m.chong@up.edu.pe

J. L. García Alcaraz et al. (eds.), *Best Practices in Manufacturing Processes*,
https://doi.org/10.1007/978-3-319-99190-0_8

8.1 Introduction

International Courier services allow importing and exporting diverse products, such as books, cell phones, clothes, medicine, or goods classified as urgent, as well as enterprise production inputs, such as automotive replacement parts, inputs for production tests, samples for marketing studies, among other things (Frontier 2015). The courier express operators offer guaranteed, fast, liable, ordered, integrated, door-to-door and around the world deliveries, which are tracked and controlled throughout the entire journey. They are the "business class" of freight services (Oxford 2009).

It is important to mention that studies and inquiries performed highlight the importance of the Express sector in the country's economy, since their international logistics operations represent the fast connection with diverse countries around the world, allowing people to also access new markets through e-commerce (GEA. ecommerce 2016a), and it will allow micro, small and large national companies to explore new suppliers for their production tests and/or marketing globally, for internal consumption sales, production and export (GEA.Policies 2016b).

This study will focus on the responses to the following questions: Do imports of international express shipments explain the external trade volume of the whole country? If so, to which extent do they do it? Which will be the future field of action of projected foreign trade based on Express shipment growth? Furthermore, the analysis takes place in Peru and it has information from primary and secondary sources, being projected in a worldwide scenario.

8.2 Literature Review

8.2.1 *Logistics Courier*

Productive enterprises, some of them exporting, use the Express service when they need machine parts urgently, avoiding production interruptions. This allows small enterprises to be part of and attend export global markets (Oxford 2009) or when they need to obtain samples to commercialize or produce on a large scale. Once approved, products mobilize commercial exchange when entering the country, as long as they are consumed in large quantities. Products, such as clothing, footwear, food, drinks, pharmaceutics, toiletries or internal production inputs, as well as subsequent export of non-traditional products, such as agricultural, textiles, fishery, and chemical products, etc., increase the country's productivity, finally permitting Gross Domestic Product expansion (Chirinos 2007).

According to interviews undertaken with the Express' sector representatives (DHL, FEDEX, APECE 2017), it was noted that Express imports represent products' exchange path on a small scale, which will later become a large volume of international commerce. In the same way, the Global Express Association asserts digital technology even allows both small businesses and individual businessmen to become small multinationals, who sell and contract products, services and ideas that go beyond borders (GEA.ecommerce 2016a).

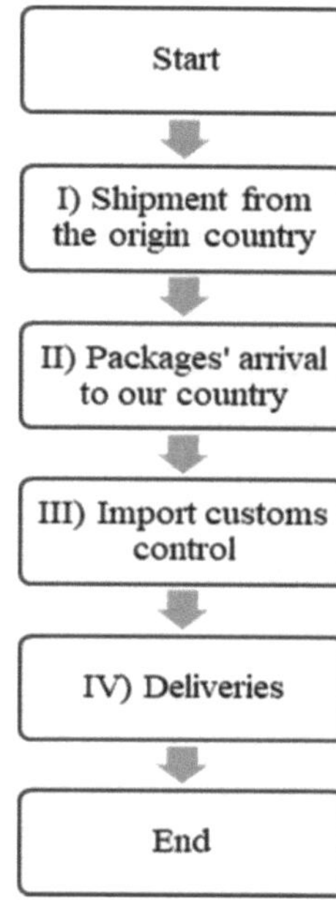

Fig. 8.1 Courier process diagram

From the literature review, only one previous study which relates Express deliveries to foreign trade, has been found and it was presented by Frontier Economics to Global Express Association (GEA), titled "Urgent shipment services and trade facilitation: Impact on the global economy" from March 2015. It was developed with data taken from members of the GEA (DHL, FedEx, TNT and UPS) at global level through a regression model.

One of the conclusions of this study is to point out that GEA plays a significant role when facilitating commercial exchange improvements, emerged from the custom's capacity indices and that urgent international shipment services, enable up to two-third of those commercial operations, being 50% a reasonable estimation of the possible effect of this sector (Frontier 2015). According to this study, an indirect estimation of supply chain's impact can be calculated, and this is developed by Express services enterprises, in relation to foreign trade of different countries.

The Trade Promotion Agreement (TPA), was signed on December 4, 2006 with the United States of America (effective as of January 2, 2009). In relation to e-commerce, it is noted that non-bureaucratic customs are needed to accelerate the commerce products that are urgent or of a fast delivery service—Express (FTA_Peru—USA 2006).

According to the regulatory customs' system of Express shipments (Supreme Decree N° 011-2009-EF, Procedure INTA-PG.28), the supply chain's process for Express shipments consists of: (i) shipment from the origin country, (ii) packages' arrival to our country, (iii) import customs control, and (iv) deliveries (MEF 2009) (Sunat.INTA-PG.28 2009). See Fig. 8.1.

8.2.2 E-Commerce

Commercial e-commerce is defined as the sale or purchase of goods or services, conducted over computer networks by methods specifically designed for the purpose of receiving or placing of orders; the payment and final delivery of the goods or services do not have to be done Online. An e-commerce transaction can be between businesses, homes, individuals, administrations, and other public or private organizations. The orders are done through the web, extranet or electronic data interchange (World Trade Organization 2013).

Technologies of information and communication have become a major factor of growth in OECD countries (David and Audretsch 2002). E-commerce has created changes in the economic activity which has been directly and conclusively impacting sectors such as the wholesale and retail distribution in Latin America and it has been apparent for many years now in The United States and Europe (Peciña 2017). Governments have an obligation to develop a process to suit the needs of the population by the quickest means possible. The use of these technologies in governments enables administrative procedures to have greater transparency, speed, efficiency and an approach to the citizens (Cárdenas 2017).

8.2.3 Economic Growth

The trade openness practiced by a large number of countries in the world reflects the implementation of the economic growth theory in the long term (Díaz-Bautista 2003). After the 2008 financial crisis, it argued that, if the governments in the biggest economies maintain their expansionary policies, the GDP contraction could be reduced by 2010 and the growth would be resumed, but at a slower pace (Naciones Unidas 2009). With the signing of Free Trade Agreements and commercial agreements, the countries have made progress in the economic and social aspects. This recognizes that, sustained economic growth is based on market opening through economic integration with other countries in the world (Berghe 2014).

The growing of the economy goes hand in hand with the reduction of poverty. Studies carried out by the OECD in that regard, point out the growth in foreign trade has a positive impact on the reduction of poverty (OECD 2014); taking that into consideration, the present work seeks to improve the logistics within customs processes, increasing the development of international trade (productive sector) and, thus, Peru's economic development.

OECD outlines some suggestions to measure the development of the country, framed in the logistics and facilitation of the same; six areas were identified: (a) the efficiency of customs clearance procedure on border; (b) the quality of business and transport infrastructure; (c) the facility of organizing shipments at competitive prices; (d) the competence and quality of logistics services; (e) the ability to track shipments; and f) the frequency with which the shipments reach their destination at the time scheduled and planned (World Trade Organization and OECD 2015); through the use of measurements such as a correlation study, Canvas or VSM (value stream mapping); an approximation has been made to the above-mentioned areas.

8.2.4 Customs Control

Since the dreadful attacks on the USA on September 11, 2001, the USA customs has strengthened its controls in an effort to protect the country from the menace of terrorism. One of the efforts in order to achieve their goals is the globalization of politics and safety practices of the international logistical chain through the SAFE framework of the World Customs Organization (WCO) (Customs and Border Protection 2008). In order to control fraudulent actions and facilitate licit movements, customs offices adopt "risk management" in their processes. It is supported by analysis technology of information database sent by customs prior to the arrival of the goods (Khalid Bichou 2014).

For more than 15 years in Peru, customs control processes are being implemented and they are supported by the risk management, disseminated by the World Customs Organization (WCO) and supported by a technological platform of information database, this is how customs administration incorporates such policies and practices in article 163 of the customs act, into national legislation (Ley General de Aduanas 2008).

This study will proceed with statistical analysis of the "Express Services Importation" and its relationship with the "Commercial Exchange" between our nation and other countries, this last one being the sum of imports and exports.

8.3 Methods and Procedures

In order to perform the study, the procedure presented below, which is composed of four consecutive phases, has been followed (see Fig. 8.2).

The methodology developed in the second phase is based on a correlation statistical analysis. A simple regression analysis can also be developed by using the method of least squares between the "Express Shipments" and "Foreign Trade" (Commercial Exchange). See Fig. 8.3.

Lastly, foreign trade projections of the country will be made. These projections are based on the performance of Express services imports, which is justified in the e-commerce growth of the country. The set of equations are summarized as follows:

Correlational coefficient:

$$r = \frac{\sigma_{xy}}{\sigma_x \sigma_y} \tag{8.1}$$

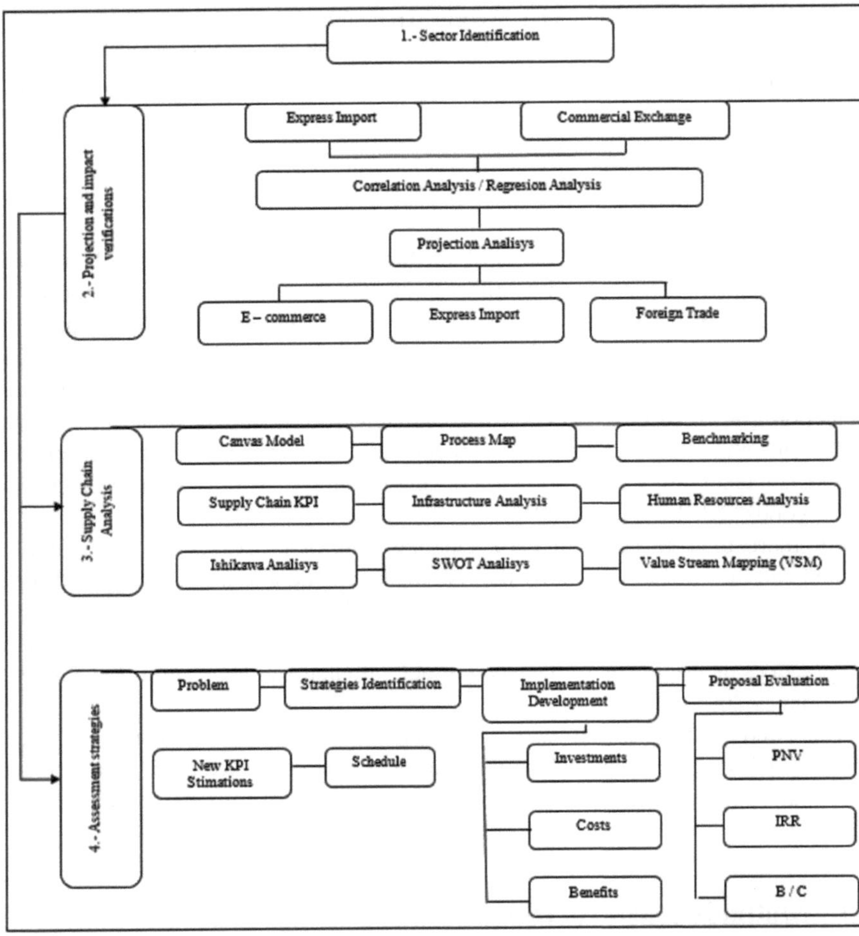

Fig. 8.2 Research procedural framework

Regression equation:

$$f(X) = Y = \beta o + \beta 1X \tag{8.2}$$

The projection of demand will be estimated using forecasting methods of: (i) time series and (ii) causal (Sunil Chopra 2013), taking seasonality of the demand into consideration in monthly ranges, being the seasonal index as follows, seasonal index for projections' analysis (Rander 2012):

$$\text{Seasonal index} = \text{Period average/total average} \tag{8.3}$$

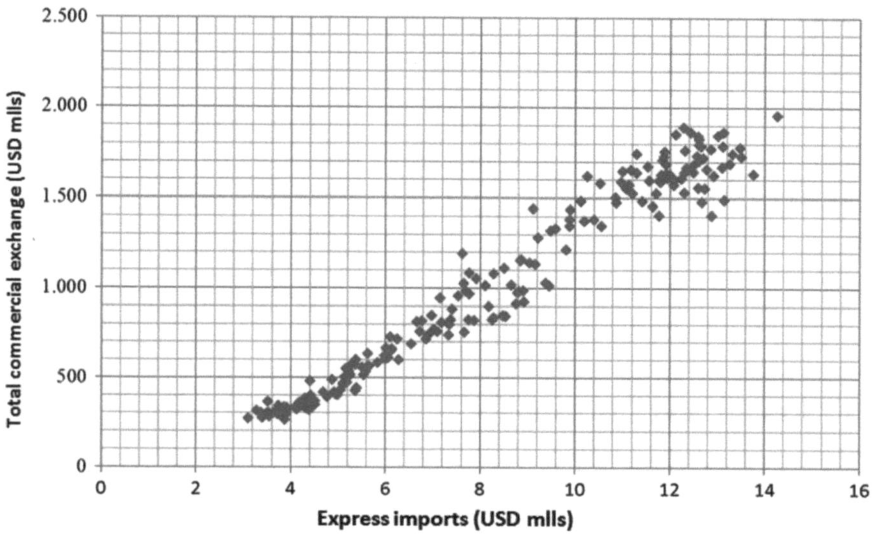

Fig. 8.3 Dispersion map, variables: "Express Imports" and "Total Commercial Exchange" (Jan 2000–Mar 2017)

8.4 Experimental/Numerical Setting

Demand's projection estimation of Express shipments is founded on e-commerce's performance. The Peruvian e-commerce growth values that have been observed are diverse. Thus, according to e-commerce news, based on Visanet data, during 2013, 2014, 2015, and 2016 it increased 8% annually (Ecommerce News 2017). It is important to note that international e-commerce trade's growth depends on the international Express shipment's proper functioning (Frontier 2015). See Table 8.1.

8.4.1 Definition of Variables to Correlate

- *Independent Variable (X)*: "Express import (CIFCourier)" measured in millions of dollars (USD), in terms of CIF, considering all the sales revenue to the country by this route. The values have been obtained from the customs administration records from 2000 to 2017, which are accumulated monthly (Sunat 2017).
- *Dependent Variable (Y)*: "Commercial Exchange (InterComer)" which is the sum of CIF Imports + FOB Exports, measured in millions of dollars (USD). It considers consumer goods imports and non-traditional products exports. The values have also been obtained from the customs' administration records from 2000 to 2017, which are accumulated monthly (Sunat.Statistical analyses 2017).

Table 8.1 Annual growth of e-commerce, data collected

Year	Increase	Source 1	Source 2
2013	8%	Ecommerce News	Visanet
2014	8%	Ecommerce News	Visanet
2015	8%	Ecommerce News	Visanet
2016	8%/9%	Visanet	Perú Retail
2017	11%	Perú Retail	–
2018	8%/16%	Visanet	Perú Retail
2020	9.6%	Visanet	–

Source Euromonitor Internacional and Visa (2016); PerúRetail (2017); Gamarra (2017)

8.5 Results and Discussion

8.5.1 Correlation and Regression Analysis

Applying the Eq. (8.1), a value of +0.982 is obtained for "r", which expresses the degree or linear relation force (Rander 2012) (Anderson 2012), being solid and positive. The coefficient of determination has a value of +0.964, this means that 96.38% of the "Commercial Exchange" values would be explained by the equation of regression, based on "Express Imports". See Table 8.2.

The regression's coefficient of the variables and the regression's equation are as follows in Table 8.3.

With this data, the simple linear regression's equation Eq. (8.2), can be described as follows:

$$Y = -321.976 + 161.414X \tag{8.4}$$

The regression's coefficient is $\beta 1$ is of +161.414, and it is interpreted as follows: *"For every 1.00 USD millions of CIF additional Express import, a commercial exchange increase of 161.4 USD million is expected in the country"*. The line of such equation could be graphed next to the data cloud, as it is observed in Fig. 8.4.

8.5.2 Impact's Estimation and Expected Projection in the Country's External Trade

With the projected demands' values from 2019 to 2028 (10 years), the commercial trade's CIF values will be estimated; these are based on the regression's analysis relation, which has already been obtained. It can be observed in Table 8.4.

Table 8.2 Model summary

Model	R	R square	R square corrected	Error tip. of estimation	Change statistics					Durbin Watson
					Change on R square	Change on F	gl1	gl2	Sig. Change on F	
1	0.982[a]	0.964	0.964	103.40478	0.964	5515.525	1	205	0.000	0.845

Source Elaborated by the authors

[a]Predictor Variables: (Constant), CIFExpress,

[b]Dependent Variable: Y3InterComer

Table 8.3 Regression model coefficients

Model		Non-standardized coefficients		Typified coefficient	*t*	Sig.	Confidence Interval to 95.0% for B		Correlations		
		B	Tip. error	Beta			Lower limit	Higher limit	Zero order	Partial	Semi partial
1	(Constant)	−321.976	18.935		−17.004	0.00	−359.308	−284.643			
	CIFExpress	161.414	2.173	0.982	74.267	0.00	157.129	165.699	0.982	0.982	0.982

Source Elaborated by the authors
[a]Dependent variable: Y3InterComer

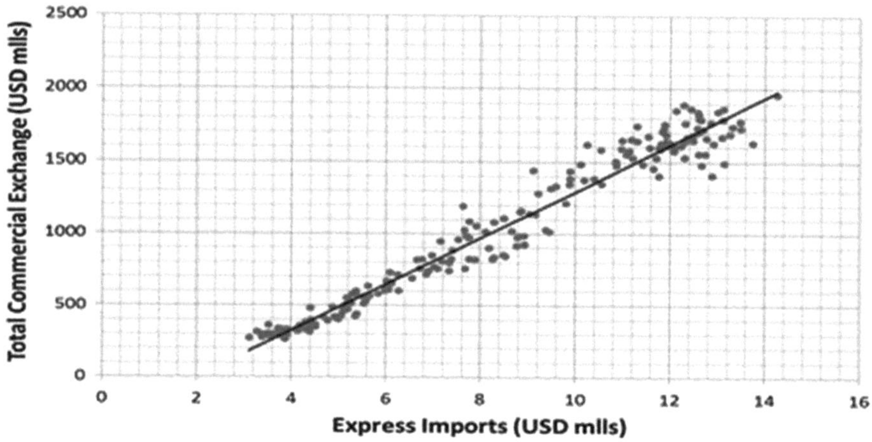

Fig. 8.4 Dispersion map and simple regression line, variables: "Express Imports" and "Total Commercial Exchange" (Jan 2000–Mar 2017)

Table 8.4 Express import projection, foreign trade impact

Year	CIF express value (USD millions) of all country		Commercial exchange value (USD millions) of all country	
	Pessimistic scenario	Optimistic scenario	Pessimistic scenario	Optimistic scenario
2019	181.28	181.28	28,944.95	28,944.95
2020	184.22	195.78	29,420.15	31,286.33
2021	187.17	211.44	29,895.40	33,815.02
2022	190.11	228.36	30,370.60	36,546.01
2023	193.05	246.63	30,845.85	39,495.48
2024	196.00	266.36	31,321.05	42,680.91
2025	198.94	287.67	31,796.25	46,121.17
2026	201.88	310.68	32,271.51	49,836.65
2027	204.83	335.54	32,746.70	53,849.36
2028	207.77	362.38	33,221.96	58,183.10

Source Elaborated by the authors

As it is observed in the Table 8.4, a pessimistic scenario is substantiated in the historic performance of shipments, making no improvements. As we observe the optimistic scenario, using e-commerce performance will lead us to estimate a purely theoretical value; supporting the implementation of best approaches in the Express supply chain. Graphically represented in Fig. 8.5.

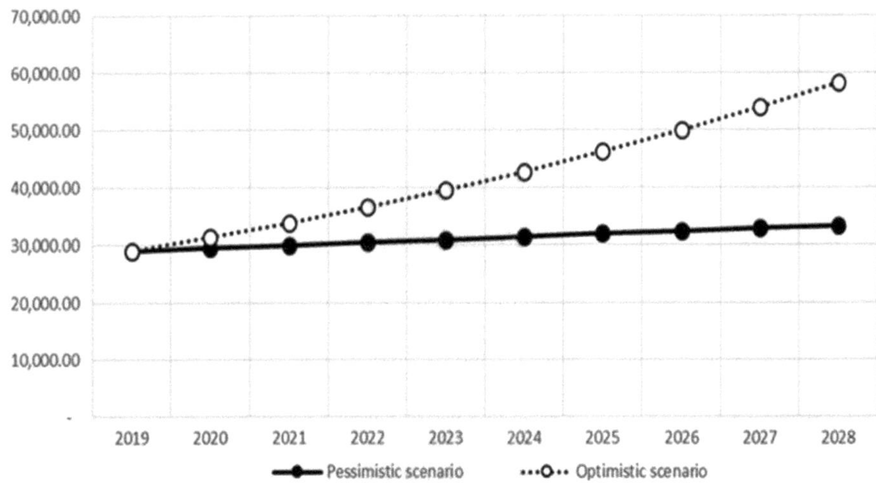

Fig. 8.5 Commercial exchange projection based on Express shipments (pessimistic and optimistic scenarios, performance gap of impact from the improvements)

8.5.3 Supply Chain Analysis

The logistical costs and time, both measured since the arrival until the package or Express deliveries have been made, these are the most important logistical Express indicators. The method of VSM has been used for the analysis and diagnosis. The details are shown in appendix 1.

- *Value of actual indicators:*
  - *User's lead time*: Is the average time that it takes shipments to be delivered and it is of 58.87 h (2.45 days) on average (Sunat 2015–2016).
  - *Storage costs*: Total storage costs were estimated in 1,091,792.64 USD in the reporting year for every 473,406 transaction, with a unit shipping cost of S/. 3.9 per day, while in storage.
  - *Total logistic cost:* In general, logistic costs incurred by Express shipments' users ascend to S/. 391.63 per transaction.

Main problem of Express' supply chain: the main problem has been identified as follows: "Express' chain supply does not count with the necessary skills, which allow them to meet the growing demand, driven by e-commerce, or its growth potential, currently having, lead times and elevated costs, which are finally moved to importers and users, discouraging its use, harming the country's external trade".

- *Identifying improvement strategies*: improvement strategies that have been identified, can be implemented to be able to meet the variable time in the Express' processes. To that effect, the two following proposals are discussed (see Table 8.5):

Table 8.5 Qualitative comparison of options

Characteristics		Weighing (%)	1st option		2nd option	
Order	Description		Score from 1 to 5	Total weighted	Score from 1 to 5	Total weighted
1	Solve the main problem (lead time)	10	4	0.40	4	0.40
2	Consider the long-term demand projection	10	5	0.50	5	0.50
3	Solve the potential growth of e-commerce	15	3	0.45	5	0.75
4	Consider the best global practices in the sector	7	2	0.14	4	0.28
5	There is a reasonable investment	10	5	0.50	1	0.10
6	Reduce process costs	10	3	0.30	5	0.50
7	Allows final customer satisfaction	10	3	0.30	5	0.50
8	Encourages the growth of the country's foreign trade	10	2	0.20	5	0.50
9	Consider the projection of demand in the short term	5	5	0.25	2	0.10
10	Contains strategic components at the country level	8	2	0.16	5	0.40
11	It allows to improve the competitiveness of the sector	5	2	0.10	5	0.25
Total		100	36	3.30	46	4.28

- 1st Option: "Optimizing Express' processing operations maintaining the same supply chain".
- 2nd Option: "Centralization of inputs' operations, implementing a distribution center, similar to Cross Docking in the main airport and redesigning Express' logistic chain".

From the evaluation of the two options, the second one has been selected, due to the fact that it has a great strategic importance. The implementation's characteristics are described in appendix 2.

- *New costs and time estimation*: this improvement will help save costs and improve timing, as follows:
  - *User's Lead Time*: This would change from 56.86 (2.37 days) to 22.49 (0.94 days), it represents a reduction of 60%.

- *Storage costs*: With the current proposal up to S/. 408,490.36 (USD 127,653.24) will be reduced, and this could represent an annual savings of 88%.
- *Total logistic cost*: In general, logistic costs incurred by Express shipment users can also decrease with the improvement's proposal. It starts with a value of S/. 391.63 per transaction to a value of S/. 262.59, which represents an order reduction of 33%.

8.5.4 Socioeconomic Evaluation

Savings in costs have been estimated, these cause a straight impact on users, in the order of USD 22 million in the first operational year of Cross Docking Center.

The social evaluation of the strategy's implementation considers marginal socioeconomic flows (cost savings), which represent benefits. These benefits are deducted at an annual rate of 11% and are established for the public sector by the ministry of economy and finances. Considering a period of 10 years for the evaluation, the total investment of USD 21,248,250.00 to implement Cross Docking's center in AIJCH (International Airport Jorge Chavez) redesigning the supply chain, a positive socioeconomic VAN of USD 123 million, is finally obtained; with a benefit/cost parameter of 6.79 and TIR of 106%.

8.5.5 In Regard to Control on the Borders

The administration of customs revenue and outgoing processes of courier shipments is done on the basis of the risk management system, implemented in the Peruvian customs, according to the rules and good practices published by the World Customs Organization. To that effect, the available control-oriented resources are addressed to those Customs transactions that proof higher risk.

In relation to courier shipments, these enter and leave Peru from the Jorge Chávez International Airport. Notwithstanding, control actions are done in a scattered way in seven different control points both inside and outside the airport. These points correspond to the temporary storage of the airport and the distribution points of the courier operators.

With the implementation proposal of a cross docking distribution center, the control actions would transfer to just one point, this center, the same that will be located inside the airport facilities, whereby customs tasks would be carried out as efficiently as possible and there would be a better control of courier processes, based on the pre-existing data analysis of the manifest and customs declarations that arrive in the country to identify high-risk shipments, facilitating the flow of traffic and customs release of low or zero risk shipments, which policies and customs risk management practices refer to.

8.6 Conclusions and Future Research

According to the analyzed information, it is stated that a considerable impact of "International Express Supply Chain" about "the country's foreign trade" exists; that is to say, a logistic Express overall improvement would help increase commercial relations of businesses and Peruvian people with their foreign providers, improving the economy.

The future foreign trade's field of action projected, based on the growth of shipments, would vary between the two estimated values for the pessimistic and optimistic scenarios (see Table 8.4 and Fig. 8.5).

The present study intends to evaluate if the international Express' services are not only fulfilling its function, but also improving its economic growth.

It may be noted that it is important to study the possibility of facilitating international express commerce of natural people, in both, imports and exports of products in small quantities, eliminating tariff and non-tariff barriers, such as internal tax charges. In our view, this would be beneficial for the access of global markets in a rapid manner and at a low cost, motivating at the same time, companies' formation through entrepreneurship until it is consolidated and formalized as a company.

Annex 1

See Fig. 8.6.

Annex 2

See Fig. 8.7.

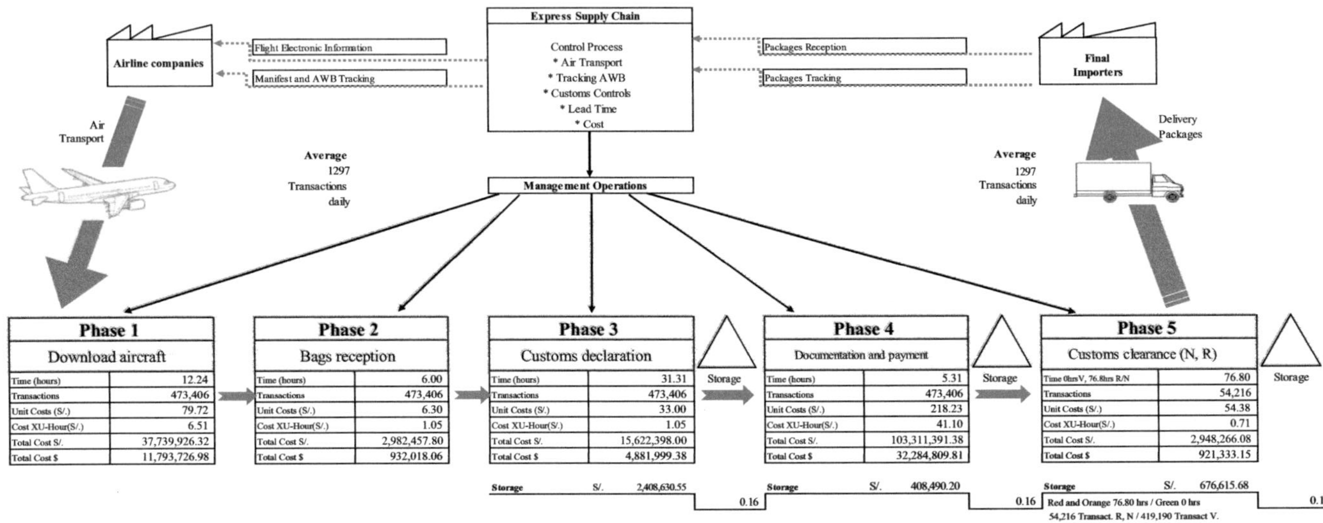

Fig. 8.6 Value stream mapping to supply chain of express shipments—current process. *Source* Elaborated by the authors

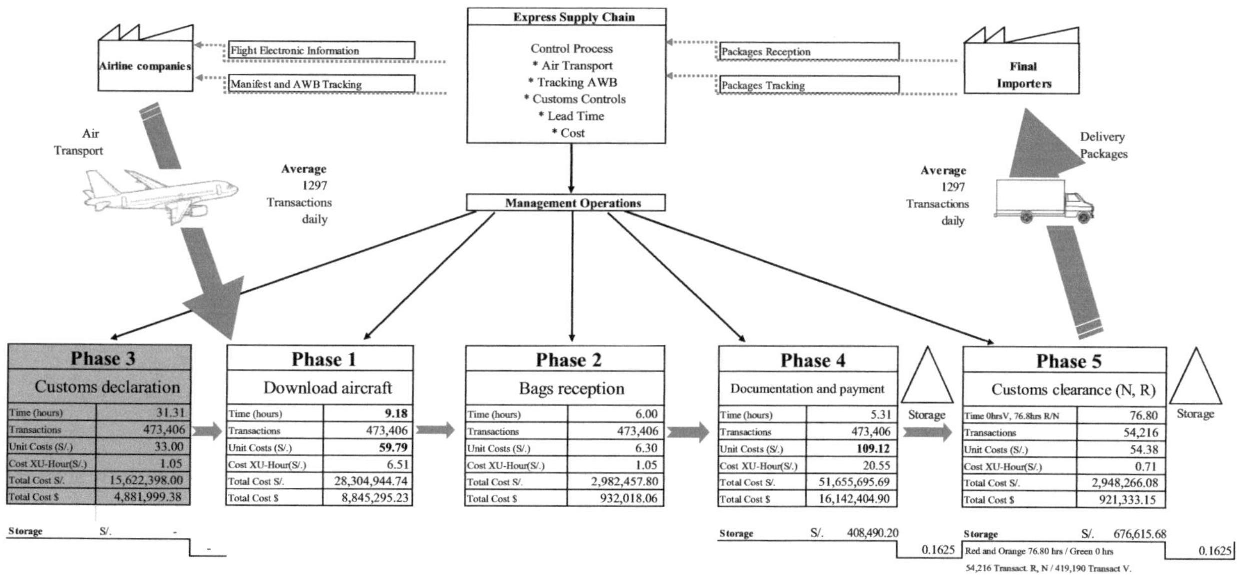

Fig. 8.7 Value stream mapping to supply chain of express shipments—proposed process. *Source* Elaborated by the authors

References

50Minutos.es (2017) The Canvas model: analyze efficiently your business model. Plurilingua Publishing, España

Anderson S (2012) Statistics for business and economy. Décimo primera edición. Cengage Learning, México DF

Arbós L (2011) Production organization and operations management, current systems of efficient and competitive management. Edtions Díaz de Santos, España

Basalisco B, Okholm H, Möller A, Wahl J (2015) Private express carriers versus posts treatment of packages inbound in US: security risks and customs concerns. A study prepared for the Coalition of Services Industries. Copenhagen Economics, Copenhagen. [PDF]. Date of consultation: 21/07/2017. Available at: https://www.copenhageneconomics.com/dyn/resources/Publication/publicationPDF/1/321/1441788333/private-express-carriers-vs-posts-treatment-of-packages-inbound-to-the-us_security-risks-and-customs-concerns.pdf

Bengt S, Östblom S (2010) Benchmarking: a signpost to excellence in quality and productivity. Wiley, Indiana

Berghe EV (2014) Free trade agreements—challenges and opportunities. Ecoe Ediciones, Bogotá

Board of Governors of the Federal Reserve System (2016) Rate of the US investment market. On: federalreserve.gov. [Online]. Date of consultation: 20/06/2017. Available at: https://www.federalreserve.gov/publications/budget-review.htm

Central Reserve Bank of Peru (BCRP) (2017) Risk free rate, On: estadisticas.bcrp.gob.pe. [Online]. Date of consultation: 20/06/2017. Available at: https://estadisticas.bcrp.gob.pe/estadisticas/series/cuadros/notasemanalmensual/NC_037

Chirinos R (2007) Trade and growth: a review of the hypothesis 'Learning by exports. DT. N° 2007–002. Working Paper series. Febrero 2007. Central Reserve Bank of Perú (BCRP) and San Ignacio de Loyola University (USIL), Lima. On: bcrp.gob.pe. [PDF]. Date of consultation: 02/04/2017. Available at: http://www.bcrp.gob.pe/docs/Publicaciones/Documentos-de-Trabajo/2007/Documento-Trabajo-02-2007.pdf

Customs and Border Protection (2008) Supply chain security—CBP works with international entities to promote global customs security standards and initiatives, but challenges remain. United States Government Accountability Office, Washington

Da Silva J (2016) Rankings de Competitividad WEF y Doing Business. National Council of Competitiveness, Ministerio de Economía y Finanzas. On: cnc.gob.pe. [Presentación en power point]. Date of consultation: 20/06/2017. Available at: https://www.cnc.gob.pe/images/cnc/Rankings/PPT-Rankings-WEF-y-DB_v.18.05.2016-_final.pdf

David B, Audretsch PJ (2002) The new economic growth in Europe and the US. Springer, New York

DHL. Freight transport. On: dhl.com.pe. [Online]. Date of consultation: 23/11/2017. Available at: http://www.dhl.com.pe/es.html

DHL. Rates. On: dhl.com.pe. [Online]. Date of consultation: 23/11/2017. Available at: http://www.dhl.com.pe/es.html

Díaz-Bautista A (2003) Economic growth determinants. International growth, convergence and institutions. Plaza y Valdés, México

EC Redaction (2017) Government signed addendum for the expansion of Jorge Chávez Airport. On: elcomercio.pe. [Online]. 25/07/2017. Date of consultation: 20/06/2017. Available at: http://elcomercio.pe/economia/peru/gobierno-firmo-adenda-ampliacion-aeropuerto-jorge-chavez-445060

eMarketer (2015) Worldwide retail ecommerce sales: eMarketer's updated estimates and forecast through 2019. On: emarketer.com. [Online]. December 23th of 2015. Date of consultation: 20/01/2017. Available at: https://www.emarketer.com/Report/Worldwide-Retail-Ecommerce-Sales-eMarketers-Updated-Estimates-Forecast-Through-2019/2001716

Euromonitor Internacional and Visa (2016) E-readiness inform in Latin américa 2016. On: visa.com.mx. [PDF]. Date of consultation: 20/07/2017. Available at: https://www.visa.com.mx/dam/VCOM/regional/lac/SPA/Default/Documents/PDFs/ereadiness-spanish-latinoamerica1.pdf
Falcón D (2017) Perú e-commerce—2017. On: blogs.gestion.pe. [Online]. March 21th of 2017. Date of consultation: 02/06/2017. Available at: http://blogs.gestion.pe/innovaciondisrupcion/2017/03/comercio-electronico-en-el-peru-2017.html
Fedex Express Information about your rates in 2017. On: fedex.com. [Online]. Date of consultation: 20/06/2017. Available at: http://www.fedex.com/pe/rates/rateinfo.html
Frontier Economics Ltd. and Global Express Association (GEA) (2015) Express delivery services and trade facilitation: impact on the world economy. Inform to the Global Express Association. March 2015. [PDF]. Date of consultation: 20/06/2017. Available at: https://global-express.org/assets/files/Trade/GEA_FinalReport_STC_15012015—es—con-cambios.pdf
Gamarra K (2016) Online consumption in Peru: country "yung" with a market full of opportunities. On: ecommercenews.pe. [Online]. Date of consultation: 16/06/2017. Available at: http://ecommercenews.pe/consumo-online-en-el-peru-un-pais-joven-con-un-mercado-lleno-de-oportunidades/
General Customs Law, Legislative Decree N° 1053 (MEF—Ministry of Economy and Finance 27/06/2008)
Gestión Redaction (2017) 41.3% of users believe that online purchases are dangerous. On: gestion.pe. [Online]. 12 de enero de 2017. Date of consultation: 20/06/2017. Available at: https://gestion.pe/tecnologia/41-3-usuarios-cree-compras-online-son-peligrosas-126486
Global Express Association (GEA) (2016a) L'e-commerce transfrontalier– Moteur de croissance. Suggestions pour favoriser la croissance économique. Exposé de position présenté par la Global Express Association (GEA). [PDF]. Date of consultation: 20/06/2017. Available at: https://global-express.org/assets/files/Trade/GEA-Position-Paper-on-e-commerce-final-2016-FR-rev2.pdf
Global Express Association (GEA) (2016b) Policies at promoting international trade in MIPYMEs: how to take advantage of the full potential of global electronic commerce. A proposal for express delivery services. Report. [PDF]. Date of consultation: 12/11/2016. Available at: https://global-express.org/assets/files/Whats%20new%20section/GEA-MSME-(final-adopted)-ESP-F(2).pdf
Google maps. Internacional Airport Jorge Chávez. On: google.com.pe. [Online]. Date of consultation: 20/06/2017. Available at: https://www.google.com.pe/maps/place/Aeropuerto+Internacional+Jorge+Chavez/@-12.0241868,-77.1127171,442m/data=!3m1!1e3!4m5!3m4!1s0x9105cdb5800881f7:0x417ef7657c8baf97!8m2!3d-12.0242576!4d-77.1110415?hl=es
Intendance of Air Customs and Postal of the Sunat (2017) Indicators of the Fast Delivery Shipments, Peru [In Spanish]
International Airport Jorge Chávez, Perú. About LAP. On: lima-airport.com. [Online]. Date of consultation: 06/07/2017. Available at: https://www.lima-airport.com/esp/lap-negocios-y-proyectos/acerca-de-lap/fundacion-y-socios
International Bank for Reconstruction and Development/The World Bank (2016) Doing business 2016. Measuring regulatory quality and efficiency. The World Bank, Washington
Khalid-Bichou MG (2014) Risk management in port operations, logistics and supply chain security. Informa Law from Routledge, New York
Liberos E (2011) The electronic commerce book. ESIC Editorial—Business Marketing School, España
Loppacher J (2013) Competing through the business value chain. Temas Grupo Editorial SRL, Buenos Aires
Manutención and Storage Magazine (2015) On: manutencionyalmacenaje.com. [Online]. 15/05/2015. Date of consultation: 20/06/2017. Available at: http://www.manutencionyalmacenaje.com/es/notices/2015/05/tnt-abre-en-madrid-un-centro-estrategico-para-exportaciones-38149.php#.Wib9D0ribIU

Michalczewsky K ¿What is electronic commerce?. On: conexionintal.iadb.org. [Online]. Date of consultation: 20/06/2017. Available at: http://conexionintal.iadb.org/2017/03/08/el-comercio-electronico-y-los-factores-de-su-desarrollo/#_ftnref2

Ministry of Economy and Finance (MEF) (2009) Supreme decret N°011-2009-EF, Approve the regulation of the special customs regime for fast delivery shipments and other provisions. "On: mef.gob.pe. [Online]. Date of consultation: 20/06/2017. Available at: https://mef.gob.pe/es/por-instrumento/decreto-supremo/3417-d-s-n-011-2009-ef/file

Ministry of Economy and Finance (MEF) (2017) Annex SNIP 10, evaluation parameters. On: mef.gob.pe. [Online]. Date of consultation: 20/06/2017. Available at: https://www.mef.gob.pe/contenidos/inv_publica/docs/anexos/new_direc/v12/05.3_Anexo_SNIP_10_modificado_por_RD_006-2012.pdf

Ministry of Foreign Trade and Tourism (Mincetur) PENX: National Plan Strategic Exporter 2025. Mincetur, Lima. [Online]. Date of consultation: 20/06/2017. Available at: https://www.mincetur.gob.pe/wp-content/uploads/documentos/comercio_exterior/plan_exportador/Penx_2025/PENX_FINAL_101215.pdf

National Superintendency of Customs and Tax Administration (Sunat) (2017a) Tax revenues of the Central Government: 2011–2016. On: sunat.gob.pe. [Online]. Date of consultation: 30/12/2016. Available at: http://www.sunat.gob.pe/estadisticasestudios/index.html

National Superintendency of Customs and Tax Administration (Sunat) (2011) Customs procedure INTA-PG.28. Fast delivery shipments. General procedure. On: sunat.gob.pe. [Online]. Date of consultation: 20/06/2017. Available at: http://www.sunat.gob.pe/legislacion/procedim/despacho/especiales/envioEntRap/procGeneral/despa-pg.28.htm

National Superintendency of Customs and Tax Administration (Sunat) (2016) Customs procedure DESPA-PE.00.06. Control of restricted goods. On: sunat.gob.pe. [Online]. Date of consultation: 20/06/2017. Available at: http://www.sunat.gob.pe/legislacion/procedim/despacho/procAsociados/despa-pe.00.06.htm

OECD (2014) OECD trade facilitation indicators calculating the potential impact of the WTO Trade Facilitation Agreement on trade costs. [Online]. Date of consultation: 20/06/2017. Available at: http://www.oecd.org/trade/facilitation/indicators.htm

Osterwalder A, Pigneur Y (2011) Generation of business models. Deusto S.A. Ediciones, Barcelona

Oxford Economics (2009) The Impact of the Express Industry on the Global Economy. Inform [PDF]. Date of consultation: 20/06/2017. Available at: http://www.apese.org/docs/BOLETIN-INFORME-OXFORD-GEA-2009.pdf

Peciña IS (2017) Electronic commerce. ESIC Business Marketing School, Madrid

PerúRetail (2017) E-commerce in Perú would grow 16% by 2018. On: peru-retail.com. [Online]. Date of consultation: 20/06/2017. Available at: http://www.peru-retail.com/entrevista/ecommerce-peru-creceria-para-2018/

Peruvian Corporation of Airports and Commercial Aviation (Corpac) The supply chain infrastructure in courier shipments of Peru. On: corpac.gob.pe. [Online]. Date of consultation: 20/06/2017. Available at: http://www.corpac.gob.pe/Main.asp?T=3763

ProInversión (2008) Preparation of the pre-investment study for the creation of a Lima-Callao multimodal high-tech logistic and services center. Volume B. Design and economic-financial evaluation of the ZAL del Callao. On: mtc.gob.pe. [Online]. Date of consultation: 20/06/2017. Available at: https://www.mtc.gob.pe/estadisticas/files/estudios/ZAL%20Callao%20-%20TOMO%20B.pdf

Ramírez E (2010) Introduction to psychology. University of Jaén, Spain

Rander S (2012) Quantitative methods for business, 10th edn. Pearson Educación, México

Rincón E, Cárdenas CV (2017) Electronic Public Administration towards the electronic administrative procedure. Editorial Universidad del Rosario, Bogotá

Risk management of corruption in the supply chain. On: esan.edu.pe. [Online]. 20/06/2017. Date of consultation: 20/06/2017. Available at: https://www.esan.edu.pe/sala-de-prensa/2017/06/gestion-de-riesgo-de-corrupcion-en-la-cadena-de-suministro/

Santos del Cerro J, García M (Coordinadores) (2006) History of Probability and Statistics (III). Delta, Madrid

Scharff. (s.f.). Distribución. On: scharff.com.pe. [Online]. Date of consultation: 23/11/2017. Available at: http://www.scharff.com.pe/#formGrow

Shohin About us? On: shohin.com.pe. [Online]. Date of consultation: 20/06/2017. Available at: http://www.shohin.com.pe/

Sunil P (2013) Administration of the Supply Chain, strategy, planning and operation. Pearson Educación, México

Superintendency of Banking, Insurance and AFP (SBS) (2017) Cost of debt according to Tamex average 2016. On: sbs.gob.pe. [Online]. Date of consultation: 20/06/2017. Available at: http://www.sbs.gob.pe/app/pp/EstadisticasSAEEPortal/Paginas/TIActivaMercado.aspx?tip=B

Supervisory Organization of Investment in Public Use Transport Infrastructure (Ositran) Infrastructure Competitiveness Ranking (WEF). On: ositran.gob.pe. [Online]. Date of consultation: 17/05/2017. Available at: https://www.ositran.gob.pe/publicaciones1/otras-publicaciones.html

Talma Infrastructure of Talma. On: talma.com.pe. [Online]. Date of consultation: 20/06/2017. Available at: http://www.talma.com.pe/infraestructura.php

Trade Agreements of Peru. (s.f.). PERÚ-EE.UU Trade Promotion Agreement. On: acuerdoscomerciales.gob.pe. [Online]. Date of consultation: 11/02/2016. Available at: http://www.acuerdoscomerciales.gob.pe/index.php?option=com_content&view=category&layout=blog&id=55&Itemid=78

United Nations (2009) United Nations Conference on Trade and Development—Report on Trade and Development. United Nations, New York and Geneva

UPS (2017) Find peace of mind in the details. On: ups.com. [Online]. Date of consultation: 23/11/2017. Available at: https://www.ups.com/pe/es/Home.page

UPS de Perú (2017) 2017 Guide to rates and services. Effective as of May 1, 2017. On: ups.com. [PDF]. Date of consultation: 20/06/2017. Available at: https://www.ups.com/media/es/ar/rate_guide_pe.pdf

Woods R (2016) Getting I.T. Right: what forwarders can learn from DHL's struggle. On: aircargoworld.com. [Online]. Date of consultation: 20/06/2017. Available at: https://aircargoworld.com/allposts/getting-i-t-right-what-forwarders-can-learn-from-dhls-struggle/

World Bank (2012) Logistic Performance Index (IDEL) per component, by country. On: bancomundial.org. [Online]. Date of consultation: 20/06/2017. Available at: https://datos.bancomundial.org/indicador/LP.LPI.OVRL.XQ?end=2016&start=2016&view=map

World Trade Organization (2009) Communication from the European Communities, Mongolia, the Separate Customs Territory of Taiwan, Penghu, Kinmen and Matsu, and Switzerland. Addendum. Customs Brokers. On: trade.ec.europa.eu. [Online]. March 18, 2009. Date of consultation: 20/06/2017. Available at: http://trade.ec.europa.eu/doclib/docs/2009/november/tradoc_145312.pdf

World Trade Organization (2013) E-commerce in developing countries. [Online]. Date of consultation: 10.10.2107 Available at: https://www.wto.org/english/res_e/booksp_e/ecom_brochure_e.pdf

World Trade Organization (WTO) Electronic commerce. On: wto.org. [Online]. Date of consultation: 20/06/2017. Available at: https://www.wto.org/spanish/tratop_s/ecom_s/ecom_s.htm

World Trade Organization and OECD (2015) Aid for trade in synthesis 2015. Reduce the costs of trade with a view to an inclusive and sustainable growth. On: WTO. [Online]. Date of consultation: 01.10.2107. Available at: http://www.oecd.org/trade/facilitation/indicators.htm

Part II
Production Management

Chapter 9
Implementation of the S&OP Process in Textile Company Case Study: Ecuadorian Textile ABA

María Angeline Bofill-Altamirano and Sonia Valeria Avilés-Sacoto

Abstract Among the different issues that many companies face, the lack of a proper Planning process deviates the company to accomplish their objectives and goals. This shortage generates miscommunication, discoordination and problems among different departments, such as Marketing, Sales, Operations, Purchases and Finances. Planning is a guide for organizations to meet these goals and objectives, while efficiently using resources, managing risks and uncertainty and, creating competitive advantages. Therefore, it is important to implement and improve an appropriate planning process, based on a supportive methodology. It is here that Sales and Operations Planning (S&OP) arises as a useful methodology. It starts with an assessment of the current planning methodology the company uses, the market in which the business operates (based on Forecasts), and the supply provided to meet the demand (based on Time Studies, Production Capacity, Safety Stocks and Optimal Batch Size calculations); later creates and implements a production plan that will guide the company by tactical and operational procedures to fulfill customers' demand monthly. This methodology will be presented by a case study applied to the Ecuadorian textile company named ABA, which gained significant results due to the usage of the S&OP process.

Keywords S&OP · Demand · Offer · Forecasts · Time studies
Production capacity · Safety stocks · Optimal batch size

M. A. Bofill-Altamirano · S. V. Avilés-Sacoto (✉)
Department of Industrial Engineering, Universidad San Francisco de Quito, Diego de Robles entre, Francisco Orellana y Pampite, Quito 170157, Ecuador
e-mail: svaviless@usfq.edu.ec

M. A. Bofill-Altamirano
e-mail: angeline.bofill@estud.usfq.edu.ec

J. L. García Alcaraz et al. (eds.), *Best Practices in Manufacturing Processes*,
https://doi.org/10.1007/978-3-319-99190-0_9

9.1 Introduction

Many of the discrepancies and problems found in many manufacturing companies are due to the lack of three factors: communication, coordination, and planning between the different departments within the company, especially among Marketing, Sales, Operations, Purchases and Finances. The result is an uncoordinated organization, with segregated, isolated, miscommunicated departments without mutual coordination in their departmental and individual objectives that go along with the global company objectives (Krajewski et al. 2008).

Therefore, a systematic management orientation is required conformed by the fundamental areas of: planning, organization, staff integration, direction, and control (Koontz et al. 2008). Management based on these areas will satisfy the objective of handling the business and its resources in an optimal procedure, thus fulfilling established goals and objectives (Druker 2003).

Every company has several and important functions, however, the planning function under the approach established by Sales and Operations Planning (S&OP), requires further analysis and study. The S&OP is a process lead by a team or a coordinator, that representing the interests of the different areas of Marketing, Sales, Operations, Purchasing and Finance, creates a set of tactical plans that allow the business to synchronize market demand with the supply of the company. In addition, it offers the possibility of controlling the business tactically, by performing as a bridge, linking the established strategic planning with daily activities while supporting continuous improvement, updating not only the value but also the supply chain in a more agile, proactive and aligned chain; thus, achieving a significant growth in the level of service and customer satisfaction (Ferreira 2013; Swaim et al. 2016; Tavares et al. 2011).

Thus, it is possible to launch an appropriate action plan as a guide to the business in the right direction. This function is the priority of many companies worldwide, as the case of the Ecuadorian company ABA, a textile company with more than 50 years in the market, dedicated to meet the needs of local and international markets by offering quality textile products. It has a family of products: threads, which face the problem that managers and supervisors, based on a subjective coordination founded on their experience, perform the S&OP process, without a theoretical basis or methodology. Additionally, the departments that participate in this planning process such as Marketing, Sales, Operations, Purchases and Finances, have ineffective communication and planning methods developed, originated on departmental goals and objectives.

These problems have led to a significant reduction in: the visibility of its processes and products, the satisfaction of local, international and internal customers, in the fulfillment and creation of established annual and monthly budgets by area, and a high increment of raw material, semi-finished and/or finished products inventories, and production costs. Similarly, the current work of machinery and personnel registers data with considerable fluctuations in their installed capacities, generating high economic and time investments in purchase and installation of state-of-the-art machinery, layoffs and/or unnecessary hiring of personnel.

ABA seeks to improve its S&OP management to direct the company towards continuous improvement, and even serve as a reference for other companies in the same situation (Cecere 2015; Ferreira 2013). This study exposes the implementation of the S&OP process, detailed for company ABA, with the aim to create a set of monthly plans of demand and production in the area of spinning, which are also aligned with the needs of the other areas involved. Formerly, the Standard Operational Procedures Manual for each of the Marketing, Sales, Operations, Purchasing and Finance departments will be analyzed and evaluated. This will lead to an in-depth behavior's evaluation of the family of threads with data from 2015 and using statistical and mathematical tools. Thus, by implementing the S&OP process and with the previous analysis, optimum standards will be proposed to achieve compliance with the demand, production, business plan and budget plans.

9.2 Literary Review

The Systematic Management approach proposed by Koonzt et al. (2008) is created by fundamental, critical and interdependent areas, such as Planning, Organization, Staff integration, Direction and Control; being Planning, the one which requires a further analysis. In the same line, Krajewski et al. (2008) propose the use of Operations Management because of its contribution to the learning of the processes within a company, due to the fact that the transformation of inputs into outputs occurs in all areas of companies. Both approaches integrate the S&OP planning because they encompass and influence several areas within a company and the decision-making process in them.

S&OP has existed since 1980. It was created as a vehicle of success of the MRP II, Manufacturing Resource Planning, a tool in vogue at the end of 1970. Then, using the concepts of the aggregated planning, it evolves from a process that considered only one component in the supply chain with purely operational or commercial relations, to a more integrative and complete process, known by some authors as Integrated Business Plan (IBP), which now involves Finances and Marketing departments, proposing a collaborative work between all the components in the supply chain to formulate courses of action beneficial for all (Wagner et al. 2013; White 2016).

S&OP is considered a bridge between long-term decisions (strategic plans) and medium- and short-term decisions (tactical and operational plans). It is a collaborative process, which feeds the Strategic plans of the company, the Annual Business Plan and it acts tactically, on a monthly or daily frequency. The decisions related to this process can be classified in three levels: long, medium and short term. Long-term decisions are those that establish capacity constraints in which medium-term planning must operate. Examples of this type of decision are: the design of a product or service, facilities, procedures, size, layout, equipment, technologies, location, all of them covering a period of time of one year or more. Medium-term decisions establish limits for short-term decision-making.

Medium-term decisions can be general levels of: people hiring, layoffs, rehiring, subcontracting, overtime, inventories and materials supply covering a period of time from 3 to 18 months. Finally, there are the short-term decisions which decide the best way to achieve desired results considering the restrictions of the long and medium-term plans, covering a period of time ranging from one day to 6 months (Chase et al. 2006; Lloret 2014; Stevenson 2012).

The planning process is based on a criterion for decision-making. The human component, the judgment, the capacity for decision-making and learning will mark the success of processes of this nature. For this reason, S&OP not only proposes a systematic methodology with steps to be followed for its implementation, but also gives a great weight to the human component of the process: the company's workforce and the S&OP team (Ambrose 2016). Some of the great benefits of implementing S&OP process include the increment in communication, teamwork and labor unification.

However, as usual, the problem of resistance to change arises, because employees avoid change in their routines that give confidence and security at work, therefore it is important to solve the problem promptly (Prokopets 2012); additionally leaders implementing this project must be innate leaders, being knowledgeable of S&OP, as well as of the processes of the company involved in planning, not just being a superficial support (Duncan 2013) (Wallace and Stahl 2004) and, finally S&OP needs to be, with respect to planning, the only way to operate and all business functions must follow the same process as the standard way (Parekh).

9.3 Methodology

The S&OP planning follows an iterative methodology that, through its planning process, will direct the organization towards a successful planning trajectory. Several authors suggest different steps, however, in this study it is presented a methodology that combines the methodology proposed by Ferreira (2013) and the one proposed by Wagner et al. (2013).

9.3.1 Step 1: Maturity Model

This first step can be applied either, at the beginning or at the end of the methodology. It depends on whether the S&OP application is being done for the first time or it has already been applied in the company. This model proposed by Ferreira (2013) cites the original model proposed by Grimson and Pyke (Ferreira 2013). It evaluates the maturity of S&OP within the company. The process uses a scoring guide that will evaluate and grade the company placing it in stages: Stage 1 is "There is no S&OP process", Stage 2 is "Reactive", Stage 3 is "Standard", Stage 4 is "Advanced", and Stage 5 is "Proactive". Each Stage lists characteristics of the company in that state allowing the evaluator to perform an in-depth assessment of the organization in the planning process. By knowing in what stage, a company is standing, it will be easier to identify the next steps that need to be followed to improve and move on to the next stage.

9.3.2 Step 2: Information Gathering

The person in charge of the planning process, the Planning coordinator, will compile relevant sales and production data. Information such as products, by-products, raw materials, processes, machines, personnel, monthly production data, monthly sales data and Key Performance Indicators (KPIs) used by the company that measure their performance in terms of production and demand. The knowledge of the company must be deep for the correct process of planning. This step will facilitate decision-making for future months and next steps (Krajewski et al. 2008).

9.3.3 Step 3: Demand Planning

The Planning coordinator with the support of the Sales department, using the information gathered in the previous step, creates a yearly forecast demand plan that will update every month. The plan must be permanently monitored and updated. The demand plan should be transformed into monetary terms to facilitate the intervention of the Finance department.

9.3.4 Step 4: Supply Planning

The Planning coordinator jointly with the Operations department, gathers production information and computes production times, efficiencies, current available production capacity, historical data of the behavior of production, products with greater rotation by area, products or processes with high production costs, optimal quantity of production, among other useful information and/or calculations. These plannings allow the S&OP process to be realistic of the supply that can be offered under certain scenarios and to establish action plans based on the demand.

9.3.5 Step 5: Comparison

The planning coordinator will compare the demand and production plans established in the previous steps, and will propose and choose at the operational level, a plan and activities to satisfy the demand with the offer (Cecere 2015). Two steps are proposed for the realization of this comparison:

- Generate different scenarios with possible alternatives of courses of action. The people involved in this process must be the S&OP coordinator and those responsible for the Sales, Operational and Financial areas. These scenarios will be cataloged as production plans. The process proposed by Stevenson (2012) will be used for the creation of this production plans:
  - The information calculated and defined in step 3 is used as the objective to be reached.
  - In Step 4 a set of possible production scenarios were defined in order to meet current demand. Now, the appropriate scenario must be selected for the period under analysis, to get the capacity that the company will have and to calculate the production index (number of units required per unit of time, considering initial and final inventory of products with higher turnovers and client requests not reached so far), level of workforce required (number of workers needed) and level of raw material required (fiber, cones, chemicals, among other supplies) (Chase et al. 2006).
  - Company or department policies relevant to the planning process should be considered (these can be security stocks, policies that prohibit hiring by hours, restriction to maintain a stable workforce at all times, work overtime, among others).
  - If applicable, alternative plans will be determined to satisfy the demand.
  - The best plan that meets the objectives is selected from these options. Otherwise go back to Step 4.
  - To create the production plan, the techniques used fall into two categories: informal test and error techniques (tables or graphs in programs such as Excel) and mathematical techniques (linear programming and simulation models) (Stevenson 2012).
- With the production plans ready, a S&OP Board will be held, where high executives together with managers of the aforementioned areas, will review the potential production plans and select the best one. This plan should be published by the S&OP coordinator in order to be an operational level guide during the month (Lapide 2005).

9.3.6 Step 6: Implementation

The implementation of the set of actions proposed in the previous step will take place. Its focus is on the Sales and Production departments but depending on whether the planning is reactive or aggressive, it will focus more on one than on the other. An aggressive planning is one in which the demand is adjusted to the capacities of the supply chain while the reactive planning does the opposite (Krajewski et al. 2008).

9.3.7 Step 7: Maturity Model

Once the implementation of S&OP has begun, the maturity model should be used again to evaluate the stage in which the company is now. Likewise, an evaluation can be made based on the indicators established in Step 2 and see the potential improvement that the process has meant. The aforementioned KPI's will be analyzed again. They can be complemented with other indicators of accomplishment, efficiency and effectiveness (Ferreira 2013).

9.4 Application to Ecuadorian Textile ABA

Following the described methodology, each of the steps are applied to the Ecuadorian Textile ABA. These steps are fed by the strategic and business plans of ABA established at the beginning of the year 2017.

9.4.1 Step 1: Maturity Model

Since ABA will apply S&OP for the first time, the model will be used as a first step. ABA in the different stages is graded based on observations and an action plan is proposed to achieve higher graded stages.

- *Collaboration and meetings*: This rubric measures the human component in the S&OP process (Ferreira 2013). ABA is identified in Stage 2: *Reactive*, based on the following observations and an action plan is proposed to reach the next and subsequent stages.
  - Observations: No personnel dedicated to the planning process, just handled at managerial levels based on orders sent by Sales department to Production department to satisfy the proposed plan. The level of integration and communication between Sales and Operations is medium because the flow of information and meetings are done at non-regular periods with non-significant contents and with a focus on financial metrics.
  - Action plan to reach next stage, Stage 3: *Standard* and subsequent stages: A greater integration between Sales and Operations is needed for the Planning coordinator to create an individual planning for area (Sales and Production), for then combine the information in a single plan in a formal S&OP meeting, including customers and suppliers, both small and large ones.
- *Organization*: This rubric measures the structure of S&OP in terms of how planning is currently used at the company (Ferreira 2013). ABA is identified in Stage 3: *Standard*, based on the following observations and a plan of action is proposed to reach the next and subsequent stages.

- Observations: A new planner position was recently hired by the company, a Planning coordinator. Since it is a new position, it is neither independent nor respected by all the staff. Planning was made at a managerial level, and it continues the same even though a planner was hired.
- Action plan to reach Stage 4: *Advanced* and subsequent stages: The process and the recently hired planner are already understood and respected by all the staff because of proposals and improvements made based on the S&OP process (these are detailed throughout this chapter).

- *Indicators*: This rubric measures the metrics used to evaluate the performance and effectiveness of S&OP (Ferreira 2013). ABA is identified in Stage 2: *Reactive*, based on the following observations and an action plan is proposed to reach the next and subsequent stages.

  - Observations: There are a limited number of indicators to measure the performance of the planning at the end of the month and are added to financial indicators. The main indicator used by the Sales department is the fulfillment of the monthly budget established by the company's strategic plan; they also use the indicator broken down by sales assistant and product family giving the department manager an idea of their monthly sales performance.
  - Action plan to reach Stage 3: *Standard* and subsequent stages: Other indicators should be used to provide measurement of the progress in the profitability and growth of the company. For example, the accuracy of sales forecasts to the reality of the market, performance of the S&OP process from the perspective of all those involved or utilization of the plant capacity. Based on the results of these indicators, improvements can be made in the planning process.

- *Information technology*: This rubric measures how related and evolved the technology tools of the company are in relation to the S&OP process (Ferreira 2013). ABA is identified in Stage 2: *Reactive*, based on the following observations and an action plan is proposed to reach the next and subsequent stages.

  - Observations: ABA currently uses Excel spreadsheets, presenting disorganization and lack of standardization. This is because people in charge or related to planning, handles their own templates without an established format. Information is consolidated manually and not automatically.
  - Action plan to reach Stage 3. *Standard* and subsequent stages: The ERP, Enterprise Resource Planning software, used by the company is a powerful tool that could have its own planning module fed by information involved in planning, such as Sales, Operations, Purchases, and information from other departments. This module could be implemented when information from these departments is centralized and online on the ERP. With the ERP supplier this option is being considered in a nearby future.

- *S&OP Integration Plan*: This rubric measures how efficient the planning processes and creation of demand and supply plans are, as well as their final integration (Ferreira 2013). ABA is identified in Stage 3. *Standard* based on the following observations and a plan of action is proposed to reach the next and subsequent stages

- Observations: Sales department is responsible for carrying out the first step in planning by sending their demand plan based on current customer orders. Operations based on this information shares important details about the successful fulfillment of this plan, which is now considered as feedback to modify it.
- Action plan to reach Stage 4. *Advanced* and subsequent stages: Permanent transferring of information between Sales and Operations is needed. It should include capacity, optimal production quantities, among other limitations provided by Operations. The planning reaches the point where the fulfillment of the demand plan is optimal, as well as the use of capacities and resources in the plant. The interaction is considered perfect, the Sales and Operations departments work as one towards the same objective.

9.4.2 Step 2: Information Gathering

Initially, the portfolio of products that ABA sells should be listed and detailed. There are 42 products, from which, 32 originate from acrylic fiber (Named from Product 1 to Product 31), 1 is a mixture of acrylic with nylon (Named Product 32), 5 are a mixture of acrylic with polyester (Named from Product 33 to Product 37) and 4 are a mixture of polyester and wool (Named from Product 38 to Product 41).

For a better understanding of the names and compositions of each product, textile terms of interest for this study are described below (Benalcazar 2010):

- ABA names its products based on a numerical list followed by a fraction. Annex 1 details this characteristic for the 42 products. The numerator of the fraction represents whether the thread is simple (1 rope) or composed (2 or more ropes).
- ABA more commonly sells threads of 1 and/or 2 ropes. This characteristic defines the path of the process that the thread takes the compound threads have more processes in the spinning.
- The denominator of the fraction represents the metric number of the thread, known as the thread title. This is the determination of a ratio index between the thickness, length and weight of it. This value or title will also be accompanied by the measuring method used. There are two measurement methods: Inverse Systems and Direct Systems. ABA uses both.
  - Inverse System measures the title based on a constant weight in a variable length, its unit is in Metric Number (Nm—from the Spanish). In the reverse system, the higher the metric number, the thinner the thickness of the thread. The title of these products or their thickness is characterized by the length in meters in 1 g of thread. One of ABA's products is a 1/16. This thread

indicates that it is a simple thread and has a title of 16 Nm, that is, 16 m of thread weigh 1 g. Another product is the 2/21, a thread composed of two ropes and has a title of 21 Nm, that is, 10.5 m of the wire weighs 1 g. The metric number of the composed yarns changes in relation to that of single yarns.

  - The Direct System measures the title based on a constant length in a variable weight, its unit is in Deniers (Den) of the raw material. The title of the raw material or its thickness is characterized by the weight in grams of 9000 m. This system, being direct indicates that the larger the Denier, the thicker the thickness of the yarn. For example, for Product 1/16 it is composed of a 2.6 Denier fiber, that is, 9000 m of fiber weigh 2.6 g.

- The torsion applied to a thread is the number of turns given per unit length. These twists are applied to the threads in the process of creation of bobbins and twisted yarns (processes will be detailed later) and define:

  - Characteristics in the production process, such as production time.
  - Characteristics in the thread as the resistance. For example, the greater the torsion, the greater the resistance, the lower the elasticity.

These terms described above define the spinning processes. Threads with higher metric numbers need more production time, are more delicate and laborious while threads of lesser metric number need shorter production times and are very easy to work. Torsions vary depending on the metric number and the characteristics of the yarn and this component of the process also defines the time and condition of production. The same happens with the threads of a single and double rope, their process lines are different, and this defines their production times.

For each spinning process, by-products will be obtained until finally reaching each of the 42 aforementioned products. The line of processes to follow will depend on the composition and the final destination of the yarn, in Annex 1 all the by-products for each final product are listed. For some products there will be shared by-products given that they are composed of the same material. In general, the processes and machinery used in ABA fall within the category of acrylic spinning long-fiber wool system (machines and processes similar to wool) (Benalcazar 2010).

The raw material market offers a wide variety of fibers, both natural and artificial. ABA specializes more in the production of yarns with synthetic fibers (acrylic, polyester or a mixture of both), but also produces mixed yarns of synthetic and natural fibers (mixture of polyester and wool).

9.4.3 Step 3: Demand Planning

Demand planning is composed by expected demand (calculated with forecasts) and by the current customers demand. The value of forecasts is reflected in exceeding the client's expectations when anticipating production yarns before the real demand of the client is known (Lloret 2014). Forecasts are a new tool in ABA but there is historical data on the demand since 2015 that facilitate its new implementation.

Historical data of the demand is broken down by the type of yarn and its color. Annex 1 in column "Color Quantities" shows the articles that have more than 300 colors. This complicates the process and reduces the accuracy of forecasts as well as increasing projection errors (Chase et al. 2006). For this reason, the data is conditioned so that sales data is grouped by-product family and not per color. As mentioned above ABA has 42 products, which are the product families considered in this study.

For forecasting a factor, in this case demand, there are qualitative (subjective, based on personal judgments) and quantitative (objective, mathematical models based on historical data) forecast methods. Within quantitative methods, there are Causal Methods or Time Series. The first method considers more than one variable that influences the forecast. Time series, on the other hand, depend only in one variable, the demand of previous periods. Within Time Series there are various methods: Moving Average (MA), Weighted Moving Average, Simple, Double and Triple Exponential Smoothing (Chase et al. 2006). Due to the existence of data and the company requesting simplicity but accurate forecasts, the second method is chosen, and within it time series are chosen.

Based on the analysis of sales frequency from 2015 to 2017 products that maintain a constant sales frequency (considerable production and billed quantities) are identified and selected for this study. Annex 2 listed the products and highlights those with this pattern, which are Product 2, Product 7, Product 11, Product 20, Product 22 and Product 36.

Neither seasonality nor trend is identifiable in the demands of the selected products, therefore all-time series are applied. Based on the estimate with the lowest MAPE, Mean Absolute Percentage Error, the best forecast method for each product is selected. Solver-Excel was used to find the optimal values of the smoothing constants that will minimize forecast error in the Simple and Double Exponential Smoothing methods (Nahmias 2007). For Double Exponential Smoothing that uses double smoothing constants: α for interception and β for trend, Solver always calculated $\beta = 0$ confirming the lack of trend in the data.

On the other hand, the Winters or Triple Exponential Smoothing method is not applied because the periodicity p needed to replace in the equations of the method is not possible to identify given that there is no seasonality in the data.

The equations to be used for Simple Moving Averages, Simple and Double Exponential Smoothing are (Nahmias 2007):

- Simple Moving Average (SMA) Eq. (9.1):

$$F_{t+1} = \frac{\sum_{i=1}^{n} X_{t-1}}{n} \tag{9.1}$$

- Simple Exponential Smoothing (SES) Eq. (9.2):

$$F_{t+1} = \alpha X_{t+1} + (1 - \alpha)L_t \tag{9.2}$$

- Double Exponential Smoothing (DES) Eq. (9.2),

$$L_t = \alpha X_t + (1 - \alpha)(L_{t-1} + T_{t-1}) \tag{9.3}$$

$$T_t = \beta(L_t + L_{t-1}) + (1 - \beta)T_t \tag{9.4}$$

$$F_{t+1} = L_{t-1} + T_{t-1} \tag{9.5}$$

where F_{t+1} is the forecast in period t, α is the smoothing constant, β is the trend constant, both constants must have a value between 0 and 1, X_t is demand in the previous period, L_t is level from the previous period and T_t is the previous period trend.

For each product, MAPE errors obtained are shown from Annex 3. The forecast for each method is also shown in each table compared to the real sales data of month 34.

As seen for each product the best forecast method is:

- *Product 2*: In spite of the SMA of 3 periods has the lower error, the SES is more consistent. Hence SES is selected.
- *Product 7*: SMA of 13 periods is the method with the lower error. Hence SMA 13 is selected.
- *Product 11*: Errors with time series are too high. Therefore, Croston method is applied (forecast method explained below) and selected for its small error.
- *Product 20*: SES has the lower error. Hence SES is selected.
- *Product 22*: Errors with time series are too high. Croston is also applied but the error is still too high. There are three atypical data (from 3 months) with assignable causes therefore those are eliminated. These assignable causes are identified with the Sales department who identified those months as problematic months for the customer because he started negotiating with suppliers from China. SMA of 14 periods is selected because it has de lower error. See Annex 3 for results.
- *Product 36*: Errors with time series are too high. There are three atypical data (from 3 months) with assignable causes therefore those are eliminated. Similarly, to Product 22 these months were identified as abnormal months because of the increment in international competition. The results show that the SMA of two periods is the most suitable forecast method for these data. See Annex 3 for results.

For products 11 and 36 traditional forecasting methods show high errors, making them poor forecasts of demand. One factor that generates these high errors is the variability within sales data. It is considered that the behavior of demand in these products is intermittent or variable. The forecast method of Croston has the ability to forecast data of this nature (periods with demand equal to 0). Despite not having demand equal to zero, sales in some months of these products are significantly lower. Equations (9.6) and (9.7), are used (Waller):

$$\text{If demand } X_t \neq 0 \begin{cases} Z_{t+1} = \alpha X_t + (1-\alpha) Z_t \\ V_{t+1} = \alpha q + (1-\alpha) V_t \\ Y_{t+1} = \frac{Z_{t+1}}{V_{t+1}} \end{cases} \tag{9.6}$$

$$\text{If demand } X_t = 0 \begin{cases} Z_{t+1} = Z_t \\ V_{t+1} = V_t, \\ Y_{t+1} = Y_t \end{cases} \tag{9.7}$$

where X_t is demand in period t, Z_t is the estimate of the mean of the size of demand when it is different from 0 in period t, V_t is the estimate of the mean of the size of the interval between demands different from 0, Y_t is the forecast of demand in period t.

9.4.4 Step 4: Supply Planning

The company manages an ERP that contains different modules or parts. All the productions, movements of stock, consumption of resources, generation of production costs, generation of production waste, etc., are recorded. The requirements and calculations necessary for the supply planning could be extracted from the information contained in this module.

In the process of recording operations information, it was not able to subtract it from the ERP. The feeding at the ERP system, is identified as altered from reality, specifically in production times including idle times for each by-product.

In ABA, the production information entered into the ERP is made by the spinning supervisors. Supervisors provide operators on each machine with forms containing including: article to be worked, time of start and end of production, time of stoppages of production or machinery based on specific coding, produced quantity and waste entry. Then this information is recorded in the production module in the ERP.

The problems identified during this step were:

- The forms do not have a standard structure or the necessary information to be recorded.
- Each supervisor handles different codes for types of production stoppages.
- The Productive hours and idle times are not being recorded correctly.
- Stoppage codes were poorly made and coded.
- Due to the lack of supervision, the recording of the information is poorly or modified.

- ABA has no supervisors during the night shift, so the information recorded by them is entered by day shift supervisors.

Improvements are proposed for enhancing production information for the supply planning:

- Standard forms for each area with specific information to be include depending on the area.
- Revision when shift changes to verify the information recorded.

9.4.4.1 Measurement of Production Times and Calculation of Production Efficiencies

Because information could not be retrieved directly from the ERP a time study is carried out to calculate time and efficiency standards in each process. Forms proposed before that includes productive and idle time of each machine are used. The article produced, the operator in charge of the production, the speed used, types of stoppages that occur and their classification, among other useful information are also collected.

For the calculation of the sample size that is representative of the population, pilot data are taken for each by-product in each machine based on the analysis of the regular cycle time. A table of recommended number of observation cycles is used to perform this initial calculation (Niebel and Freivalds 2009). Based on this recommendation, it is assumed that at least three samples should be taken per article.

Depending on the behavior of the cycle time, this sampling calculation can increase its size or stay in three cycles. After taking the pilot samples, the true sample size is calculated using the Eq. (9.8), used in (Niebel and Freivalds 2009).

$$n = \left(\frac{ts}{k\bar{x}}\right)^2, \tag{9.8}$$

where n is the sample size, $\bar{x}$ is the sample mean, s is the sample deviation, the acceptable fraction of error k established in 5% and the cumulative probability of distribution t with variable degrees of freedom depending on the by-product and the machine analyzed. A confidence level of 95% is used for calculations (Niebel and Freivalds 2009).

The results are shown in Annex 4. In the column "Recommended number of cycles" the number of measurements that were taken (based on the first analysis) and on the column "# of samples calculated" the sample size calculated subsequently are shown by machine and by-product. As can be seen, in most of the cases calculations taken at the beginning exceed the ones calculated with the formula. But, in some cases this amount is inferior. In these cases, an Anderson Darling normality test is performed to verify the normality of this data and accept the samples taken as representative of the population. The hypotheses of this assumption are:

H_0= data follows a normal distribution
H = data does not follow a normal distribution

Using a confidence level of 95%, $\alpha = 5\%$, to have the certainty not to reject incorrectly the hypothesis 95 of the times, it is concluded that for all the cases where the plot data was less than the suggested data calculated by the formula p is greater than α, therefore there is not sufficient statistical evidence to reject h_0, thus the times taken initially follow a normal distribution and are representative of the sample and population (Montgomery 2013). Around 24 tests were performed in the 24 processes that showed this behavior. Annex 5 shows the test result from Minitab for the by-product Con11.

On the other hand, Annex 6 shows by final product a breakdown for each by-product in each machine of: average time of direct stoppages allowed (cleaning, preparation, conditioning, among others), average production time, real efficiency calculated and the adjusted efficiency (considering the elimination of times of indirect or not allowed stoppages assigned or justified by a cause). With this data, the cycle time for each by-product is calculated, as well as for the final product.

The removal of indirect or non-permitted stoppages assigned to a cause was a process carried out jointly with the supervisors and managers of Production and Maintenance departments of the company. These stoppages include mechanical damages, activities made to improve the operation of the machine or abnormal conditions in the produced or fed by-product. Additionally, there are non-permitted stoppages that happen more frequently and are not eliminated from the calculation. An analysis of the "5 Whys" is carried out. This is a tool proposed by the Toyota Production System for helping analyst to find the root causes of problems and potential solutions for each identified sub-cause (Ries 2012). Classified into two the main stoppage causes are classified into Raw-Material Machinery and Administration.

The conclusion of this analysis for Administration is that the main cause for machines to stop work because of administration is the workload gave to operators. The solution proposed is to carry an organization based on methodological analysis of workload in each line of machines.

9.4.4.2 Machine Capacities Calculations

Using equations provided by machine suppliers, as well as external audits carried out in the company, capacities for each by-product in each machine are calculated. Two types of equations are used:

- For all the machines until Spinning Machines Eq. (9.9) is used:

$$\text{Capacity}[\text{kg}] = \text{Velocity}\left[\frac{\text{m}}{\text{min}}\right] \times \text{wick weight}\left[\frac{\text{kg}}{\text{m}}\right] \times \text{shift}[\text{min}] \times \text{efficiency}[\%] \tag{9.9}$$

- From Spinning Machines until Packaging the Quantity of spindles is taken into account, then Eq. (9.10) is used:

$$\text{Capacity}[\text{kg}] = \frac{\text{Velocity}\left[\frac{\text{m}}{\text{min}}\right] \times \#\,\text{spindles} \times \text{Shift}[\text{min}] \times \text{Efficiency}[\%]}{\text{Metric Number of the yarn}\left[\frac{\text{m}}{\text{kg}}\right]} \tag{9.10}$$

Results of these calculations are summarized in Annex 7. Two capacities are calculated, the first with the real efficiency of production and the other with the adjusted efficiency (without stoppages not allowed with assignable causes). It also shows the increase in production that can be achieved with this increase in efficiencies.

9.4.4.3 Calculation of Optimal Production Batch, Reorder Point and Safety Stocks

The demand for products in ABA follows a *stochastic* behavior because they are not constant over time. For these cases, the calculation of the optimal order batch (production) must be calculated using stochastic or random reorder models. It is also identified that the inventory control of the company is continuous, which means that the inventory level is always known and controlled. It is established that the model (Q, R) using Service Level is the most appropriate model based on the information provided by the company and the need of knowing this production details, the optimal production batch Q, the reorder point R and the safety stock ss. The assumptions of this model are (Nahmias 2007):

- Continue audit to the system
- Random demand, which follows a normal distribution with mean and standard deviation
- Fixed delivery time τ
- Existence of: Preparation costs K ($/order), cost of maintaining inventory ($/year), c purchase cost ($/unit), p penalization cost for unities not satisfied or backordered ($/unit)

The company does not have information on penalty costs, but they have been managing service levels, specifically the Type 1 Service. This service specifies α probability of not having backorders during the time of delay. Equations (9.14) and (9.15), to be used are (Nahmias 2007):

$$F(R) = \alpha \tag{9.11}$$

Table 9.1 Results of (Q, R) model

Details	Product 2	Product 7	Product 11	Product 20	Product 22	Product 36
Q	2360.43	1645.93	1661.84	1853.76	2453.18	1393.42
R	4514.31	2641.38	2023.91	2113.92	3346.34	1097.29
ss	1300.32	788.34	752.72	796.03	1285.74	465.12

$$R = \sigma z + \mu \tag{9.12}$$

$$Q = EOQ = \sqrt{\frac{2K\lambda}{h}} \tag{9.13}$$

$$h = c \times i \tag{9.14}$$

$$ss = R - \mu \tag{9.15}$$

where z is a standardize variable obtained from the function value $F(R)$, λ average of annual demand, i the % of cost assigned to inventory management. This percentage should be 20% (Durlinger 2012; REM 2010).

$$\mu = \text{media} \times \tau \tag{9.16}$$

$$\sigma = \text{standard deviation } \times \tau \tag{9.17}$$

Months where warehouses have initial inventory μ:

- If $ss < \mu$ (initial inventory) Q^* must be ordered
- If $ss \geq \mu$ Q^* must not be ordered (Table 9.1).

9.4.4.4 Production Requirements Calculations

The calculations previously presented are in terms of final product, these must be translated to each of the by-products that make up this final product. This calculation uses the Eq. (9.18) (Tompkins et al. 2010):

$$I_k = \frac{O_k}{(1 - d_k)}, \tag{9.18}$$

where, I_k is the quantity to be produced, O_k is the final quantity of the product demanded (optimal production quantity) and d_k waste generated. d_k for each by-product is the average registered since 2015. There are some processes where waste equals to 0%. Annex 8 shows the analysis for each by-product and process.

9.4.5 Step 5: Comparison

This step consolidates the results of step 3 and step 4. Now the information available is: customers current demand, monthly forecasts by family of products, limitations, requirements and variations the plant has and the supply it can offer (Ferreira 2013).

9.4.6 Step 6: Implementation

The proposed plan built on previous steps can be modified and adapted as the time goes. Everything will depend on the fulfillment achieved on the demand plan or any limitations or troubles it has encountered.

9.4.7 Step 7: Maturity Model

As applied at the beginning of the methodology, this process will allow to measure progress in the S&OP process. The comparison of the initial maturity model with the final one will allow evaluate the effectiveness of this new implementation on the planning process.

- *Collaboration and meetings*: After applied the methodology, now ABA has an independent position dedicated only to planning. This position works together with the Sales and Operations department. This position integrates the business needs and the supply offered in a single integrated plan. Customers and suppliers are much more considered in the plans made. ABA advances from Stage 2 to Stage 5.
- *Organization*: The planner has more independence, greater respect among the personnel involved in the planning process. There is a collaborative process with the sales and operational department. ABA advances from Stage 3 to Stage 5 where a formal planning team with the participation of executives is already identified.
- *Indicators*: Although ABA has not improved from Stage 2 given that forecasts have not been yet implemented, indicators used now have more relation to planning with no longer a financial focus.
- *Information technology*: ABA has not improved from Stage 2 given that the same technological tools are kept. There is less manual consolidation, but the ERP has not changed in relation to a planning module.
- *S&OP integration plan*: ABA has not improved from Stage 3 given that the same level of integration is maintained. Being a process of change, adequate integration has had some resistance.

9.5 Conclusions and Recommendations

- The S&OP planning process can be a bridge for internal and external communication. Helping companies to have a realistic and accurate demand planning.
- It is needed the use of systematic tools that improve and focus planning: Production times, Capacities, Efficiencies, Forecasting, Model (Q, R), Calculation of production requirements.
- Replacement of calculations and decision-making based on experience by theoretical and methodological foundations.
- With the comparison of the Maturity Model before as well as after implementing S&OP, improvement is evident in the rubrics related to the structure of the planning, the planner position, the process itself and indicators used.
- Proper planning will allow ABA to be competitive in price, delivery times and quality in the product and service.
- It is recommended to:
  - Have commitment between every employee in the company and the position of planner so that S&OP is correctly implemented.
  - Constant feeding and updating of information. S&OP needs to constantly update its calculations as the company, products, personnel, customers, departments change.
  - Compliance with S&OP processes, metrics and indicators. Even though the planning process under this methodology is flexible, processes, metrics and indicators will better guide the correctly implementation.
  - Apply the complete methodology for a better knowledge of their customers, products, departments and supply and a guide on how to merge each of this information for properly calculate standards, apply them and measure their effectivity and necessity.
  - Adapt ERP to Planning to create a Planning module so that everyone involved have unlimited access to information regarding to that module and decision-making is properly done.
  - Update and adapt S&OP to the needs and changes that the company presents throughout its implementation in order to achieve a correct implementation.
  - Present improvements from S&OP to the members of the company. It helps to acknowledge benefits, challenges, difficulties, advantages, and disadvantages.
  - Spread basic S&OP concepts to the entire company, so that the proposed working culture of S&OP and the benefits obtained with it like teamwork, communication, follow ups, use of standards and definition of responsibilities and processes can be applied among everyone.

Annex 1 Breakdown of by-Products (Done by the Author)

Product Name	Color Quantities	Composition	Dyed Thread Title	Raw Thread Title	Skeins	Twisted Cone 3 ropes	Twisted Cone 2 ropes	Simple Cone	Bobbins	Rovings	Sliver Cans TP1 or TP2	Sliver Cans SP1 o SP2	Sliver Cans SR1 o SR2	Top	Botes SR1, GC1, C, GC2	SCs R1	Sliver Cans T1 y T2	Raw Material
Product 1	Raw	Acrylic		1/16			1 rope cone	Con1	Bob1	Rov1	SCTP1	SCSP1	SCSelfR1	Top1		No Sliver Can	SC RMA N SC RMA S	RMA
Product 2	300		1/29	1/35	Sk2		1 rope cone	Con2	Bob2	Rov2	SCTP2	SCSP2	SCSelfR2	Top2		SC B HB	SCRMB N SCRMB S	RMB
Product 2a			1/29	1/35	Sk2a		1 rope cone	Con2a	Bob2a	Rov2a	SCTP2a	SCSP2a	SCSelfR2a	Top2a		SC BC HB	SCRMB N SCRMC S	RMB RMC
Product 3	Raw			1/37			1 rope cone	Con3	Bob3	Rov3	SCTP3	SCSP3	SCSelfR3	Top3		SC DE HB	SCRMD N SCRME S	RMD RME
Product 4	5		1/29	1/37	Sk4		1 rope cone	Con4	Bob4	Rov4	SCTP4	SCSP4	SCSelfR4	Top4		SC DE HB SC F N	SCRMD N SCRME S SCRMF N	RMD RME RMF
Product 5	14		1/29	1/37	Sk5		1 rope cone	Con5	Bob5	Rov5	SCTP5	SCSP5	SCSelfR5	Top5		SC F N SC E HB	SCRMF N SCRME N SCRME S	RMF RME RME
Product 6	7		1/29	1/37	Sk6		1 rope cone	Con6	Bob6	Rov6	SCTP6	SCSP6	SCSelfR6	Top8		SC F N SC E S	SCRMF N SCRME S	RMF RME
Product 7	316		1/36	1/42	Sk7		1 rope cone	Con7	Bob7	Rov7	SCTP7	SCSP7	SCSelfR7	Top9		SC DaB HB	SCRMDa S SCRMB N	RMDa RMB
Product 8				1/42			1 rope cone	Con8	Bob8	Same as Product 3								
Product 9	Raw			2/21			Ret9	Con9 Con9a	Bob9 Bob9a	Same as Product 4 Same as Product 3								
Product 10				2/25.5			Ret10	Con10	Bob10	Same as Product 3								
Product 11	285		2/27	2/30	Sk11		Ret11	Con11	Bob11	Rov11	SCTP11	SCSP11	SCSelfR11	Top10		No SC	SCRMH N	RMH
Product 12	108		2/30	2/30	Sk12		Ret12	Con12	Bob12	Rov12	SCTP12	SCSP12	SCSelfR12	Top11		SC F N	SCRMF N	RMD
Product 13	122		2/30	2/30	Sk13		Ret13	Con12	Bob12	Rov12	SCTP12	SCSP12	SCSelfR12	Top11			SCRMF N	RMD RMI
Product 14	3			2/32			Ret14	Con14	Bob14	Same as Product 4								
Product 15				2/32			Ret15	Con15	Bob15	Same as Product 5								
Product 16				2/32			Ret16	Con16	Bob16	Same as Product 6								
Product 17	2			2/32			Ret17	Con17	Bob17	Same as Product 3								
Product 18	7		2/25	2/32	Sk18		Ret18	Con18	Bob18	Rov18	SCTP18	SCSP18	SCSelfR18	Top12		No SC	SCRMB N SCRMB S	RMJ RMB
Product 19	3		2/36	2/34	Sk19		Ret19	Con19	Bob19	Rov19	SCTP19	SCSP19	SCSelfR19	Top13		No SC	SCRMC N	RMC
Product 20	357		2/29	2/35	Sk20		Ret20	Con20	Bob20	Igual al Producto 18								
Product 21	91		2/31	2/37	Sk21		Ret21	Same as Product 3										
Product 22	Raw			2/37			Ret21	Same as Product 3										
Product 23			2/29	2/37			Ret22	Same as Product 4										
Product 24			2/29	2/37			Ret23	Same as Product 5										
Product 25			2/29	2/37			Ret24	Same as Product 6										
Product 26			2/29	2/37			Ret26	Con26	Bob26	Rov26	SCTP26	SCSP26	SCSelfR26	Top3 Top6		SC DE HB SC K N	SCRMD N SCRME S SCRMK N	RMD RME RMK
Product 27				2/38			Ret27	Con27	Bob27	Same as Product 3								
Product 28				2/42			Ret28	Con28	Bob28	Same as Product 3								
Product 29				3/18		Twst29rope3	Ret21	Same as Product 3										
Product 30				3/18		Twst30rope3	Same as Product 26											
Product 31				3/18		Twst31rope3	Same as Product 23											
Product 32	9	Nylon	2/28	2/28	Sk32		Ret32	Con32	Bob32	Rov32	SCTP32	SCSP32	SCSelfR32	Top14		SC D N Purchased	SCRMD N	RMD RML RMM
Product 33	Raw	Polyester-Acrylic		2/48			Ret33	Con33	Bob33	Rov33	SCTP33	SCSP33	SCSelfR33	Top16	SC33	SC F N Purchased	SCRMF N	RMF RMN
Product 34	27			2/48			Ret34	Con34	Bob34	Rov34	SCTP34	SCSP34	SCSelfR34	Top17	SC34	Purchased SC E N	SCRME N	RMM RME
Product 35	Raw			2/48			Ret35	Con35	Bob35	Rov35	SCTP35	SCSP35	SCSelfR35	Top19	SC35	Purchased		RME RMM
Product 36				2/48			Ret36	Con36	Bob36	Rov36	SCTP36	SCSP36	SCSelfR36	Top20	SC36	Purchased SC E N	SCRME N	RMN RME
Product 37				2/48			Ret37	Con37	Bob37	Rov37	SCTP37	SCSP37	SCSelfR37	Top21	SC37	Purchased		Same as Product 36
Product 38		Polyester-Wool		1/28			1 rope cone	Con38	Bob38	Rov38	SCTP38	SCSP38	SCSelfR38	Top22	SC38			RMO RMP
Product 39				2/40			Ret39	Con39	Bob39	Rov39	SCTP39	SCSP39	SCSelfR39	Top23	SC39			RMO RMP
Product 40	32			2/48			Ret40	Con40	Bob40									RMO RMP
Product 41	46			2/56			Ret41	Con41	Bob41									RMO RMP

Annex 2 Sales Frequency Analysis from 2015–2017 (Done by the Author)

2015		2016		2017	
Product	Annual sales frequency [month]	Product	Annual sales frequency (month)	Product	Annual sales frequency (month)
Product 2	12	Product 2	12	Product 2	9
Product 7	12	Product 7	12	Product 20	9
Product 22	11	Product 22	11	Product 7	9
Product 20	11	Product 20	11	Product 22	9
Product 11	9	Product 36	8	Product 11	8
Product 36	9	Product 11	7	Product 36	7
Product 27	5	Product 27	7	Product 17	6
Product 17	2	Product 23	5	Product 27	4
Product 37	1	Product 17	3	Product 37	3
Product 23	1	Product 32	2	Product 38	1
Product 41	1	Product 25	2	Product 25	1
Product 5	1	Product 37	1	Product 24	1
Product 9	1	Product 3	1	Product 23	1

Annex 3 Error MAPE Forecasts Products (Done by the Author)

PRODUCT 2

	MAPE		Forecast	Real Demand	Forecast	Real Demand
MA 2	35		25,033.50		35,046.80	
MA 3	30		22,125.60		30,975.80	
MA 4	34		21,817.10		30,543.90	
MA 5	35		20,967.70		29,354.80	
MA 6	35		21,173.80		29,643.30	
MA 7	38		21,539.00	26,178.59	30,154.60	36,650.00
MA 8	38		21,189.30		29,665.00	
MA 9	39		20,385.80		28,540.10	
MA12	44		18,755.60		26,257.90	
SES	31	α 0.48	23,623.30		33,072.50	
DES	32	α 0.34 β 0.005	22,705.40		31,787.50	

PRODUCT 7

	MAPE		Forecast	Real Demand	Forecast	Real Demand
MA 2	26		11,823.60		16,553.00	
MA 3	28		12,277.10		17,189.90	
MA 4	28		12,135.60		16,989.90	
MA 5	31		12,427.10		17,397.90	
MA 6	33		12,233.80		17,127.30	
MA 7	34		11,907.50		16,670.50	
MA 8	36		11,480.80		16,073.20	
MA 9	39		11,057.00	14,695.74	15,479.90	20,574.02
MA12	23		9,898.76		13,858.30	
MA13	21		9,922.22		13,989.10	
MA14	23		10,106.30		14,148.90	
MA15	24		10,050.50		14,070.70	
SES	24	α 0.8001	12,505.80		17,508.10	
DES	26	α 0.3391 β 0.005	11,527.30		16,138.20	

PRODUCT 11

	MAPE		Forecast	Real Demand	Forecast	Real Demand
MA 2	84		9,018.73		12,626.20	
MA 3	89		8,041.79		11,258.50	
MA 4	105		8,156.37		11,418.90	
MA 5	111		7,318.99		10,246.60	
MA 6	114		7,542.33		10,559.20	
MA 7	121		7,833.34		10,966.70	
MA 8	128		8,204.24		11,485.90	
MA 9	131		7,659.02	9,521.09	10,722.60	13,329.51
MA12	128		7,079.99		9,911.97	
MA13	129		7,487.10		10,481.90	
MA14	51		7,501.12		10,501.60	
MA15	53		7,318.84		10,246.40	
SES	106	α 0.005	7,384.96		10,338.90	
DES	99	α 0.005 β 0.005	7,024.94		9,434.90	
Croston	26	α 1.00			7,000.00	

PRODUCT 20

	MAPE		Forecast	Real Demand	Forecast	Real Demand
MA 2	36		15,856.60		22,199.30	
MA 3	40		16,229.40		22,721.11	
MA 4	48		15,967.90		22,355.00	
MA 5	56		15,130.50		21,182.80	
MA 6	61		14,522.80		20,331.90	
MA 7	64		14,286.30		20,000.80	
MA 8	64		13,661.50		19,126.10	
MA 9	64		12,682.60	13,768.43	17,755.60	19,275.78
MA12	54		10,909.30		15,272.90	
MA13	51		10,774.10		15,083.70	
MA14	41		10,784.00		15,097.60	
MA15	37		10,586.20		14,820.60	
SES	34	α 0.836	17,483.20		24,476.50	
DES	34	α 0.85 β 0.005	17,786.90		24,901.70	

PRODUCT 22

	MAPE		Forecast	Real Demand	Forecast	Real Demand
MA 2	135		7,213.85		10,099.40	
MA 3	122		13,023.50		18,232.90	
MA 4	132		13,143.00		18,400.00	
MA 5	135		13,538.30		18,953.70	
MA 6	132		13,576.20		19,006.60	
MA 7	129		13,238.40	15,307.13	18,533.70	21,429.96
MA 8	132		12,953.40		18,134.80	
MA 9	138		11,803.20		16,524.50	
SES	163	α 0.005	19,847.90		27,787.10	
DES	102	α 0.005 β 0.00475	10,646.20		14,904.70	
Croston	105.26	α 1.00			3,725.00	

PRODUCT 22 MODIFIED

	MAPE		Forecast	Real Demand	Forecast	Real Demand
MA 2	38		17,672.70		24,741.80	
MA 3	38		16,282.20		22,795.10	
MA 4	39		15,991.70		22,388.30	
MA 5	38		15,546.40		21,764.90	
MA 6	38		14,823.90		20,753.50	
MA 7	42		14,271.80		19,980.50	
MA 8	46		14,787.80	15,307.13	20,702.80	21,429.96
MA 9	49		15,597.50		21,836.40	
MA12	41		15,868.40		22,215.70	
MA13	32		15,176.80		21,247.50	
MA14	31		15,927.10		22,298.00	
SES	59	α 0.005	20,106.80		28,149.60	
DES	38	α 0.00454 β 0.005	13,675.20		19,145.30	

PRODUCT 36

	MAPE		Forecast
MA 2	47		11,558.00
MA 3	52		9,589.63
MA 4	58		8,757.54
MA 5	66		7,926.21
MA 6	74		7,231.78
MA 7	80		6,647.62
MA 8	83		6,181.56
MA 9	86		5,977.09
MA12	50		5,124.85
MA13	44		5,163.63
MA14	40		5,373.64
MA15	38		5,433.59
SES	64	α 0.62	8,997.31
DES	60	α 0.642 β 0.005	9,055.02

PRODUCT 36 MODIFIED

	MAPE		Forecast
MA 2	38		11,558.00
MA 3	42		9,589.63
MA 4	46		8,757.54
MA 5	53		7,926.21
MA 6	59		7,231.78
MA 7	62		6,647.62
MA 8	62		6,181.56
MA 9	61		5,977.09
MA12	44		5,124.85
SES	37	α 0.92	7,780.78
DES	36	α 1 β 0.005	7,126.47

Annex 4 Sample Size Analysis (Done by the Author)

Section	Machine	Analyzed Product	By-Product	Cycle Time [min]	Recommended # of samples	# taken of samples	# of samples calculated	Normality Test
Cutting and Mixing Line	SBC1	Product 2, 7, 20	RMB N	> 40	3	11	2.00	
		Product 2, 20	RMB S	> 40	3	5	1.00	
	SBC2	Product 2, 7, 20	RMB N	> 20	5	10	14.00	Test performed
	SBC1	Product 7	RMD S	> 45	3	9	8.00	
	SBC2		RMD S	> 20	5	5	4.00	
	SBC1	Product 22	RMDa N	> 40	3	6	8.00	Test performed
	SBC2		RMDa N	> 25	5	10	39.00	Test performed
	SBC1	Product 11	RME N	> 40	3	3	1.00	
		Product 22	RME S	> 40	3	9	11.00	Test performed
	SBC2	Product 11	RME N	> 20	5	10	29.00	Test performed
	R1	Product 2	SC B HB	> 4	15	15	7.00	
		Product 7	SC DaB HB	> 3	15	15	2.00	
		Product 22	SC DE HB	> 5	10	15	1.00	
	W1	Product 22	Top 3	> 3.6	15	15	5.00	
		Product 7	Top 9	> 4	15	15	2.00	
		Product 2	Top 2	> 2.5	15	15	1.00	
	W2	Product 22	Top 3	> 2.5	15	12	6.00	
		Product 7	Top 9	> 3	15	16	7.00	
		Product 2	Top 2	> 3.4	15	27	4.00	
	RW2	Product 20	Top 12	> 3	15	20	9.00	
		Product 11	Top 10	> 4	15	17	1.00	
		Product 7	Top 9	> 3.5	15	12	11.00	
Preparation Line 1	SelfRegulator Line 1	Product 2	SCSelfR2	> 4.5	15	10	1.00	
		Product 7	SCSelfR7	> 5	10	5	1.00	
		Product 20	SCSelfR18	> 5	10	10	1.00	
		Product 22	SCSelfR3	> 5	10	23	1.00	
	SP 1	Product 2	SCSP2	> 9.5	10	15	1.00	
		Product 11	SCSP11	> 9	10	18	6.00	
		Product 20	SCSP18	> 9.5	10	15	1.00	
		Product 22	SCSP3	> 9.5	10	15	1.00	
	TP 1	Product 2	SCTP2	> 10	8	15	6.00	
		Product 11	SCTP11	> 9.5	10	15	1.00	
		Product 20	SCTP18	> 11	8	15	1.00	
		Product 22	SCTP3	> 10	8	15	1.00	
	Rub-Rover 1	Product 2	Rov2	> 13	8	15	1.00	
		Product 11	Rov11	> 11	8	15	1.00	
		Product 20	Rov18	> 11	8	6	1.00	
		Product 22	Rov3	> 13	8	15	1.00	
Preparation Line 2	SelfRegulator Line 2	Product 7	SCSelfR7	> 5	10	16	1.00	
		Product 20	SCSelfR18	> 7	10	25	1.00	
		Product 36	SCSelfR36	> 7	10	15	1.00	
	SP 2	Product 7	SCSP7	> 13	8	10	1.00	
		Product 20	SCSP18	> 13	8	22	16.00	
	TP 2	Product 2	SCTP2	> 19	8	15	1.00	
		Product 7	SCTP7	> 19	8	15	1.00	
		Product 20	SCTP18	> 19	8	15	1.00	
	Rub-Rover 2	Product 2	Rov2	> 19	8	10	2.00	
		Product 7	Rov7	> 19	8	15	1.00	
		Product 20	Rov18	> 23	5	15	6.00	
		Product 36	Rov36	> 23	5	9	1.00	
Combing Line	SR1	Product 36	SC36	> 5.5	10	12	1.00	
	GC1			> 9.5	10	10	1.00	
	C1			> 120	3	2	3.00	Test performed
	C2					1		
	C3					2	226.00	Test performed
	C4					4	31.00	Test performed
	GC2			> 10	8	8	1.00	
	GC3			> 6	10	10	1.00	
	WC1		Top20	> 5	10	10	1.00	
Spinning Machines	SM3	Product 36	Bob36	> 40	3	3	60.00	Test performed
	SM4	Product 7	Bob7			2	36.00	Test performed
	SM5	Product20	Bob20			2	1.00	
	SM6	Product 20	Bob20			6	2.00	
	SM7	Product 22	Bob3			3	79.00	Test performed
	SM8	Product 2	Bob2			6	7.00	Test performed
	SM9	Product 11	Bob11			3	3.00	
	SM10	Product 2	Bob2			6	5.00	
	SM11	Product22	Bob3			3	9.00	Test performed
Winding Machine	WM3	Product11	Con11	> 40	3	24	55.00	Test performed
		Product36	Con36			11	74.00	Test performed
		Product20	Con20			24	75.00	Test performed
		Product7	Con7			12	6.00	
	WM4	Product2	Con22			36	35.00	
		Product22	Con22			46	4.00	
Skeins	Skeins1-R	Product2	Sk2	> 40	3	2	28.00	Test performed
	Skeins2-R					4	14.00	Test performed
	Skeins2-L					4	21.00	Test performed
	Skeins3-R					5	9.00	Test performed
	Skeins4-R					5	16.00	Test performed
	Skeins5-L					5	5.00	
	Skeins6-R					2	45.00	Test performed
	Skeins3-R	Product7	Sk7			2	59.00	Test performed
	Skeins1-L	Product11	Sk11			5	149.00	Test performed
	Skeins3-L					5	2.00	
	Skeins4-L	Product20	Sk20			5	72.00	Test performed
	Skeins5-R					5	2.00	
	Skeins6-L					4	46.00	Test performed

Annex 5 Anderson Darling Normality Test of the by-Product Con11 in Machine WM3 (Done by the Author)

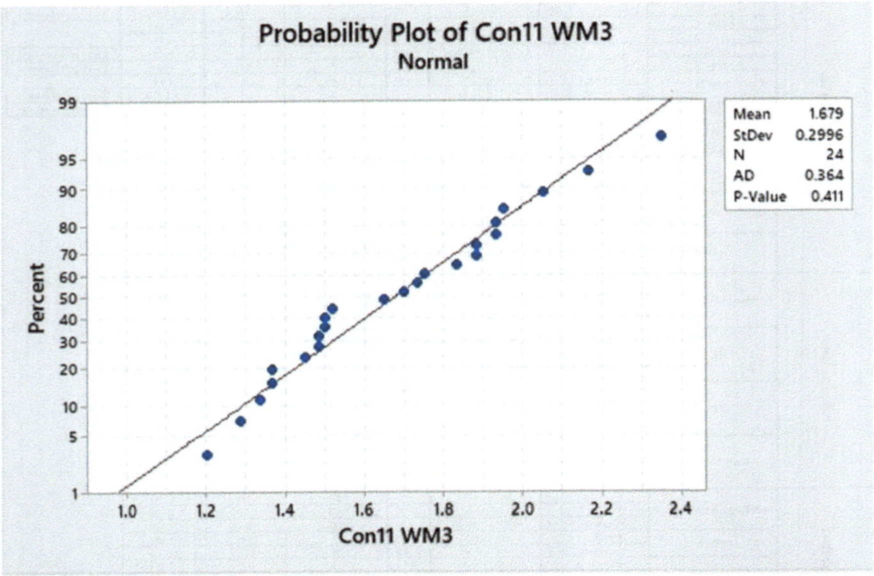

Annex 6 Standard Production, Allowed Stoppages and Efficiencies for Each by-Product and Machine

Product	By-Product	Machines	Total production time [min]	Average productive time [min]	Average allowed stoppages [min]	Efficiency with not allowed stoppages	Real Efficiency
Product 2	Sk2	Skein1-R	77.00	68.00	9.00	89.02%	89.02%
		Skein2-R	62.25	54.75	7.50	95.40%	95.40%
		Skein2-L	65.00	55.25	9.75	97.74%	97.74%
		Skein3-R	72.00	58.6	13.40	100.00%	100.00%
		Skein4-R	64.40	53.60	10.80	99.08%	99.08%
		Skein5-L	66.20	53.20	13.00	96.22%	96.22%
		Skein6-R	66.00	54.00	12.00	100.00%	100.00%
	Con2	WM4	110.08	109.862	0.217	100.00%	100.00%
	Bob2	SM8	212.13	205.21	6.92	98.30%	98.30%
		SM10	227.97	221.69	6.28	98.35%	98.35%
	Rov2	RM1	16.37	13.24	3.13	98.29%	98.29%
		RM2	20.68	20.15	0.53	90.69%	60.29%
	SCTP2	TP1	12.77	10.43	2.34	98.63%	98.63%
		TP2	20.25	19.19	1.06	88.38%	63.06%
	SCSP2	SP1	11.50	9.66	1.84	94.72%	87.45%
	SCSelfR2	SelfR2	24.66	5.00	19.66	100.00%	100.00%
	Top2	W1	4.66	3.61	1.05	98.59%	100.00%
		W2	7.81	3.47	4.34	95.57%	92.50%
	SCB HB	R1	4.68	4.49	0.19	97.30%	92.14%
	RMB S	SBC1	46.80	44.30	2.50	100.00%	93.35%
	RMB N	SBC1	60.65	50.65	10.00	89.44%	89.44%
	RMB N	SBC2	25.77	23.42	2.35	96.08%	92.63%
Product 7	Sk7	Skein3-R	78.50	70.50	8.00	94.77%	94.77%
	Con7	WM3	117.80	117.58	0.22	100.00%	100.00%
	Bob7	SM4	245.48	241.00	4.48	100.00%	100.00%
	Rov7	RM2	25.82	23.79	2.03	98.47%	98.47%
	SCTP7	TP2	21.40	19.94	1.46	98.74%	97.87%
	SCSP7	SP2	17.49	13.91	3.58	100.00%	94.08%
	SCSelfR7	SelfR1	9.14	5.19	3.95	100.00%	84.57%
		SelfR2	10.11	6.98	3.13	100.00%	65.30%
	Top9	W1	5.79	4.25	1.54	90.38%	90.38%
		W2	6.12	3.45	2.67	92.45%	89.09%
		RW2	5.90	3.87	2.03	44.46%	40.42%
	SCDaB HB	R1	5.50	4.98	0.52	65.92%	63.94%
	RMB N	SBC1	60.65	50.65	10.00	89.44%	89.44%
		SBC2	25.77	23.42	2.35	96.08%	92.63%
	RMD S	SBC1	63.67	47.18	16.49	97.25%	97.25%
		SBC2	24.80	24.09	0.71	97.64%	97.64%
Product 11	Sk11	Skein1-L	37.40	28.60	8.80	86.98%	86.98%
		Skein3-L	32.60	22.80	9.80	94.77%	94.77%
	Con11	WM3	100.97	100.75	0.22	99.79%	99.79%
	Bob11	SM9	220.17	209.67	10.50	99.55%	99.55%
	Rov11	R1	16.36	11.92	4.44	97.86%	93.28%
	SCTP11	TP1	10.98	10.06	0.91	95.05%	95.05%
	SCSP11	SP1	11.18	9.80	1.38	96.63%	96.63%
	Top10	RW2	9.11	4.03	5.07	78.66%	78.66%
	RME S	SBC1	51.11	47.11	4.00	97.72%	97.72%
	RME N	SBC1	70.68	44.68	26.00	81.64%	63.53%
		SBC2	27.05	24.15	2.90	96.78%	83.88%
Product 20	Sk20	Skein4-I	38.60	29.60	9.00	91.90%	91.90%
		Skein5-R	38.20	29.00	9.20	100.00%	100.00%
		Skein6-I	41.00	31.25	9.75	97.04%	97.04%
	Con20	WM3	64.93	64.71	0.22	99.67%	99.67%
	Bob20	SM6	202.54	196.70	5.83	98.70%	98.70%
	Rov18	R1	16.73	13.08	3.65	97.76%	93.01%
		R2	21.81	21.00	0.81	98.74%	98.74%
	SCTP18	TP1	14.38	11.19	3.19	87.20%	87.20%
		TP2	20.59	19.86	0.73	98.84%	98.84%
	SCSP18	SP1	13.26	9.72	3.54	88.42%	84.05%
		SP2	15.68	13.96	1.72	100.00%	97.15%
	SCSelfR18	SelfR1	6.19	5.15	1.04	100.00%	100.00%
		SelfR2	7.79	7.01	0.78	100.00%	91.47%
	Top12	RW2	3.82	3.57	0.25	82.13%	80.17%
	RMB S	SBC1	46.80	44.30	2.50	100.00%	93.35%
	RMB N	SBC1	60.65	50.65	10.00	89.44%	89.44%
	RMB N	SBC2	25.77	23.42	2.35	96.08%	92.63%
Product 22	Con3	WM4	59.69	59.47	0.22	98.44%	98.44%
	Bob3	SM7	234.50	229.33	5.17	97.64%	97.64%
		SM11	216.12	211.08	5.04	100.00%	100.00%
	Rov3	R1	16.51	13.02	3.49	99.60%	98.78%
	SCTP3	TP1	13.04	10.00	3.04	99.49%	93.02%
	SCSP3	SP1	11.42	9.67	1.75	95.01%	95.01%
	SCSelfR3	SelfR1	7.05	5.16	1.89	100.00%	80.23%
	Top3	W1	5.64	3.67	1.97	97.30%	97.30%
		W2	5.22	2.92	2.30	97.69%	97.69%
	SCDB HB	W1	5.54	5.26	0.28	82.20%	82.20%
	RMDa N	SBC1	66.69	56.69	10.00	100.00%	100.00%
		SBC2	33.19	32.60	0.59	81.29%	79.33%
	RME S	SBC1	51.11	47.11	4.00	97.72%	97.72%
Product 36	Con36	WM3	139.22	139.00	0.22	99.48%	99.48%
	Bob36	SM3	364.99	360.33	4.66	100.00%	100.00%
	Rov36	RM2	15.93	15.09	0.84	91.90%	69.26%
	SCSelfR36	SelfR2	7.02	7.022		100.00%	100.00%
	Top20	BP1	5.47	5.47		100.00%	100.00%
	SC36	GC3	6.61	6.61		100.00%	100.00%
		GC2	10.06	10.06		100.00%	100.00%
		C4	124.77	124.77		100.00%	100.00%
		C3					
		C2					
		C1					
		GC1	10.21	9.52	0.69	100.00%	100.00%
		WC1	6.31	5.94	0.37	100.00%	100.00%

Annex 7 Capacities

Machine	Prod.	By-Product	Vel. [m/min]	Wick Weight [g/m]	Wick Weight [kg/m]	Shift [h]	Shift [min]	Real Efficiency	Adjusted Efficiency	Perfect Efficiency	Real Capacity	Theoretical Capacity	Real Machine Capacity	Theoretical Machine Capacity		
SBC1	2, 7, 20	RMB N	179.71	32.65	0.0327	11.50	690.00	89.44%	89.44%	95.00%	3,621.13	3,621.13	3,583.31	3,641.99		
	2, 20	RMB S	179.04	32.20	0.0322	11.50	690.00	93.35%	100.00%	95.00%	3,713.38	3,977.91			Efficiency Increase	8.61%
	22	RMDa N	178.71	30.30	0.0303	11.50	690.00	82.20%	82.20%	95.00%	3,070.99	3,070.99			Production in kilos increased	58.68
	7	RMD S	179.65	28.50	0.0285	11.50	690.00	97.25%	97.25%	95.00%	3,435.52	3,435.52			% increase in production	1.64%
	11	RME N	178.76	34.20	0.0342	11.50	690.00	97.72%	97.72%	95.00%	4,122.08	4,122.08				
	22	RME S	178.25	36.25	0.0363	11.50	690.00	79.33%	81.29%	95.00%	3,536.77	3,624.29				
SBC2	2, 7, 20	RMB N	250.00	28.30	0.0283	11.50	690.00	92.63%	96.08%	95.00%	4,521.96	4,690.59	3,852.97	4,052.84	Efficiency Increase	21.56%
	22	RMDa N	250.00	27.00	0.0270	11.50	690.00	100.00%	100.00%	95.00%	4,657.50	4,657.50			Production in kilos increased	199.87
	7	RMD S	250.00	23.86	0.0239	11.50	690.00	97.64%	97.64%	95.00%	4,018.63	4,018.63			% increase in production	5.19%
	11	RME N	250.00	20.20	0.0202	11.50	690.00	63.53%	81.64%	95.00%	2,213.80	2,844.65				
R1	2	SliverCanB HB	288.00	25.42	0.0254	11.50	690.00	92.14%	97.30%	95.00%	4,653.79	4,914.63	3,860.10	3,975.00	Efficiency Increase	7.14%
	7	SliverCanDaB HB	254.00	24.25	0.0243	11.50	690.00	63.94%	65.92%	95.00%	2,717.68	2,801.52			Production in kilos increased	114.89
	22	SliverCanDB HB	254.00	24.58	0.0246	11.50	690.00	97.69%	97.69%	95.00%	4,208.83	4,208.83			% increase in production	2.98%
W1	2	Top2	230.00	24.03	0.0240	11.50	690.00	100.00%	98.59%	95.00%	3,813.29	3,759.64	3,438.96	3,672.71	Efficiency Increase	18.36%
	7	Top9	230.00	23.98	0.0240	11.50	690.00	90.38%	90.38%	95.00%	3,439.21	3,439.21			Production in kilos increased	233.76
	22	Top3	230.00	24.07	0.0241	11.50	690.00	80.23%	100.00%	95.00%	3,064.36	3,819.29			% increase in production	6.80%
W2	2	Top2	234.67	24.03	0.0240	11.50	690.00	92.50%	95.57%	95.00%	3,598.74	3,718.33	3,622.97	3,706.72	Efficiency Increase	6.44%
	7	Top9	236.00	24.05	0.0241	11.50	690.00	89.09%	92.45%	95.00%	3,489.44	3,621.10			Production in kilos increased	83.75
	22	Top3	235.00	23.96	0.0240	11.50	690.00	97.30%	97.30%	95.00%	3,780.73	3,780.73			% increase in production	2.31%
RW2	7	Top9	251.00	25.07	0.0251	11.50	690.00	40.42%	44.46%	95.00%	1,754.98	1,930.22	2,851.67	3,022.32	Efficiency Increase	12.57%
	11	Top10	251.50	23.38	0.0234	11.50	690.00	78.66%	78.66%	95.00%	3,190.69	3,190.69			Production in kilos increased	170.66
	20	Top12	250.50	22.83	0.0228	11.50	690.00	91.47%	100.00%	95.00%	3,609.33	3,946.05			% increase in production	5.98%
Self1	2	SliverCanSelfR2	276.00	22.14	0.0221	11.50	690.00	100.00%	100.00%	95.00%	4,216.98	4,216.98	3,962.01	4,154.69	Efficiency Increase	18.28%
	7	SliverCanSelfR7	276.00	22.18	0.0222	11.50	690.00	84.57%	100.00%	95.00%	3,572.22	4,223.96			Production in kilos increased	192.68
	20	SliverCanSelfR18	276.00	21.92	0.0219	11.50	690.00	97.15%	100.00%	95.00%	4,055.94	4,174.92			% increase in production	4.86%
	22	SliverCanSelfR3	276.00	22.12	0.0221	11.50	690.00	95.01%	95.01%	95.00%	4,002.90	4,002.90				
SP1	2	SliverCanSP2	260.30	23.71	0.0237	11.50	690.00	87.45%	94.72%	95.00%	3,723.57	4,033.21	3,983.04	4,130.59	Efficiency Increase	13.75%
	11	SliverCanSP11	260.20	24.00	0.0240	11.50	690.00	96.63%	96.63%	95.00%	4,164.02	4,164.02			Production in kilos increased	147.55
	20	SliverCanSP18	261.30	22.52	0.0225	11.50	690.00	98.84%	98.84%	95.00%	4,013.10	4,013.10			% increase in production	3.66%
	22	SliverCanSP3	265.20	23.69	0.0237	11.50	690.00	93.02%	99.49%	95.00%	4,031.46	4,312.03				
TP1	2	SliverCanTP2	206.70	12.20	0.0122	11.50	690.00	98.63%	98.63%	95.00%	1,715.55	1,715.55	1,832.30	1,836.23	Efficiency Increase	0.81%
	11	SliverCanTP11	205.20	14.88	0.0149	11.50	690.00	95.05%	95.05%	95.00%	2,002.57	2,002.57			Production in kilos increased	3.93
	20	SliverCanTP18	206.30	12.12	0.0121	11.50	690.00	98.74%	98.74%	95.00%	1,703.47	1,703.47			% increase in production	0.22%
	22	SliverCanTP3	204.30	13.70	0.0137	11.50	690.00	98.78%	99.60%	95.00%	1,907.61	1,923.34				
RM1	2	Rov2	193.00	28.00	0.0280	11.50	690.00	98.29%	98.29%	95.00%	3,664.99	3,664.99	3,697.73	3,745.58	Efficiency Increase	4.58%
	11	Rov11	193.00	31.37	0.0314	11.50	690.00	93.28%	97.86%	95.00%	3,896.94	4,088.35			Production in kilos increased	47.85
	20	Rov18	193.00	28.00	0.0280	11.50	690.00	98.70%	98.70%	95.00%	3,680.30	3,680.30			% increase in production	1.33%
	22	Rov3	193.00	26.65	0.0266	11.50	690.00	100.00%	100.00%	95.00%	3,548.67	3,548.67				
SelfAuto2	7	SliverCanSelfR7	235.00	22.18	0.0222	11.50	690.00	65.30%	100.00%	95.00%	2,348.59	3,596.49	2,773.54	3,452.91	Efficiency Increase	57.34%
	20	SliverCanSelfR18	235.00	21.92	0.0219	11.50	690.00	100.00%	100.00%	95.00%	3,554.73	3,554.73			Production in kilos increased	679.36
	36	SliverCanSelfR36	235.00	21.52	0.0215	11.50	690.00	69.26%	91.90%	95.00%	2,417.32	3,207.51			% increase in production	24.49%
SP2	7	SliverCanSP7	235.00	20.20	0.0202	11.50	690.00	94.08%	100.00%	95.00%	3,080.91	3,274.78	3,075.08	3,251.69	Efficiency Increase	10.28%
	20	SliverCanSP18	235.00	22.52	0.0225	11.50	690.00	84.05%	88.42%	95.00%	3,069.24	3,228.60			Production in kilos increased	176.61
															% increase in production	5.74%
TP2	2	SliverCanTP2	270.60	12.20	0.0122	11.50	690.00	63.06%	88.38%	95.00%	1,435.90	2,012.46	1,763.10	1,960.77	Efficiency Increase	26.19%
	7	SliverCanTP7	271.00	10.09	0.0101	11.50	690.00	97.87%	98.74%	95.00%	1,845.59	1,862.03			Production in kilos increased	197.67
	20	SliverCanTP18	275.33	12.12	0.0121	11.50	690.00	87.20%	87.20%	95.00%	2,007.81	2,007.81			% increase in production	11.21%
RM2	2	Rov2	132.30	28.00	0.0280	11.50	690.00	60.29%	90.69%	95.00%	1,540.92	2,318.03	2,250.80	2,474.94	Efficiency Increase	35.16%
	7	Rov7	130.50	24.18	0.0242	11.50	690.00	98.47%	98.47%	95.00%	2,144.22	2,144.22			Production in kilos increased	224.14
	20	Rov18	130.10	28.00	0.0280	11.50	690.00	93.01%	97.76%	95.00%	2,337.82	2,457.27			% increase in production	9.96%
	36	Rov36	180.00	24.00	0.0240	11.50	690.00	100.00%	100.00%	95.00%	2,980.23	2,980.23				

Annex 8 Production Requirements Calculations for Each Product

Product	By-Product	Ok	dk	Ik
Product 2		2,360.43		
	Packing		0.00%	2,360.43
	Winding		0.00%	2,360.43
	Drying		0.00%	2,360.43
	Centrifuge		0.00%	2,360.43
	Dyeing		0.00%	2,360.43
	Skeins		10.73%	2,644.15
	Twisting		3.08%	2,728.18
	Winding		2.06%	2,785.56
	Spinning			
	Roving		0.31%	2,794.33
	Third Passage			
	Second Passage			
	Self Regulator			
	Winding		10.14%	3,109.65
	Stretch Breaking			
	Converter			

Product	By-Product	Ok	dk	Ik
Product 20		1,853.76		
	Packing		0.00%	1,853.76
	Winding		0.00%	1,853.76
	Drying		0.00%	1,853.76
	Centrifuge		0.00%	1,853.76
	Dyeing		0.00%	1,853.76
	Skeins		3.69%	1,924.78
	Twisting		0.86%	1,941.48
	Winding		3.08%	2,003.18
	Spinning		1.47%	2,033.06
	Roving			
	Third Passage		0.22%	2,037.55
	Second Passage			
	Self Regulator			
	Winding			
	Stretch Breaking		9.27%	2,245.72
	Converter			

Product	By-Product	Ok	dk	Ik
Product 22		2,453.18		
	Packaging		0.00%	2,453.18
	Twisting		4.14%	2,559.13
	Winding		1.70%	2,603.39
	Spinning		2.11%	2,659.51
	Roving			
	Third Passage		0.86%	2,682.66
	Second Passage			
	Self Regulator			
	Winding			
	Rebreaker		1.36%	2,719.64
	Stretch Breaking			
	Converter			

Product	By-Product	Ok	dk	Ik
Product 7		1,645.93		
	Packing		0.00%	1,645.93
	Winding		0.00%	1,645.93
	Drying		0.00%	1,645.93
	Centrifuge		0.00%	1,645.93
	Dyeing		0.00%	1,645.93
	Skeins		1.79%	1,675.93
	Twisting		11.23%	1,887.95
	Winding		2.53%	1,936.95
	Spinning			
	Roving		0.39%	1,944.54
	Third Passage			
	Second Passage			
	Self Regulator			
	Winding		1.58%	1,975.75
	Stretch Breaking			
	Converter			

Product	By-Product	Ok	dk	Ik
Product 11		1,661.84		
	Packing		0.00%	1,661.84
	Winding		0.00%	1,661.84
	Drying		0.00%	1,661.84
	Centrifuge		0.00%	1,661.84
	Dyeing		0.00%	1,661.84
	Skeins		0.65%	1,672.72
	Twisting		3.06%	1,725.52
	Bending		3.74%	1,792.56
	Winding		1.33%	1,816.72
	Spinning		1.25%	1,839.72
	Roving			
	Third Passage		0.25%	1,844.33
	Second Passage			
	Self Regulator			
	Winding		0.61%	1,855.65

Product	By-Product	Ok	dk	Ik
Product 36		1,393.42		
	Packaging		0.00%	1,393.42
	Twisting		0.95%	1,406.78
	Bending		0.00%	1,406.78
	Winding		1.71%	1,431.26
	Spinning		1.81%	1,457.64
	Roving			
	Third Passage		0.42%	1,463.79
	Second Passage			
	GC3			
	GC2			
	Combing		2.27%	1,497.79
	GC1			
	SR1			

References

Ambrose S (2016) Achieving S&OP success: how principles of team effectiveness can help. Foresight: Int J Appl Forecast (43):25–31

Benalcazar D (2010) Practical theory manual of acrylic spinning. Bachelor thesis. Universidad Técnica del Norte, Ecuador [In Spanish]

Cecere L (2015) What is the value proposition of sales and operations planning? Retrieved from http://supplychaininsights.com/what-is-the-value-proposition-of-sales-and-operations-planning/. Accessed on 16 June 2018

Chase R, Jacobs R, Aquilano N (2006) Operations management: production and supply chain, 12th edn. McGraw-Hill, México [In Spanish]

Druker P (2003) The management: tasks: responsibilities and practices. El Ateneo editorial, Buenos Aires, Argentina [In Spanish]

Duncan A (2013) The journal of business forecasting: S&OP and Strategy: building the bridge and making the process stick

Durlinger P (2012) Inventory and holding costs: a white paper approach for managers. Durlinger Consultancy

Ferreira G (2013) Sales and operations planning: design and implementation of S&OP process in a multinational company. M.sc thesis, Business Administration, At the Católica Lisbon School of Business and Economics [In Portuguese]

Koontz H, Weihrich H, Cannice M (2008) Administration: a global and business perspective, 4th edn. McGraw Hill, México [In Spanish]

Krajewski L, Malhotra M, Ritzman L (2008) Operations management. Processes and value chains, 8th edn. Pearson Education, Mexico [In Spanish]

Lapide L (2005) Sales and operations planning part III: a diagnostic model. J Bus Forecast 13–16 (Spring)

Lloret J (2014) Proposal to implement a production planning and control model in the company Isollanta Cía. Ltda. Bachelor degree thesis. Universidad Politécnica Salesiana de Cuenca, Ecuador [In Spanish]

Montgomery D (2013) Design and analysis of experiments, 8th edn. Wiley, New York, USA

Nahmias S (2007) Production and operations analysis, 6th edn. McGraw Hill, Mexico [In Spanish]

Niebel B, Freivalds A (2009) Industrial engineering: methods, standards and work design, 12th edn. McGraw Hill, Mexico [In Spanish]

Prokopets L (2012) S&OP: what can you learn from the top performers. Supply Chain Management Review. Retrieved from http://www.scmr.com/article/SOP_What_You_Can_Learn_from_the_Top_Performers. Accessed on 18 June 2018

REM (2010) Methdology of calculating inventory carrying costs. Retrieved from http://www.remassoc.com/portals/0/remprecc.pdf. Accessed on 20 May 2018

Ries E (2012) The 5 Whys. Harvard Business Review. Retrieved from https://hbr.org/2012/02/the-5-whys.html. Accessed on 12 Apr 2018

Stevenson W (2012) Operations management, 11th edn. McGraw Hill, New York, USA

Swaim J, Maloni M, Bower P, Mello J (2016) Antecedents to effective sales and operations planning. Ind Manage Data Syst 116(6):1279–1294. https://doi.org/10.1108/IMDS-11-2015-0461

Tavares A, Scavarda L, Fernandez N, Scavarda A (2011) Sales and operations planning and the firm performance. Int J Prod Perform Manage 61(4):59–381. https://doi.org/10.1108/17410401211212643

Tompkins J, White J, Bozer Y, Tanchoco JMA (2010) Facilities planning, 4th edn. Wiley, Edison, New Jersey, USA

Wagner S, Ullrich K, Transchel S (2013) The game plan for allying the organization. Bus Horiz 57 (2):189–201. https://doi.org/10.1016/j.bushor.2013.11.002

Wallace T, Stahl R (2004) Sales and operations planning: the how-to handbook, 3rd edn. Steelwedge Software, San Mateo, CA, USA

White O (2016) Transitioning from S&OP to IBP. Retrieved from Oliver White Paper Series http://georgepalmatier.com/white-papers/transition-sop-ibp-palmatier.pdf. Accessed on 16 June 2018

Chapter 10
Reflecting on Industrial Business Models: A History of Tradition, Challenges, and Potential Innovations

George Leal Jamil, Antonio Juan Brioñes-Peñalver and Domingo García-Perez de Lema

Abstract This chapter aims to contribute to evaluate market options for business models to be adopted by industrial agents. To achieve this object, a methodological approach for a critical observation was pursued. First, it was conducted a theoretical review, regarding the business model (BM) concept and related definitions, as its relationship with strategy, tactical, and operational level of a generic company, innovation, innovation management, structure, and components, among others. Then an analysis of experiences of practical business model adoption by economy actors, typical industrial arrangements or those related to industrial actions, productive chains or practices, were studied, to produce an overlook of its classical, traditional and influential aspects. Finally, in this analysis, reflections around these findings are presented, contributing for BM practice for industrial organizations and associated value-aggregation partners, as from service sectors, which, nowadays, even pressure industries to adopt or adapt to their business models.

Keywords Industry 4.0 · Business models · Industrial business models Innovation

G. L. Jamil (✉)
Informações em Rede Consultoria e Treinamento Ltda, Belo Horizonte, Brazil
e-mail: gljamil@gmail.com

A. J. Brioñes-Peñalver
Department of Economy and Business, Universidad Politecnica de Cartagena, Cartagena, Spain
e-mail: aj.briones@upct.es

D. García-Perez de Lema
Financial Economics and Accounting, Universidad Politécnica de Cartagena, Cartagena, Spain
e-mail: Domingo.garcia@upct.es

J. L. García Alcaraz et al. (eds.), *Best Practices in Manufacturing Processes*,
https://doi.org/10.1007/978-3-319-99190-0_10

10.1 Introduction

It is possible to understand the industrial sector as the most relevant proponent of business fundamentals along economic markets development history. Just considering the period of Industrial Revolution, 1760 to, at most, 1840 (Guile 1987), it is possible to understand a remarkable change not only in emergent productive theories, but also in influences in mankind's way of life. The structural changes posed by the transformation, when manual production, sometimes improvised and without rules was changed—by many ways—to new systems, seeking basic planning control and forecasts, issuing an expressive fact on how the conception of business models impacted lives this way. This was one event which, through the comprehension of BMs influences, can define a valid context for this discussion.

The development of industrial chains, their political and strategical influences were some of the undeniable factors to decide critical historical happenings, such as last phases of colonial era (the end of the cycle, started in the fifteenth century in the West), world wars, national association in blocks and conflicts, emerging of commercial and industrial powerful associations, among several other implications which delimited and influenced organizational positioning and effective works from centuries. These productive chains reached, sometimes, a level in which they become as powerful as some governments, producing influences for market and regulatory decisions.

Ranging from pre-Industrial revolution era, with manual handling, imprecise, unplanned and geographically restricted competition to modern complexes, where massive automation, new wave of robotization, application of emerging technologies, such as internet of things and its relation to data and knowledge management (KM), result in a multifaceted opportune context for research (Schumpeter 1942; Venkatraman 1989; Utterback 1996; Cano 2012).

Thinking as a provocation, present phenomena still pressure industrial connections to adopt and propose new solutions (Zalewska-Kurek et al. 2016; Wirtz and Daiser 2017) in business models. Sometimes, these demands are simply presented to industries, calling for a fast change in a traditional, value-set chain. In alternative events, industries must face options do strategically align to service-oriented designs which constitute platform BM that will attempt to answer massive "scalable" needs, like some startups propositions (Mullins 2014; Nielsen and Lund 2018).

Conceptually, business models (BM) are difficult definitions to restrict, observing the actual competitive context. This conceptualization intends to relate to various processes, methods, analysis and strategic to tactical positionings proposed during a larger period, a conceptual life cycle. This way, this concept needs to compose a base where it refers on how they produce their plans and, effectively, how it competes or offer its final value to customers or agents. This concept will be approached, from several points of view, in the theoretical background developed in the following, but it is initially taken as the design of organizational relations which states or report its ways of thinking and planning its action towards its customers and market actions, referring to intermediate-level and operational works, providing a whole comprehension on how one company plans and performs its proposed actions, resulting in a contextual alignment (Magretta 2002; Casadesus-Masanell and Ricart 2007, 2010; Osterwalder and Pigneur 2010; George and Bock 2011).

From modern cases, it is possible to identify that businessmen and investors, sponsors, and proponents must pay attention to their planning capabilities and resources spent for their firm success, specially composing and updating their business models. This happens when it is observed the apparent definition of some new standards, as the SaaS (platform) models, marketplace ("Uber-like") models, signature, metamodels and various market conceptions—(Mullins 2014; Ovans 2015)—which are facing the true competition for less than a decade, still in their conceptual "youth". These models, however, have an estimated huge number of applications and alternate practical instances, even reaching the possible level of becoming a standard proposition, or conceptual paradigms in near future (Massa and Tucci 2014).

While this dynamic competition emerges, supported by various other almost uncontrollable phenomena, such as globalization, world economy fast development and changes and, more noteworthy, impressive surge of new technologies, such as those from data analysis, information and knowledge processing and internet of things, business models became harder to define, plan and implement, leading them to become a more strategic component of sectors and derived strategies. In an apparent paradox, a more challenging paradigm turns to be a potential differentiation, unfolding in a source of competitive advantage (Porter 2008).

The industrial sector has shown an impressive path for the Economy, as it supported events of the human historical evolution, from the urban formation, national competition, towards today's technology application arena. This chapter aimed an analysis of BM influence, contribution and development in this scenario, aiming to reach their importance for actual competitive arrangements. Among the results of this exploratory, initial and superficial research, business models perceived in an informal market observation were detailed, from an analysis model, consolidated from the literature review. This allowed to consider these alternatives, some of them still risky positionings, being tested by entrepreneurs and practitioners, from a more structured point of view, producing the first level of outcomes in this study.

For this purpose, we started with the theoretical background approach, where basic concepts were discussed, exploring also their relationships. This initial development aimed to define a common understanding of their dynamic interactions with many concepts. In the conclusion of this part, together with an integrative, relational study for BM concept along with strategy and further business-oriented framework concepts, a basic analytical definition, that proposed a view of components was reached, was to be applied in an exploratory exam of practical cases. Finally, these discussions permit to formulate basic reflections around proposition and its practice, allowing to the development of further studies, taking this chapter as a basic, initial and contributive base.

10.2 Theoretical Background

This section aims to discuss some of main concepts, being business model the essential, and produce a demanded view of its relationships towards a practical understanding. Based on this development, it is possible to observe, along with the criteria adopted for the contextual focus, or an exploratory analytical model, to be applied in the real cases analysis, at the ending remarks.

10.2.1 Business Models

The discussion around business models (BM) definition is endless and, usually, tends to follow cultural and sectorial delimitations. Magretta (2002) presented an attempt for conceptual study and definition, in a period which strategy and strategic planning were regarded by several authors and managers as old structures for organizational thinking, related to exhausted competitive arenas. At that time, in a controversial approach, some stated that corporative conception should reside in "well designed business models for the Internet age".

As that market trend proved, dramatically, with disastrous commercial, financial and strategic outcomes, business models, taken as a structural ordering, are essential, although not infallible, way to project and plan aligned processes. Analyzing cases, observing it from critical points of view and referring to authors, Magretta stated that business models relate, basically, on how a company makes money, understanding and fulfilling customer's expectations.

This way, she divided her understanding about BMs in two main aspects: (1) How companies produce something and (2) How they offer what they produce. Interestingly, this author also perceives how the introduction of information technology and its associated resources enabled strict guidance of alignment, resulting in potential optimization, but also posing excessive control and bureaucracy. Joan Magretta also pointed out the importance of BMs as "stories that explain how companies work". This remark leads us to demands on comprehend what was proposed, what was done and what was the consequence of the whole interaction on the market, a dynamic and concise functioning that will require strategic thinking (Porter 2008; Porter and Magretta 2014).

Johnson et al. (2008) defined an observation about components that oriented various studies and experiments throughout following years, regarding the topic itself and its relationships with innovation and its management processes. For these authors, main components for business models conceptions are: (a) Customer value propositions—consolidating future customer-oriented strategic positioning; (b) Profit formulas—which set potential directives and controls for financial administration, and (c) Key resources and processes—as a dynamic and systemic approach to analyze and value resources and processes that can be strategically positioned, through deliberated and/or reactive actions. This last item is to propose a correspondence with Penrose (1959) studies that relate resources, examined through a Resource-Based View (RBV) analysis, to a potential sustainable competitive advantage definition and practice.

For Casadesus-Masanell and Ricart (2010), business models are the reference to the "logic" of the firm, how it operates towards value creation for customers. Authors introduce business model as a consolidating thought produced at the strategic thinking phase, which will enlighten how corresponding tactics and operational efforts will be planned, serving both as an orientation and contextual definition for effective strategy implementation and monitoring. These authors, adopting a simple analogy of different models of car conceptions, production and operation, exemplified how the same type of organization can perform what was expected by its final customers, when (a) Components were known; (b) Context of application was known; (c) An overall coordination existed—which can be considered the corporative strategy.

As an additional, opportune reflection, they also sought to separate strategy from business models (although, as it was pointed out by them, it is difficult). Strategy, for them, "is a contingent plan of action as to what business model to use", developing an understanding that BM composition and the related choices on their adoption ("changing the car parts") are details from strategic decisions.

Almost all authors presented so far report to Peter Drucker's former work (Drucker 1993a, b), where this researcher, without referring specifically to BM concept, affirmed that a design should be composed to answer basic corporative questions, related to customer knowledge, their perception of offered value, value's relation to resources and, finally, correspondent costs. His definition for BMs, as "stories that explain how organizations work", can be broadly explored, for example, to understand graphic tools (for example, the Business model canvas, proposed by Alex Osterwalder and his research team) as means of building histories, materializing decisions and corresponding alternatives taken, among several other outcomes.

It can also be related to knowledge management principles, when we observe how they correspond to companies' evolution, like innovation alternatives (Nonaka and Takeuchi 1995). Opportunely, these points are to be addressed in the following development, because they are targets of several emerging technological applications, change, by this way of thinking, some fundamentals of concept reasoning (Grönlund et al. 2010; Edgett 2015; Christensen et al. 2016; Gatautis 2017).

Usually, works and researches try to produce an understanding of the concept, searching the relation of factors, such as: Customer behavior comprehension; Market segmentation; Value and Costs to offer; Decision-making and Market monitoring (Henderson and Clark 1990; Deshler and Smith 2011; Kim and Mauborgne 2015; Ovans 2015; Anunciação and Peñalver 2017). Attempts vary from diagrams propositions, textual definition for processes, relationship maps—specially to strategic planning processes—among alternative choices. The comprehension of a model, however, leads to a conceptual issue, defined by Silva (2011), when he argues about the real role for a model in the development of systems and implementable solutions. A model, for this author, is a conceptual, high-level outcome which must be, thereafter, adopted as a base for the real application, serving as the conceptual base, a wider theoretical reference, which can provide the contextual delimitation for real solution proposition. His thoughts can be validated, when we assess Duin et al. (2013) and Singleton-Green (2014), in which authors state that a business model is a "simplified version of the reality". It is an assumption that can be considered for this conceptual production, as BMs are far from a consensual convergence, especially when addressing industrial evolution.

One of these successful tools was the "canvas" diagnosis and hypothesis proposition stated by Osterwalder and Pigneur (2010)—the business model canvas—which synthetized nine mains aspects for one effective model. The application of this easy-to-use method for planning cycles is a significant perception of the "model concept", defined in the previous discussion. It is an important supporting tool to adjust competitive alignment through further work, building its plans and related processes towards its coordinated and monitorable operation. Confirming this assumption, Ovans (2015) calls attention to the role of the BM canvas as a "hypothesis generator" for firms proposition and actions.

For Casadesus-Masanell and Ricart (2010), we have that business models' components are divided in two major sets: "(1) The concrete choices made by administration about how the organization will operate and, (2) The consequences of these choices". Again, one initially naïve affirmation can lead to understand and explore several interactions among critical fundamentals like corporative structure, strategic planning resources availability, technological learning and application, culture, social relationships, knowledge management, among others. These definitions enlighten the complexity and power of business models, to be applied not only in corporative conceptions but, complementarily, as strategic alternatives that can propose competitive advantage (Ries 2011; Mullins 2014).

Structuring these conceptions into a concise definition, essential points can be determined for the business model conceptualization that will be used for the remaining of this chapter:

- Value proposition is to be negotiated with the market. This element, frequently, overshadows the remaining ones, deserving special attention on its analysis and delimitation. It has a potential relationship which covers from strategic decisions and plans to tactical definitions, such as strategic marketing planning or financial project validation, from many others.
- It is a description of practical ideas to achieve strategy fundamentals: structure, intangible (financial, informational, patents and licensing, etc.) and tangible (buildings, vehicles, laboratories, and information technology computer platforms) resources mapping and associated classification.
- Internal processes, methods, and aspects definition are likely to be derived from the strategic planning process, taken as the critical and fundamental assessment, that will identify methods and techniques adopted by the organization for customer perception of its performance. (These two last points can be combined, eventually, in one aspect focusing the internal environment projection).

- Basic mapping of external competitive environment, pointing and classifying competitors, market regulators and agents, customers, potential partnerships, among other.
- Key indicators (KI) classification, hierarchy and definition, which enable basic controlling and monitoring of each strategic goal and action executed to achieve that goal.

Dynamics of business models exercise in the market arise from Teece (2009) and Teece and Linden (2017), in an approach that permits a historical review, as David Teece, a remarkable author, evolved his appreciation around this topic. As it was previously stated, BMs were defined to produce a view of operational work and its integration, aligned to value offer, in a coherent development from the thoughts of Peter Drucker. In an updated study, both authors approached the hypothesis related to value creation, retention, share and protection, advancing conceptions to detail both for customer-related knowledge and operational market actions in the new scenario of emerging technologies, and their implication to new strategic marketing actions.

Two significant research and practice trends come from the conceptual application: (1) Study its application for sectors, markets, regions and national systems; (2) Understand the intended "business model innovation" context, somewhat derived from important and disseminated studies, such as the "Blue Ocean strategy" definition, remarkably composed, proposed and studied by Kim and Mauborgne (2015). These will be addressed in the following sections, as some concepts must be evaluated before, composing an intended situation for its improved discussion.

10.2.2 Strategy, Strategic Planning, Alignment and Business Models

Strategy can be considered as the basic formulation for corporative future, designed with the help of systems, researches and accumulated knowledge, providing conditions for offer positioning to final customers with the goal of competitive survival and development (Penrose 1959; Porter 2008; Porter and Magretta 2014).

Strategy can be understood as materialized through plans or orientation that explicit goals to be achieved in one determined period, together with the coordination for its corporative branch tactics and operations. It will, in a more traditional way of thinking, be formulated, developed, projected, implemented and modeled through the process of strategic planning, associating it to the specifications and coordination for organizational middle-level detail and its final alignment to the operational tasks (Mintzberg et al. 2008; Porter 2008; Hitt et al. 2011). Modern views also approach the continuous learning and emerging market signals as ways to adjust, promote, change, and update strategies dynamically (Mintzberg et al. 2008), contextualizing strategy in several levels, and analyzing how it can be detailed in planning actions, in a more flexible, fast, and adaptive way.

The relationship between strategy and business models has been theme of several researches. For Casadesus-Masanell and Ricart (2007, 2010), in two remarkable and related works, business models evolved in the last years, being expressively influenced by information technology, although not restricted to its usage to be developed and implemented. As cited in their work, pressures, as the adoption of "e-business", identified as services based on internet infrastructures and interfaces, and the wave of "it must be innovative—an innovation", took several firms to an imprecise path of strategic—tactical—operational alignment. This fact led to implementation of new rules, processes, methods, and connections to partners, suppliers, and customers, developing unstable alternatives to compete in their markets. The development from strategy already shows a potential bi-directional motivation, where BMs are taken into consideration as planning resources to develop and execute strategic movements and configuration, but are also strategic resources by itself (Chesbrough 2010).

A competitive, differential model can be a strategic item to be positioned, searching for a position when it will achieve a competitive advantage in the market. The cyclic dynamism of a business model and strategy can be verified when these authors comment about the "aggregation" as a method to understand a "group of strategic choices" to relate to one "consequence", aligning with the following approach of "decomposability" criteria, addressing (strategically, indeed) components for competitive advantage positioning, in a typical relationship between these two important topics. This observation is completely confirmed by the "virtuous cycle" concept, practiced by these authors on analyzing market cases, such as Ryanair, presented in the same study, considering the dynamics of strategic planning process.

These findings can also be found in the works built around resource-based view of firms, when authors such as Penrose (1959) and Wernerfelt (1984) sought to discuss, study, and define relations to evaluate corporative resources through metrics which could identify their potential usage to build a sustainable competitive advantage. As pointed out by Barney (1991), a criterion named VRIN was proposed to identify the Value, Rarity, Inimitability and "Non-substitutive" grades of each corporative resource, aiming to value it for a potential sustainable competitive advantage positioning.

Business model components, like human resources management, abilities on developing and implementing information systems and processes (not restricted to HR policies, but also relating to IS methods and analysis itself), knowledge generation processes, specialized machinery, communication, are always cited in their original studies and in the extensive post-publication thread. This discussion generated a productive approach regarding strategic planning process, its focus on resources and its market statement and, finally, dynamics regarding the continuous cycle of monitoring and executing strategies through levels alignment.

Interestingly, HBSP (2006) also validated this way of thinking, identifying basic business model components such as (a) Revenue resources; (b) Cost drivers; (c) Investment size and (d) Critical success factors. Except for the last one, considered too wide and complex, the remaining can be regarded as points of strategically aligned operation, executed according to corporative strategy. This affirmation can be also found when authors declare organizational strategy as the complete analysis for planning and operation in a specific market, explaining their view for this alignment.

From this starting point, it is opportune to recall Johnson et al. (2008) study, when these authors propose a definition previously discussed for business models, typically adopting strategic elements: (a) Value positioning (Customer value proposition)—which can be considered, for example, one essential component of a possible "Blue Ocean" strategy definition; (b) Profit formulas—like the tactic-to-operation definition of financial control and implementation, taken as a fundamental quantitative report; (c) Key resources mapping—resources identification (for example, the application of VRIN criterion, presented above) and, finally, (d) Key processes definition and implementation—which involves structuration, coordination, and operation methods and goals setting disciplines. This strategy—business model evaluation confirms, details, and validates this relationship.

Eyring et al. (2011) analyzed the opportunity to address and develop a BM for middle markets, presenting a typical and generic strategic decision to define and negotiate with the focus for a well-defined segment. In this work, authors also approach its update, studying the demand to expose a competitive and dynamic strategic value and, finally, compose items, such corporative structure (redefined), that enables this company to jump from one choice to another, for example, in a reaction to the competitor's movement. This is a contribution that leads to the trend "Business model innovation" to be discussed in the following section.

In what can be considered a closing scenario for this objective development, Sachsenhofer (2016) described a context where business models represent the coherent design of various components, that can be addressed strategically. He approached, from works like Chesbrough (2010) and Zott et al. (2011), the constitution of a business model and how these components can be managed strategically. It was analyzed how various of these elements, cited in the above development can be integrated and corresponding, forming the desired business model.

Among these studied components are: technology, funding sources, knowledge management, value-offer techniques, administrative methods (Marketing strategies, for example), profit sources, customers negotiation instances—market channels, for instance (Lee and Ho 2010; Anunciação and Peñalver 2017).

Nielsen and Lund (2018) analyzed how BMs can constitute into a scalable factor. This study observes the modern competitive era, when they can be implemented, changed ("pivoted") from managerial and technological constituents, also integrating corporative networks. Factors announced by authors, relating distribution channels, improving abilities around scarce resources, sharing investments requirements and operative tasks and, finally, the adoption of the "platform" concept—which is one of the promises about innovative integration, propelled by information technology—show how intricated, appealing and motivating this theme is considered today.

Clearly, industrial complexes, with its huge economic and social importance, compose a significant part on this arena, demanding review of its strategic fundamentals, to understand how corporative arrays, networking, production and commercial efforts sharing and several trends are to be applied and implemented with the dynamics with business models (Castells 1996; Slávik and Hanák 2017).

10.2.3 Innovation and Business Model Innovation

Innovation became one organizational image for almost all competitive companies in any market that try to keep a modern characteristic for its customers and agents. However, increasingly, it left the advertising trend to become one strong strategic factor for competitive survival. To understand its propositions and possible types, it is opportune to assess the seminal works from Joseph Schumpeter—Schumpeter (1942)—and James Utterback—Utterback (1971)—which can be considered consolidated in the conceptualization proposed by OECD (2005).

In this valuable, historical path, the theoretical background around innovation concept resulted, first, in a practical differentiation from inventions. Inventions are unstable, unfinished, incomplete versions or propositions of a socioeconomic market solution, which was originally designed to answer one final-customer problem (Henderson and Clark 1990). Innovations, on the other hand, are practical solutions that propose, through a service and/or product project, to solve one practical problem, with the insertion of newness. It happens, at least, with a starting version which presents this initial new offer, called by Ries (2011) the "minimum viable product" (MVP).

From those references cited, six types of innovations can be defined:

- *Product*: new or updated technological device or implement, a physical, tangible asset.
- *Process*: new or updated method to do something, to conduct or make a service possible, a new way to propose, plan, and implement the execution of a practical task.
- New supply or material to be applied to assemble, produce, or support an assembly.
- New market to be created or defined by one innovative position.
- New structure, when a company changes its way of serving and negotiating with external agents and customers and
- New strategic positioning, from market competitive pressures, which demand, sometimes, continuous innovative depiction, as it occurs in sectors which are strictly or predominantly based on technology.

These types of innovation are to be considered along with a classification around their degree, referring to the sources cited, analyzing how two fundamental dimensions—business model and technology—were proposed to be modified. This way, it is possible to define:

- *Incremental innovations*: for those which implement small modifications and/or technology in a new offer from the original proposition.
- *Semi-radical innovations*: when an expressive change in one of these two dimensions is provided, compared to the other. This way, one new positioning of a service which changes expressively its associated BM with few or no changes in technology is to be considered a semi-radical of business model. Alternatively, one new product which results in a new device, format, conception, among tangible characteristics (not exclusively), can be considered an innovation semi-radical of technology.
- *Radical*: when it is possible to identify that both dimensions were expressively changed in a new offer, compared to the older provision.

With these refinements in the innovation concept—basic definitions, types, and degrees—it is possible to advance on initial analysis of roles for innovative efforts and how innovation of BM is one of the most attractive actual topics in this sector's literature and for practitioners, investors, and economy agents.

Business models, by itself, are source of one perceived dimension of innovation (Bashir and Verma 2017). When a new way to produce, distribute, share, store, commercialize, negotiate, and keep one offer in one market, it is possible to understand BM function for innovation (Bereznoi 2014; Khanaga et al. 2014). For example, when films entertainment options are available through catalog-supported streaming services, such as Netflix and others, we understand that a new one, practiced with customer association methods (capture, retention, etc.) and information technology support, a new set of choices emerged and is adopted for one's lifestyle. This is a change for movies and TV shows, series watching, a remarkable market case.

From this conceptual base, it is also possible to understand how components of business models, discussed in the previous subsection, can become factors for different types of innovation. A new method to be approached by one company, in its relationship with its chain, can become a process-like innovation (Kavadias et al. 2016). Changes in a product or service, which can be an offer to new customers and markets, can potentialize innovation of market, even creating new segments and profiles to be analyzed by strategic marketing studies and implementations. Process and market innovation, by themselves, are a potential change for a design, which, finally, is one innovation based on its planning conception and practice. As it is possible to understand these concepts as unrestricted, not exclusive basics, this intricacy can lead to observe definition types as complementary and even cooperative.

Christensen et al. (2016) studied the evolution of BMs in "stages" that constitute a matter of learning and, unfortunately, a factor that poses difficulties to change over time. These authors, after a reflection based on methodological analysis, understood that "Business models, by their nature, are not to change, and they become less flexible and more resistant as they develop over time". These findings were compared with the "stages" model, in the same reference are composed by following elements: (a) Value creation; (b) Conceptual fundamentals; (c) Sustainability and, finally, (d) Efficiency.

This set of elements reinforces some of the findings for business model concept, innovation and its relation, pursuing a way where managerial maturity play a remarkable role. Interestingly, this approach recalls findings from Mintzberg et al. (2008), when they considered the composition of deliberated and emerging strategies, as one organization must comprehend its strategic coordination process not like a behavior, but a continuous process of high-level analysis learning, motivating to relate the managerial maturity to innovation abilities and capacities comprehension and overall coordination.

In the conclusions of their work, Christensen et al. (2016) report about conditions to define business models that address competitively new market opportunities: real opportunity analysis (analytical knowledge), conceptual independence of the new to the previous one and effective partnerships or networks. From Pfeifer et al. (2017), it is possible to understand how new markets and alternatives provided by new propositions show these evidences, also enabling to understand a persistent factor when assessing BMs, strategy and innovation demands.

Wirtz and Daiser (2017) argued about the potentials of the association between business model innovation and strategic planning, although observing the lack of sufficient theoretical formulation regarding this important base, offering an integrative framework for its future exploration in practical conception processes which can constitute a strategic asset for managers. Their integrative framework is based on: (a) macro (external) factors; (b) micro (internal) factors and (c) areas definitions with its possible KM processes. Analyzing these three components, it is possible to perceive one way where an applied, practiced business model can be changed, eventually producing an innovative arena, as a competitive BM innovation.

These studies illustrate, over various years, trends and thoughts from different points of view—emerging technologies adopted by final customers, market-pull and technology-push innovations, demanded strategic changes, external and internal causes and resources management aspects, etc—produced a pressured context, which eventually became a strong pressure for BM innovation.

From the sources analyzed above, two important evidences must be added for this study:

- Strategic relationship and importance (Chesbrough 2003; McAfee and Brynjolfsson 2012; Khanaga et al. 2014; Kavadias et al. 2016; Wirtz and Daiser 2017). This approach relates how innovation can be positioned strategically, producing a two-way strategy—business model.
- How the contextual interaction between innovation and business models structure and pressures for changes in components and its consolidation into an effective model occur at the strategic level (Davila et al. 2006; Deshler and Smith 2011; Diaconu 2011; El-Bashir et al. 2011; Kavadias et al. 2016; Gatautis 2017; Wirtz and Daiser 2017).

Reaching this level of understanding, it is opportune to evaluate the technological factor on new pressures and propositions, aiming to discuss its integration to result in the "Industry 4.0" market trend, that will be studied in the following subsection.

10.2.4 Emerging Technology Trends and Industry 4.0 Concept

This subsection does not intend to promote a deep debate of emerging technologies, otherwise permitting enough support to define a level of perception regarding Industry 4.0 perspective, a strong sectorial tendency. Obviously, as it happens with any market trend, the endless debate promoted now is controversial and even sometimes conflicting, when authors, researchers and practitioners still argue about Industry 4.0 real evolution, conceptual distortion pressured by commercial competition, innovation potential, social and economic implications, among many influences.

A brief overlook of industrial evolution shows that technology development and its related adoption are fundamental factors for remarkable historical events. As industrial complexes still hold a significant part of world's economy as integrators and final producers of end-user solutions, its development reaches levels of social, demographic and economic impacts for societies (Drucker 1993a, b).

Nowadays, with the usual evolution of tangible and intangible assets, it is expected that several ways of consecutive and accumulative scientific knowledge, usually in a random fashion applied in devices, software and market-oriented solutions, reach people's lives, an amazing succession of new products and services. As illustrated in the previous subsection, this is a situation where innovations can be announced, keeping the unstable and challenging relation with corporative strategies, in which business models play an expressive role on implementing controls, methods, processes, and structures for organizations to keep their competitive presence in markets. With the definitions studied so far, it is possible to understand that BMs serve as strategic partners to address strategies coherently to operational and implementation levels. Some of these trends are to be addressed in the following, aiming to promote a basic level of understanding around technological potential application for Industry 4.0 conceptualization, which is the convergent point for this subsection.

Data generation became one of the most massive processes in human living (McAfee and Brynjolfsson 2012; Kavadias et al. 2016; Pfeifer et al. 2017). With this expressive content continuously generated, it is comprehensible that some companies successfully start to offer analysis and additional informational products and tools for final customers, including companies which apply these analyses for strategic and corresponding tactical planning (Yonce et al. 2017). In this arena, two efforts affirmed in the last years: Big data and Analytics.

Big data is understood, basically, as the potential on producing knowledge from structured data and associated unstructured sources, aiming to compose scenarios and simulations that is an advanced comprehension regarding real phenomena and its combined implications (Chen et al. 2012; Mahrt and Scharkow 2013; El-Gayar and Timsina 2014). Its potential is shown by fast adaptations in their customer-oriented sale, applying, for example, automated internet interfaces, new design and final presentation, among others. These offers are taking into consideration factors like instantaneous competition, comprehensive changes of customer commercial interests ("humors" and "behaviors"), bringing a somewhat chaotic dynamic on final transactions with clients (Kartik et al. 2017), when market negotiations change rapidly. Data production is regarded as an elementary process for knowledge management efforts, improving conditions for innovation successful positioning (Jamil and Magalhães 2015; Jamil and Silva 2016).

Additionally, "Analytics" are related to the application of specific algorithms to generate knowledge from information, strictly applied to online transactions—such as productive systems monitoring, electronic commerce retailing or internet search engines usage—for fast understanding about customer reactions (Chen et al. 2012; SAS 2017; Yonce et al. 2017; Jamil et al. 2018). Analytics services are, sometimes, offered as open, easy-to-use services, reaching to complex, deep sense-formulation that can be applied for strategic and tactical immediate decisions, typically describing and pointing potential end-user reactions to mobile and web-based interactions (SAS 2017; Yonce et al. 2017). It results in a turbulent, challenging and fast-changing situation to propose innovations, both in the business model or in products and services.

One of the most referential emerging technologies that can produce immediate effect on industrial capabilities, resources, and systems is "Internet of things" (IoT). As presented by Chui et al. (2010), this technological composition integrates devices and its integrated services through Internet addresses, provided by Internet protocol connections. With this simple approach, a new availability of apparatus, platforms and a level of scheduling, intelligent and autonomous operation and, moreover, inter-device communication with programmable level of decision, arouse in the last years, leading to market solutions which are now definitely being absorbed by customers. Clearly, it is possible to include, in this customer base, all industrial complexes that can apply IoT to implement robots, auto-driven vehicles, machines with adaptation techniques ("decision capabilities") and inter-communication.

IoT is a new context where projects can be implemented, producing different production schemes, where human operators can be replaced by machines, in a new wave of operational tasks substitution, incorporated with a level of inter-relationship still to be effectively understood by managers and implementors (Gatautis 2017; Jamil et al. 2018). It is possible to argue that basic business model components, related to operational efficiency, alternatives, and tactical integration are immediately affected by the introduction of IoT, signaling a potential change that will propose an alternative for industrial arrangements and complexes. Concluding this brief overlook regarding IoT, it is opportune to affirm that its costs of implementation fell, in a period of few years, to small amounts of its original values, enabling not only its standalone usage, but also its implementation in domestic devices, such as TV sets, air conditioning devices and smartphones.

At this point, oriented mainly by market efforts and facts, a new version of automated industrial integration, called "Industry 4.0" emerged, announced as the implementation of automated machinery, Internet of things, consistent flow with external and internal corporative decision contexts (promoted by the usage of analytics and big data), association with distributed processing and infrastructural and supporting elements. In the focus of this chapter, the specific attention is not to promote a deep comprehension on how these solutions are composed.

Otherwise, potential changes in the business models and the emerging new ones, which can be elaborated, designed and implemented are facts that will deserve better attention (Turban et al. 2002; Teece and Linden 2017). This theoretical background discussion was initially promoted to appreciate how these technological context and economy demands motivate the real implementation and their potential innovation. Along with strong market movements, sectorial leadership definitions and competitive actions, there is a remarkable market pressure to innovate and business models are, as developed above, tools and ways to promote this expected level of innovation.

This way, it is the final proposition of this theoretical review, to understand that BMs have a straight and dynamic relationship with organizational strategy and they can also be favored by the introduction of actual advances, producing, potentially, conditions to innovate. It is not a stable, definite level of conceptual base, as it can offer to managers and designers new contextual perspectives in which they have conditions to propose new business models, which, by their way, produce strategic alternatives for value positioning.

10.2.5 Supporting Concepts

This subsection closes the planned theoretical review, presenting a base of additional, supporting concepts that will enable better comprehension on aspects of business model elements and their application.

Organizational structure, according to George and Bock (2011) and Jamil and Magalhães (2015), are formal or intuitive perceptions, describing personal and functional delegations. From previous studies, it is possible to perceive the continuous presence of hierarchical structure, describing controlling and commanding levels, also defining instances for communications and decision-making. It is possible also to understand characteristics of new propositions, like the matrix or process-oriented details, usually driven by external factors—like customer behavior changes, market opportunities, functional redistribution, regulatory pressures—orient their alignment to optimize all its internal processes and interventions, aiming to maximize the performance and customization levels.

Another remarkable influence over structure details is originated by project management works and researches, which proposed a definition that can be comprehended as a fusion of hierarchical and process-oriented configurations, called "projectized" structure. Centered in the personal figure of the project manager—institutionalized by the efforts of Project Management Institute PMBoK (2013)—this structure is based on his professional interventions in project planning, execution and coordination, structuring his leadership among groups, combining a hierarchy and matrix-oriented works.

Information and knowledge governing are conceptualized as processes designed to allow the potential evolutive knowledge generation from data (Tuomi 2000; Jamil and Magalhães 2015; Jamil and Silva 2016). In the context of this study, information and knowledge processes can be defined as resources used to propose new business models, supply market and corporative knowledge to build new value-offer positionings. Among these cases, it is possible to identify innovation conception in the core of a possible BM, as those adopted by companies which deal with "data analysis" and "data science" markets, with great evidence nowadays in sectors like market research, entertainment, informational services and communication (Akbar 2003; Setia et al. 2013).

Finally, the strategic alignment can be understood as the contextual fit, from the highest conceptual level of one specification, to its operational services and production, on how a strategy is conceived, planned, executed and followed, enabling actions like optimization, risk management, fast and consistent market movements (Chesbrough 2003; Porter 2008; Casadesus-Masanell and Ricart 2010; Massa and Tucci 2014; Christensen et al. 2016).

This review aimed to detail a fair level of coherence among a complex network of concepts and aspects that will allow the intended analysis for business models strategic planning and performance, to be developed in the remaining, to the intended context of reflections.

10.3 Examining Business Model Propositions for Industries

An initial attempt to evaluate the immense availability of approach from literature unavoidably led to works that observe what and how they were planned and what are the outcome, from one basic focus point, taken, for example, from defined BM components (Osterwalder and Pigneur 2010; Eisenmann 2014; Ali Mahdi et al. 2015).

This way, the orientation followed by this study is to develop an analysis, which will attain on components, as defined in the previous theoretical background, and, in this section, examine classical, innovative (theoretical), and implemented business models studied, inherited (as cultural traditional influences) and practiced by industry leaders and competitors. In the following section, trends or perspectives for each model are analyzed, proposing a development that can unfold in various themes for future theoretical and empirical studies.

BM components adopted for this study, from the conceptualization worked before, are, basically:

- *Value Proposition (VP)*: The context to be negotiated, offered to final customers and its perception.
- *Internal structure proposition (IS)*: The definition of elements, managed with high degree of independence by a company, from their managerial decisions.
- *External structure configuration (ES)*: Description of channels, distributors, partners, networked cooperative elements and other agents who help planning and positioning implementations.
- *Key indicators (KI)*: Factors, predominantly quantitative (costs, operational numbers, times measured, etc.) or that result in scales or relative measurement (customer satisfaction, quality) which can be collected, checked and verified to be compared, a comprehension on how goals and planned checkpoints are to be reached.

Some typical occurrences for each one of these components are (based on the sources already presented in this literature review):

- *VP*: Customer-oriented processes; Product and services lines and its supervision (i.e., How it is possible to keep your offer valued by customers); value-support channels (distribution, storage, logistics, etc.) planning; information technology ("digital" or "digital transformation") support for customer optimal negotiations; high-level processes of identifying and classifying customers and oriented processes; innovation handling principles and some others.
- *IS*: organizational structure; personnel profile details; human resources management capabilities; decision-making processes; tangible infrastructure (buildings, accesses, laboratories, workspaces, communication systems, conventional machinery; automated or robotic infrastructure, etc.), linked with internal intangible support, as motivational activities, satisfaction level of employees and working personnel, for example.
- *ES*: integrated distribution systems, like logistics planning techniques along with information technology support; transportation and storage systems, relationship with suppliers, distributors and commercialization channels; components of a distributed or shared processing system (as integrated automatized industrial plants which can be dynamically coordinated and managed with human or automated intervention), etc.
- *KI*: costs of production (partial and final activities), offer, transportation, storage, moving, ordering, placement, negotiation, volumes of supply and factors sampled for costs calculation; productivity forecast and compared final values; time-related variables—physical processes times, intervals, operational, transportation; values of negotiation; customer satisfaction/rejection levels; channel performances, costs and productivity, as examples.

Recalling that this chapter is proposed for reflections around the main subjects and with an exploratory approach, these definitions stated for a starting analytical approach, can help identify and classify business model experiences and cases, to be discussed in the following. Both, this briefly defined base and its analysis can justify and provide demands for further studies which can validate, review, detail, apply for further detail.

10.4 An Overview of Practical Business Models: Cases and Analysis

From the categorization defined above, it is possible to produce the intended reflections of some business models, collected from the literature (see Table 10.1) and from practical cases, observing it as the design proposed, adopted or, simply, intuitively implemented by some industrial agents—factories, complexes, supporting facilities, associations and other members or components of these chain arrangements. For each case listed in the following, those VP, IS, ES, and KI aspects are to be elaborated to reach those intended reflections around business models.

It is important to recall from studies that observed how BM concept was approached in the literature, reported views regarding the original focus, methodological approach aid considerations regarding the objectives and results of these

Table 10.1 Collected cases from the literature review

Case	Literature source
Disney Studios	Magretta (2002)
Apple and Tata Motors	Johnson et al. (2008)
Dell, E-Bay, Amazon	Osterwalder and Pigneur (2010)
Ryanair	Casadesus-Masanell and Ricart (2010)
TDC and Telmore	Casadesus-Masanell and Ricart (2010)
Chunghwa Telec. Company	Lee and Ho (2010)
Godrej and Boyce	Eyring et al. (2011)
Tata Motors	Sako (2012)
Nespresso	Matzler et al. (2013)
ICBC (Intl. Commerce Bank of China) and South Africa Standard Bank	Deloitte and Huawei (2015)
BMW (car and parts manufacturing)	Sachsenhofer (2016)
Boeing	Christensen et al. (2016)
Spain and Portugal tourism initiatives	Anunciação and Peñalver (2017)
Intuit	Colvin (2017)
Uber	Casumano (2018)

studies, reaching definitions considered for this base for reflections. Taking Osterwalder et al. (2005) as a valid parameter, from this analysis, it is possible to fit this text as a combination of their second type of research—authors describe abstractions of BMs relating it to firms' common characteristics—and the third, also —authors present a conceptualization from real-world implementations.

Along with a literature review, visits were conducted in industrial complexes and associated services. These cases included an automated top-level plant of an automotive industry, a food processing industrial complex, some startup supporting co-working installations (which promote public and private incubator/accelerator programs), taken as an opportune, dependent and guided sample, where that initial level of perception for components was applied and studied, producing the following panorama and corresponding reflections.

As a starting point for analysis, it is opportune to define one additional question around industrial business models: who was the "author" or "source of demand" to define what business model was adopted by one organization? It is possible to understand that, some centuries or decades ago, market leadership could allow industries to define and position their business models, formatting productive chains and even imposing some conditions for customers, as they faced restricted offers and competition, because of technologic, geographical and market conditions. This way, industries could project or even improvise market arrangements because, mainly, their control of competitive advantage negotiated.

Nowadays, with the emerging services markets, increase in competition, globalization, economy dynamics, necessity of fast reconfiguration of design, among other factors, industries had, sometimes, to accept internal and external pressures when considering their choices, losing the comfortable condition of an isolated and controlling leadership that characterized some markets and sectors. This question presents one basic aspect to be added, not as a conceptual definition for BM adoption itself, but as one external pressure or position factor for industrial complexes, considering the competitive dynamics throughout time, composed in the following study.

A traditional, *hierarchical model* for industrial complexes can be found when companies consider their functional actions with a higher priority level, instead of customer or market needs, for example. For hierarchical, function-oriented business models, command and control from "outside the factory" define structures that will be implemented "inside the factory". Typically, sectorial standards are adopted and implemented, in a classical homogeneity effect. VP is usually obtained from operational efficiency, leading to immediate objective of costs reduction, higher performance levels and consequent resources positioning. IS element is characterized by classical specifications for manufacturing and continuous processes structures.

As we observe the productive chain, a rigid, structured plan, with low-degree of flexibility, is usually found, in which times and volumes are set and sought by all components. A "continuous line" (a predominance of a rigid structure that conducts the operational level) can be identified and provide overall coordination for sub-tasks and processes instances. ES is almost a reproduction of the internal structure, as the leading organization defines and controls operative adjustments, serving as the main reference. Centralization is another factor that will probably emerge, for example, when information and knowledge management are conceived and held with this controlling/coordination fashion also. As expected, KI will focus on rigorous quantitative data measurements, serving for immediate control over production levels and end-customers negotiation. Although related to typically old-fashioned industrial sector, this conception has a strong cultural contribution, still with expressive influence in situations of new implementation reaching services sectors, unexpected risky situations, market uncertainty, pressured associations from merge and acquisition processes, etc.

As thinking from a historical, traditional point of view, it is undisputable to cite the business model *process-oriented*, like the solution remarkably developed by organizations which aimed to dynamically realign its operational design towards answering processes demands. Historically, the automotive industrial complex Toyota is one of the main references, as some companies and economy actors, ranging from services-oriented (like some e-commerce retailers, as parts of Amazon, E-bay, among many others) to NGOs, attempted to adapt these principles (Liker 2004; Deloitte and Huawei 2015). It is a competitive context where some companies started to face competitive advantage factor, also observing some risky conditions never experienced before. This alternative, which defines a fast-answer pressure for industrial players, requires a thorough preparation, as financial control implementations, human resources preparation and management, automation, overall control, goal-seeking culture.

Analyzing the conceptual components: VP is the main objective, as the complete arrangement is made in function of customer, end-process perception. It demands high-level of value perception by industrial managers and strategic control of the remaining components, as to provide the fast alignment of them to promote a productive process orientation towards maximum value answer. Pressures over IS elements are severe, when fast reorientation of groups and teams structure, application of operational and tactical knowledge towards optimal production limits, efficiency, etc., and their associated control, provided by integrated information systems and contributive information technology applications, must be implemented to allow the intended dynamism.

It is possible to understand this alternative for a business model like the first type in which ES elements can eventually format the IS arrangement, when it happens with the service-oriented models, to be approached in the end of these reflections. Here, the market strength, competitive facts, regulatory phenomena, technological advances and changes and, clearly, customer behavior changes, require adaptations, producing a correspondence between ES components and IS elements, illustrating how industrial strategies are affected by new competitive implementations. KI expand those cited in the previous case to encompass quantitative signals, from the value-aggregation chains, admitting to complete control arrangement performance, dynamically set goals (in terms of production levels, costs optimization, profitability, products acceptance, etc.).

It is important to recall, when to seek for industrial best practices, about the presence of external forces, originated from business model adoption, such this one, that imposes some conditions, parameters, and strategic demands for industries, leading to a situation where the industrial sector is a component of the process-oriented complex, somewhat in an unfavorable condition of being led by other sectors.

Another opportune trend which provoke business model conception and implementation is the reorganization towards project management disciplines, the *projectized* BM. Efforts of the PMI, documented by the referential Project Management Body of Knowledge publication (PMBoK 2013), defined an interesting level of understanding about projects specification and managerial efforts, works and communications, reaching structures, factors such as performance administration and, mainly, human resources definition. This solid conceptual framework defined stances where several companies that relate strictly to industrial complexes were proposed. These players act in sectors like civil construction, traffic coordination, capital goods supplying, implemented their specifications considering "project" as the focus element and fundamental concept.

The following analysis for components does not have the objective on approaching project coordination itself. Otherwise, it attempts, like the previous analysis, to reflect about components when thinking about projects and project governing principles. VP is highly related to the project consequences, relating its definition processes—such scheduling, costs forecast, human resources performances supervision—to the levels obtained.

Perception of values being reached, answering the quality levels of maturity (Prado and Archibald 2009) and producing the expected results for final users, constitute main compositions for value positioning. Specifications for project demanded resources map, almost completely, the internal structure, IS. Along the definitions of "knowledge areas" and "project management processes", defined by PMBoK (2013), it is possible to understand several elements of IS definitions, which may be detailed in the organization's business model, aligning it to this conceptual understanding. Almost the same occurs with the ES elements, added by the disruptive context of innovative technologies to be applied, such as Internet of Things (Chui et al. 2010), studied before. Finally, KI component is expressively defined by the main framework conceptualization and remaining conceptual and practical compositions, which implement project coordination activity.

Considering quality as one of the main objectives in these tasks, precise measurement of all quantitative items of a project conception, planning, execution and final monitoring (for example, the usage of a construction, delivered for its normal operation) are among the quantitative indicators that must be addressed in its design. This reflection also shows how a business model is affected and pressured by external conditions and processes, when it happened with the process-oriented type and is observed for the next cases.

An interesting alternative, with intense discussion is a general model which can be identified as *platforms*. This alternative can be considered from fundamentals of some different areas, as industrial management, production engineering and information technology. It generally describes a basic assembly context where components and parts are dynamically mounted, installed or logically configured, aiming to differ the final goal. It can be considered, taking the managerial principal outcome, as a business model derived from the process-oriented, but with the addition of component change, adaptation of parts of the business model, which is a dynamic associated with fast-moving industries and its partners of the productive and value-aggregation chain.

Gatautis (2017) showed how the concept platform evolved in the last years, mainly affected by the perspectives of digital transformation, which are found in information technology-based platforms, offered by companies like Google, Amazon, Microsoft, Apple, among many others. In this approach, those platforms are considered as one opportune type, but it is proposed an advance towards an industrial integration (and perspectives of componentization) where its components—machinery, administration, controls, etc—are to be changed in a way similar to "Lego" components, with objective to dynamize processes and actions oriented to market opportunities.

Analyzing the BM components: VP analysis shows the connection of external scenario comprehension (potentially provided by information and knowledge management support from information systems—Nonaka and Takeuchi (1995) and Jamil and Silva (2016)—producing knowledge-based conditions) to business modularity expertise as the most attractive aspect to obtain the desired value positioning adaptability. This must be intensively related to ES, when the value-aggregation chain must show the desired transparency, effective connection, knowledge and information capabilities to admit such fast-moving plans and actions.

For IS elements, it is important that "Lego" oriented connections must prevail, as process input/output specifications, with predicted figures, data collection, and deliveries, permitting one specific planning component to be changed with complete support of management view. Indicators context, KI, shall inform precisely changing-time-related impacts, results and allow the follow-up about the new chain configuration. Platforms, although a well-known alternative, are gaining consistence as networks can now be scheduled, connected, and changed with technology support and cultural acceptance, becoming more accepted.

Finally, the *service-oriented* trend is transforming in a real pressure over industrial complexes, as the productive chain, sometimes, is gaining a different, competitive aspect that shows services leading these network relations. It is important to note that some decades ago, services were complementary strategic value increment options, when, for example, commercial organizations usually were driven by industrial manufacturers to offer value for final customers, in cases such automotive or electrical apparatus.

For example, marketplaces business models, or the "Uber-like" configurations are driving some plans and actions of automotive industries, not only the car manufacturing, but also with important strategic services, like maintenance, product line update and redesign, among several others. Is important to mention, to reinforce this notion, investments done by leading automotive manufacturers on buying or controlling car-sharing services—like the BMW investment in DriveNow and Daimler in Car2Go, which are now merging the operation and value positioning at customer-level (Bloomberg 2018). These signals are reinforced by the worries, in some countries, of "des-industrialization", in which national economies are more influenced by services sectors (Bresser-Pereira 2008; Oreiro and Feijó 2010; Cano 2012). Interestingly, business models like *marketplaces, signature-frequent demand, pay as you serve or use, scarcity consumption, pay in advance and service-to-product* (this last one being a strong evidence of service dominance)—Mullins (2014)—are, nowadays, becoming increasingly adopted by final customers for their usual needs, as buying food, transportation (even international), finance negotiations, goods purchasing, among many others.

The analysis of business model components is, at this moment, too superficial, as strategic and market movements are still turbulent, leading to a situation where reached analytical method reached through literature review must be applied in following studies, even to understand its sufficiency on really addressing the evidences from these remarkable and huge actions.

Business model alternatives continue to be proposed, associated and practiced by market agents. With the Economy advances, flexibility and dynamics, faced recently, analytical methods for research, planning, and coherent implementation of business models become the most critical tool for professionals involved in decisions towards BM real implementations.

10.5 Conclusions

This chapter proposition was to approach an opportune topic which demands severe thinking about industrial organizations nowadays: Business models. For these initial studies, some reflections, supported by an analytical support developed from theoretical review was conducted, observing some of the most typical market alternatives adopted so far. As competition evolves, in one undeniable signal of economy development, business models are still on study, planning, adaptations, turbulent implementations showing a requisite for market actions, substantially affected by strategic results that must be comprehended by any scholar and practitioner.

For this purpose, the proposition was stated, followed by a literature review around the main topic—business models—and its associated analysis to essential related concepts—with specific approach of strategic and tactical implications—and supporting contextual items, that provide structures and tools to implement business model in real cases. Following, an analytical proposition, constituted by four main aspects or elements—Value Proposition (VP), External structures (ES), Internal structures (IS) and Key indicators (KI)—was discussed and superficially applied on some market-adopted choices, to produce the desired level of reflexive thoughts, motivating a basic comprehension on BM planning and application.

It is possible to understand business model importance in nowadays competitive scenarios and industrial complexes face another tough front of strategic pressures, even reaching a fast-changing picture where services sectors are forcing reviews for industrial alternatives. This chapter intended to bring some light to these researches and discussions, proposing an initial theoretical framework which allows to develop analytical methods to support further conceptions and practices.

For further studies, among several perspectives, detailed case study analysis (although already explored in the literature, follow-ups always can provide more detail on business model practical implementations), key success factors on business models adoptions, exploration around VP, ES, IS, and KI internal aspects to evaluate their details and relate it to results from practical BM applications are among important alternatives.

This chapter has the potential on contributing to these and studies to approach business model real importance for strategic decision-making for industries and their related value-aggregation partners.

References

Akbar H (2003) Knowledge levels and their transformation: towards the integration of knowledge creation and individual learning. J Manag Stud 40(8):1997–2021

Ali Mahdi HA, Abbas M, Mazar TI, George S (2015) Comparative analysis of strategies and business models of Nike, Inc. and Adidas Group with special reference to competitive advantage in the context of a dynamic and competitive environment. Int J Bus Manag Econ Res 6(3):167–177

Anunciação PF, Peñalver AJB (2017) Information urbanistic perspective in the context of Blue economy: analysis of Setubal and Cartagena tourism offer. Int J Sociotechnol Knowl Dev 9(3): 65–83

Barney J (1991) Firms resources and sustained competitive advantage. J Manag 17(1):99–120

Bashir M, Verma R (2017) Why business model innovation is the new competitive advantage. IUP J Bus Strategy 14(1):7–17

Bereznoi A (2014) Business model innovation in corporate competitive strategy. Probl Econ Transit 57(8):14–33. https://doi.org/10.1080/10611991.2014.1042313

Bloomberg (2018) Daimler, BMW reach a deal to merge car-sharing units. Available at https://www.bloomberg.com/news/articles/2018-03-28/daimler-bmw-are-said-to-reach-deal-to-merge-car-sharing-units. Accessed on Mar 2018

Bresser-Pereira LC (2008) The Dutch disease and its neutralization: a Ricardian approach. Rev Econ Polít 28(1):47–71

Cano W (2012) A desindustrialização no Brasil. Econ Soc 21:831–851

Casadesus-Masanell R, Ricart JE (2007) Competing through business models. IESE Business School Working Paper No. 713. Available at SSRN: https://ssrn.com/abstract=1115201 or http://dx.doi.org/10.2139/ssrn.1115201

Casadesus-Masanell R, Ricart JE (2010) From strategy to business models and to tactics. Long Range Plan 43(2–3):195–215, April–June. https://doi.org/10.1016/j.lrp.2010.01.004

Castells M (1996) The rise of the network society, the information age: economy, society and culture, vol I. Blackwell, Oxford, United Kingdom

Casumano MA (2018) The sharing economy meets reality. Commun ACM 81(1):26–28

Chen H, Chiang RHL, Storey VC (2012) Business intelligence and analytics: from big data to big impact. MIS Q 36(4):165–1188

Chesbrough HW (2003) Open innovation. The new imperative for creating and profiting from new technology. Harvard Business School Press, Cambridge, USA
Chesbrough HW (2010) Business model innovation: opportunities and barriers. Long Range Plan 43(2):354–363
Christensen CM, Bartman T, van Bever D (2016) The hard truth about business model innovation. Manag Rev 58(1):21–40
Chui M, Loffler M, Roberts R (2010) The Internet of Things. McKinsey Q. Available at https://www.mckinseyquarterly.com/ghost.aspx?ID=/The_Internet_of_T. Accessed on May 2016
Colvin G (2017) How Intuit reinvents itself. Fortune Magazine, November 2017, pp 77–83
Davila T, Epstein MJ, Shelton R (2006) Making innovation work: how to manage it, measure it and profit from it. Pearson Education Inc, New York, USA
Deloitte, Huawei (2015) Transformation and reconstruction of Banks in the digital era. Deloitte Touche Tohmatsu Limited and Huawei Technologies. Available at http://e.huawei.com/en/publications/global/ict_insights/201512291113/analyst/201512301703. Accessed on Feb 2017
Deshler R, Smith K (2011) Making business model innovation stick. People Strategy 34(4):18–23
Diaconu M (2011) Technological innovations: concepts, process, typology and implications in economy. Theor Appl Econ 10(563):127–144
Drucker PF (1993a) Post-capitalist society. HarperCollins, New York, USA
Drucker PF (1993b) Innovation and entrepreneurship. HarperCollins, New York, USA
Duin PAV, Ortt JR, Aarts WTM (2013) Contextual innovation management using a stage-gate platform: the case of Philips shaving and beauty. J Prod Innov Manag 31(3):489–500
Edgett SJ (2015) Idea-to-launch (Stage-gate ® model): an overview. Available from http://www.stage-gate.com/resources_stage-gate_wp10.php. Accessed on Feb 2017
Eisenmann T (2014) Business model analysis for entrepreneurs. Harvard Business School Background Note 812-096, Original on Dec 2011, Revised Oct 2014. Available at https://www.hbs.edu/faculty/Pages/item.aspx?num=41268
El-Bashir MZ, Collier P, Sutton SG (2011) The role of organizational absorptive capacity in strategic use of business intelligence to support integrated management control systems. Acc Rev 86(1):155–184. https://doi.org/10.2308/accr.00000010
El-Gayar O, Timsina P (2014) Opportunities for business intelligence and big data analytics in evidence based medicine. In: Annals of 47th Hawaii international conference on system science, IEEE. https://doi.org/10.1109/hicss.2014.100
Eyring MJ, Johnson MW, Nair H (2011) New business models for emerging markets. Spotlight on business model innovation. Harvard Business Review, Jan–Feb, pp 1–9
Gatautis R (2017) The rise of platforms: business models innovation perspectives. Inz Ekon-Eng Econ 28(5):585–591. https://doi.org/10.5755/j01.ee.28.5.19579
George G, Bock AJ (2011) The business model in practice and its implications for entrepreneurship research. Entrep Theory Pract 35(1):83–111. https://doi.org/10.1111/j.1540-6520.2010.00424.x
Grönlund J, Sjödin DR, Frishamar J (2010) Open innovation and stage-gate process: a revised model for new product development. Calif Manag Rev 52(3):106–131
Guile BR (1987) Technology and global industry. National Academy Press, Washington DC, USA
HBSP—Harvard Business School Publishing Corporation (2006) Building a business model and strategy: how they work together in entrepreneur's toolkit: tools and techniques to launch and grow your new business, chapter 4. Cambridge, Massachusetts, USA, pp 47–62
Henderson RM, Clark KB (1990) Architectural innovation: the reconfiguration of existing product technologies and the failure of established. Adm Sci Q 35(1):9–30
Hitt MA, Ireland RD, Hoskisson RE (2011) Strategic management concepts: competitiveness and globalization, 10th edn. Cencage Learning, New York, USA
Jamil GL, Magalhães LFC (2015) Perspectives for big data analysis for knowledge generation in project management contexts. In: Jamil GL, Lopes SM, Silva AM, Ribeiro F (eds) Handbook of research on effective project management research through the integration of knowledge and innovation. IGI Global, Hershey, USA, pp 1–18. https://doi.org/10.4018/978-1-4666-7536-0.ch001

Jamil GL, Silva AM (2016) Inteligência de mercado como um processo de gestão do conhecimento. Editora Media XXI, Porto, Portugal

Jamil LC, Vieira AAP, Xavier AJD (2018) Reflecting on analytics impacts on information architecture contexts as a source of business modelling for healthcare services. In: Jamil GL (ed) Handbook of research on expanding business opportunities with information systems and analytics. IGI Global, Hershey, USA. https://doi.org/10.4018/978-1-5225-6225-2

Johnson MJ, Christensen C, Kagermann H (2008) Reinventing your business models. Harvard Business Review, Dec 2008, pp 2–11

Kartik K, Tarun K, Tracey AS (2017) The differential impact of new product development "make/buy" choices on immediate and future product quality: insights from the automobile industry. J Mark 81(6):1–23. https://doi.org/10.1509/jm.14.0305

Kavadias S, Ladas K, Loch C (2016) The transformational business model. Harvard Business Review, Dec, pp 91–98

Khanaga S, Volberda H, Oshi I (2014) Business model renewal and ambidexterity: structural alteration and strategy formation process during transition to a Cloud business model. R&D Manag 44(3):322–340

Kim WC, Mauborgne R (2015) Blue ocean strategy: expanded edition. Harvard Business Review Press, Cambridge, USA

Lee CS, Ho JC (2010) A framework for analyzing business model innovation in mobile commerce. J Int Technol Inf Manag 19(4):37–60

Liker J (2004) The Toyota way: fourteen manufacturing principles from the world's greatest manufacturer. McGraw Hill, New York, USA

Magretta J (2002) Why business models matter. Harvard Bus Rev 80(5):86–92

Mahrt M, Scharkow M (2013) The value of big data in digital media research. J Broadcast Electron Media 57(1):20–33. https://doi.org/10.1080/08838151.2012.761700

Massa L, Tucci (2014) Business model innovation. In: Dogdson M, Gann D, Nelson P (eds) The Oxford handbook of innovation management. Oxford University Press, Oxford, United Kingdom, pp 420–441

Matzler K, Bailom F, Von den Eichen SF, Kohler T (2013) Business model innovation: coffee triumphs for Nespresso. J Bus Strategy 34(2):30–37. https://doi.org/10.1108/02756661311310431

McAfee A, Brynjolfsson E (2012) Big data: the management revolution 90(10):60–68. Harvard Business Review, Cambridge, USA

Mintzberg H, Ahlstrand B, Lampel J (2008) Strategy safari. Pearson Education, New York, USA

Mullins J (2014) Customer-funded business: start, finance or grow your company with customer's cash. Wiley, Hoboken, USA

Nielsen C, Lund M (2018) Building scalable business models. Sloan MIT Manag Rev 59(2):65–79

Nonaka I, Takeuchi H (1995) The knowledge-creating company. Oxford Press, Oxford, United Kingdom

OECD—Organization for Economic Cooperation and Development (2005) Oslo manual. Available at https://www.oecd.org/sti/inno/2367580.pdf. Accessed on Mar 2017

Oreiro JL, Feijó CA (2010) Desindustrialização: conceituação, causas, efeitos e o caso brasileiro. Rev Econ Polít 30(2):219–232. https://doi.org/10.1590/S0101-31572010000200003

Osterwalder A, Pigneur Y (2010) Business model generator: a handbook for visionaries, game changers and challengers. Wiley, Hoboken, USA

Osterwalder A, Pigneur Y, Tucci CL (2005) Clarifying business models: past, present and future of the concept. Commun Assoc Inf Syst 15(1):1–25

Ovans A (2015) What is a business model? Harvard Business Review blog. Available at https://hbr.org/2015/01/what-is-a-business-model. Accessed on May 2017

Penrose ET (1959) The theory of the growth of the firm. Wiley, New York, USA

Pfeifer S, Peterka SO, Stanić M (2017) Business models of micro business: empirical evidences from creative industries. J Contemp Manag Issues 22(Special issue):1–19

PMBoK (2013) Project Management Body of Knowledge, 5th edn. Project Management Institute

Porter M (2008) On competition. Harvard Business School Press, Cambridge, USA

Porter M, Magretta J (2014) Strategy and competition: the Porter collection. Harvard Business Press Review, Boston, USA
Prado D, Archibald R (2009) Gerenciamento de Projetos para Executivos. INDG Tecnologia e Serviços Ltda, Nova Lima, Brazil
Ries E (2011) The lean startup: how today's entrepreneurs use continuous innovation to create radically successful businesses. Crown Business, New York, USA
Sachsenhofer W (2016) Leveraging business model components as drivers of business model portfolios. J Bus Models 4(3):37–47
Sako M (2012) Business models for strategy and innovation. Commun ACM 55(7):22–24
SAS (2017) What is analytics? Available at https://www.sas.com/en_us/insights/analytics/what-is-analytics.html. Accessed on Nov 2017
Schumpeter JA (1942) Capitalism, socialism and democracy. Harper-Collins, New York, USA
Setia P, Venkatesh V, Joglekar S (2013) Leveraging digital technologies: how information quality leads to localized capabilities and customer service performance. MIS Q 37(2):565–590
Silva AMB (2011) Modelos e Modelizações em Ciência da Informação: O Modelo eLit.pt e a investigação em literacia informacional. J PRISMA.COM 13(1):1–56
Singleton-Green B (2014) Should financial reporting reflect firms' business models? What accounting can learn from the economic theory of the firm. J Manage Governance 18(3):697–706. https://doi.org/10.1007/s10997-012-9240-7
Slávik Š, Hanák R (2017) Identification of relation between business model and business strategy and measurements of its tightness. Probl Manag 21st Century 12(1):54–68
Teece DJ (2009) Dynamic capabilities and strategic management: organizing for innovation and growth. Oxford Press, Oxford, United Kingdom
Teece DJ, Linden G (2017) Business models, value capture and digital enterprise. J Organ Des 6 (8):2–14. https://doi.org/10.1186/s41469-017-0018-x
Tuomi I (2000) Data is more than knowledge: implications of the reversed knowledge hierarchy for knowledge management and organizational memory. J Manag Syst 16(3):103–117
Turban E, Mc Lean E, Wetherbee J (2002) Information technology for management: transforming business in the digital economy, 3rd edn. John Wiley and Sons, Hoboken, USA
Utterback JM (1971) The process of technological innovation of the firm. Acad Manag J 14(1):75–88
Utterback JM (1996) Mastering the dynamics of innovation, 2nd edn. Harvard Business Press Review, Cambridge, USA
Venkatraman N (1989) Strategic orientation of business enterprises: the construct, dimensionality and measurement. Manage Sci 35(8):942–962
Wernerfelt B (1984) A resource-based view of a firm. Strateg Manag J 5(1):171–180
Wirtz BW, Daiser P (2017) Business model innovation: an integrative conceptual framework. J Bus Models 5(1):14–34
Yonce C, Taylor J, Kelly N, Gnau S (2017) BI Experts' perspective: are you ready for what's coming in analytics. Bus Intell J 22(3):36–42
Zalewska-Kurek K, Kandemir S, Englis BG, Englis PD (2016) Development of market-driven business models in the IT industry. How firms experiment with their business models? J Bus Models 4(3):48–67
Zott C, Amit R, Massa L (2011) The business model: recent developments and future research. J Manag 37(4):1019–1042

Chapter 11
The Constrained Joint Replenishment Problem Using Direct and Indirect Grouping Strategies with Genetic Algorithms

Julian Andres Zapata-Cortes, Martín Darío Arango-Serna and Víctor Jaime Saldarriaga-Romero

Abstract The Joint replenishment problem (JRP) is a model for inventory optimization when ordering multiple products is required. This model allows reducing total inventory costs compared to the practice of performing individual optimization of each product. The saving cost is produced by sharing the fixed ordering costs between several products. The JRP model can be solved using two strategies: the direct grouping strategy (DGS) and the indirect grouping strategy (IGS), which vary in the way the products are grouped. As it can be seen in this chapter, several authors use the JRP model in its simplest version, but it can be easily extended to include restrictions for approximate the model to more realistic applications. Due to the combinatorial nature and the restriction inclusion to the JRP model, it is necessary to use advanced techniques that allow obtaining good solutions in reasonable computing times. This chapter presents the JRP with resource and capacity constraints, which is solved using two genetic algorithms for the evaluation of the Indirect and the Direct Grouping Strategies (IGS and DGS).

Keywords Inventory management · Joint replenishment problem
Constraints · Grouping strategies · Genetic algorithm

J. A. Zapata-Cortes (✉) · V. J. Saldarriaga-Romero
Escuela de Negocios, Institución Universitaria CEIPA, Calle 77 Sur No. 40-165, Sabaneta, Antioquia, Colombia
e-mail: julian.zapata@ceipa.edu.co

M. D. Arango-Serna
Departamento de Ingeniería de la Organización, Universidad Nacional de Colombia, Carrera 80 No 65-223, Medellín, Antioquia, Colombia

J. L. García Alcaraz et al. (eds.), *Best Practices in Manufacturing Processes*,
https://doi.org/10.1007/978-3-319-99190-0_11

11.1 Introduction

The growing competitive pressure in which organizations are immersed leads to innovate in their processes seeking for new mechanisms that allow them to optimize their operations and so use their resources and capabilities more efficiently. Logistics has become more important in different sectors, due to its impact in aspects such as customer satisfaction, operating costs, relationships with suppliers and distributors and other actors in the supply chain (Arango et al. 2013). Logistics and supply chain management are essential to contribute to the value generation and competitiveness in organizations, since they are responsible for delivering the products to customers. This is why logistics and supply chain management play a fundamental role in customer satisfaction and in reduction in operating costs by allowing a better resources assignment.

Inventory management is one of the main business logistics functions, since it is responsible for ensuring the adequate amount of stocks in order to accomplish the required service level while trying to bring the inventory costs to their minimum value. Many strategies and models have emerged in inventory management seeking to reduce costs associated with this logistics function (Arango et al. 2011; Arango et al. 2013).

Decisions around Inventory Management have to do with the definition of the quantity and the right moment in placing an order of products, looking to avoid shortages and ensuring an appropriate supply to customers, seeking to minimize the ordering total cost and keeping the required products in the warehouse (Nagasawa et al. 2015). One of the most implemented models for inventory optimization is the Economic Order Quantity (EOQ) that establishes the quantity and the right moment of replenishment. This model is applied to optimize the inventory for only one product at a time, which is no practical in many real situations that require ordering several products at the same time.

Performing inventory optimization models for a single product at a time, can bring additional costs if it is compared with multiple products optimization models (Chopra and Meindl 2008; Moon and Cha 2006). The optimization of a jointly inventory process for multiple products can be done through the Joint Replenishment Problem (JRP), which is responsible for defining the moments and quantities in which a product or a set of them must be ordered to the supplier. This allows that the total inventory cost for all products, as well as the ordering cost will be as low as possible (Bastos et al. 2017). Porras and Dekker (2008) mention that the JRP generates solutions in which the reduction of total cost, compared with the results of the EOQ model for a set of 20 products, is around 13%.

The problem of solving the JRP model can be addressed with two different strategies: the Direct Grouping Strategy (DGS) and the Indirect Grouping Strategy (DGS). Several techniques can be used for solving the mathematical problem (Bastos et al. 2017). This chapter presents a review of the JRP model in which the direct and indirect grouping strategies for its solution is analyzed. It is also presented a JRP model which is extended with capacity and resources constraints and two genetic algorithms are used for solving this model.

11.1.1 The Joint Replenishment Problem (JRP)

The joint replenishment of several products can be used for organizing the management inventory process with multiple products in a company. It seeks that the products can be grouped in the best way when placing orders, pursuing to reduce the total storage and ordering cost (Moon and Cha 2006; Bastos et al. 2017).

There are several inventory management models for the joint replenishment of products in companies that allow deciding the optimal quantities to be ordered from the same supplier (Khouja et al. 2000). One of these models is the JRP, which seeks to determine the replenishment policy of different products, so that the total cost of keeping products, placing new orders and out of stocks are reduced (Olsen 2005).

In its simplest version, the JRP model only considers the costs of ordering and keeping inventory to generate supply plans without product shortages. The model is subjected to several assumptions, which are (Olsen 2005):

- Demand is known and constant.
- Time is finite.
- Costs of storage are constant.
- No quantity discounts.
- Out of stocks are not allowed.
- Instantaneous supply.
- Required amounts of products are completely available in the suppliers.
- There are no constraints on the storage capacity.
- There are no constraints on the available budget to place orders.

There are several variations of the JRP model that basically are focused on solving situations of capacity and resources constraints (Bastos et al. 2017), in which the ordering cost of the items are dependent on each other (Olsen 2008) and in which demands are presented with non-deterministic behavior, as in the case of stochastic and dynamic demand (Bastos et al. 2017).

Other extensions of the JRP model consider multi-step networks (Bastos et al. 2017; Yang et al. 2012) and the inclusion of routing and delivery decisions (Joint Replenishment and Delivery Problem—JRD) (Cha et al. 2008; Coelho and Laporte 2014; Qu et al. 2013; Wang et al. 2012; Wang et al. 2013; Zeng et al. 2014; Nagasawa et al. 2015).

The JRP model can be solved with simple methods, such as the algorithm presented in Chopra and Meindl (2008). However, this procedure is useful for a low number of products and do not ensure the optimum solution in reducing total cost

(Chopra and Meindl 2008). Due to its combinatorial nature, the JRP model is considered a NP-Hard problem, even in its simplest version (Hong and Kim 2009).

In the JRP model, the ordering cost is composed of two elements. A fixed cost of order preparation, considered as the higher cost, in which companies must incurred each time an order is placed. The second one, considered as the lower cost of ordering, is associated with each product *i* and must be included each time the product *i* is placed in an order, which increases the total cost (Olsen 2005). By finding the best combination of order cycles it is possible to reduce total inventory costs, since the fixed ordering cost is shared among all products (Olsen 2005).

Establishing the replenishment policy of the JRP model can be carried out by two ways: the Direct Grouping Strategy (DGS) and the Indirect Grouping Strategy (IGS) (Khouja et al. 2000). In DGS, the products are organized into several groups. An ordering Cycle time is calculated for each group, and the products in each group are jointly ordered.

In the IGS, a single cycle time and the multiple time in which every product must be ordered is calculated. This way, some products are ordered each time the cycle time is completed, and other products will be ordered every two cycles, every three cycles, etc. (Khouja et al. 2000). In general terms and based on the results reported by authors such as Olsen (2005), Porras and Dekker (2006), Khouja and Goyal (2008) and (Bastos et al. 2017), the IGS is more efficient when there is a low relation between Higher/Lower ordering costs. Mathematically, the total cost is the sum of the annual costs of keeping and order products, as can be seen in Eq. (11.1), which corresponds to the JRP objective function.

$$\mathrm{TC} = C_\mathrm{h} + C_\mathrm{o} \tag{11.1}$$

where C_h is the holding inventory cost and C_o is the ordering cost. These terms differ if the Direct or Indirect grouping strategy is used. The notation for both formulation is:

- n: number of ordered products.
- i: index of products, with $\mathbf{1 \leq i \leq n}$.
- D_i: annual demand of the product i.
- T: order cycle (time between orders), in years.
- T_i: cycle time of consecutive orders of product i.
- hi: holding cost of product i.
- S: fixed cost of placing an order.
- S_i: variable cost of including product i in an order.

The mathematical formulation for each JRP grouping strategy is presented below (Olsen 2005).

11.1.2 Indirect Grouping Strategy (IGS)

In order to find the time and quantity to be ordered in this strategy, it is necessary to define an optimal common cycle time T^* and a set of integers k_i which are multiples of the cycle time and indicates the moment when the product i must be ordered from the supplier. In this way, products with the same value of k are ordered at the same time and share the major cost of ordering S. For example, a product with $k = 1$ is ordered each time the cycle time is finished, a product with $k = 2$ is ordered every two cycles and so on (Olsen 2008). According to Olsen (2005), this is a relatively simple strategy to be calculated and implemented in most of the cases.

The order quantity of each product in each cycle time is calculated with Eq. (11.2).

$$Q_i = T^* k_i D_i \tag{11.2}$$

The total annual cost incurred for keeping each product is given by Eq. (11.3) and the total keeping cost of all products is calculated with Eq. (11.4):

$$Tk_i \frac{D_i}{2} h_i \tag{11.3}$$

$$C_{\mathrm{h}} = \frac{1}{2} T \sum_{i=1}^{n} h_i k_i D_i \tag{11.4}$$

The variable ordering cost, which must be added to the major fixed costs every time the product i is included in an order, is calculated with Eq. (11.5). The total annual ordering cost, including both the fixed and variable cost, is expressed in Eq. (11.6)

$$\frac{s_i}{k_i} \tag{11.5}$$

$$C_{\mathrm{o}} = \frac{1}{T}\left(S + \sum_{i=1}^{n} \frac{s_i}{k_i}\right) \tag{11.6}$$

The total replenishment cost for the IGS strategy is calculated with Eq. (11.7), which results from adding Eqs. (11.4) and (11.6).

$$\mathrm{TC} = \frac{T}{2} \sum_{i=1}^{n} h_i k_i D_i + \frac{1}{T}\left(S + \sum_{i=1}^{n} \frac{s_i}{k_i}\right) \tag{11.7}$$

The optimal cycle time T^* is calculated for the set of integers k_i, as presented in Eq. (11.8).

$$T^* = \sqrt{\frac{2\left(S + \sum_{i=1}^{n} \frac{s_i}{k_i}\right)}{\sum_{i=1}^{n} k_i h_i D_i}} \quad (11.8)$$

The minimum total cost for the IGS can be calculated by finding the k_i values for the n products to be ordered, using Eq. (11.9).

$$TC = \sqrt{2\left(S + \sum_{i=1}^{n} \frac{s_i}{k_i}\right) \sum_{i=1}^{n} k_i h_i D_i} \quad (11.9)$$

11.1.3 Direct Grouping Strategy (DGS)

This strategy is based on forming m groups of products to define the moment and the quantity to order. For each group a single cycle time is assigned. The objective is to find the adequate grouping (which products are part of each group) and the optimal cycle time for each group, in order to produce a minimal holding and ordering cost. For the DGS, the total replenishment cost is given by Eq. (11.10).

$$\text{TC} = \sum_{j=1}^{m} \left(\frac{1}{T_j} \left(S + \sum_{i \in G_j} s_i \right) + \frac{1}{2} T_j \sum_{i \in G_j} h_i D_i \right) \quad (11.10)$$

In Eq. (11.10), m is the number of groups, j is the index of the groups with $j = 1, 2, \ldots, m$, G_j is the group j and T_j is the cycle time for group j. When the first derivative of the TC equation with respect to T_j is obtained, the optimum cycle time for each group j is obtained, as presented in Eq. (11.11)

$$T_j^* = \sqrt{\frac{2\left(S + \sum_{i \in G_j} s_j\right)}{\sum_{i \in G_j} h_i D_i}} \quad (11.11)$$

Once the optimal cycle time for each group T_j^* has been calculated, the optimal total cost TC* can be obtained by substituting the values of T_j^* in Eq. (11.10).

Authors such as Olsen (2005) mention that most of the works in the specialized literature are focused on solving the JRP model using the indirect grouping strategy and without considering constraints, which ignores the real conditions of many practical situations (Moon and Cha 2006). This chapter presents the IGS and DGS to solve the JRP model with capacity and resource constraints.

11.2 Constraints in the Joint Replenishment Problem (JRP)

In real situations there are several situations in which the simple JRP model cannot be applied, due to the existence of conditions that restrict its application. These constraints are related with capacity, such as the limited space in the warehouses, the maximum production of suppliers, the maximum weight allowed in vehicles (Moon and Cha 2006) and maximum number of supply operations to be carried out (Taleizadeh et al. 2011). Resource restrictions may be included as well, such as limitations of budget to execute purchases or to keep products in inventory (Moon and Cha 2006; Taleizadeh et al. 2011).

Goyal (1975) was the first author to include constraints on the JRP model, introducing constraints in resources by limiting the maximum amount of money (budget) that can be invested in inventory. In his proposal, Goyal (1975) developed constraints for the IGS as presented in Eq. (11.12), which has been adopted by multiple authors when including constraints on the JRP model, as in the case of Khouja et al. (2000), Moon and Cha (2006), Li et al. (2014) and (Taleizadeh et al. 2011).

$$\sum_{i=1}^{n} Tk_i D_i C_i \leq R \tag{11.12}$$

In this equation C_i represents the purchase cost of each product i and R is the maximum budget that can be invested for all products in each order cycle. In the case of the DGS, the constraint is written in a similar way, this time assuring the budget for each group, as expressed in Eq. (11.13)

$$\forall j : T_j \sum_{i \in G_j} D_i C_i \leq R \tag{11.13}$$

Equations (11.12) and (11.13) restrict the financial resources in the company. The capacity constraints can be used to restrict the use of space and load of vehicles. The total amount that can be stored in a certain cycle time can be restricted by using Eq. (11.14) for the IGS (Taleizadeh et al. 2011) and with Eq. (11.15) for the DGS strategy.

$$\sum_{i=1}^{n} Tk_i D_i f_i \leq F \tag{11.14}$$

$$\forall j : T_j \sum_{i \in G_j} D_i f_i \leq F \tag{11.15}$$

In Eqs. (11.14) and (11.15), f_i represents the volume of each product i, and F the available storage capacity in the warehouse (Taleizadeh et al. 2011). The mathematical formulation for other constraints is similar and can be found in the works of

Moon and Cha (2006), Hoque (2006), Taleizadeh et al. 2011, Chen et al. (2016) and Ongkunaruk et al. (2016).

This way, the JRP model becomes the minimization of the total replenishment cost, using Eq. (11.9) or (11.10) for the IGS or DGS strategies respectively, taking into account resource and capacity constraints of Eqs. (11.11–11.15) depending of the selected strategy. Despite the importance of including these restrictions in the formulation of the JRP model, in the scientific literature the number of works that include restrictions on the JRP is not very extensive (Moon and Cha 2006).

11.3 Solution of the Joint Replenishment Problem

Several authors have developed different solution methods for the JRP model, seeking to find the optimal solution to the joint inventory replenishment. However, due to its NP-Hard nature and the possibility of including constraints, the use of new heuristics and metaheuristics techniques has been the area of major study in recent years around the JRP model (Nagasawa et al. 2015, Li et al. 2014; Bastos et al. 2017). As mentioned in the second section of this chapter, there are two solution strategies for the JRP Model, which strongly influenced the heuristic and metaheuristic technique to be used.

For the IGS strategy, the first works published were the studies of Shu (1971), Goyal (1973), Goyal (1974), Silver (1976), (Goyal SK Belton 1979) and Kaspi and Rosenblatt (1983). One of the most successful algorithms in the IGS solution is the RAND, proposed by Kaspi and Rosenblatt (1991), which improved the results of all the algorithms used for the JRP solution at this moment. Goyal and Deshmukh (1993) proposed an improvement to the method presented by Kaspi and Rosenblatt (1991) using a better estimate of the lower limits used in the original RAND. This method is still valid today, yielding acceptable results compared with more recently proposed algorithms (Moon and Cha 2006, Khouja et al. 2000). Other authors who have proposed exact methods for the JRP are Van Eijs et al. (1993) and Viswanathan (1996).

The DGS strategy has been used by authors such as Rosenblatt (1985), Van Eijs et al. (1992), Strijbosch et al. (2002), Olsen (2005) and Olsen (2008), which in addition to solve the JRP model they have evaluated their performance against the IGS, finding that the IGS strategy outperformances DGS results when the ordering costs are high, because many products can be ordered using IGS and the fixed total cost can be distributed among more products. Literature review on the JRP can be found in the works of Goyal and Satir (1989), Khouja and Goyal (2008) and Bastos et al. (2017).

Due to the need of finding faster algorithms, capable of solving larger problems, as well as including constraints into the model, several heuristic and metaheuristic methods have been used to solve the JRP model. Genetic and Evolutionary Algorithms are widely used for solving the JPR model (Li et al. 2014, Bastos et al. 2017). Genetic algorithms are searching algorithms based on the evolutionary concept of natural selection, in which the fittest individuals are the ones who survive in the evolution process. These algorithms represent a random exploration of the search space, in which it is expected that after a number of evolutions it can be found a solution close to the optimum of the problem (Nagasawa et al. 2015). Genetic algorithms, as well as evolutionary algorithms and evolution strategies are part of evolutionary computing, which are useful mechanisms to find good solutions in large search spaces (Olsen 2008; Arango-Serna et al. 2014).

Some works that use genetic algorithms and other evolutionary methods for the JRP model solution are: Khouja et al. (2000) presented a genetic algorithm to solve the JRP and its result was compared with the RAND. Olsen (2005) used a genetic algorithm to solve the JRP through DGS. Moon and Cha (2006) presented a genetic algorithm to solve the JRP with budget constraints. Olsen (2008) introduced an evolutionary algorithm to solve the JRP in which the relationship between ordering costs are interdependent with each other. Moon et al. (2008) proposed a genetic algorithm for the JRP considering discounts for quantities from multiple suppliers. Hong and Kim (2009) used a genetic algorithm to solve the JRP with exact inventory costs. Li et al. (2009) developed a Genetic Algorithm to solve the JRP with deteriorate items.

Yang et al. (2012) applied a genetic algorithm for the JRP with multiple retailers and discounts. Nagasawa et al. (2015) solved the JRP using the Can-order level policy using a multi objective genetic model. Li et al. (2014) presented an evolutionary differential algorithm to solve the JRP with budget restrictions. Ongkunaruk et al. (2016) used a genetic algorithm to solve the JRP model with defective products and shipping restrictions. The same authors presented a genetic algorithm for the JRP with multiple vehicles and constraints on capacity and quantity of shipments (Chen et al. 2016).

One of the main characteristics of the genetic algorithms used to solve the JRP model are the easy application and ability to work with complex problems and the possibility of including restrictions into the model (Khouja et al. 2000; Moon and Cha 2006). In order to include restrictions in the genetic algorithms different possible methods could be used, which are penalty functions, repair procedures, specialized operators and decoders (Khouja et al. 2000). Each of previous methods can be developed in different ways, such as the case of static, dynamic and adaptive penalties. The static penalty is the most commonly used case, in which the objective function is increased by a constant factor v multiplied by the unfeasible factor p_r, which takes a value of 1 for those solutions that do not comply with the restrictions, and zero (0) for those that do. The evaluation formula considering the non-viability penalty is expressed in Eq. (11.16)

$$\text{Fitness} = \text{TC} + vp_r \tag{11.16}$$

Although the penalization methods are useful for their application in genetic algorithms, Moon and Cha (2006) demonstrated that this practice is not applicable for the capacity and resource constraints in the JRP, since it does not generate the optimal solution to the problem. Instead of using the penalty, it is necessary to determine the cycle time using every restriction and compare those with the cycle time calculated with the cost equation. The optimal supply cycle is given by Eq. (11.17), with T^0 the cycle time obtained with the total cost equation and T^w the cycle time obtained with the wth constraint (Moon and Cha 2006; Chen et al. 2016).

$$T^* = \min\left(T^0, T^1, T^2, \ldots, T^w\right) \tag{11.17}$$

In this chapter the JRP model with budget and capacity constraints in the warehouse is presented, so the optimal cycle time T^0 must be calculated from Eqs. (11.8) and (11.11) for each IGS or DGS respectively. The cycle times T^1 and T^2 for the budget and maximum capacity constraints must be calculated with Eqs. (11.18) and (11.19) for IGS, and (11.20) and (11.21) for the DGS. For DGS, as presented in Eqs. (11.20) and (11.21), there are as many restrictions as number of groups j.

$$T^1 = \frac{R}{\sum_{i=1}^{n} k_i D_i C_i} \tag{11.18}$$

$$T^2 = \frac{F}{\sum_{i=1}^{n} k_i D_i f_i} \tag{11.19}$$

$$T_j^1 = \frac{R}{\sum_{i \in G_j} D_i C_i} \tag{11.20}$$

$$T_j^2 = \frac{F}{\sum_{i \in G_j} D_i f_i} \tag{11.21}$$

As previously mentioned, the JRP model with constraints is solved efficiently using genetic algorithms, and it is the reason why this technique is used to solve both the IGS and DGS strategies.

11.3.1 Genetic Algorithm for Solving the JRP Model

Genetic algorithms are random search techniques based on biological process of evolution of the species, in which a set of possible solutions, represented as individuals, are combined with the objective of producing new individuals with different characteristics in order to obtain better solutions to the problem (Arango-Serna et al. 2015). Genetic algorithms are used to solve complex mathematical combinatorial problems (Vergidis et al. 2012; Zapata-Cortés 2016; Arango-Serna et al. 2016) and they have the ability to find good solutions in problems with characteristics of discontinuity, multimodality, noisy functions (Fonseca and Fleming 1995; Tiwari et al. 2002), nonlinear and non-convex search spaces (Wang et al. 2004).

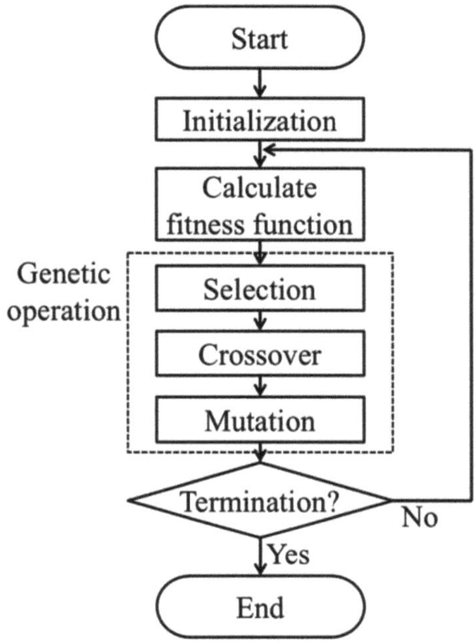

Fig. 11.1 Solving process of a genetic algorithms. *Source* Nagasawa et al. (2015)

The set of individuals that represent the solutions to the model is known as the population and these individuals are produced randomly or by following a specific procedure, in order to help the algorithm to generate seed solutions, with the aim of helping the random process to produce feasible solutions and finding the best solution to the problem more quickly. This initial population is evaluated to determine the quality of the solutions and to guide the following populations through the evolution process, in which some individuals are selected for their recombination (Crossover) and mutation.

Subsequent populations include the best individual of previous populations (Elitism) and the new individuals produced in the crossover and mutation process. Each population is evaluated, and it again goes through the evolution process. At the end of a certain number of evolutions, it is expected to find individuals that represent the optimum or at least a close solution to the problem (Arango-Serna et al. 2015; Zapata-Cortés 2016; Arango-Serna et al. 2018). Figure 11.1 represents the process of a genetic algorithms (Nagasawa et al. 2015).

The individuals' representation, known as the chromosome, is a critical element in the success of the genetic algorithm, since this is the coding of real solutions to the genetic language (Nagasawa et al. 2015). In the algorithms developed to solve the JRP model, the individuals' representation is done using a vector of values, which can be integers (Olsen 2005), real (Moon and Cha 2006) or binary numbers (Khouja et al. 2000). Each value in the vectors are known as a gene, and for the algorithms presented in this chapter, based on the work of Olsen (2005), integers are used in the chromosome of the DGS and real numbers for the IGS strategy, following the work of Moon and Cha (2006). Depending on the strategy to be followed (IGS or DGS), these genes have a different meaning and behavior.

In the case of the IGS, a gene located at position i of the chromosome, represents the value k_i for product i. In other words, it is the multiple of the cycle time in which this product must be ordered. This chromosome of real values is easily coded to the k_i values, by using the procedure presented in Moon and Cha (2006). With this coding, it is possible to determine upper and lower limits for k_i, that allows reducing the search space, as proposed by Moon and Cha (2006). Figure 11.2 shows the used chromosome in the representation of a problem of 10 products for the IGS.

The integer numbers of Fig. 11.2 is the codification of the chromosome into the k_i values. For the JRP of 10 products presented in this figure, products 1, 2, 3, 7, 8, and 9 are ordered every cycle time T; products 4, 5, and 10 are ordered every two cycle times; and product 6 is ordered every 5 cycle times. The genetic algorithm internally calculates the cycle time and the total cost of the solution. For the chromosome in the DGS, a vector of integers is used, where the position i represents the product i, and the gene value corresponds to the group of which the product belongs. The representation of the individual for the DGS strategy is shown in Fig. 11.3.

The individual for the DGS presented in Fig. 11.3 indicates that products 2, 3, 5, and 10 belongs to group 1; group 2 includes products 1, 7, and 9; group 3 includes products 4 and 8; and group 4 includes only product 6. Similar to the IGS strategy, both the cycle time and the total cost are calculated internally by the genetic algorithm.

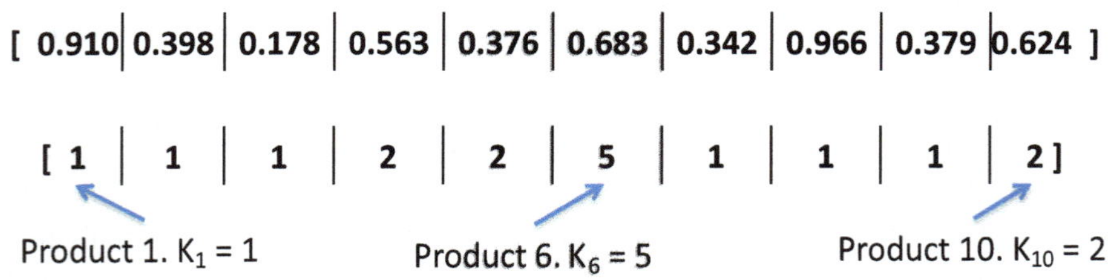

Fig. 11.2 Chromosome representation for the IGS strategy. *Source* Own Source

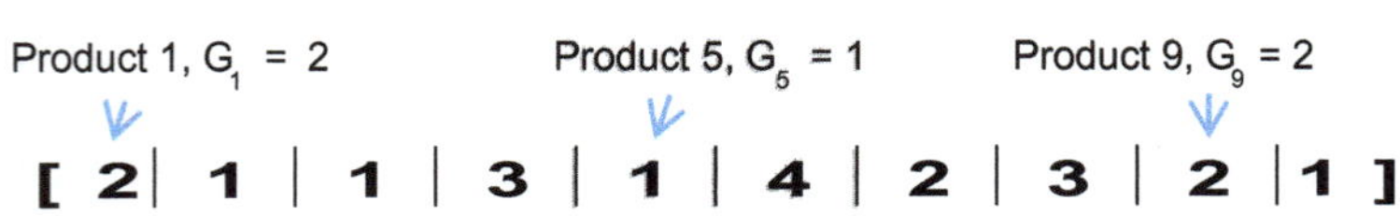

Fig. 11.3 Representation of the chromosome for DGS strategy. *Source* Own Source

Since each grouping strategy has a different representation, it is necessary to use two different genetic algorithms. The evaluation of the individuals is done through the fitness function, which determines how good the individual is in relation to the real problem. The fitness evaluation of the individuals is done using the cost equations, which are obtained by replacing the T^* found by Eq. (11.17) in Eqs. (11.7) and (11.10) for the IGS and DGS strategies respectively. Budget and storage capacity constrains are include to the JRP model and they are evaluated with Eqs. (11.7–11.21).

Crossover of individuals for the two algorithms is done by the two-point method. In this method, two different points are randomly selected. The crossover points indicate the fraction of chromosomes that are combined between individuals to interchange genetic information and produce new solutions. Figure 11.4 presents the crossover process of two individuals using the IGS, for an operation in which the random numbers produced automatically are 3 and 7. Figure 11.5 presents the same crossing process for the DGS.

The mutation process is performed using the single-point method. This point is randomly selected. The gene that corresponds to the mutation point is also randomly modified, changing the value k_i by a value between the minimum and the highest real value that represents the k_i of each product for the IGS. In the DGS the

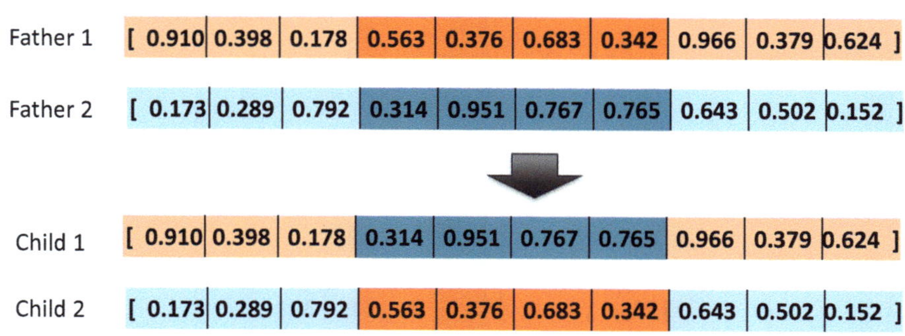

Fig. 11.4 Two-point crossover operator for the IGS strategy

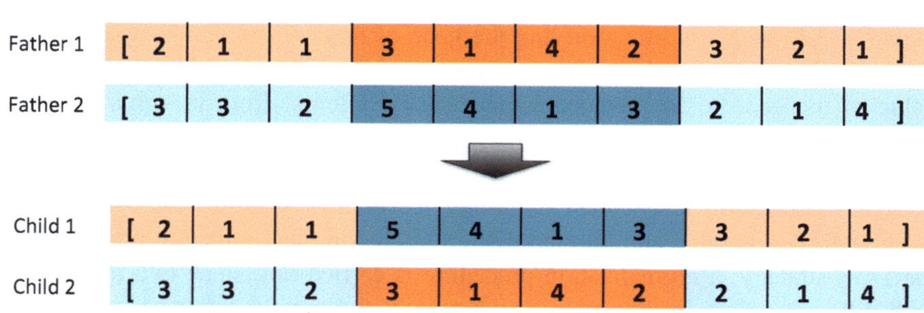

Fig. 11.5 Two-point crossover operator for the DGS strategy. *Source* Own Source

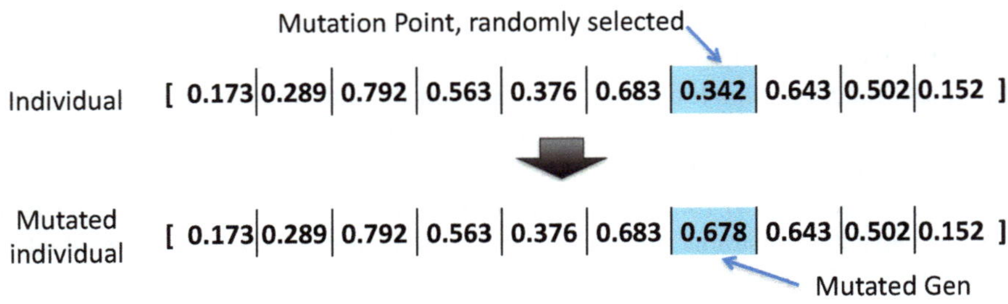

Fig. 11.6 Mutation operator for the IGS strategy. *Source* Own Source

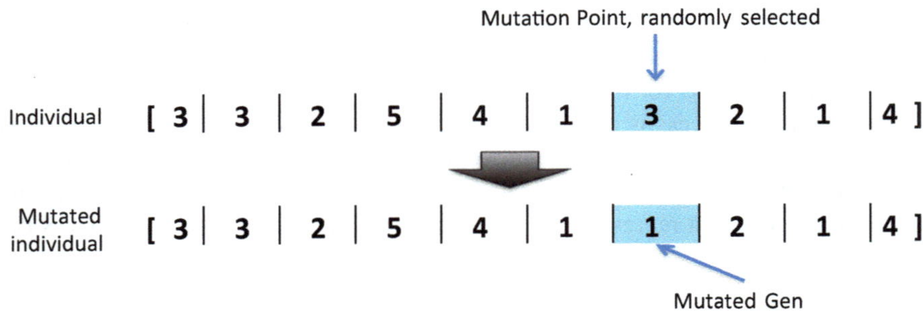

Fig. 11.7 Mutation operator for the DGS strategy. *Source* Own Source

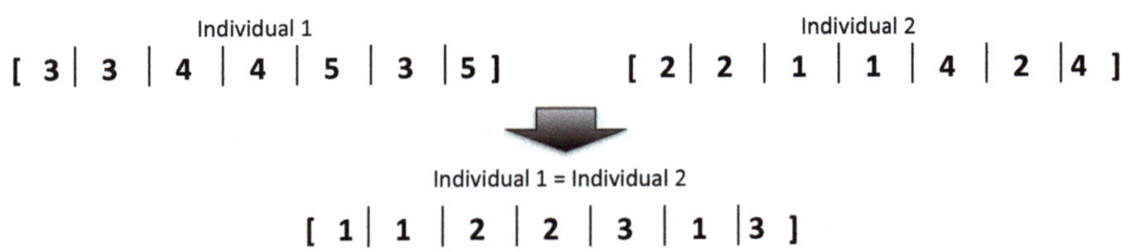

Fig. 11.8 Different individuals representation for the same solution. *Source* Adapted from Olsen (2005)

mutation is carried out by modifying the gene of the mutation point by a random value between 1 and the largest group number of the individual. In both cases, the value of the gene to be replaced is necessarily different from its original value. Figure 11.6 represents the mutation operator for the IGS and Fig. 11.7 for the DGS strategy.

Initial population is obtained randomly for both algorithms. Once the initial population is generated and after applying the crossing and mutation operators for the DGS strategy, a normalization procedure is carried out, since two individuals with different values in the genes may represent the same solution, as is the case with the individuals presented in Fig. 11.8 (Olsen 2005).

In both algorithms, the selection of individuals is carried out using tournament, in which individuals are randomly selected until they complete groups of capacity equivalent to 10% of the population. The algorithm is stopped after running a certain number of evolutions. The stopping condition, the population size and the percentage of mutation are determined by experimentation and adjustment (tuning of the algorithm).

11.4 Application of the JRP Model with Constraints

The JRP model is applied by solving the multiple product inventory assignment problem using the DGS and IGS strategies in a medium size company (Bancoldex 2013) dedicate to the underwear clothing production in the city of Medellín, Antioquia, Colombia. The company replenishes all the products from the same supplier, which is also located in Medellin. In the supply process, 24 different products must be ordered, which have different demands, costs and volumes. Table 11.1 presents the annual basis data of the company for the joint inventory policy definition.

The higher fixed cost of ordering is equal to 220,000 cost units, the maximum budget for the products is 25,000,000 and the maximum volume units of merchandise that can be received in the warehouse is 100,000. The calculation of the EOQ is implemented in the company for the individual product optimization in order to reduce inventory cost. Since only the optimal quantity for a single product is considered with the EOQ method, budget and storage restrictions are ignored in the company. The calculation of the EOQ and the cycle time for each product, as well as the total costs of this policy are presented in Table 11.2.

The joint inventory policy is defined using the JRP using both the IGS and DGS strategies considering the budget and maximum capacity constraints. Tables 11.3 and 11.4 show the best individual obtained by running the genetic algorithms for the IGS and DGS respectively. In both cases, the JRP model was solved with and without constraints. The algorithms used a population of 200 individuals, 500 evolutions as stop condition and a mutation percentage of 0.2.

The joint replenishment plan generated by the best individual for the IGS strategy with restrictions, presented in Table 11.3b, means that products 1, 7, 8, 13, 14, 19 and 20 must be ordered every cycle time of $T = 0.1135$ years; products 2, 3, 4, 9, 15, 21, and 22 every two cycle times; products 10, 16, and 17 every three cycles; products 5, 11, and 23 every four cycles; product 24 every five cycles; 12 every six cycles; the product 6 every seven cycles and the product 18 every 8 cycles. In a similar way, the replenishment plan can be deduced for the solution of the IGS without restrictions.

In the DGS strategy, the joint replenishment plan generated by the best individual with restrictions (Table 11.4b) indicates that the products 5, 6, 11, 12, 17, 18, 23, and 24 belong to group one and must be ordered together every 0.6432 years. The products 2, 3, 4 9, 10, 15, 16, 21 and 22 belong to group two and

Table 11.1 Data for the joint inventory policy definition

Item	1	2	3	4	5	6	7	8	9	10	11	12
$d_i =$	9100	4000	2700	1160	630	178	9200	4350	2700	870	582	196
$s_i =$	48,150	51,060	40,890	42,680	52,650	52,640	36,900	40,020	54,990	37,840	53,100	54,050
$h_i =$	1090	1020	860	1110	930	1050	990	1010	1110	820	820	1200
$b_i =$	5250	5625	6375	7125	7187.5	6687.5	5000	6125	5625	6187.5	5250	5000
$f_i =$	3	5	6	5	6	2	4	3	6	4	3	5
Item	13	14	15	16	17	18	19	20	21	22	23	24
$d_i =$	9200	4400	3360	1170	666	164	9900	6000	3090	930	516	206
$s_i =$	46,350	45,540	42,300	37,400	47,250	51,700	46,350	51,060	47,470	39,600	50,400	40,890
$h_i =$	1060	980	1100	820	1060	820	1100	910	1060	1190	880	1120
$b_i =$	5375	6750	6125	5375	5750	5250	5937.5	6812.5	6750	5250	6562.5	6250
$f_i =$	3	5	4	5	7	2	4	3	5	4	3	5

Table 11.2 Amounts to be ordered and total cost using the EOQ method

Item	1	2	3	4	5	6	7	8
EOQ	2116	1458	1280	741	608	304	2185	1497
T	0.23	0.36	0.47	0.64	0.96	1.71	0.24	0.34
Cost	2,306,417	1,487,229	1,100,715	822,468	565,236	319,238	2,163,259	1,511,554
Item	9	10	11	12	13	14	15	16
EOQ	1157	740	623	299	2150	1544	1266	857
T	0.43	0.85	1.07	1.53	0.23	0.35	0.38	0.73
Cost	1,283,857	606,536	510,557	359,045	2,279,230	1,513,280	1,392,452	702,780
Item	17	18	19	20	21	22	23	24
EOQ	580	330	2190	1891	1249	637	563	310
T	0.87	2.01	0.22	0.32	0.40	0.68	1.09	1.50
Cost	614,277	270,327	2,408,548	1,720,458	1,323,685	758,023	495,547	346,966
Total cost	26,861,682							

Table 11.3 Best individual for IGS with and without constraints

(a) Without constraints	(b) With constraints
[0.757, 0.282, 0.194, 0.459, 0.54, 0.537, 0.318, 0.396, 0.456, 0.599, 0.67, 0.527, 0.83, 0.377, 0.212, 0.542, 0.572, 0.766, 0.105, 0.989, 0.01, 0.533, 0.686, 0.443]	[0.681, 0.702, 0.725, 0.555, 0.77, 0.779, 0.062, 0.71, 0.987, 0.913, 0.854, 0.649, 0.876, 0.06, 0.791, 0.671, 0.726, 1.0, 0.262, 0.444, 0.921, 0.464, 0.971, 0.611]
[1, 1, 1, 2, 3, 5, 1, 1, 1, 2, 3, 5, 1, 1, 1, 2, 3, 6, 1, 1, 1, 2, 3, 4]	[1, 2, 2, 2, 4, 7, 1, 1, 2, 3, 4, 6, 1, 1, 2, 3, 3, 8, 1, 1, 2, 2, 4, 5]
Cycle time: 0.1468	Cycle time: 0.1135
Total cost: 13,046,149	Total cost: 13,306,436

Table 11.4 Best individual for DGS with and without constraints

(a) Without constraints	(b) With constraints
[3, 3, 3, 3, 1, 1, 3, 3, 3, 1, 1, 1, 3, 3, 3, 3, 1, 2, 3, 3, 3, 3, 1, 1]	[3, 2, 2, 2, 1, 1, 3, 3, 2, 2, 1, 1, 3, 3, 2, 2, 1, 1, 3, 3, 2, 2, 1, 1]
Cycle time group 1: 0.5824	Cycle time group 1: 0.6432
Cycle time group 2: 2.0102	Cycle time group 2: 0.2239
Cycle time group 3: 0.1556	Cycle time group 3: 0.1135
Total cost: 13,808,461	Total cost: 14,730,043

must be ordered every 0.2239 years and products 1, 7, 8 13, 14, 19, and 20 belong to group 3 and must be ordered every 0.1135 years.

The costs for all the replenishment plans obtained for the JRP model with the IGS and DGS strategies are lower than the inventory policy reported in the

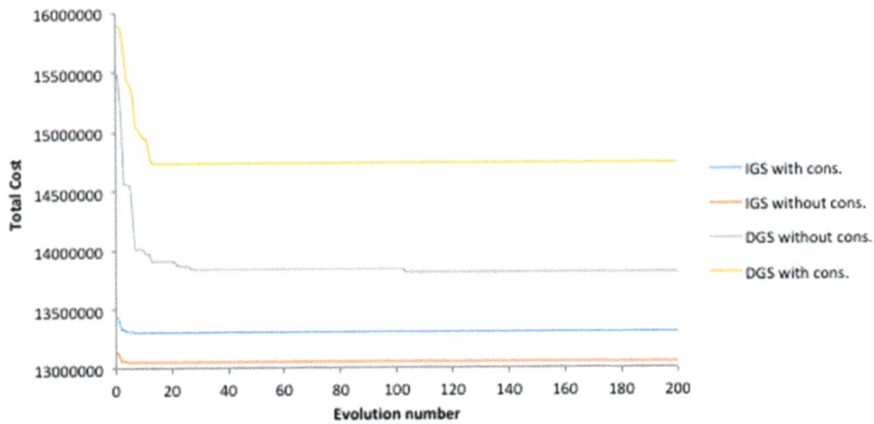

Fig. 11.9 Best individual fitness value evolution for the IGS and DGS strategies. *Source* Own Source

company when the EOQ model is used to individually optimize the inventory costs of each product. The cost reduction is produced by achieved a suitable combination of the fixed ordering costs.

As can be observed from Tables 11.3 and 11.4, the replenishment costs in the IGS strategy are lower than the DGS, in accordance with the specialized literature. The replenishment costs for both strategies are greater when the restrictions are included. This occurs because it is necessary to decrease the cycle times, which reduces the quantities to be ordered in order to accomplish the budget and the storage capacity.

The genetic algorithms developed to solve the JRP model for both for the IGS and the DGS strategies allow including the budget and storage capacity constraints. The algorithms solutions produced the expected results for the JRP and showed a fast convergence to find the best solutions to the model, as it depicted in Fig. 11.9. In this figure it is possible to observe that for the 4 cases analyzed (The IGS and DGS strategies with and without restriction), the best individuals are obtained before the 110 evolutions.

11.5 Conclusions

In this chapter it is presented an application of the JRP using the indirect and the direct grouping strategies (IGS and DGS), including budget and storage capacity constraints. The JRP model allowed generating replenishment plans that reduce logistics costs compared with the individual optimization of all the products using the EOQ model. The cost reduction was produced by a better distribution of the fixed ordering cost, which was shared among the products jointly ordered.

The results obtained with the JRP model using the direct and indirect grouping strategies indicate that the IGS strategy generates lower replenishment costs than those produced by the DGS strategy, as it is reported in the specialized scientific literature. The effect of including the budget and capacity constraints to the JRP model was also analyzed, finding that when working with the constrained model, the costs are increased due to the necessity of decreasing the orders cycle time, what reduces the replenishment quantities ordered from the supplier. This way, the number of orders in a year is increased and the amount of products in every order is reduced, allowing fulfilling the budget and the storage capacity restrictions.

For the solution of the JRP model using both the IGS and DGS strategies, a genetic algorithm was used in each case. Those algorithms allowed obtaining good solutions and including the resource and capacity constraints. The solutions of the algorithm are acceptable and allow to adequately define the replenishment plans for the case of the studied company. Every algorithm also presents a rapid conversion to its best individual, which indicates its ability to solve this kind of problems.

As future research lines, it is recommended to use new restrictions that allow a closer approach to more realistic situations, such as the case of including maximum production and load vehicle capacities. It is also recommended to integrate this approach with the vehicle routing processes, seeking for a joint optimization of transport and inventory processes. It is also interesting to analyze the possibility of studying multiobjective optimization models that allow not only to consider the costs associated with the supply processes, but also includes variables such as service level and the collateral impacts of logistic operations on society and the environment.

References

Arango-Serna MD, Zapata-Cortes JA, Serna-Uran CA (2018) Collaborative multiobjective model for urban goods distribution optimization. In: García-Alcaraz J, Alor-Hernández G, Maldonado-Macías A, Sánchez-Ramírez C (eds) New perspectives on applied industrial tools and techniques. Management and industrial engineering, pp 47–70. Springer, Cham

Arango-Serna MD, Andrés-Romano C, Zapata-Cortes JA (2016) Collaborative goods distribution using the IRP model. DYNA 83(196):204–2012

Arango-Serna MD, Gómez-Montoya RA, Zapata-Cortes JA (2013a) Measurement and improvement of the coal dispatch operation through statistical models R & R. Bol Cienc Tierra 33:135–146 (in Spanish)

Arango-Serna MD, Adarme W, Zapata-Cortes JA (2013b) Collaborative inventories in the optimization of the supply chain. Dyna 80(181):71–80 (in Spanish)

Arango MD, Zapata-Cortes JÁ, Adarme W (2011) Application of the inventory model managed by the seller in a company of the Colombian food sector. Rev EIA 15:21–32 (in Spanish)

Arango-Serna MD, Zapata-Cortes JA, Gutierrez D (2015) Modeling the inventory routing problem (IRP) with multiple depots with genetic algorithms. IEEE Lat Am Trans 13(12):3959–3965

Arango-Serna MD, Serna-Uran CA, Zapata-Cortes JA, Alvarez AF (2014) Vehicle routing to multiple warehouses using a memetic algorithm. Procedia Soc y Behav Sci 160(19):587–596

Bancoldex (2013) Classification of companies in Colombia. Available in: https://www.bancoldex.com/Sobre-microempresas/Clasificacion-de-empresas-en-Colombia315.aspx. Retrieved on Feb 2017 (in Spanish)

Bastos LSL, Mendes ML, Nunes DRL, Melo ACS, Carneiro MP (2017) A systematic literature review on the joint replenishment problem solutions: 2006–2015. Production 27 e20162229:1–11. https://doi.org/10.1590/0103-6513.222916

Cha BC, Moon IK, Park JH (2008) The joint replenishment and delivery scheduling of the one-warehouse, n-retailer system. Transp Res Part E Logist Transp Rev 44(5):720–730. https://doi.org/10.1016/j.tre.2007.05.010

Chen T, Wahab MIM, Ongkunaruk P (2016) A joint replenishment problem considering multiple trucks with shipment and resource constraints. Comput Oper Res 74:53–63

Chopra S, Meindl S (2008) Supply chain management, 3rd edn. Prentice Hall, Mexico (in Spanish)

Coelho LC, Laporte G (2014) Optimal joint replenishment, delivery and inventory management policies for perishable products. Comput Oper Res 47:42–52. https://doi.org/10.1016/j.cor.2014.01.013

Fonseca CM, Fleming PJ (1995) An overview of evolutionary algorithms in multiobjective optimization. Evol Comput 3(1):1–16

Goyal S (1975) Analysis of joint replenishment inventory systems with resource restriction. Oper Res Q 26:197–203

Goyal SK (1973) Determination of economic packaging frequency for items jointly replenished. Manag Sci 20(2):232–235

Goyal SK (1974) Determination of optimum packaging frequency of items jointly replenished. Manag Sci 21(4):436–443

Goyal SK Belton AS (1979) A simple method of determining order quantities in joint replenishment under deterministic demand. Manag Sci 25(6):604

Goyal SK, Deshmukh SG (1993) A note on 'The economic ordering quantity for jointly replenished items. Int J Prod Res 31(12):2959–2961

Goyal SK, Satir AT (1989) Joint replenishment inventory control: deterministic and stochastic models. Eur J Oper Res 38:2–13

Hong SP, Kim YH (2009) A genetic algorithm for joint replenishment based on the exact inventory cost. Comput Oper Res 36(1):167–175. https://doi.org/10.1016/j.cor.2007.08.006

Hoque MA (2006) An optimal solution technique for the joint replenishment problem with storage and transport capacities and budget constraints. Eur J Oper Res 175:1033–1042

Kaspi M, Rosenblatt MJ (1983) An improvement of Silver's algorithm for the joint replenishment problem. IEEE Trans 15(3):264–267

Kaspi M, Rosenblatt MJ (1991) On the economic ordering quantity for jointly replenished items. Int J Prod Res 29(1):107–114

Khouja M, Goyal SK (2008) A review of the joint replenishment problem literature: 1989-2005. Eur J Oper Res 186(1):1–16. https://doi.org/10.1016/j.ejor.2007.03.007

Khouja M, Michalewicz Z, Satoskar S (2000) A comparison between genetic algorithms and the RAND method for solving the joint replenishment problem. Prod Plan Control 11:556–564

Li CY, Gao J, Zhang TW, Wang XT (2014) Differential evolution algorithm for constraint joint replenishment problem. In: 2014 8th international conference on future generation communication and networking. IEEE, New York. https://doi.org/10.1109/fgcn.2014.23

Li C, Xu X, Zhan D (2009) Solving joint replenishment problem with deteriorating items using genetic algorithm. J Adv Manuf Syst 8(1):47–56. https://doi.org/10.1142/S0219686709001626

Moon IK, Goyal SK, Cha BC (2008) The joint replenishment problem involving multiple suppliers offering quantity discounts. Int J Syst Sci 39(6):629–637

Moon IK, Cha BC (2006) The joint replenishment problem with resource restriction. Eur J Oper Res 173(1):190–198. https://doi.org/10.1016/j.ejor.2004.11.020

Nagasawa K, Irohara T, Matoba Y, Liu S (2015) Applying genetic algorithm for can-order policies in the joint replenishment problem. Ind Eng Manag Syst 14(1):1–10. https://doi.org/10.7232/iems.2015.14.1.001

Olsen AL (2005) An evolutionary algorithm to solve the joint replenishment problem using direct grouping. Comput Ind Eng 48(2):223–235. https://doi.org/10.1016/j.cie.2005.01.010

Olsen AL (2008) Inventory replenishment with interdependent ordering costs: an evolutionary algorithm solution. Int J Prod Econ 113(1):359–369. https://doi.org/10.1016/j.ijpe.2007.09.004

Ongkunaruk P, Wahab MIM, Chen Y (2016) A genetic algorithm for a joint replenishment problem with resource and shipment constraints and defective items. Int J Product Econ 175:142–152
Porras E, Dekker R (2006) An efficient optimal solution method for the joint replenishment problem with minimum order quantities. Eur J Oper Res 174(3):1595-1615. https://doi.org/10.1016/j.ejor.2005.02.056
Porras E, Dekker R (2008) A solution method for the joint replenishment problem with correction factor. Int J Prod Econ 113(2):834–851. https://doi.org/10.1016/j.ijpe.2007.11.008
Qu H, Wang L, Zeng YR (2013) Modeling and optimization for the joint replenishment and delivery problem with heterogeneous items. Knowl Based Syst 54:207–215. https://doi.org/10.1016/j.knosys.2013.09.013
Rosenblatt MJ (1985) Fixed cycle, basic cycle and EOQ approaches to the multi-item single-supplier inventory system. Int J Product Res 23(6):1131–1139
Shu FT (1971) Economic ordering frequency for two items jointly replenished. Manag Sci 17(6): B406–B410
Silver EA (1976) A simple method of determining order quantities in joint replenishments under deterministic demand. Manag Sci 22(12):1351–1361
Strijbosch LWG, Heuts RMJ, Luijten MLJ (2002) Cyclical packaging planning at a pharmaceutical company. Int J Oper Product Manag 22(5):549–564
Taleizadeh AA, Akhavan-Niaki ST, Nikousokhan R (2011) Constraint multiproduct joint-replenishment inventory control problem using uncertain programming. Appl Soft Comput 11:5143–5154
Tiwari A, Roy R, Jared G, Munaux O (2002) Evolutionary-based techniques for real-life optimisation: development and testing. Appl Soft Comput 1(4):301–329
Van Eijs MJG, Heuts RMJ, Kleijnen JPC (1992) Analysis and comparison of two strategies for multi-item inventory systems with joint replenishment costs. Eur J Oper Res 59:405–412
Vergidis K, Saxena D, Tiwari A (2012) An evolutionary multi-objective framework for business process optimization. Appl Soft Comput 12(8):2638–2653
Viswanathan S (1996) A new optimal algorithm for the joint replenishment problem. J Oper Res Soc 47(7):936–944
Wang K, Salhi A, Fraga ES (2004) Process design optimisation using embedded hybrid visualisation and data analysis techniques within a genetic algorithm optimisation framework. Chem Eng Process 43(5):657–669
Wang L, Dun CX, Bi WJ, Zeng YR (2012) An effective and efficient differential evolution algorithm for the integrated stochastic joint replenishment and delivery model. Knowl Based Syst 36:104–114. https://doi.org/10.1016/j.knosys.2012.06.007
Wang L, Qu H, Li Y, He J (2013) Modeling and optimization of stochastic joint replenishment and delivery scheduling problem with uncertain costs. Discrete Dyn Nat Soc 2013:1–12. https://doi.org/10.1155/2013/657465
Yang W, Chan FT, Kumar V (2012) Optimizing replenishment polices using Genetic Algorithm for single-warehouse multi-retailer system. Expert Syst Appl 39:3081–3086
Zapata-Cortés, JA (2016) Optimization of merchandise distribution using a multiobjective genetic model of collaborative inventory of m suppliers with n customers. Doctoral thesis, Universidad Nacional de Colombia—Sede Medellín (in Spanish)
Zeng YR, Wang L, Xu XH, Fu QL (2014) Optimizing the joint replenishment and delivery scheduling problem under fuzzy environment using inverse weight fuzzy nonlinear programming method. Abstr Appl Anal 2014:1–13. https://doi.org/10.1155/2014/904240

Chapter 12
Improvement of the Demand Planning of Imported Seeds in the Company Agro Perú SA

Leonilda Coba, Cesar Shimabuku and Jesús González-Feliu

Abstract This chapter analyzes the current structure of the demand planning in Agro Peru SA and the potential of rethinking the demand planning structure and decisions on the whole supply chain performance. Using a methodology that combines the CSAR framework for rethinking supply chain strategies, a multi-criteria Analytic Hierarchy Process method, and a Sales and Operation Planning methodologies. The present work proposes a diagnosis of demand planning operations, strategy and a management plan to improve them. First, the chapter presents the company and the agricultural sector in Peru. Second, the methodological framework is described. Third, the main results are presented and discussed. Results indicate that it is necessary to elaborate two internal projects to improve the forecast of demand and thus reduce inventories without affecting sales.

Keywords Demand planning · Management · Forecasting

12.1 Introduction

According to the report published by the National Institute of Statistics and Informatics, the agricultural and fishing sector accounted for 5.17% of Peru's GDP in 2017. It also indicates that the industry has a high use of unskilled labor, with livelihood in direct to 2.3 million households. Within these 2.3 million are the customers of the agricultural sector which are served by companies such as Agro Perú S.A.

L. Coba (✉) · C. Shimabuku
Department of Supply Chain Management, Universidad Del Pacifico,
Av. Gral. Salaverry 2020, Jesús María, 15072 Lima, Peru
e-mail: l.cobaarribasplata@alum.up.edu.pe

J. González-Feliu
Institut Henri Fayol, 158 cours Fauriel, CS 62362,
42023 Saint-Etienne Cedex 2, France

J. L. García Alcaraz et al. (eds.), *Best Practices in Manufacturing Processes*,
https://doi.org/10.1007/978-3-319-99190-0_12

Agriculture activities are essential contributions to a country's economy, mainly in developing countries. For example, 11% of Peruvian territory is considered to be farmed areas (CAT 2017), which includes both agricultural and livestock activities, becoming one of the most important sectors for financial support of millions of homes. Moreover, this sector is mainly composed of small farmers, who have between 3 and 5 ha used for agricultural production (CAT 2017). In that context, the agricultural supply chains are mainly related not to organized agroindustry groups but too small producers, with local and national markets dominating the agrarian economy (although some exportations are seen) and a low level of integration in supply chain practices.

Providing seeds to those small agriculture producers is a key economic activity that needs to be developed efficiently, as the seed supply is an important market and their customers (agriculture producers of small size) rely on the efficiency of seeds providers for their effectiveness. To do that, a supply chain strategy viewpoint seems essential. Although agri-food logistics systems are already studied in the literature, they deal mainly with end-consumer agricultural products (like fruits and vegetables) or agroindustry supply chains, but the field of seeds remains in general an input of those supply chains. However, to ensure the efficiency of the whole supply chain, it is important to verify the efficiency of seeds supply. And as observed in the Peruvian field, the seed supply chains present several improvement points that need to be identified and processed.

To address those issues, but also in a generalization aim to propose a framework to analyze and improve the seeds supply chains, this paper proposes an analysis of the imported supply chain system of Agro Peru S.A., a Peruvian agricultural company, in a generalization perspective. To do that, and on the base of a case study, the main supply chain strategies are identified, and analysis of improvement is made on both supply chain and project management viewpoints. First, the background, context, and hypotheses are presented, including a presentation of Agro Peru S.A. Then, the methodological framework used here is described. After that, the main results are presented and discussed. Finally, as a conclusion, managerial implications, are presented.

12.2 Background, Problem and Hypotheses Statement

12.2.1 Background and Context

In recent years the pace of growth of the company has increased, so it is continuously evaluated whether it is possible to make improvements in its strategic, key and support processes. One of the objectives is to review the procedures of inventory management linked to the planning of demand and the supply of products.

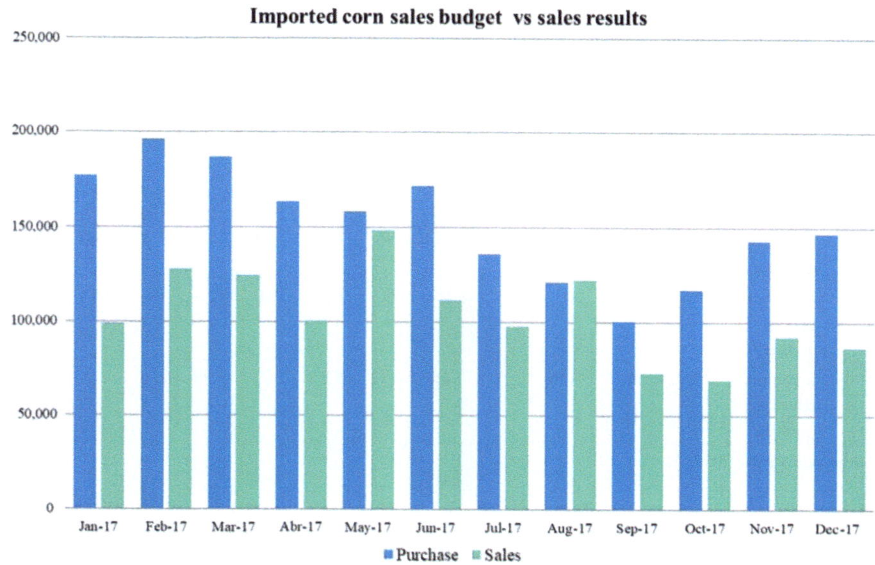

Fig. 12.1 Imported corn sales budgets versus sales results 2017

According to the analysis of the investigation, in 2017 the company has had an overstock of the product under study which has increased the handling costs of the inventory. For example, the behavior of sales versus the budget that the company manages from January to December of the year 2017 was the following:

Figure 12.1 presents for each month two columns representing respectively forecasted and real sales. As we can see in each right column, the behavior of sales was inaccurate compared to the projection of their budgets. Table 12.1 shows the measurement of the inventory coverage that is by the current purchase planning, where it was observed that it has an average of 3.5 months, and should be according to the policy of the company a maximum of two months:

Also, the average accuracy of the sales budget versus sales was 70% and must be at least 90% according to the company's policy and nature of the business.

Having identified the problem of the overstock, we seek to review the expenses of maintaining an inventory because of not having more detailed planning as follows in Table 12.2.

In 2016 the total cost of maintaining inventory in 2016 was USD 1,570,686. In the context of imported corn seeds, an vital bias in forecasting (which, for the proposed case, often overestimates sales) leads to an increase of costs since unsold corn seeds are however imported and activate logistics costs which are non-negligible (mainly for transport and inventorying).

For that reason, the imported corn supply division needs to improve both the forecasting accuracy but also their supply strategies and tactics to deploy a more flexible and reactive supply chain. Thus, the focus of this chapter resides on imported corn supply planning and management operations.

Table 12.1 Inventory report of imported corn 2017

Unit	Description	Jan	Feb	Mar	Apr	May	Jun	Jul	Ago	Sep	Oct	Nov	Dec
(t)	Initial inventory	596	497	557	429	305	354	506	628	531	408	336	267
(t)	Purchase	177	196	187	163	158	172	136	121	101	117	143	147
(t)	Buy			159			150	300	234				
(t)	Sales	99	128	124	100	148	112	98	122	73	69	92	86
(t)	Final inventory	497	369	592	329	156	393	708	740	458	339	240	180
	Accuracy versus budget (%)	56	65	67	61	94	65	72	101	72	59	65	59
	Inventory coverage (months)	3.97	2.97	5.16	2.70	1.46	4.32	6.23	5.91	3.99	2.76	1.74	0.86

Source Agro Peru 2017

12.2.2 Literature Review

The supply chain can be defined as a group of entities which are directly involved in product flow, services, money, and information provided by the customer (Mentzer et al. 2001). While the supply chain strategy, is defined as a collection of general and specific objects, as well as, policies and decisions made in the supply chain to align their operations to the overall strategy of the company (Perez-Franco et al. 2016; Perez-Franco 2016).

However, and although the definition of supply chains can be done with different frameworks (Lambert et al. 1998; Beamon 1999; Christopher and Towill 2001; Lambert 2008), those frameworks address physical and informational flows, but do not catch the issues of supply chain strategies. To deal with that lack, Perez-Franco et al. (2016) developed the CSAR methodology (Conceptual System Assessment and Reformulation). Authors propose it to develop a preliminary understanding (at a theory level) of supply chain strategy at the level of business units, then to have a methodology applicable also in practice. The method allows identifying supply chain strategies, assessing them and proposing improvement elements. Authors applied it to a laminates' supply chain.

However, agricultural supply chains do not yet follow a supply chain strategy analysis vision, and among them, seed supply chain is little studied as a central part of agricultural supply chains. Indeed, it is often seen as a part of the providers and supply strategies but often remains seen as an input for supply chain optimization or as a fixed or observed environment to which few can be changed (Lazzarini et al. 2001; Bevilacqua et al. 2009; Morganti and Gonzalez-Feliu 2015; Palacios-Argüello et al. 2017).

Some works focusing on supply chain design and strategic planning optimization consider the seed production and distribution stages (Ahumada and Villalobos 2009)

Table 12.2 Cost without demand planning in 2016

Item/Month	Cost	Jan	Feb	Mar	Apr	May	Jun
Final inventory (units)		497,034	369,144	403,794	303,324	154,974	193,254
Unit cost (USD)	5.43	2,698,895	2,004,453	2,192,602	1,647,050	841,510	1,049,370
Weighted average cost of capital	5.80%	156,536	116,258	127,171	95,529	48,808	60,863
Maintaining inventory costs (insurance, taxes, handling) (USD)	0.09%	2510	1864	2039	1532	783	976
Warehouse cost × m^2 (USD)	0.02	9788	7269	7952	5973	3052	3806
Inventory risks costs (handling damage, etc.)	0.84%	22,563	16,757	18,330	13,769	7035	8773
Maintenance total		191,396	142,149	155,492	116,803	59,677	74,418
Total inventory cost (inventory cost + maintenance)		2,890,292	2,146,602	2,348,094	1,763,853	901,187	1,123,788
Item/Month	Cost	Jul	Ago	Sep	Oct	Nov	Dec
Final inventory (units)	395,446	507,466	434,596	365,596	270,346	183,916	395,446
Unit cost (USD)	2,147,270	2,755,539	2,359,855	1,985,185	1,467,977	998,662	2,147,270
Weighted average cost of capital	124,542	159,821	136,872	115,141	85,143	57,922	124,542
Maintaining inventory costs (insurance, taxes, handling) (USD)	1997	2563	2195	1846	1365	929	1997
Warehouse cost × m^2 (USD)	7787	9993	8558	7199	5324	3622	7787
Inventory risks costs (handling damage, etc.)	17,951	23,036	19,728	16,596	12,272	8349	17,951
Maintenance total	152,277	195,413	167,353	140,783	104,104	70,822	152,277
Total inventory cost (inventory cost + maintenance)	2,299,547	2,950,952	2,527,208	2,125,967	1,572,081	1,069,484	2,299,547

Source Agro Peru 2017

but remain either theoretical or optimization-based, not relying on the definition of strategies but only on the combinatorial optimization visions. Other works deal with tactical and operational aspects (Albrecht 2010), mainly on coordination issues, but do not necessarily address the strategy identification and assessment issues. For those reasons, we aim to apply and extend the CSAR methodology to the seeds distribution field via the proposed case of Agro Peru S.A.

12.3 Methodologies

12.3.1 CSAR Methodology

This methodology was used to capture the supply chain strategy, based on the supply area of Agro Peru S.A. Consider the following steps to achieve the objective of rethinking and studying the processes as shown in Fig. 12.2.

To identify the critical area in the supply chain of Agro Perú SA (from the perspective of the supply or logistics area), the CSAR methodology was used, following the procedures starting from capturing the processes throughout the chain

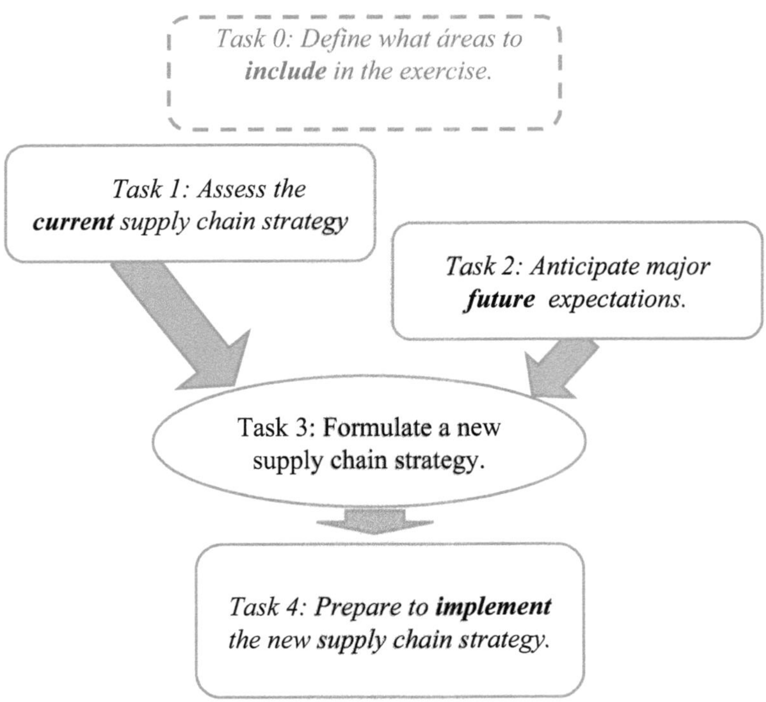

Fig. 12.2 Introduction to supply chain

to the rethinking of the conceptual map with the processes and main sub-processes. All areas of the operation must be analyzed before rethinking a new strategy in the supply chain in this case supply.

It is essential to comply with the four tasks to achieve a correct analysis; work teams must be formed, committed to meeting the objectives for this, the following team must be defined:

- Project leader
- Main team (people who participate in all activities)
- Evaluation team
- Advisory team

This project included the participation of the central unit in the logistics area of the Agro Peru Company, which was led by the Operations and Logistics Manager.

First, four pillars were identified in conjunction with those responsible for the logistics area; we sought to determine the strategic objective and found the following: "Contribute to the growth of all our customers by providing quality products, at a lower price and in the expected time." The pillars found in this first capture were those indicated below, each one aligned with its sub-processes:

- Manage purchasing planning.
- Comply with the delivery of the products.
- Control and verify purchases with quality.
- Generate savings in the supply area.

Each of the pillars was named with a letter and a number starting from A1, A2, A3, and A4, and the next level with the letter B1, B2 and so on. For the correct identification and at the same time allow the evaluation of each level with the following methodology (AHP) under the defined criteria. This methodology allows the integration of all the areas involved and seeks to review the processes and strategies openly to contribute to the decision-making of the strategic areas of the company.

12.3.2 AHP Methodology

For the determination and validation of the alternatives, the AHP methodology was used, which helped us to focus on the most important option to address. It concluded that we should focus on the demand planning that is the beginning of all the supply in the company, it was not done efficiently, and there could be a reduction of costs.

With this methodology, we developed the criteria to be applied in the validation of each alternative level, through a score of greater importance that was developed by the team. This methodology is designed to review problems with multiple criteria, based on subjective evaluations regarding the importance of each criterion,

Table 12.3 Scale criteria and description

Number	Scale of verbal criteria	Description
1	Equally important	Both options are equal
3	Moderately important	Slightly preference between options
5	Strongly important	Preference for one option
7	Very important	High preference for one option
9	Extremely important	Absolute preference for an option
2, 4, 6, 8		Intermediate of the previous values

Source Saaty (1990)

to then show the preference of each decision alternative. After having identified the pillars of the first capture, we continue with the assignment of the following criteria as shown in Table 12.3.

After identifying the criteria and the pillars, we start with the hierarchy for the evaluation and selection of the pillar that leads us to the second analysis of the recapture of the capture according to the CSAR methodology.

12.3.3 Oliver Wight Checklist

The evaluation of the demand planning area was conducted with a survey of the leading managers according to the Oliver Wight Class A checklist obtaining the following results shown in Table 12.4.

- In this evaluation, an average of 1.33 was obtained.
- To be considered acceptable, the diagnosis must have a minimum average of 4.5; therefore, the company needs to have a demand planning area.

12.3.4 S&OP Methodology

This methodology allows balancing the demand and supplies in the integration process with the integration of the sales, finance, marketing, and operations areas in a short and medium-term horizon.

This methodology was used in the present work to complement the solution in one of the pillars aligned with the change management that was found in the stakeout of the pillars that we will see in the next chapter.

It was found that there was no communication integration between all the areas for the correct development of the operations, starting from the strategic areas to the functional areas which show as an opportunity to integrate with the S&OP methodology.

Table 12.4 Oliver wight checklist in Agro Perú

	Not done	Poor	Regular	Good	Very good	Excellent
1. Processes are effectively controlled						
Inventory and service levels			X			
Planning from start to finish			X			
Resource management			X			
Integrated planning processes		X				
Fulfillment of the planning (complete and on time)			X			
2. Use of technical tools						
Advance planning	X					
Scenario analysis			X			
Options to meet the demand profitably and quickly		X				
Product delivery options		X				
3. Inventory control processes						
Accuracy of inventory management		X				
Process control		X				
Accuracy (at least 95%)		X				
Punctuation	0	6	10			

Source Wight (2005)

12.4 Diagnosis Results

12.4.1 Stakeout of the Capture of the Processes of the Supply Area

Through the CSAR methodology we carry out the rethinking of the catch in the operations of the supply area together with those responsible for the area obtaining the following result (see Fig. 12.3).

The main pillars are

- Manage supply planning.
- Comply with the delivery of products.
- Control and verify purchases with quality.
- Generate savings in the supply area.

To begin with, the selection of the pillars to be developed, we perform the hierarchy with the AHP methodology. From everything evaluated, it can be concluded that the critical process to study in this research work is A, as shown in Table 12.5 that corresponds to "Manage supply planning."

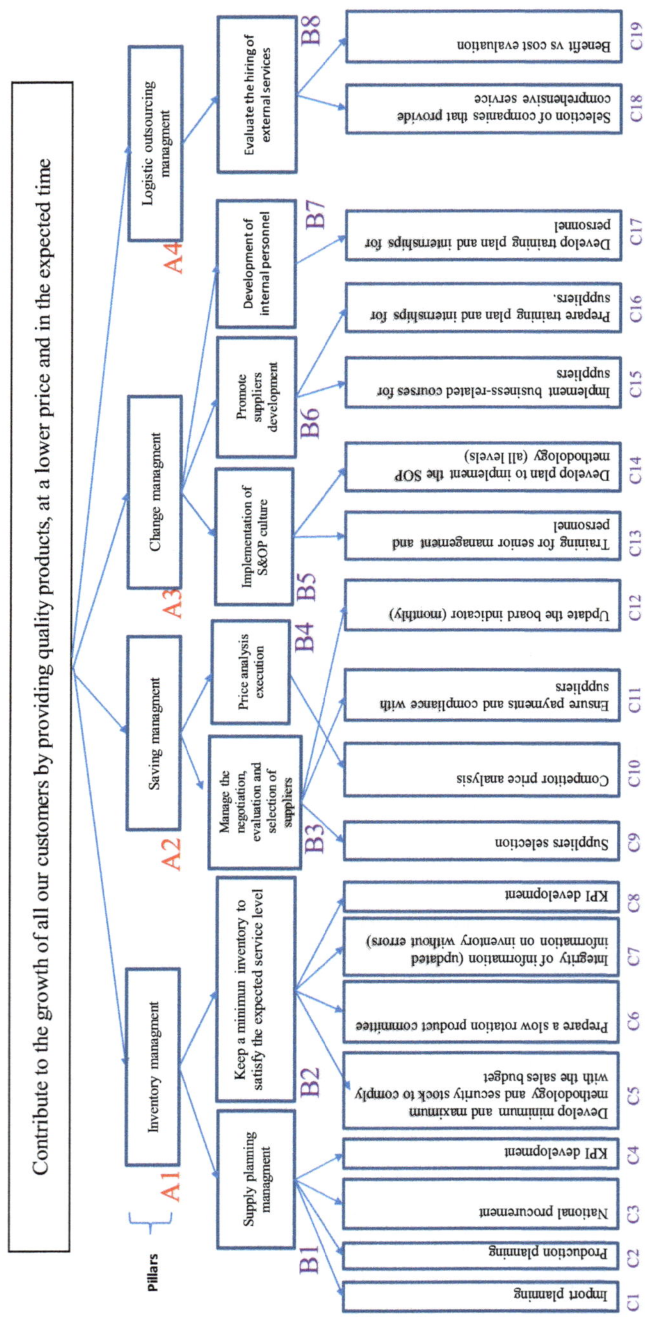

Fig. 12.3 Improved conceptual map

The result of the analysis reveals that the critical area of the supply chain is the A1 because it has a higher score on the average of the study followed by the A3, as shown in the following table of standardization. (Table 12.6).

In conclusion, the analysis sustains that the outstanding option is "managing inventory management." followed by "managing change management," which is the orientation of this research work. For this it has been considered important to start with the following objectives:

- Implementation of an area of demand planning to achieve a good "manage inventory management" that will have as an objective the planning of supply and the reduction of inventories.
- Start with the proposal of the implementation of the SO & P methodology aimed at "managing change management" that will strive to mobilize people inside and outside the company, seeking an integration that generates value as a whole to the organization and strengthens its growth.

12.4.2 Analysis of the Impact on the Business Line of Imported Corn

After having identified the need for a demand planning area, the impact that inventory management currently has and then what is necessary to implement it is

Table 12.5 Matrix to manage supply planning

Alternatives	A1	A2	A3	A4
A1	1.000	9.000	5.000	5.000
A2	0.111	1.000	1.000	3.000
A3	0.200	1.000	1.000	3.000
A4	0.200	0.333	0.333	1.000
Total	1.511	11.333	7.333	12.000

Source Saaty (1990)

Table 12.6 Standardization

Alternatives	A1	A2	A3	A4	Score	%	Position
A1	0.662	0.794	0.682	0.417	0.639	63.86	1
A2	0.074	0.088	0.136	0.250	0.137	13.70	3
A3	0.132	0.088	0.136	0.250	0.152	15.17	2
A4	0.132	0.029	0.045	0.083	0.073	7.26	4

Source Saaty (1990)

Table 12.7 Seed line ABC classification

	Zone	Items	Goods percentage	Accumulated percentage	Investment percentage	Percentage of accumulated investment
0–90%	A	9	20	20	78	78
80–95%	B	12	26	46	17	95
95–100%	C	25	54	100	5	100
	Total	46	100		100	

Source Agro Perú (2017)

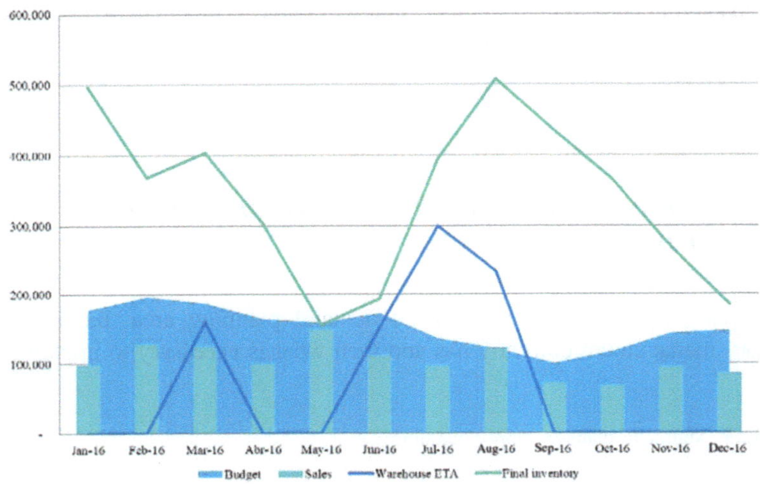

Fig. 12.4 Inventory behavior during 2016

evaluated. The analysis of Pareto ABC products has been carried out under the investment criterion, to identify which products have the highest investment in the company and the rotation of these, shown in Table 12.7.

As can be seen, "A zone" is made up of nine products that represent 20% of the product line of the seed line that represents imported yellow corn seed, and these are responsible for 78% of the company's investment.

Where we find as a result of the planning curve the graph in Fig. 12.4 that points out the need for the company to have correct planning.

As you can, observe the behavior of the inventory is high compared to the budget or sales that the company has in the indicated products; that is, it is immobilized money.

12.4.3 *Justification of Selection of Critical Processes*

The company recognized various factors when forecasting the sale of the fiscal year/period. Among the objective factors it is considered:

- Previous year's sales
- Replenishment waiting time (imports)
- Price discounts for buying full containers
- Country's economy (exchange rate)
- Competitor shares (new products)

However, the problem is that there is no demand forecasting method based on any theory and formula or simulation. The company makes a qualitative simulation, applies time series (if possible), and includes causes that may have altered the demand in a period. There is no record of when, how, or why decisions were made to correct errors or deviations presented.

After having made the diagnosis and evaluation in the present research work has taken the initiative to focus on the development of the process of demand planning from the supply chain point of view, considering that:

- The company is growing; there is an increase in unexpected purchases to meet demand.
- Demand planning process is not precisely executed by product divisions (different marketing approaches).
- There is no documentation of the execution of the sale or statistical methodology for generating the forecast.
- Budget per month is executed annually and revised monthly making slight adjustments.
- There are 990 SKUs at the company level and 46 SKUs in the Seeds division.
- The dependence on cash flow in the Credit and Collections area.
- The forecast has been calculated with the seasonal variation with the trend of imported corn to find the demand for 2017 as shown in Annex

Why "Demand Planning" was chosen as part of the pillar "Manage Inventory Management."

According to the diagnosis made of Agro Peru in the crucial area of the supply chain carried out, it is observed that the capture of the first pillar focus on "Managing supply planning." Then in the rethinking of the process according to the CSAR methodology, it becomes "Manage Inventory Management," this is because the scheduling of the supply becomes a sub-process of the same as seen in Fig. 12.5.

The sub-pillars or sub-processes are adapted to have a balance with demand planning for a correct supply, both in imports and national purchases. That is why it was necessary to initiate the change from the creation of the Demand Planning area that directly impacts the first pillar of the Administration of the inventory management that has the company as "supply planning."

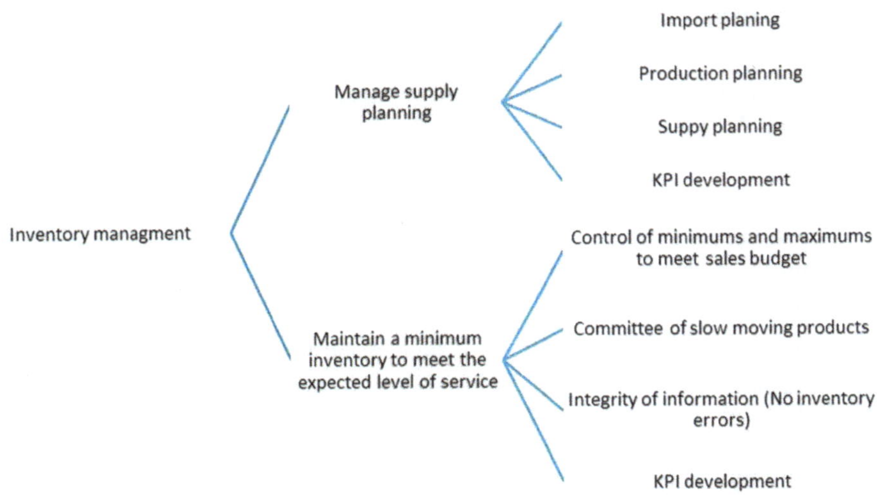

Fig. 12.5 Inventory management pillar detail

According to Suni Chopra's Supply Chain Management book, it tells us that a company must have a correct supply management to improve the synchronization with the supply chain, so it must correctly manage the demand to maximize the supply. Profitability as a company, therefore, the planning of the supply chain works together with inventory management and the alignment that exists between supply and demand.

12.5 Proposed Improvement Project and Expected Results

After having applied the methodologies, the results obtained show us that the outstanding options are: "managing the inventory management," followed by the "managing change management," which is the orientation of the present research work. Concluding that they should start with implementing the following projects (see Fig. 12.6):

- Implementation of a demand planning area to achieve a good "manage the inventory management" that will have as an objective the sourcing planning and the reduction in inventory management.
- Start with the proposal of the implementation of the SO & P methodology aimed at "managing change management." The primary objective is to mobilize people inside and outside the company, seeking an integration that generates value as a whole to the organization and strengthens its growth.

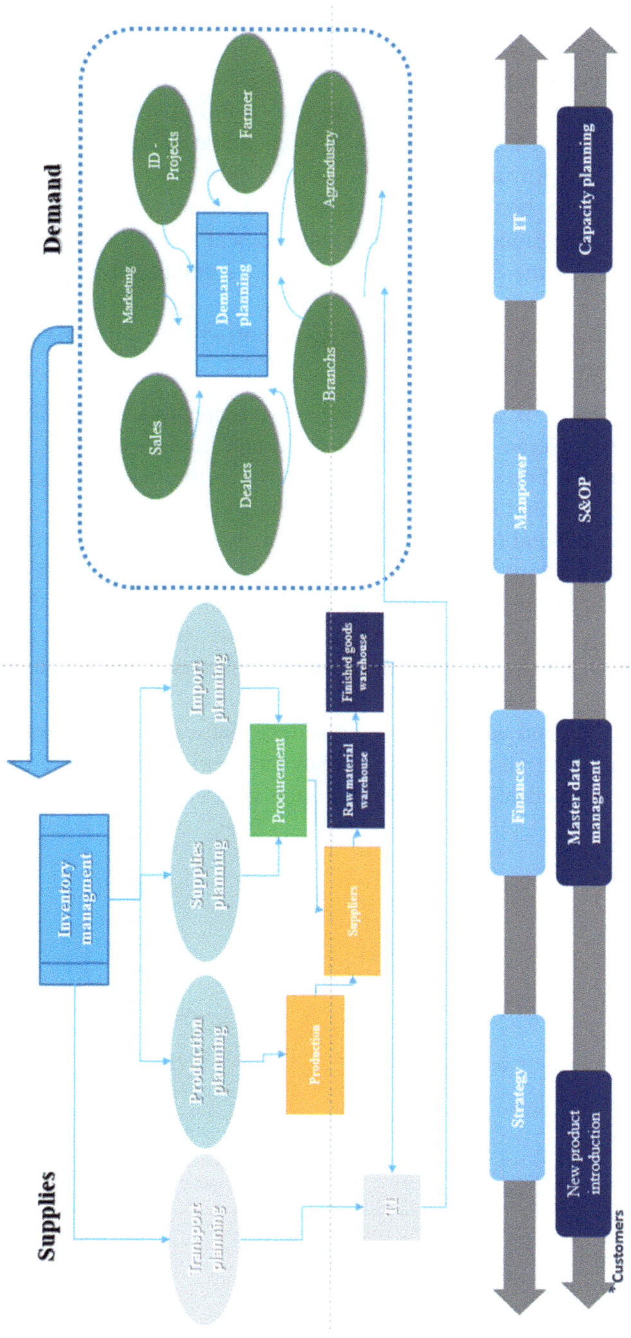

Fig. 12.6 Ideal process map

12.5.1 Project 1: Development of the Area of Demand Planning

The area of demand planning proposed will be within the structure of the Operations and Logistics Management as shown in Fig. 12.7.

The area must have a demand planner for each line of business; in this case, one is being considered for the seed division of the corn line.

According to the book "Best practices of demand management" (Crum and Palmatier 2003), the process of demand management is not the responsibility of the Commercial area or the Supplies area. Instead is a collaborative work between all them.

It is considered that the demand management will be able to work more efficiently when they can count on:

- The commitment of senior management (participation), especially of the business lines considered to be the most important in each division, and it is they who decide the horizon. Also, you must have the close participation of the general manager and the operations and logistics manager.
- Sales leadership and marketing management, sales management and marketing must have definite plans for the sales and launch budgets that they have during the projection of the following year. The Seeds line works directly with a team of developers that act as marketers in the fields; Also, there are demonstration internships that help generate the loyalty of farmers with the company Agro Peru.
- A full-time demand administrator considers that a person is hired to approach and integrate with all areas and the company begins to generate the change.
- Effective communication between the sales and supplies area.

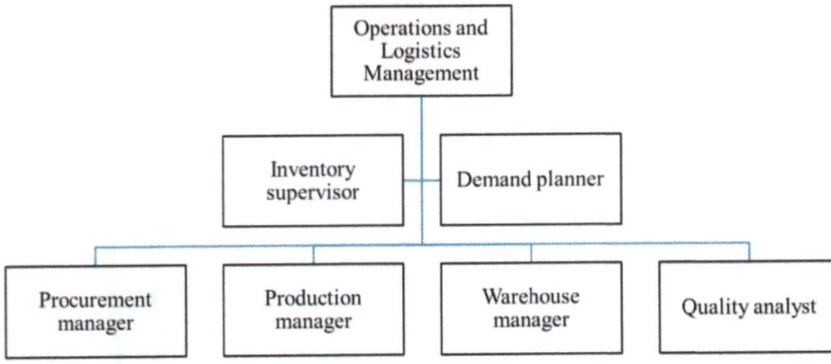

Fig. 12.7 New operations and logistics structure

12.5.2 *Project 2: Foster the Culture of S&OP Methodology*

Sales and Operations Planning is a process that must be carried out on a monthly basis between the Operations, Sales and Finance areas for the review of supply projections, demand and financial results. It will be led by the demand planner of the corn seed line. The S&OP seeks a balance between demand and the supply chain.

12.5.2.1 Steps for the Implementation of the S&OP

The book "Enterprise Sales and Operations Planning: Synchronizing Demand, Supply and Resources for Peak Performance" by Palmatier and Crum (2002) states that for efficient management, sales and operations planning processes require the integration of:

- People
- Processes
- Tools

The book sustains that people perform the processes, so they should know and understand what they are doing and understand what is expected of them. Operations represent the need for defined methods with inputs, outputs, methods, results, and measurements. The tools represent the need to provide the means to people to comply with their procedures.

According to the Kodak model, according to Palmatier and Crum (2002), the S&OP process is considered strategic and is driven by demand (see Fig. 12.8).

Where:

- *Product review*: it refers to product development and new launch plans; this is related to the Marketing area.
- *Demand review*: it refers to the updated forecast along with the offer. It is the work of the Demand Planning area.

Fig. 12.8 S&OP process

- *Review of the chain or supply*: It's related.
- *Financial Projections*: related to the financial review, investment, cash flow from a horizon of 12–18 months.
- *S&OP meeting:* that is where the decisions of the actions are taken.

Table 12.8 shows some activities that must be carried out in the S&OP meetings. In the meeting the demand planner must have a clear agenda, for example:

- Review of main indicators.
- Confirm a monthly validation of the forecast in units and value with the horizon for the full year.
- Evaluation of adjustments for the forecast.
- The estimated time of the meeting should not exceed two hours.
- Participants: Demand planner, head of the Corns Line, Administration and Finance Manager, General Manager, Commercial Manager and Logistics and Operations manager (optional).

12.6 Conclusions

- This chapter presented the assessment and analysis of an agro-industrial supplier company supply strategy, via the combination of the CSAR methodology an AHP method and an S&OP framework. The research focused on the business division of seeds, maize line, specifically corn imported from Agro Perú S.A.
- Currently, the company does not have an area of demand planning or demand management that is responsible for consolidating all the information of the commercial area, development area and marketing for the proper preparation of the supply area, both nationally and internationally.
- The Rethinking methodology evaluates the company's need for better inventory management and integration of the areas through the S&OP methodology.
- Through the methodology of the Checklist for the excellence of Oliver Wight, the current demand planning process was evaluated, obtaining. As a result an average of 1.33 of 4.5 (acceptable data), which demonstrates the need to have a demand planning area in the company Agro Perú SA
- Based on the need to create a demand planning area, the company Agro Perú S. A. must acquire Forecast Pro software that allows the interaction and analysis of the data.
- The results showed the potential of the framework, which can be generalized to other agribusiness companies. Indeed, CSAR methodology being business unit (BU) oriented, it can be extended to other supply chains if BUs can be identified. That supposes identifying the main activities in the supply chain and the primary stakeholders (internal and external). Frameworks like that of Lambert et al. (1998) or Mentzer et al. (2001) can be used to identify supply chain processes and relate them to BUs.

Table 12.8 Activities established for S&OP meetings

Task	Action	Detail
Initial forecast adjustment	Forecast prepared according to:	Previous sales
		For each sales channel
		By division (seeds, corn, etc.)
		By SKU (corn): national and imported
Incorporation of commercial and finances input	Commercial	Changes in the client portfolio
		Sector behavior
		Changes in market trends
		Sell-out information
	Finance	Price changes
		Return of investment of promotions, special sales
	Marketing corn division	Portfolio changes
		Discounts modifications
		Meteorological information
Review with division management	Review of previous months	
	Recommendation of promotions, inventory reduction of slow-moving goods	
	Review of forecast proposal by month and corn line	
Review with commercial team	Review of previous months	
	Monthly forecast review of the corn line	
	Proposed aggregate forecast review per channel per month	
Final review with general management	Prognosis review by division per month	
	Identification of differences against the initial budget	
Capacity assessment and communication of the final forecast	Final update of the forecast	
	Delivery of forecast by SKU to the planning team of procurement and production	
	Review of capacities and inventories	
	Update and communication of adjusted forecast	

Source Agro Perú (2017)

- Moreover, S&OP, being already in a standard form, is used by various companies and is able to be adapted to different cases. AHP and other decision support frameworks can complete the analysis framework to prioritize and evaluate possible strategies and projects, but a more in-depth analysis of those methods seems to be still made to state on the most suitable structure to choose among proposed improvement alternatives.

References

Ahumada O, Villalobos JR (2009) Application of planning models in the agri-food supply chain: a review. Eur J Oper Res 196(1):265–266

Albrecht M (2010) Supply chain coordination mechanisms: new approaches for collaborative planning. Springer, Berlin, pp 265–266

Beamon BM (1999) Measuring supply chain performance. Int J Oper Product Manag 19(3): 275–292

Bevilacqua M, Ciarapica FE, Giacchetta G (2009) Business process reengineering of a supply chain and a traceability system: a case study. J Food Eng 93(1):13–22

CAT Project (2017) The agricultural and livestock sectors in Peru. Climate, agro and risk treatment. Accessed at: http://www.agroyriesgo.pe/informacion/cifras-y-datos/los-sectores-agricola-y-pecuario-en-el-peru/ (in Spanish)

Christopher M, Towill D (2001) An integrated model for the design of agile supply chains. Int J Phys Distrib Logist Manag 31(4):235–246

Clima, Agro y Transparencia de Riego (CAT) (2017) The agricultural and livestock sectors in Peru. Agroyriesgo.pe. Accessed at: http://www.agroyriesgo.pe/informacion/cifras-y-datos/los-sectores-agricola-y-pecuario-en-el-peru/ (in Spanish)

Crum C, Palmatier G (2003) Demand management best practices: process, principles and collaboration. J. Ross Publishing, Incorporated, Estados Unidos

Lambert DM (2008) Supply chain management: processes, partnerships, performance. Supply Chain Management Institute

Lambert DM, Cooper MC, Pagh JD (1998) Supply chain management: implementation issues and research opportunities. Int J Logist Manag 9(2):1–20

Lazzarini S, Chaddad F, Cook M (2001) Integrating supply chain and network analyses: the study of netchains. J Chain Netw Sci 1(1):7–22

Mentzer JT, DeWitt W, Keebler JS, Min S, Nix NW, Smith CD, Zacharia ZG (2001) Defining supply chain management. J Bus logist 22(2):1–25

Morganti E, Gonzalez-Feliu J (2015) City logistics for perishable products. The case of the Parma's Food Hub. Case Stud Transp Policy 3(2):120–128

Palacios-Argüello L, Morganti E, Gonzalez-Feliu J (2017) Food hub: Una alternativa para alimentar las ciudades de manera sostenible. Rev Transp Territ 17:10–33

Palmatier G, Crum C (2002) Enterprise sales and operations planning: synchronizing demand, supply and resources for peak performance. J. Ross Publishing Incorporated, Estados Unidos

Perez-Franco R (2016) Rethinking your supply chain strategy. A brief guide. Massachusetts Institute of Technology, Cambridge

Perez-Franco R, Phadnis S, Caplice C, Sheffi Y (2016) Rethinking supply chain strategy as a conceptual system. Int J Prod Econ 182:384–396

Porter M (2004) Competitive advantage: creation and maintenance of superior performance. Mexico, D.F. Continental Editorial (in Spanish)

Saaty T (1990) How to make a decisión: the analytic hierarchy process. Eur J Oper Res 48, 1990. North Holland: Elsevier Science Publishers B.V. Accessed at: http://linkinghub.elsevier.com/retrieve/pii/037722179090057I
Wight O (2005) Class A checklist for business excellence, 6th edn. Oliver Wight Editorial EAME LL, United States

Chapter 13
Numerical Analysis in a Beverage Can Utilizing Tube Hydroforming Process

Jorge Carlos León-Anaya, Juan Carlos Cisneros-Ortega, Eloy Dimas-Celestino and José Antonio Juanico-Lorán

Abstract Numerical analysis for Tube Hydroforming (THF) was developed in this work to predict the behavior of extruded aluminum tube in the forming die for beverage can applications. THF is a metal-forming process dependent on three parameters: friction between the tube and the die, internal pressure, and material properties of the tube. For a proper simulation, experimental determination of the mechanical properties of aluminum 6061-T5 was conducted, test specimens were obtained directly from the aluminum tube. Numerical simulation and design of experiments were performed to obtain interactions between factors.

Keywords Tube hydroforming · Numerical · Analysis · Design of experiments
Finite element method · Aluminum

13.1 Introduction

Tube hydroforming (THF) offers advantages over the traditional forming process, and in recent years it has become an alternative manufacturing process for medium range quantity production. Some of the advantages over traditional forming process are: (a) dimensional accuracy, (b) reduced spring back, (c) weight reduction, (d) uniform wall thickness, and (e) reduced scrap (Singh 2003).

J. C. León-Anaya (✉) · E. Dimas-Celestino · J. A. Juanico-Lorán
División de Ingeniería Industrial, Universidad Politécnica de Valle de México, Av. Mexiquense s/n, Col. Villa Esmeralda, Tultitlán, Estado de México
C.P. 54910, Mexico
e-mail: jleon_mx@yahoo.com

J. C. Cisneros-Ortega
Universidad Popular Autónoma del Estado de Puebla, Posgrado, 21 sur 1103 Barrio Santiago, C.P. 72410 Puebla, Mexico

J. L. García Alcaraz et al. (eds.), *Best Practices in Manufacturing Processes*,
https://doi.org/10.1007/978-3-319-99190-0_13

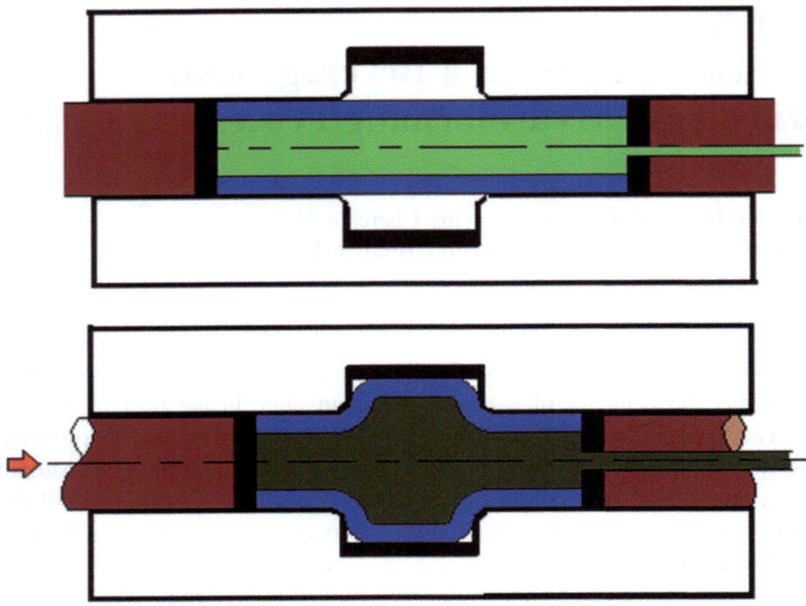

Fig. 13.1 Tube hydroforming process. *Source* León et al. (2014)

At the beginning of the century, several studies referents to the numerical simulation of hydroforming of sheet and tubes were developed. Most authors use Finite Element Analysis (FEA) for simulation, due to the advantages of computer simulation over laboratory test (Shinde et al. 2016).

However, computer resources are finite, optimization of those resources are possible by applying statistical techniques. Experimental Design is a method in which a controlled experimental factor is subject to special treatment for purposes of comparison with a factor that is kept constant.

Due to the nature of the hydroforming process, some drawbacks can be found; the most common is the slow cycle time, limited experience in the design of tools and equipment, and lack of specialized literature, especially in Spanish language.

A brief description of the tube hydroforming process is shown in Fig. 13.1.

The tube is held between two die halves with the desired form. The first step is to fill the tube with a low-pressure fluid ,after removing the air trapped inside, a combination of axial force and internal high pressure is applied, tube expands and is forced to attain the shape of the die. Axial feeding and internal pressure can be applied simultaneously to improve material formability (Langerak et al. 2004).

Three main parameters are involved in THF. The first parameter is the friction between the die and the tube to be processed; to reduce the buckling, bursting, surface wrinkling, and tool wearing (Jun-Jang et al. 2011) a lubricant must be utilized. There are three identified friction zones: the guided, transition, and expansion zones.

The second parameter to be controlled in THF is applied pressure in the internal wall of the tube and axial feeding. There are four process related, low pressure, high pressure, multi-pressure, and hydro-bulge.

Low-pressure hydroforming is limited to 83 MPa (12,000 psi); this value was determined by the Tube and Pipe Fabricators Association. Expansion on the tube is up to 3%, minimal change in thickness is present, and cycle time and investment in equipment are lower than other hydroforming processes.

High-pressure hydroforming is up to 414 MPa (60,000 psi); higher plastic deformations can be achieved, due to the stresses in the process, for welded tubes, quality, and bursting can be a decisive factor, thickness variation can be present, especially in corners. The pressure allows complex geometry; dimensional stability and low spring back are advantages.

Finally, the third parameter is the correct determination of material characteristics, with this in consideration; the designer can improve accuracy in the prediction of process parameters for THF applications (Yeong-Maw and Yi-Kai 2006). Deformation mechanism in tube hydroforming is under biaxial strain deformation and biaxial stress state, applying plasticity, membrane, and thin-walled theories, limits, and parameters for hydroforming can be achieved (Koc and Taylan 2002), obtained analytical models to predict buckling, wrinkling, and bursting as well as axial force, internal pressure, counterforce, and thinning. In the work developed by Djavanroodi et al. (2008), in equilibrium conditions and applying the von Mises, the yielding pressure in the tube is

$$Py = \sigma_y \frac{2t_0}{D_0 - t_0} \tag{13.1}$$

where

Py yielding pressure (MPa)
σy yield stress of material (MPa)
t_0 initial tube wall thickness (m)
D_0 initial tube outside diameter (m) (Fig. 13.2).

13.2 Materials and Method

The physical properties of the material are usually the result of standardized tests, previous to a manufacturing process, such as cold forming for sheets, and cold rolling for tubes; grain orientation and homogeneity are affected, different values in different directions of the structure are present, making necessary new mechanical tests to obtain those modified values.

Initial material is an aluminum 6061-T5 tube with a diameter of 31.75 mm (1 1/4″) and 1.4 mm of thickness. In Sect. 6.9 of the ASTM-E8/E8 M-09 “Standard

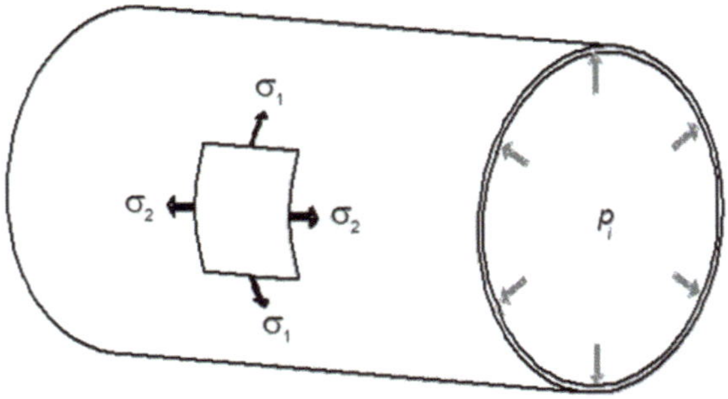

Fig. 13.2 Surface stress in tube subject to internal pressure (*Source* Author)

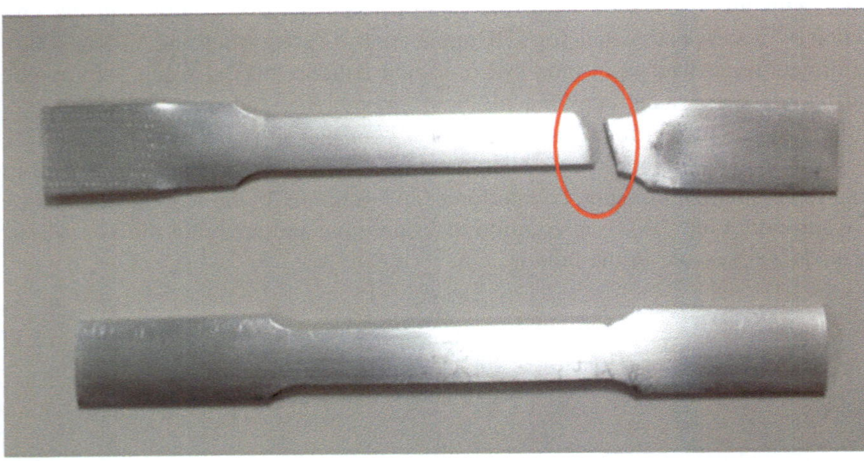

Fig. 13.3 Test specimens according to ASTM E8/E8 M-09 Standard. *Source* Authors

test methods for metallic materials", the dimension of the specimen and the manufacturing way are specified, as shown in Fig. 13.3 (ASTM E8/E8 M-09 Standard 2016).

The tensile tests were performed at room temperature using a Shimadzu Universal Testing Machine; model AG—X Plus, with a velocity test of 3 mm/min (see Fig. 13.4). The values of force and displacement were obtained with Trapezium Materials Testing Operation software in the materials characterization laboratory at Universidad Politécnica del Valle de Mexico (UPVM). The data obtained is shown in Table 13.1.

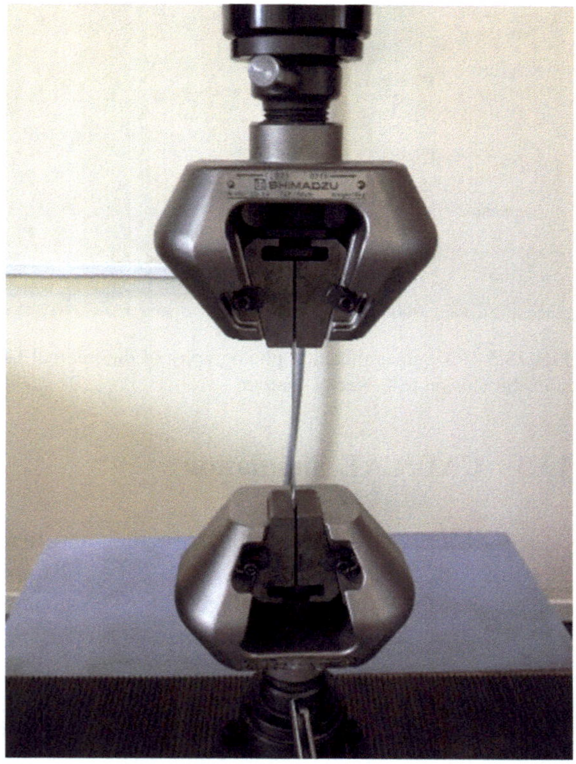

Fig. 13.4 Tension test for mechanical characterization

Table 13.1 Comparison of experimental versus literature mechanical properties

Property	Experimental	Literature
σ_y	95 MPa	110 MPa
σ_u	218 MPa	210 MPa
% Elongation	13	16
n	0.365	0.20
E	92.8 GPa	69.8 GPa

Source ASM Aerospace Specification Metals, Inc.

Optic images of macrostructures were obtained by using a Carl Zeiss microscope model Axio Vert AX10 in the materials characterization laboratory at UPVM. Several pictures were taken for the analysis. No substantial changes were observed in the structure of the specimens, nevertheless, mechanical properties obtained had significant differences with the ones reported in the literature (see Fig. 13.5).

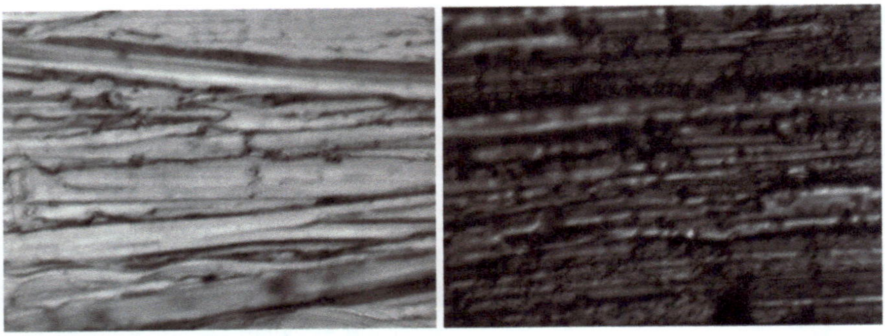

Fig. 13.5 100× magnification photography of the internal face of the specimen test before and after the tension test. *Source* Author

13.3 CAD/CAE Simulation

A geometrical 3D CAD model of the dies and the tube to be hydroformed was developed in SOLIDWORKS® 2014–2017 software, and the rendered model of one of the dies is shown in Fig. 13.6. The general dimensions on the die are 100 × 110 × 50 mm, and the mechanical properties for simulation purposes are of 4140 steel for the dice.

The tube has the following dimensions: 66 mm in external diameter, a length of 95 mm, and a wall thickness of 0.50 mm. The mechanical properties assigned were the ones obtained in the experimental testing for aluminum 6061-T5.

The model was exported in IGS format which is to be utilized in the structural module of ANSYS® workbench 16.2 for meshing and the application of boundary conditions. The model is an assembly consisting of two dies and one tube as shown in Fig. 13.8.

Once imported into ANSYS® 16.2; several geometries of solid meshing were attempted and the final model had 22,080 nodes and 11,871 elements, friction in

Fig. 13.6 Rendered model of one of the dies

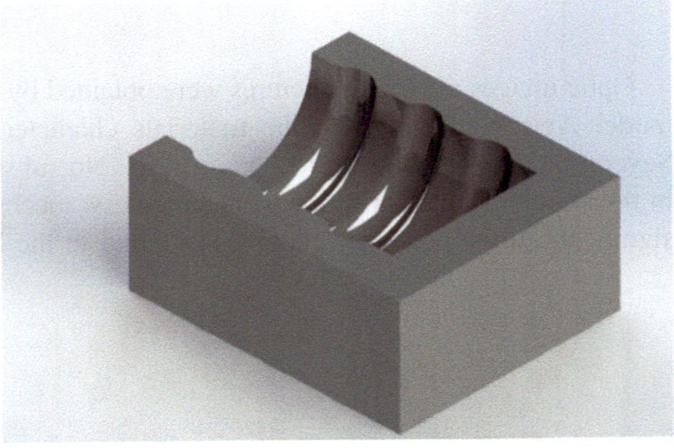

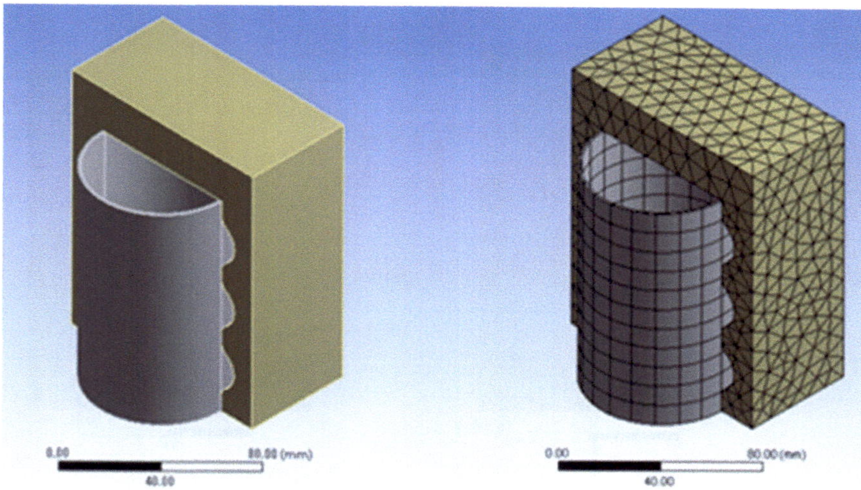

Fig. 13.7 Geometric model and meshed model of the tube (*Source* Author)

contact the external face of the tube and the internal faces of the dies were considered, the faces between the dies were considered as no contact surfaces (Fig. 13.7).

Four models for simulation were made, with friction coefficient (μ) of 0, 0.05, 0.08, and 0.10, an incremental internal pressure was applied in the internal face of the part. Three incremental steps were utilized (Ceretti et al. 2013) in the first one, a pressure of three MPa was applied, in the second step; it was seven MPa, in the third step the pressure was withdrawn. The models with a different number of steps were tested, however, three steps offered the optimal relation between simulation time and accuracy, models with four and five steps required a substantial extra processing time and no appreciable improvement in the results were appreciated.

Frictionless model suffered greatest plastic deformation in radial and longitudinal direction, spring back was with an average 4% in both the directions. However, a frictionless case was utilized only as a reference; in real experimentation, friction is always present, lubricants, especially solid ones like graphite are suggested.

Permanent deformation and maximum stresses were present at step two of the simulation. Spring back was always present. In the model with the highest friction coefficient, the final von Misses equivalent stresses are shown in Fig. 13.8, in all the cases, the maximum stress in step two was below the ultimate stress of the material, so we can conclude that there was no fracture on the tube (Fig. 13.9).

As shown in Figs. 13.10 and 13.11, we can assume that there is a direct relation between friction coefficient and deformation; both radial deformation and longitudinal deformation are dependent on friction between the surface of the part and the die. At a constant volume of material, in the longitudinal deformation, a shortening occurs, due to the material expansion in the radial direction.

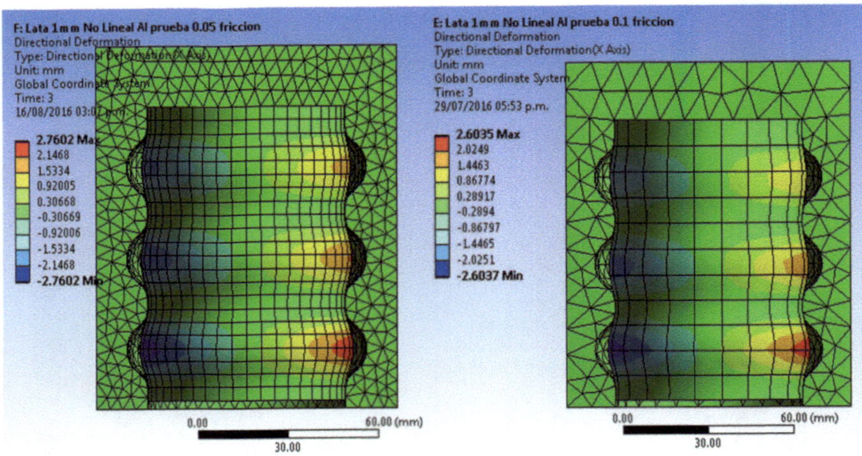

Fig. 13.8 Radial deformation, friction of $\mu = 0.05$ and $\mu = 0.10$

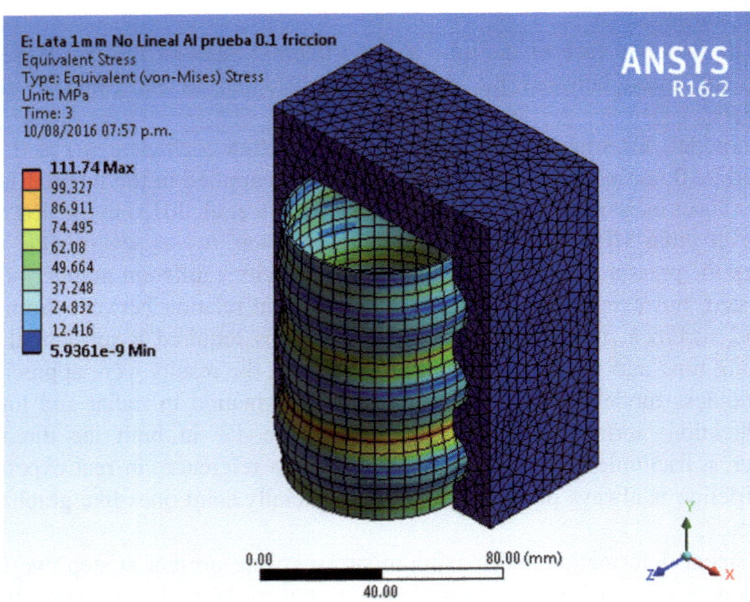

Fig. 13.9 Von Mises stress, $\mu = 0.10$

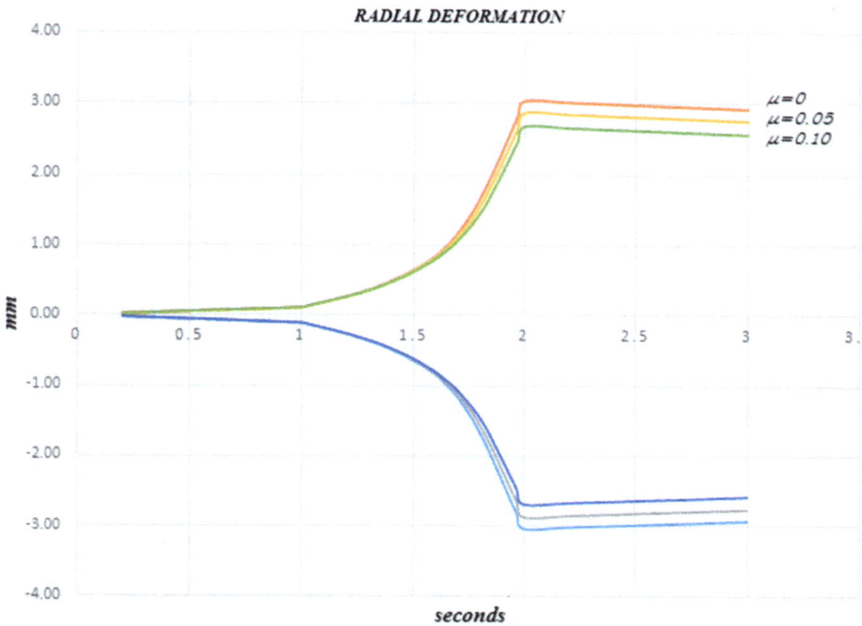

Fig. 13.10 Radial deformation versus time

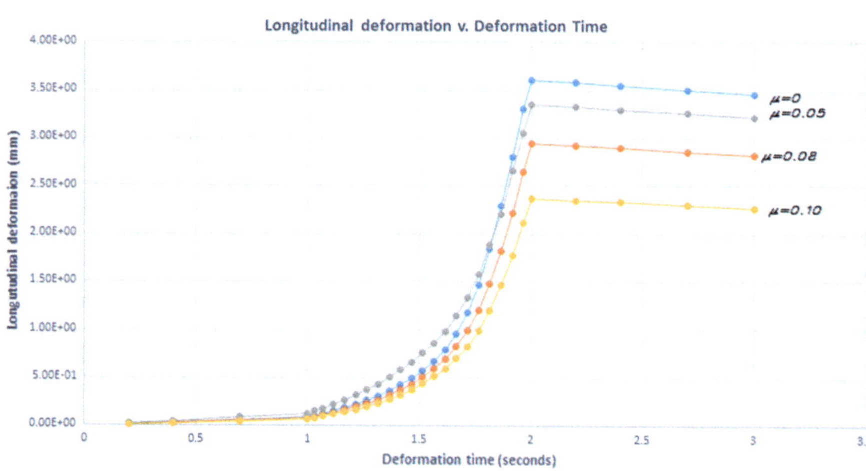

Fig. 13.11 Longitudinal deformation versus time

13.4 Design of Experiments (DOE)

Computer numerical simulation offers several advantages over experimental testing; research time, preparation of experimental equipment, and material are reduced. However, trial-error simulation, (changing one parameter at each simulation), consumes valuable computation time, and this can be optimized by applying a systematic methodology. Design of experiments (DOE) is a statistical tool that provides information about the interactions of factors and the way they behave in the test.

With the results of the simulation, several computer experimental design analyses were conducted. Notations for the simulations were $X1$ for pressure, $X2$ for time as independent variables, and coefficient friction (μ) of 0, 0.05, 0.08, and 0.10 as dependent variables $Y1$, $Y2$, $Y3$, and $Y4$.

Initially, hierarchical linear regression, for the frictionless case, with two models were generated, in the first model only pressure was considered as an independent variable, in the second both, pressure and time are considered. As shown in Table 13.2, in the second model, the correlation coefficient obtained was 0.893 ($R > 0.7$ above this value is considered a high correlation) and a higher explained variance ($R2$) with a determination coefficient of 0.797.

Hierarchical regression is utilized in the study for the reason that deals with how predictor (independent) variables are selected and entered into the model. Specifically, hierarchical regression refers to the process of adding or removing predictor variables from the regression model in steps.

The correlation coefficients are used in statistics to measure how strong a relationship is between two variables. There are several types of correlation coefficient: Pearson's correlation (also called Pearson's R) is a correlation coefficient commonly used in linear regression (reference).

The standard error of the estimate is a measure of the accuracy of predictions.

Considering the values in the ANOVA analysis, shown in Table 13.3, the significance should be less than 0.5, with the finality to know that the model tested presents a significant improvement in the prediction of the result in the dependent variable (Table 13.4).

Good collinearity has an acceptance value of Variance Inflation Factor (VIF) <5, the model presents a value of 1.121, so we can observe that a strong relation between pressure and time is present. Further simulations were performed for the coefficient friction of 0.05, 0.08, and 0.10, applied pressure and material properties were kept constant, and the results are shown in Tables 13.5 and 13.6.

Table 13.2 Models of linear regression for frictionless conditions

Model	R	R^2	Adjusted R^2	Estimation standard error
1	0.423[a]	0.179	0.149	1.02893787
2	0.893[b]	0.797	0.782	0.52106208

[a]Predictors: pressure
[b]Predictors: pressure, time

Table 13.3 ANOVA[a] table for the frictionless model

Model		Square addition	Gl	Quadratic mean	F	Sig.
1	Regression	6.242	1	6.242	5.896	0.022[b]
	Remnant	28.585	27	1.059		
	Total	34.828	28			
2	Regression	27.768	2	13.884	51.138	0.000[c]
	Remnant	7.059	26	0.272		
	Total	34.828	28			

[a]Dependent variable: Radial deformation frictionless
[b]Predictors: (Constant), pressure
[c]Predictors: (Constant), pressure, time

Table 13.4 Regression coefficients for the frictionless model

Model	Collinearity statistics coefficients[a]		β	t	Sig.	Collinearity Statistics	
	B	Standard error				Tolerance	VIF
Pressure and time	−0.794	27.768		−1.966	0.060		
	−0.069	7.059	−0.150	−1.601	0.121	0.892	1.121
	1.506	34.828	0.832	8.904	0.000	0.892	1.121

[a]Dependent variable: Radial deformation frictionless

Table 13.5 Models of linear regression, μ = 0.05

Model	R	R^2	Adjusted R^2	Estimation standard error
1	0.423[a]	0.179	0.149	0.96468205
2	0.895[b]	0.802	0.786	0.48312605

[a]Predictors: pressure
[b]Predictors: pressure, time

For each of the previous models, a linear equation model was developed, and four equations were obtained as shown in Eqs. (13.1–13.4)

$$y = -0.794 - 0.069x_1 + 1.506x_2 \tag{13.1}$$

$$y = -0.741 - 0.064x_1 + 1.417x_2 \tag{13.2}$$

$$y = -0.668 - 0.065x_1 + 1.336x_2 \tag{13.3}$$

$$y = -0.692 - 0.057x_1 + 1.322x_2 \tag{13.4}$$

Table 13.6 ANOVA[a] table for the model with $\mu = 0.05$

Model		Square addition	Gl	Quadratic mean	F	Sig.
1	Regression	5.478	1	5.478	5.886	0.022[b]
	Remnant	25.127	27	0.931		
	Total	30.605	28			
2	Regression	24.536	2	12.268	52.559	0.000[c]
	Remnant	6.069	26	0.233		
	Total	30.605	28			

[a]Dependent variable: Radial deformation frictionless
[b]Predictors: (Constant), pressure
[c]Predictors: (Constant), pressure, time

13.5 Conclusions

Material characterization was necessary, mechanical properties were modified by the intermediate process of extruded tube, determination of yield stress, elastic modulus and strain index were made, significant differences between the literature and experimental values were observed, yield stress in the test was 14% below, and elastic modulus was 33% above, strain hardening is almost 182.5% of the one reported in the literature.

Tube hydroforming process depends on the best combination of material properties, process parameters, process sequence, and die geometry. Finite Element Analysis is capable of predicting the accurate behavior of tube hydroforming process and can be used to reduce experimental efforts.

In all the simulations, the applied pressure was the same; the results show that the friction affects longitudinal and radial deformation; spring back and von Mises stress in the tube. In order to obtain the desired forming profiles with a higher friction coefficient, higher pressures should be necessary.

Numerical simulation is a helpful tool for initial prediction in the design and development of specific problems in THF in initial stages, shortening the lead times in manufacturing and reducing costs in this stage. Design of experiments and the experimental test can be developed for future works.

The regression models allow us to predict multiple behaviors in the part, if we apply a different combination of pressure–time. In combination with FEA, the development of models can be optimized reducing the computation time (León et al. 2014).

The determination of the linear regression equations with two factors (pressure and time), with four levels of friction, for prediction of the radial deformation (dependent variable), shows a correlation factor above 0.89, indicating the strong relation between the variables. The determination coefficient with a level of 80% of explained variance confirms that a prediction with high accuracy of the radial deformation is feasible with different combinations of time and pressures for this model. In the prediction of the plastic deformation in the part, the pressure is not the only variable considered; time is fundamental to the THF process. Initially in the frictionless model, if we only account pressure as an independent variable, the regression equations show that the correlation with the deformation is poor, with a value of 0.423, and determination coefficient (R^2) of 0.179, that means that only 17.9% of the deformation changes is explained.

In contrast, in the second model both pressure and time should be considered independent variables, in all of the following simulations; high correlation coefficients are obtained, with values above 0.80, according to (Walpole et al. 2007), a strong correlation between deformation, pressure, and time.

A combination of FEA and DOE allows the generation of multiple configurations and the generation of useful information.

13.6 Discussions

Studies of tube hydroforming simulation started at the beginning of the century; however, almost no investigations of commercial aluminum tubes developed in Latin America. The THF allowed controlling the necessary time demanded by a typical hydroforming work cycle.

In this work, ANSYS program was used for the simulation of THF process. It obtains results that agree with the analysis of the literature, presenting low significant errors. Future work could extend the simulation to various commercial tubes in Mexico and help the design and manufacture of them. DOE allows the development of mathematical models, combining both a shorter time of research and design is expected.

References

ANSYS® Academic Research, Release 16.1, Help System, Mechanical advanced nonlinear material, Guide, ANSYS, Inc

ASTM E8/E8 M-09 (2016) Standard test methods for tension testing of metallic materials. ASTM International, West Conshohocken, PA, 2009, U.S.A. www.astm.org

Ceretti E, Attanasio A, Fiorentino A, Giorleo A, Giardini A (2013) Aluminum can shaping by hydroforming: simulative feasibility study and prototype production. Int J Adv Manuf Technol 68:1797–1807. https://doi.org/10.1007/s00170-013-4978-x

Djavanroodi F, Gheisary M, Zoghi-Shal H (2008) Analytical and numerical analysis of free bulge tube hydroforming. Amer J Appl Sci 5(8):972–979

Jun-Jang P, Wen-Da Z, Gang L, Shi-Qiang Z, Shi-Jian Y (2011) Effect of internal pressure distribution on thickness uniformity of hydroforming Y-shaped tube. Trans Nonferrous Met Soc China 21(S2):423–428. https://doi.org/10.1016/S1003-6326(11)61618-X

Koc M, Allen M, Jiratheranat S, Taylan A (2000) The use of FEA and design of experiments to establish design guidelines for simple hydroformed parts. Int J Mach Tools Manuf 40(1):2249–2266

Koc M, Taylan A (2002) Prediction of forming limits and parameters in the tube hydroforming process. Int J Mach Tools Manuf 42(1):123–138. https://doi.org/10.1016/S0890-6955(01)00048-7

Langerak N, Rout DK, Verma R, Manikadan G, Haldar A (2004) Tube hydroforming in automotive applications. Tata Steel Research Development & Technology, Ijmuiden, The Netherlands, India

León J, Pérez C, Cisneros J (2014) Stress analysis in an automotive part manufactured by hydroforming 1:175–184 [In Spanish]

Shinde RA, Patil BT, Joshi KN (2016) Optimization of Tube Hydroforming Process. Proc Technol 23:398–405. https://doi.org/10.1016/j.protcy.2016.03.043

Singh H (2003) Fundamentals of hydroforming 1st edn. Society of Manufacturing Engineers SME, U.S.A.

Walpole RE, Myers RH, Myers S, Keying Y (2007) Probability and statistics for engineering and science. Pearson Educación, México [In Spanish]

Yeong-Maw H, Yi-Kai L (2006) Analysis of tube bulge forming in an open die considering anisotropic effects of the tubular material. Int J Mach Tools Manuf 46(15):1921–1928. https://doi.org/10.1016/j.ijmachtools.2006.01.025

Chapter 14
The E-Strategy for Lean-Sigma Solutions, Latin American Case Study in a New Product Validation Process

Francisco J. Estrada-Orantes, Abimael H. García-Pérez and Noé G. Alba-Baena

Abstract The international competition has challenged managers in Latin America to implement state-of-the-art methodologies for problem-solving and continuous improvement, and specifically, the production process validation has become an issue when it has to be completed in a short time span. Typically, the validation activities for critical equipment are performed at least a week prior to the official production launch. If, for any reason, the equipment gets into the plant late, any time taken by the validation process may impact the startup of the line. On the other hand, if production is started without validating the equipment, and performance is not as expected, then, the plant must start a cycle of process improvement activities to get the performance to the expected levels. In general, the continuous improvement activities are organized around two major methodologies: Lean Manufacturing and Six Sigma. While the Lean approach tends to be of a quick-fix type, it gets shorthanded when the causes are not so obvious, and deeper statistical analyses are required. On the other hand, Six Sigma works better when there is plenty of time to conduct all types of tests and analyses to achieve a good cost-effective solution. This chapter explores a combination of the Lean Manufacturing speed with the Six Sigma power of analysis, arranged as a set of sequential steps, for solving industrial problems and giving cost-effective solutions in a short time span. This is achieved by following a strategy identified by these authors as the E-Strategy, which is divided into two phases, the diagnostic and the solution phase. It uses a hierarchical approach of analysis for identifying the root cause of the problem. From the most frequent causes, the problems are eliminated adapting and using the most efficient set of tools. In the E-Strategy, as the complexity of the problem increases, the tools used get more specialized and elaborated. In this chapter, a case study is included as an example of the use of this methodology. The case study shows that the use of the Lean-Sigma approach is effective when following the E-Strategy sequence, and leads to improvements in overall

F. J. Estrada-Orantes (✉) · A. H. García-Pérez · N. G. Alba-Baena
Department of Industrial Engineering and Manufacturing, Universidad Autónoma de Ciudad Juárez, Av. Del Charro 450 Norte. Col. Partido Romero, Juárez, Chihuahua, Mexico
e-mail: frestrad@uacj.mx

J. L. García Alcaraz et al. (eds.), *Best Practices in Manufacturing Processes*,
https://doi.org/10.1007/978-3-319-99190-0_14

performance. The description focuses on the efforts for increasing the conforming outcomes from a crimping process. A pull test is used for performance evaluation of the outcomes. Initial data shows an overall performance of a sigma level of 1.1. However, after following the E-Strategy, running a Taguchi experiment, and performing a series of adjustments a final process evaluation shows an increment to a sigma level of 5.5 for the performance, and a three times reduction in the variation of the process, achieving this solution in a short period of time of 3 days.

Keywords Lean-Six Sigma · E-Strategy · Taguchi · Process validation

14.1 Introduction

The main characteristics of the first decades of the twenty-first-century industrial production and market products is that product lifetimes are short and are likely to become shorter, also the fast delivery, low cost, and higher expectations are the reality and represent current challenges. This means that actual factories need to be able to frequently reconfigure and validate their processes to modify procedures or replace operations.

Also, there is the need for reducing the time for factory workers and managers to learn about the changes, adaptations, and additions to production requirements and processes reconfigurations. Consequently, managers face more possibilities to test the robustness of their operations, and to highlight the weaknesses and failures of their systems. At the same time, they need to keep enhancing the competitiveness and value of their products. Manufacturing systems have to be designed to be flexible or reconfigurable while the operation can resize as the market requires. Moreover, in today's manufacturing the product diversity, shorter delivery, and stretching in pricing are key factors for keeping a successful operation (Gershwin 2017).

Particularly, the validating step of an equipment design algorithm, typically is used to provide a rapid, and smooth confirmation that such equipment performs to customer expectations. The validating step covers the installation of the equipment and the manufacture and evaluation of a significant amount of pieces done under regular production conditions. All unanticipated installation and validation concerns, as well as their corrective actions are implemented prior to the launch of mass production, which is also used to confirm the equipment design process (Yang and El-Haik 2003).

If the results of the validation are not as expected, and problems arise, a structured root cause investigation process is required to create a solution that may solve the problems in a short time so as to minimize the delay for mass production. The way the steps are structured for properly investigating an incident is critical for achieving the expected results (Heuvel et al. 2008).

In short, the main challenges experienced by industrial manufacturing companies, concentrate in keeping the production process running within the expected lead-times, with a predictable quality, a reliable delivery, and cost-effective initiatives that allow their products stay competitive in the market. In general, manufacturing companies face conditions that involve the identification and correction of problems, and have a tendency to look into the continuous improvement paths. In general, improvement programs are not in the daily agenda, but, when the burden of a company becomes higher than expected, or when the production launching of a new product is at risk, the continuous improvement initiatives arise as the most important programs in the organizations.

The mentioned challenges are generally addressed by using several engineering methodologies such as Six Sigma, Lean Manufacturing, Design for Six Sigma (DFSS), and Re-Engineering approaches. In the case of Latin America, the use of such applications and solution processes are boosted by the strong competition among Latin American enterprises, the established international corporations and the competition with other commercial regions such as the Asian and European zones. Making use of the mentioned methodologies and decision processes, the key for survival and success of the local industrial companies.

14.1.1 Latin American Scenario

When facing the described challenges, managers and industrial owners prefer a simple and fast solution with the minimum investment and fast return. For finding such solutions, researchers in Latin America have been struggling for efficient and clear strategies that fit the needs and requirements of the local industries, but also to overcome the restrictions of the Latin America scenarios. A way of doing this is by adapting tools and approaches known and successfully used in other regions and different economic platforms. In this way, Latin American researchers have been challenged to find solutions and working philosophies that can be adopted and adapted to organizational structures and behavior requirements (Contreras et al. 2006) from other regions as has been occurring at the USA–Mexico border zone for the past six decades (Wilson 2010).

In such experience, researchers have learned that there are conditions and scenarios where the adopted strategy may require additional modifications to adapt it to specific regional needs. That includes incorporating some tools and methods developed locally or importing them from other strategies. Especially, in the USA–Mexico border region, where a high percentage of the manufacturing plants are the USA owned, and most of the equipment come from outside the region, the equipment validation process takes a significant role during the production ramp up of new products. Very often, the equipment validation process has to be conducted in parallel, or very close to the mass production ramp up. Additionally, in some instances, some complexity is added when the results of the validation are not as expected, and a Root Cause Analysis, and a Problem-Solving Technique needs to be used. In such cases, the selection, combination, and adaptation of the tools, to suit the specific needs of a company, will enhance the resulting methodology to respond to the economic and process restrictions, or, in other instances, to the time constraints and subjective conditions set by some environmental aspects of the territory, as previously reported by these authors and others, who have verified the results during the development of different applications in this region. Reports have also shown that these integrations are more successful when the process is led by experienced engineers, who are flexible to adapt to the complex conditions and restrictions of this economic region.

14.1.2 Approaches Used in Latin America

Considering the described scenario, some engineering approaches have proven to be effective for solving this type of problems, and understanding the characteristics of such procedures, adapted guidelines may be developed for making decisions under the mentioned conditions. For instance, in solving short-term problems the Lean manufacturing approach, the PDCA cycle, and the 8Ds methodology have proven their value. For process improvement, Six Sigma is the most concurred choice, and for optimization projects, Lean-Six Sigma has been the most favored for the efficient results, and moreover for solving short term but complex problems, Lean-Sigma has been used more often because it is efficient and reliable.

14.1.2.1 The "Just-in-Time" or Lean Manufacturing Path

Many companies in the US and Mexico solve critical situations using either, the "quality path", or the "Just-in-Time" (JIT) production path (Schonberger 1986). The JIT ideas were first introduced by Taiichi Ohno in 1978 and are based on the foundations of the Toyota Production System (TPS). This path gains more attention after 1990 when James Womack introduces some ideas and tools for increasing the efficiency of this path (Womack et al. 1990). By redefining the objective focusing in achieving a rapid and continuous series of improvements in the production systems, Womack renames the JIT path to "Lean Manufacturing path". The new Lean Manufacturing path was keeping its identity by using the "just do it" and "keep it simple" strategy. On the other hand, the "Quality" path based on Deming's statistical approach to quality, integrates the tools that use data generated by the process (through an effective in-depth statistical analysis) in order to identify and eliminate the root cause of problems (Deming 2000). Later, the quality path takes a more relevant importance with the development of the Six Sigma approach, which will be addressed in a later paragraph.

However, when the solutions require a flexible approach, or the focus is on keeping the system working as it is, a methodology, such as Lean Manufacturing is preferred. The lean methodology focuses on identifying all immediate threats to the flow of the production process. These threats are identified as waste, and classified as muda, mura, and muri, for their names in Japanese. Muda are all aspects of waste related to the operation of the production process. Mura is related to fluctuations in the production schedule, and muri has to do with all aspects of the workstation structure and design. Lean manufacturing focuses on reducing the Lead-Time using any available tool to minimize the waste that impacts the production process performance.

According to Womack and Jones (1996) the Lean Manufacturing methodology is based on five principles: (I) **Specify** the Value added of the product. Provide exactly what the customer wants, at the right time, and at the right price. (II) **Identify** the main Value Stream. From the customer's perspective, identify all the activities in the production process that add value to the product. (III) **Develop Flow**. Make sure the process flows without interruptions, delays, or accumulations. (IV) **Use Pull Production** scheduling. The Production should be customer driven, or oriented to fulfill the customer demands. (V) **Strive for Perfection**. Continue looking for perfection by eliminating the different types of waste.

In the Lean manufacturing approach, there are 8 types of waste classified as "Muda". They are described as follows:

1. *Defects*: Parts or products that do not meet the customer's requirements and or specifications.
2. *Over production*: Producing more than what the customer requires, or producing before the customer needs it, or producing faster than the customer's consumption.
3. *Over processing*: Adding processing activities to a production process that are not necessary, and the customer is not willing to pay for. For instance, having someone remove excess material on a molded part, when it was supposed to be perfect going out of the molding machine.
4. *Transportation*: Moving materials or products around unnecessarily.
5. *Motion*: People moving around during the production process unnecessarily.
6. *Inventory*: Keeping more inventory than strictly necessary to keep the process flowing continuously.
7. *Waiting time*: Time that an operation stops production flow waiting for an input (material, machine, people, and order)
8. *Talent*: Waste for not using people's talent to improve the process.

The waste classified as Mura by Lean Manufacturing is described as follows:

1. *Changes*. Waste generated by sudden changes in the production schedule.
2. *Fluctuations*. Flow interruptions and waste generated due to fluctuations in the production schedule.
3. *Inconsistencies*. Waste caused by inconsistencies between the production schedule and the availability of materials.

Additionally, the waste classified as muri include the following descriptions:

1. *Layout*. Waste caused by a poor layout of the workstation.
2. *Tooling*. Waste generated by inadequate tooling.
3. *Work Instructions*. Waste generated by confused work instructions.

Further developments and enhancements to the Lean manufacturing path have incorporated some techniques in an effort to identify and correct problems directly on the production line, whenever they arise. The Toyota Kata approach delineates a structured strategy for identifying causes of problems and improving processes. The Kata approach uses the following steps:

1. Understand the current condition
2. Establish the next target condition
3. Conduct PDCA cycles toward the target condition

Rother (2010) describes in his work the details, and mechanics of the strategy. The procedure is oriented to solve problems on the shop floor, rather than in a meeting room.

In the case of Latin America, several reports have proven that it is possible to successfully adapt the Lean manufacturing approach to production systems in this region.

14.1.2.2 The Six Sigma or Quality Path

In the USA–Mexico border region, continuous improvement methodologies like Six Sigma have proven to be effective when a process improvement requires a deep statistical analysis of the potential causes of a condition in a production system. Six Sigma is defined as a continuous improvement methodology that focuses primarily on the identification and reduction of the process variation. As a structured methodology, Six Sigma was first introduced and implemented in Motorola during the 1990s (Pande et al. 2000).

The power of the statistical analysis makes this methodology ideal for complex improvement challenges where the root cause is hidden deeply in the process parameters or there is an intricate relationship among the variables of such process. The tools proposed for each phase in the Six Sigma methodology have been proven and adapted in many applications. Figure 14.1 shows a relationship between the phases of the Six Sigma improvement process using the DMAIC cycle (Define–Measure–Analyze–Improve and Control), the objectives of each phase and the tools most frequently used for each phase. Six Sigma aims to identify and reduce variation and looks to achieve a performance level of 3.4 defective parts per million (Rath & Strong 2000).

Moreover, the table shown in Fig. 14.1 serves also as a general guidance during the analysis for the improvement teams. Reports provide evidence that in the USA–Mexico region, the use of Six Sigma has been successful for production processes improvement (Camacho et al. 2016; Coy et al. 2016) and for product design cases as reported by Lopez et al. (2016) and Romo et al. (2016).

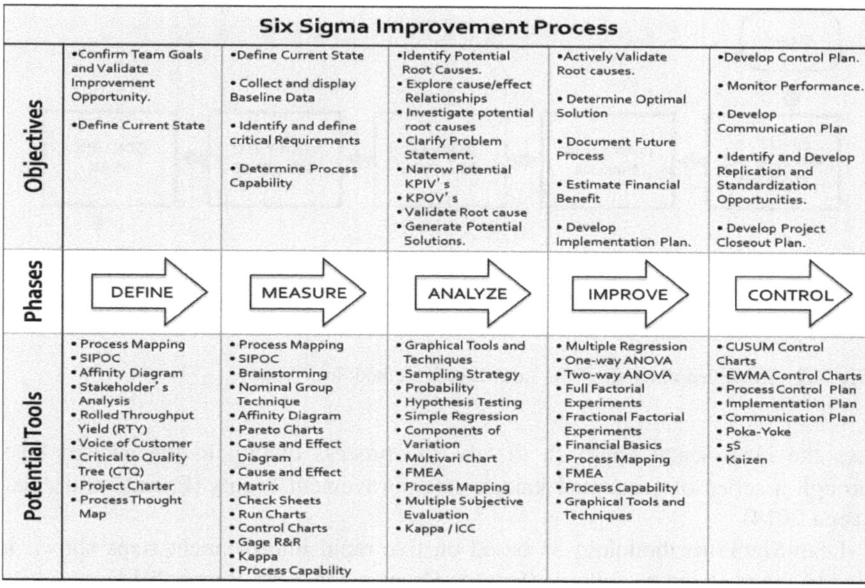

Fig. 14.1 Image of a roadmap of DMAIC phases and the potential tools for each phase

14.1.2.3 Merging Lean Manufacturing and Six Sigma

The need for combining elements used by Lean Manufacturing and those used by Six Sigma was originally identified and proposed by George (2002) and George & George (2003). For some projects addressing primarily organizational wastes affecting the production lead-time, the solutions are more related to the use of the Lean Manufacturing approach, while for long-term projects addressing quality, cost, or delivery issues that required a deep analysis, the Six Sigma approach is preferred as the solving method. However, in many production processes, the reality is that a specific scenario may involve both types of challenges, that is, the lead-time needs to be improved, but it requires a deep statistical analysis to identify and eliminate the hidden root cause of a complex problem.

For such conditions, several combinations of tools coming from Lean manufacturing and Six Sigma have been developed for different scenarios and specific needs. The original combination is known as Lean-Six Sigma (George 2002; George et al. 2005), and its evolution has taken this initiative to a methodology known as Lean-Sigma. While Lean-Six Sigma is seen as a methodology that takes advantage of the structured DMAIC roadmap, and integrates additional tools from Lean manufacturing to each phase, and is primarily used as a long-term process improvement methodology, Lean-Sigma is seen as a short-term problem-solving methodology. Solving a problem using Lean-Sigma will not always bring a process' performance to a six sigma level, but instead, the Lean-Sigma methodology

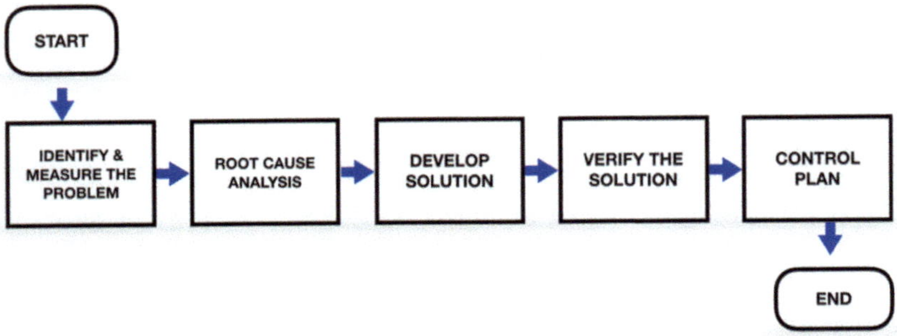

Fig. 14.2 Simplified diagram of the Lean-Sigma methodology steps

uses the incremental approach to take the process quality to the desired level through a series of sequential kaizen, or improvement events (Estrada and Alba-Baena 2014).

Lean-Sigma methodology is based on five rapid improvement steps shown in Fig. 14.2 and stated as follows (Estrada-Orantes and Alba-Baena 2014):

1. *Identify* and measure the problem. How big is it?
2. *Root cause analysis:* What is the root cause of the problem?
3. *Develop a solution:* Identify the alternative solution that best solves the problem.
4. *Verify* the solution: Make sure that the problem is eliminated by the proposed solution.
5. *Control Plan:* Make a quick and effective plan so that the previous situation does not come back.

The synergy between these different approaches was discussed by Estrada and Alba-Baena (2014) and several researchers have reported the use of Lean-Sigma for solving industrial problems in the Latin American environment, among them Gracia et al. (2016) and De la Cruz et al. (2016). Also, Alba-Baena et al. (2016) reported the use of Lean-Sigma during a product Ramp-Up event. However, the use of Lean-Sigma has shown that there are opportunities for defining new strategies in order to make more efficient the use of Lean-Sigma in industrial settings. Recently Alba-Baena et al. (2016) show that the Lean-Sigma strategy of "do it at the speed of Lean with the depth of Sigma", can be structured for solving situations in the restrictive environment of Latin America and give solutions in a short-term span.

14.1.3 The E-Strategy

The Lean-Sigma approach has been proven to be efficient all over the globe; however, at the managerial level more efficient strategies have to be developed in order to take advantage of the experiences in the use and implementation of Lean-Sigma in the context of the Latin America environment. This chapter releases and describes the E-strategy as a strategy for efficiently solving industrial problems. This strategy is the result of the analysis of the practical application of the Lean-Sigma approach in companies at the USA–Mexico border region. The "E-strategy" is a strategy that follows a series of hierarchical and logical steps in the decision process for solving a problem.

The E-Strategy is based on the Lean-Sigma methodology previously tested by the authors. In the context of the Lean-Sigma methodology, the E-Strategy concentrates on developing specific steps to conduct the Root Cause Analysis, as well as the development of the solution. The E-Strategy uses Diagnostic Phase to the steps used to identify the root cause of the problem, and the Solution Phase to the steps used to create and test the solution. In order to minimize the elapsed time from the manifestation of the problem, through the creation and test of the solution, the E-Strategy quickly evaluates six potential causes in the Diagnostic Phase, and five additional steps to develop and validate the solution.

14.1.3.1 The E-Strategy, Diagnostic Phase

Figure 14.3 depicts the E-Strategy Diagnostic Phase with six potential and basic causes that may be responsible for the problem. The Diagnostic phase may be seen as the quick identification and elimination of common obstacles that interrupt the normal flow of the production process. This phase starts immediately after the problem manifestation. Each cause is analyzed, one at a time, following the order in the diagram. Each potential cause is addressed as a question in order to filter the complexity of the problem. Each question is followed by an action *(make corrections)* and a decision *(Is the problem solved?)* up to the end of the six consecutive steps, and are as follows:

1. *Are prints and drawings correct?* Check and compare the actual prints and drawings used on the floor for all components, sub-assemblies, final products, and equipment involved in the problem, to determine if they are up to date to the most recent revision, and are used correctly.
2. *Are the tools and operating conditions correct?* Review the actual operating tools used on the floor, the working conditions of assembly and fabrication stations, and the parameter values of the equipment, to be in accordance to the official work instructions.
3. *Is raw material correct and to specifications?* Check for compliance to specifications of the raw materials being used.
4. *Is the measuring system correct?* Review the calibration status of testing equipment, and conduct a new analysis for the complete measuring system.
5. *Is process variation in statistical control?* Take additional samples to verify the statistical stability and predictability of the process.

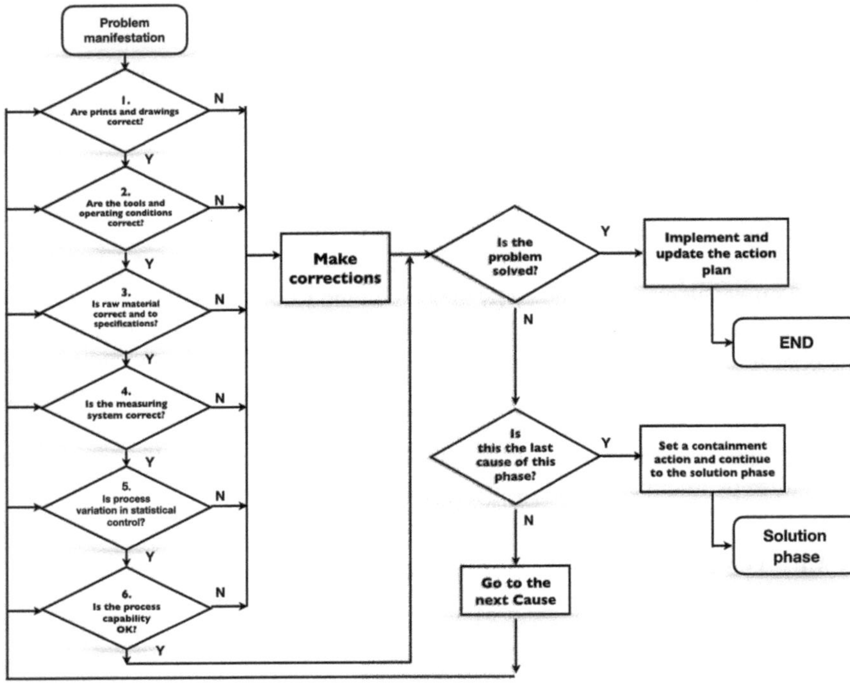

Fig. 14.3 Schematic diagram of the E-Strategy Diagnostic Phase

6. *Is the process capability OK?* Take additional samples, and review the current capability of the process.

If after reviewing the described filters, and performing any corrections required, the problem persists, a containment activity is set at the process, the condition is classified as a complex problem, and the team moves on to the Solution phase, as depicted in Fig. 14.3.

14.1.3.2 The E-Strategy, Solution Phase

After the use of the diagnostic phase, the common causes are eliminated, and basic data from the product, the process, and the working conditions have been collected. If the problem persists, it means that it is a complex problem with a hidden root cause, and the need for a deeper analysis arises. For this situation, the E-Strategy solution phase is used and consists of five sequenced activities that have to be completed in order to solve the problem. Figure 14.4 depicts the Solution phase. The first step is identified as "*let the process speak (data)*", which means to collect current data from the operations involved in the problem.

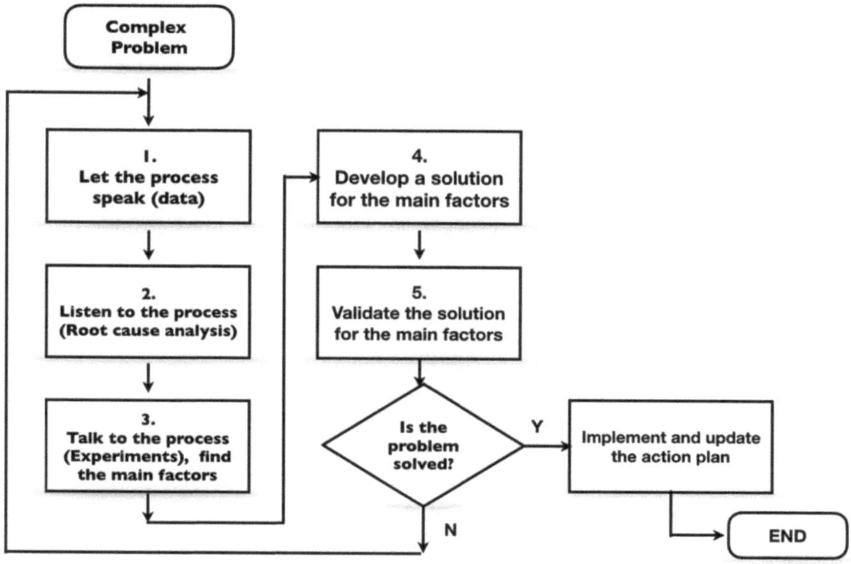

Fig. 14.4 Schematic diagram of the E-Strategy Solution Phase

The second step, "*listen to the process (root cause analysis)*" focuses on finding the potential causes of the problem and identifying the most likely to be the root cause. The next step "*talk to the process (Experiment), Find main factors*" requires running experiments to quantify the effect of the process variables in the critical characteristics of the output variable. The relationship between them and the possibilities for level adjusting and solving the problem.

Step four, "*develop a solution for the main factors*", deals with the integration of possible solutions and the selection of the most promising for solving the problem. In step five "*Validate the solution for the main factors*", the chosen solution is evaluated obtaining additional data from the process, using the chosen values for each factor, and performing statistical tests, and comparisons between the initial and final values. If such solution solves satisfactorily the problem, then the last step "*implement and update the action plan*" is documented, and the methodology concludes.

For illustrating the use of the E-Strategy, this chapter includes a case study of the integration of the E-Strategy to the Lean-Sigma approach. The example describes a condition where the deliveries cannot wait, and the quality levels are not as expected from the process. In this case, with the aggravate that it is a new product in this facility at Juarez, Mexico and the equipment and production setup are under validation for launching. During the solution process, these authors used the most efficient tools and methodology of Lean-Sigma for following the two phases of the E-Strategy and give a solution in 3 days after the problem was described by the management.

14.2 Case Study: Equipment Validation Process for New Product Introduction

14.2.1 Case Study Scenario

A manufacturing facility that produces midget fuse blocks is starting production under a tight-time scenario. The process is scheduled to start producing, and shipping to the customer in parallel to the installation and validation of the equipment. Up to the second day of the week, all the quality assurance tests of the equipment comply to specifications, but the crimping force at the fuse end caps, which show values below the specifications.

After several attempts to correct the problem without a positive outcome, the production line is stopped, and classified as not ready for mass production. The case is presented to the continuous improvement team to find the causes and create a feasible solution, with the clear objective to have the production of conforming products as soon as possible.

14.2.2 Problem Description

The crimping machine used to perform the task is a four-station turn-table (as seen in Fig. 14.5a). The stations are sequenced as follows: in the first station the fuses are loaded; the second station is used for crimping the upper side of the fuse, the third one for crimping the other side of the fuse, and the last station is used to unload the fuses. The crimping stations consist of a set of clamps which is closed by the force of a pneumatic cylinder, create notches on the end caps of the fuses to keep them in place during use. The quality of this operation is checked by measuring the retention force using the instrument shown in Fig. 14.5b. This instrument includes a load cell in the upper portion, and a motor that creates the separation movement.

The basic procedure for this test is described as follows: (1) The lower end cap of the fuse is held to the fixed base of the instrument, (2) The upper end cap of the fuse is attached to the load cell, (3) When the motor starts an upper movement of the load cell, a tension force is created on the fuse. As this force increases the fuse breaks, or either one of the end caps fall out when the product crimping force is overpassed. The maximum force used before breakage, or detachment of the fuses, is registered by the load cell, and is identified as the "maximum disassembly force".

The validation documentation of the crimping machine, submitted by the manufacturer, shows evidence that the machine is in compliance to the pull test specification, which suggests that the problem may come from the actual setup of the machine.

(a)

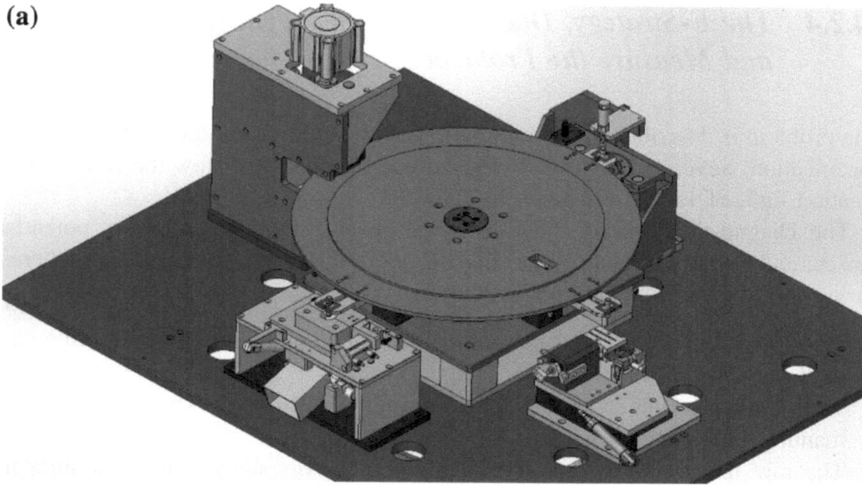

(b)

Fig. 14.5 **a** Partial drawing of the crimping machine, **b** Pulling testing equipment used for these experiments

14.2.3 Methodology

The E-Strategy, as depicted in Figs. 14.3 and 14.4, is used as the guiding methodology for addressing the problem. Details of the application of all the steps of the methodology are provided below.

14.2.4 The E-Strategy, Diagnostic Phase. Identify and Measure the Problem

The problem is described as the pulling force test of the fuses does not conform to specification. Several data points from current samples are below the lower specification limit of 150 N, as shown in Table 14.1.

The Diagnostic Phase of the E-Strategy, specifically focuses on six potential causes, and reviews one at a time. For this case, the critical information gathered during this stage is summarized as follows:

1. Prints and drawings for all the components and equipment are up to date and used correctly.
2. Tools and operating conditions for the crimping machine are set up according to manufacturer's specifications.
3. The raw material is within specifications, however shows some variation in three critical characteristics: the diameter and length of the component identified as melamine tube, and the diameter of the end caps.
4. The calibration of the testing equipment is found to be acceptable.
5. The pull force testing data shows no evidence of special cause variation (see Fig. 14.6), showing that the process is stable, predictable, and has a normal distribution as shown in Fig. 14.7a.
6. The capability of the process is found not acceptable, with a Cpk value of −0.15 as shown in Fig. 14.7b. The process has an overall performance of 683,576 ppm, which is equivalent to a sigma level of 1.1.

Since the process cannot be corrected to achieve the target using the manufacturer's recommended operating parameters, the situation is diagnosed as a complex problem, and requires the Solution phase of the E-Strategy.

Table 14.1 Data for samples after crimping and pulling testing

No.	Max pulling force (N)	No.	Max pulling force (N)
1	185.10	11	110.35
2	120.85	12	111.55
3	113.10	13	185.05
4	96.05	14	165.05
5	169.90	15	89.50
6	75.10	16	130.40
7	102.95	17	185.73
8	114.95	18	126.13
9	202.85	19	103.98
10	143.50	20	113.24

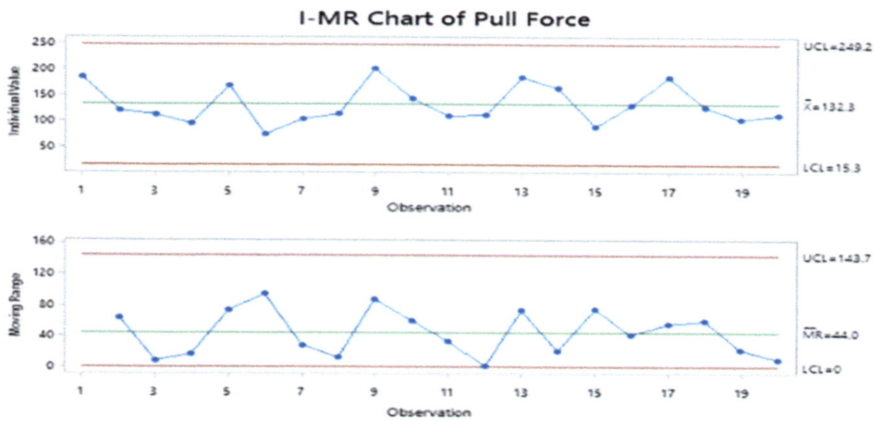

Fig. 14.6 Control chart of data after pulling force testing

14.2.5 The E-Strategy, Solution Phase. Root Cause Analysis

The solution phase of the E-Strategy requires the collection of additional information in order to allow the process to speak, understand its behavior, and identify all potential causes of the problem. Three principal components are involved when performing the pulling test: A melamine tube, and two end caps, the latter are identified with the same part number. Two of the characteristics that may potentially influence the pulling force resistance are identified in the melamine tube as: the diameter and the tube length, and one is identified in the end cap as the diameter. For the initial characterization of the variability of the incoming components, data from 100 pieces of each is used to calculate a capability analysis for each characteristic.

Figures 14.8 and 14.9 depict the measured behavior for the three characteristics. As seen in the mentioned figures, the values observed (Cpk) are 0.81, 0.68, and 1.01 for the tube diameter, tube length, and the cap diameter, respectively. These capability studies show that the components are coming from processes that are generating parts under and above the specification limits with expected defectives as: 1.3, 2.8, and 2.4% for the tube diameter and length and the cap diameter, respectively, however, these components are filtered in the incoming inspection, so most of the received parts are within the specification limits when received at the process.

From the data shown in these figures and the incoming inspection for these components, it is accepted that these are within specifications. Accepting also that the variation within the component's dimensions is considered as part of the common cause variation, even though some combinations at extreme values of the components may fall out of specifications and may create undesirable results. Also, it is accepted that such conditions are not the main and root cause of the crimping problem.

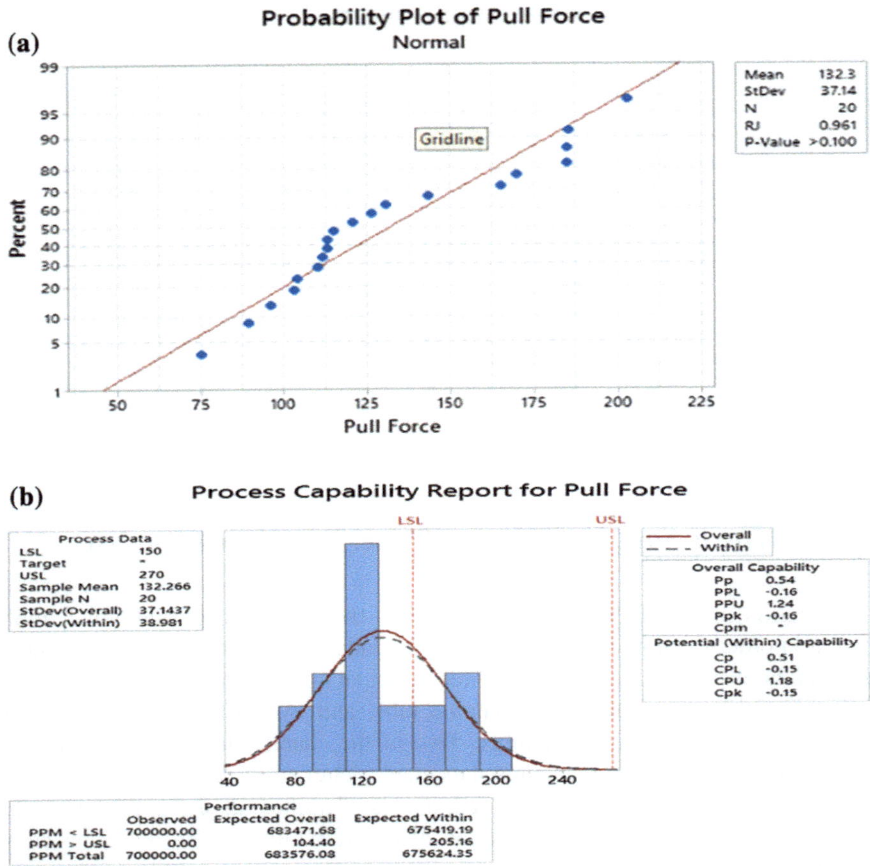

Fig. 14.7 **a** Normality test after pulling force testing, **b** Initial capability analysis after pulling force testing

As mentioned before, the designer and integrator of the crimping equipment recommended operating conditions that includes options for clamping with clamps according to the end cap model. Also, the crimping machine has a control for adjusting the distance traveled by the piston before closing the clamp over the cap. Also, the time the clamp remains closed and the air pressure can be adjusted.

In summary, the potential factors affecting the outcome of the crimping process, as shown in Table 14.2 are: the type of clamp, the piston travel distance, the time the clamp remains closed, and the air pressure. All these factors may be adjusted and controlled during the operation. Additionally, the characteristics of the components may also affect the outcome, but such variation, as long as it remains within specification, is beyond the control of the operation, for that reason they are considered as non-controllable (see Table 14.2).

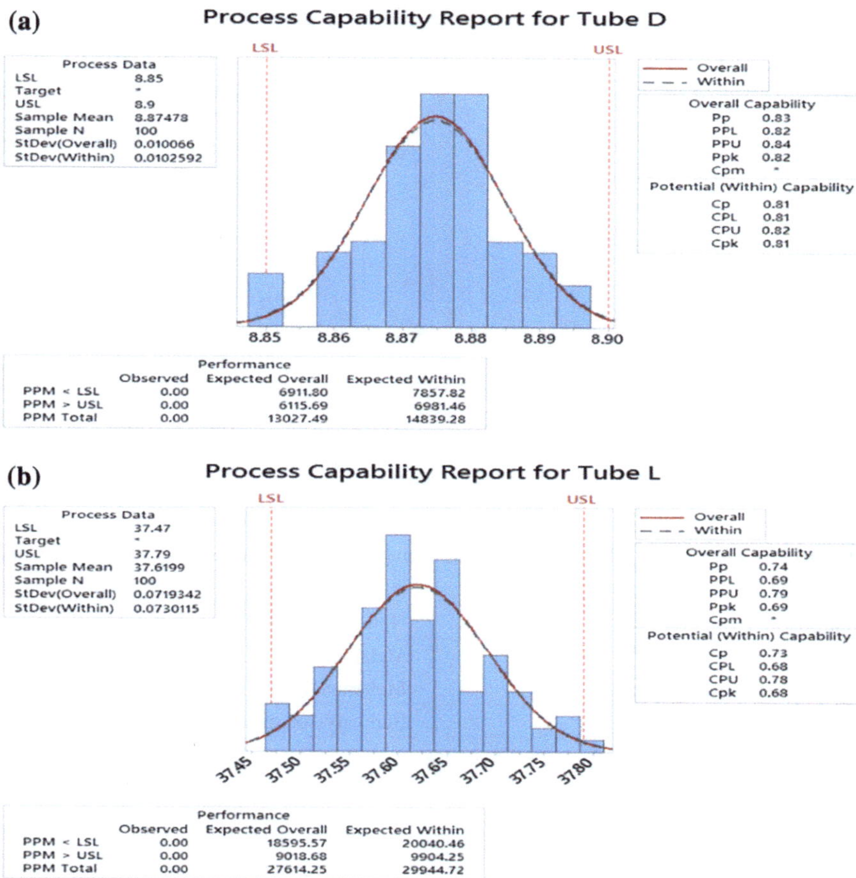

Fig. 14.8 **a** Capability analysis for tube diameter, **b** Capability analysis for tube length

14.2.6 Developing a Solution

Once the main factors have been identified, the next step in the E-Strategy methodology is to develop a solution. For this case, an experiment that includes both, controllable and non-controllable factors are designed to explore which factors are primarily impacting the pull force, and at the same time investigate the operating conditions that may help maintain the process robust against the non-controllable factors. A Taguchi Robust Parameter experiment is designed using an L9 array for the three controllable factors (see Table 14.3) and an L4 array for the three noise factors shown in Table 14.4.

Based on the crimping process characteristics and mechanical tolerances given, the experiment includes different levels, for the controllable factors three levels are assigned (see Table 14.3) and for the non-controllable factors two levels as shown

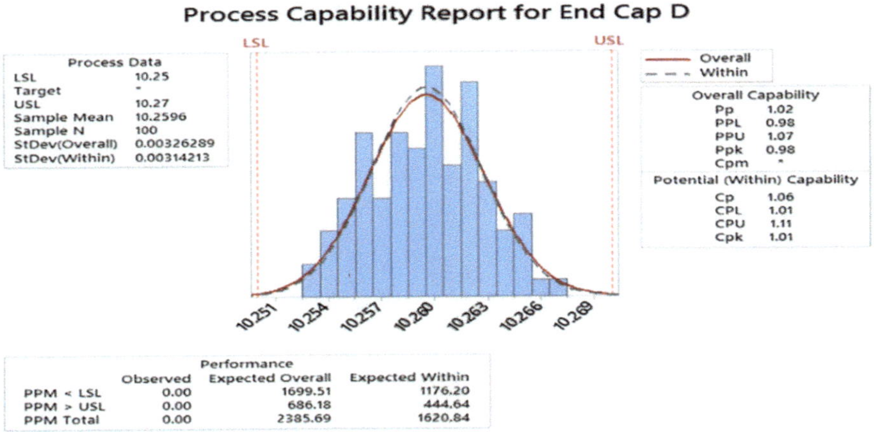

Fig. 14.9 Capability analysis for the end cap diameter

Table 14.2 Experimental identification of affecting variables divided as controllable and non-controllable

Controllable	Non-controllable
Type of clamp	Melamine tube diameter
Piston travel distance	Melamine tube length
Time clamp remains closed	End cap's diameter
Air pressure	

Table 14.3 Controllable factors and levels proposed for the experiment

Factor	Level 1	Level 2	Level 3
(*A*) Piston travel distance (mm)	15	17.5	20
(*B*) Time clamp stays closed (s)	0.5	1.5	2.5
(*C*) Air pressure (psi)	40	75	90

Table 14.4 Levels proposed for the non-controllable factors

Factor	Level 1	Level 2
(*O*) Melamine tube diameter (mm)	8.85	8.90
(*P*) End cap diameter (mm)	10.25	10.27
(*Q*) Melamine tube length (mm)	37.47	37.79

in Table 14.4. The final experimental arrangement for both, the L9 for the controllable and the L4 for the non-controllable variables are presented in Table 14.5.

From such arrangement and by crossing the two arrays, a total of 36 different combinations are generated and were tested obtaining the results as shown in Table 14.6. With the data summarized in Table 14.7, the Signal-to-Noise (*S*/*N*) table

Table 14.5 L9 array for controllable factors

Run	(*A*) Piston travel	(*B*) Time clamp close	(*C*) Air pressure
1	15	0.5	40
2	15	1.5	75
3	15	2.5	90
4	17.5	0.5	75
5	17.5	1.5	90
6	17.5	2.5	40
7	20	0.5	90
8	20	1.5	40
9	20	2.5	75

Table 14.6 L4 array combinations for non-controllable factors

Factor	Combinations			
	1	2	3	4
(*Q*) Tube length	37.47	37.79	37.79	37.47
(*P*) End cap diameter	10.25	10.27	10.25	10.27
(*O*) Tube diameter	8.85	8.85	8.90	8.90

Table 14.7 Results of the experiment for the pulling force testing (values in Newtons, N)

Run	Combination			
	1	2	3	4
1	199.30	168.00	178.80	176.05
2	173.45	179.25	185.30	159.30
3	177.80	159.30	177.15	151.10
4	176.30	192.50	188.70	205.30
5	210.00	190.60	164.15	186.00
6	185.10	169.90	202.85	185.05
7	120.85	75.10	143.50	165.05
8	113.10	102.95	110.35	89.50
9	96.05	114.95	111.55	130.40

is calculated, as shown in Table 14.8, containing the levels for each controllable factor, then, considering for the decision the combination that reports the highest values during the pulling force testing for choosing as the possible solution (see Table 14.8).

In this case, the proposed solution is based on the Signal-to-Noise (*S*/*N*) Table shown in Table 14.8, and the levels of the controllable factors are selected from those with the highest value s in this table, as follows: for factor *A*, level two is selected, for the factor *B*, level one is chosen, and for factor *C*, level two is the best option. The corresponding values for the levels selected for each controllable factor, as shown in Table 14.9 are: for the piston traveling at a distance

Table 14.8 Signal-to-noise ratio (*S/N*) values

Level	(*A*) Piston travel	(*B*) Time clamp close	(*C*) Air pressure
1	44.74	43.82	43.55
2	45.42	43.46	43.76
3	40.66	43.54	43.52

Table 14.9 Selected values for controllable factors

Factor	Selected value
(*A*) Piston travel distance (mm)	17.5
(*B*) Time clamp remains closed (s)	0.5
(*C*) Air pressure (psi)	75

value of 17.5 mm, for the time of keeping the clamp closed 0.5 s, and finally keep the pneumatic (air) pressure in 75 psi.

14.2.7 *Verify the Solution*

Once a solution is selected, in order to verify that the selected values for the controllable factors solve the problem, the E-Strategy requires the crimping machine to be run using the selected values. In this case, 40 pieces are run, and the corresponding pull force test data is collected. The verification is performed by comparing the pull force results obtained with the original settings versus the results obtained with the proposed settings. This comparison is validated by using a statistical hypothesis testing. Table 14.10 shows the results for the confirmation run.

The I-MR chart shown in Fig. 14.10 shows a graphical representation of the data for the pulling test before and after using the solution obtained with the

Table 14.10 Data of the confirmation run after the pulling force testing

No.	Force	No.	Force	No.	Force	No.
1	199.8	11	194.1	21	229.5	31
2	203.3	12	192.8	22	203.6	32
3	193.2	13	203.8	23	226.2	33
4	179.2	14	204	24	210.6	34
5	219.7	15	208.5	25	221	35
6	226.4	16	217.8	26	191.3	36
7	184.8	17	213.3	27	214.4	37
8	198.7	18	179.2	28	208.4	38
9	218.5	19	219.3	29	189.7	39
10	186.3	20	199.3	30	207.2	40

methodology. As seen in the figure, the left portion of the chart represents the behavior of the data before applying the solution and the right portion of the chart shows the behavior of the data after applying the selected values for the controllable factors. A visual comparison shows that the mean values increase, and the variation is reduced.

A noticeable shift in the process is observed in the I-MR Chart (Fig. 14.10), and it also shows that the process is still stable and predictable, and therefore, it is considered to be in a state of statistical control. The pulling force mean value, as shown in the upper portion of Fig. 14.10, moves from 132.3 to 202.7 N after using the selected solution. Meanwhile, the UCL and LCL (Upper and Lower control limits) shift from 15.3 and 249.2 N to 161.6 and 243.9 N, respectively. The moving range of the data, as depicted in the lower portion of Fig. 14.10, also shows a considerable reduction in the variability of the data. The mean value of the range goes from 44.0 to 15.5, which represents a 64.7% reduction in the variation of the process outcomes. At the same time, the UCL and LCL limits shift from 0 and 143.7 N to 0 and 50.5 N, respectively.

A capability analysis for the pulling force after applying the selected values for the controllable factors of the crimping machine is elaborated and depicted in Fig. 14.11. This study shows that the proposed solution puts the process within the specification limits achieving a Cpk value of 1.28 and a Ppk value of 1.33, which represents an expected overall performance of 33.35 ppm.

A Capability Analysis comparison for the pulling force, using the data before and after applying the solution, is depicted in Fig. 14.12, facilitating the discussion and comparison of the performance before and after the implementation.

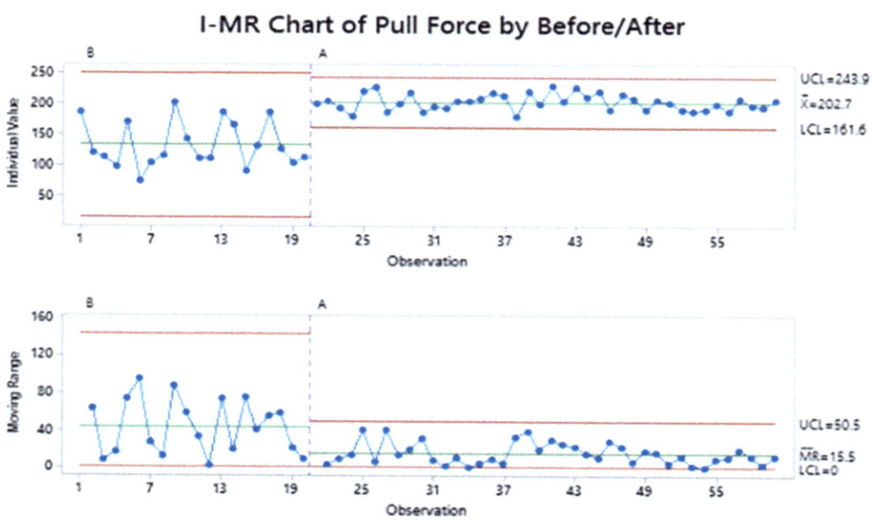

Fig. 14.10 I-MR chart comparing the before versus after implementation data for the pulling force testing

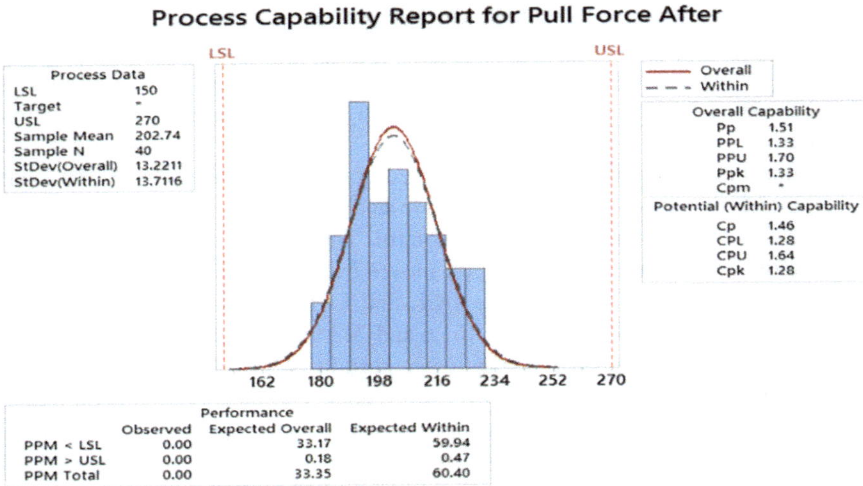

Fig. 14.11 Capability analysis for pull force after changes

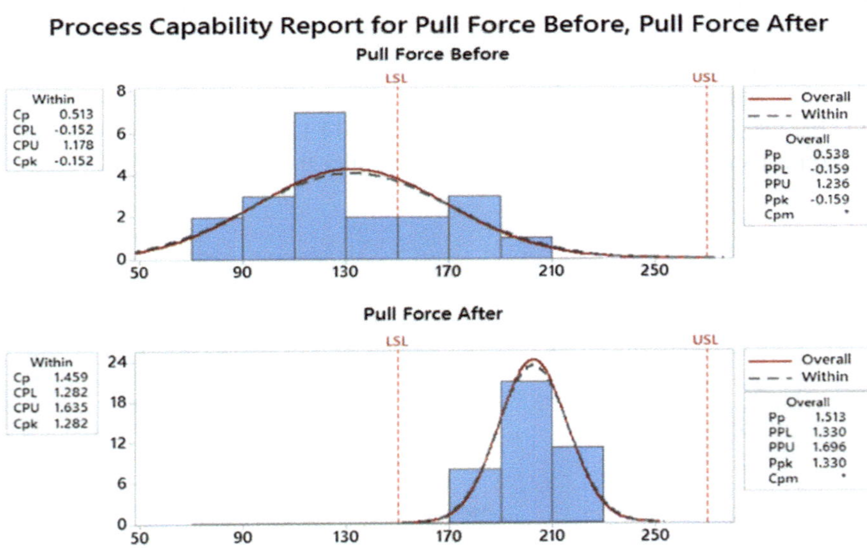

Fig. 14.12 Before versus after comparison of the capability analysis for the pulling force testing

The upper portion of the chart shows the process capability of the data before applying the solution, and the bottom portion of the chart shows the behavior of the data after applying the selected values for the controllable factors in the crimping process. The capability analysis shows that before applying the solution, over 60% of the parts had values below the LSL, with a predicted overall performance of 683,576 defective parts per million (see Fig. 14.12). While, after applying the

One-Sample T: Confirmation

```
Test of μ = 150 vs > 150


Variable       N    Mean  StDev  SE Mean  95% Lower Bound      T      P
Confirmation  40  202.74  13.22     2.09           199.22  25.23  0.000
```

Fig. 14.13 Pulling-force testing data for the confirmation run using the hypothesis test

Two-Sample T-Test and CI: Pull Force Before, Pull Force After

```
Two-sample T for Pull Force Before vs Pull Force After

                    N   Mean  StDev  SE Mean
Pull Force Before  20  132.3   37.1      8.3
Pull Force After   40  202.7   13.2      2.1


Difference = μ (Pull Force Before) - μ (Pull Force After)
Estimate for difference:  -70.47
95% upper bound for difference:  -55.74
T-Test of difference = 0 (vs <): T-Value = -8.23  P-Value = 0.000  DF = 21
```

Fig. 14.14 Pulling-force testing data comparing before versus after implementation using the hypothesis test

solution, the performance reduces to a value of 33.35 defective parts per million. Data also shows that the process capability shifts from a Cpk with a negative value of −0.152 to a positive value of 1.282. In other words, the mean of the process is now within the specification limits and it is centered.

As part of the verification step, the mean value of the response variable (pulling force) is compared to the lower specification limit using a one-sample t-hypothesis test. Figure 14.13 shows the results of the hypothesis test. With a p value <0.0001, the test suggests that there is enough evidence to support that the mean value of the crimping process will pass the pulling test with values larger than 150 N.

A second hypothesis testing is used for supporting the graphical evidence that there was an increase in the mean of the process after applying the solution. For this, a Two-Sample t test is conducted, and the results are shown in Fig. 14.14. With a p value <0.0001, the test suggests that there is enough evidence to support that the mean value of the pulling testing of the product coming from the crimping process is greater after using the selected values to the controllable factors of the crimping process.

14.2.8 Control Plan

The last step of the E-Strategy focuses on the update, and implementation of the control plan. In this case, the team updated the working instructions, the visual aids, and the control plan to include the adjusted operating parameters. This step also includes transferring the knowledge, and lessons learned to the rest of the organization. This is achieved by posting the results, and making available a document describing the details of all the steps used to achieve the final improvement.

14.3 Conclusions

Several researchers and practitioners have proven that in a restrictive environment it is possible to successfully implement state-of-the-art methodologies for problem-solving and continuous improvement, as exemplified by several technical reports. Lean-Sigma has become a fast response and effective methodology for problem-solving. The Lean-Sigma strategy of "do it at the speed of Lean with the depth of Six Sigma," can be structured for solving situations in the restrictive environment of Latin America and give solutions in a short-term span. This chapter describes the E-Strategy which is divided into two phases, the diagnostic and solution phases, and speeds up the process to reach a solution by hierarchically eliminating the most frequent causes of problems in the industries. It helps to solve the problems by adapting and using the most efficient set of tools, and depending on the complexity of the problem, the sequence of tools gets more specialized and elaborated.

The case presented in this chapter is an example of how the E-Strategy is used. The case scenario describes a situation where a manufacturing process is installing and validating assembly and test equipment in parallel to starting production of a new product. The situation gets complicated when a crimping machine does not achieve the expected value for the pulling force test using the manufacturer's recommended settings. At that point, the E-Strategy is used to solve the problematic situation, and allow the process to continue with the mass production.

Initial process data shows an overall performance of 683,576 ppm, which is equivalent to a sigma level of 1.1. This is, over 68% of the pieces out of specifications having a mean of 132 N and a standard deviation of 37 N.

After following the E-Strategy and running a Taguchi experiment, which includes an L9 array for the controllable factors and an L4 array for the uncontrollable ones, the pulling testing data permits to identify as potential solution the following values for the controllable factors: (*A*) Piston Travel distance 17.5 mm, (*B*) Time clamp remains closed 0.5 s, and (*C*) Air pressure 75 psi. After using the selected solution, data shows that, the process is stable, predictable, and, therefore, considered to be in a state of statistical control, with a mean value of 202.7 N (which is above the LCL of 150 N) and a noticeable variation reduction as seen in the change of the standard deviation (from 37 to 13 N). Also, the quality of the outcome measured in 64% defective (a sigma level of 1.1) was drastically reduced to an overall performance of 33.35 ppm, which is equivalent to a sigma level of 5.5. As a highlight of this case study, the whole solving process was completed over a period of 3 days from the problem description to the control plan trying to keep the lead-time as short as possible.

Finally, it can be concluded that the use of the Lean-Sigma approach in Latin America's competitive environment is an efficient methodology. Especially, if it is oriented as a problem-solving technique instead of a project-based methodology. Complemented with the use of the E-Strategy, Lean-Sigma is a straightforward solution process that keeps the characteristic flexibility of Lean manufacturing, but flexible enough to move quickly to the deep statistical analysis characterized by the Six Sigma methodology. The combination of Lean-Sigma and the E-Strategy helps in achieving efficiently the quality goals, while providing a fast solution, and maintaining a short lead-time, as proven by this case, during initial validation of manufacturing equipment.

References

Alba-Baena N, Estrada FJ, Sierra Torres OO (2016) Using Lean-Sigma for the integration of two products during a ramp-up event. In: Giner A-H, Cuauhtémoc S-R, Jorge Luis G-A (eds) Handbook of research on managerial strategies for achieving optimal performance in industrial processes. IGI Global, Hershey, pp 405–427. https://doi.org/10.4018/978-1-5225-0130-5.ch019

Camacho J, Baena NA, Hernández A, Molina J, Barrera MFR, de León DZ (2016) Improvement of productivity in a production line through the implementation of a specialized program to capture the metric of overall efficiency of the "OEE" equipment. CULCyT 55(12):146–154 (in Spanish)

Contreras ÓF, Carrillo J, Garcia H, Olea MJ (2006) Labor performance of the maquiladoras: an assessment of safety at work. Front norte 18:55–86 (in Spanish)

Coy MR, Salazar JM, Martínez LFF, Baena NA (2016) Application of the Six Sigma Methodology to control the significant parameters in the production of a Nylon glove with Latex coating. CULCyT 58(13):271–283 [In Spanish]

Deming WE (2000) Out of the crisis. 1st edn. The W. Edwards Deming Institute. MIT Press 2000, ©1982, 1986

De la Cruz Rodríguez MI, Estrada-Orantes FJ, Mendoza MD, Saldaña JFE, Rodríguez RR (2016) Methodology for the continuous improvement of manufacturing processes based on lean sigma and applied to the process of manufacturing automotive harnesses. CULCyT 56(12):94–106 (in Spanish)

Estrada-Orantes FJ, Alba-Baena NG (2014) Creating the Lean-Sigma synergy. In: García-Alcaraz JL, Maldonado-Macías AA, Cortes-Robles G (eds) Lean manufacturing in the developing world: methodology, case studies and trends from Latin America. Springer International Publishing, Cham, pp 117–134. https://doi.org/10.1007/978-3-319-04951-9_6

Gracia OC, Estrada-Orantes FJ, Pérez FH (2016) Application of the Lean-Sigma methodology in the solution of problems in manufacturing processes: case study. CULCyT 57(12):199–211 (in Spanish)

George ML (2002) Lean six sigma: combining six sigma quality with lean speed. McGraw-Hill

George ML, George M (2003) Lean six sigma for service. McGraw-Hill, New York
George ML, Rowlands D, Price M, John M (2005) The lean six sigma pocket tool book. Mc-Graw-Hill
Gershwin SB (2017) The future of manufacturing systems engineering. Int J Prod Res 56(1–2):224–237
Heuvel, LNV, Donal KL, Jackson LO, Hanson WO, Rooney JJ, Walker DA (2008) Root cause analysis handbook, 3rd edn. ABS Consulting, Rothstein Associates, Inc.
López DR, Gaytán MCC, Baena NA, Orantes FE (2016) Design of a station for measurement and analysis of the cable crimping process. CULCyT 58(13):203–212 (in Spanish)
Ohno T (1988 June) Toyota production system—beyond large-scale production. Productivity Press, pp 25–28. ISBN 0915299143
Pande PS, Neuman RP, Cavanach RR (2000) The six sigma way. McGraw-Hill, New York, NY, USA
Rath & Strong (2000) Six sigma pocket guide. Rath & Strong Management Consultants, Sixth Printing 2001
Romo JS, Baena NA, González JR, Pérez DMA, Estrada-Orantes FJ (2016) Redesign and characterization of an electrical test equipment in the Jack Bam product assembly. CULCyT 59 (13):26–37 (in Spanish)
Rother M (2010) Toyota Kata. Managing people for improvement, adaptiveness, and superior results. McGraw-Hill, New York
Schonberger RJ (1986) World class manufacturing, the lessons of simplicity applied, 1st edn. The Free Press, New York
Wilson PA (2010) Maquiladoras. Texas State Historical Association. http://www.tshaonline.org/handbook/online/articles/dzm02. Accessed 26 Oct 2017
Womack JP, Jones DT (1996) Lean thinking, 1st edn. Simon & Schuster. ISBN 0-684-81035-2
Womack JP, Jones DT, Roos D (1990) The machine that changed the world, First Harper Perennial edition, 1991. ISBN 0-06-097417-6
Yang K, El-Haik B (2003) Design for six sigma: a roadmap for product development. McGraw Hill, New York

Part III
Manufacturing and Technology

Chapter 15
Manufacturing's Strategic Role and Management Practices: Evidence from Colombian Companies

Jorge A. Vivares and William Sarache

Abstract Operations Strategy (OS) is a long-term plan for Manufacturing/ Operations Systems (MOS), oriented toward the generation of distinct organizational advantages. Typically, OS studies have fixated on two components: content and process. However, Manufacturing's Strategic Role (MSR), despite its importance to MOS performance, has remained as a missing link in the OS field. As such, in the present chapter, MSR is explored, and the management practices adopted by a group of companies located in the Colombian coffee region are analyzed. Survey research is conducted for the verification of MSR influence on MOS performance, and the level of management practices used in manufacturing as well as their relationship with MSR is evaluated. Investigative results permit the conclusion that MSR is a new variable that should be integrated into the OS formulation process. The proposed MSR registered internal consistency and impact on MOS performance; therefore, in this study, MSR is highlighted as a fundamental component of the OS. The case study revealed that manufacturing management practices are different than those typically addressed in the investigative literature. These findings contribute to the advancement of OS study and report empirical evidence from the Colombian context.

Keywords Manufacturing's strategic role · Management practices
Operations strategy · Manufacturing strategy

J. A. Vivares (✉) · W. Sarache
Facultad de Ingeniería y Arquitectura, Departamento de Ingeniería Industrial,
Universidad Nacional de Colombia, Carrera 27 No. 64-60, Campus Palogrande,
Bloque F, Oficina F-705, Manizales, Colombia
e-mail: javivaresv@unal.edu.co

J. L. García Alcaraz et al. (eds.), *Best Practices in Manufacturing Processes*,
https://doi.org/10.1007/978-3-319-99190-0_15

15.1 Introduction

Operations Strategy (OS) is a long-term plan for Manufacturing/Operations Systems (MOS), with which the concrete actions that support a company's general strategy are defined. As a field of study, OS originates with Skinner's (1969, 1966) seminal work. With his first publications, he indicated the importance of overcoming the merely technical role assigned to MOS, in order to convert it into a strategic competitive weapon. OS seeks to create lasting competitive advantages for companies through strategic MOS management (Miltenburg 2008; Vivares-Vergara et al. 2015).

Typically, OS has been divided into two components (Leong et al. 1990): process and content. The process establishes strategy formulation procedures, while content defines competitive priorities (MOS objectives) and the group of decision areas (products, processes, facilities, etc.) to be intervened for MOS performance improvement. Although OS investigation has concentrated on these two components, the systematic literature review conducted in this study established the existence of a third component, which has remained absent in OS study: Manufacturing's Strategic Role (MSR).

The MSR concept was introduced by Hill (1983) and further advanced by Hayes and Wheelwright (1984) in the so-called four-stage model. In the said model, MOS is categorized in accordance with the role to be completed in a company's competitive strategy, thus: (1) internally neutral, (2) externally neutral, (3) internally supportive, and (4) externally supportive. However, Barnes and Rowbotham (2004, p. 613) reported the existence of limited empirical evidence on this topic. In fact, in the review of the literature, which supports this study, the knowledge gap detected by the said authors remains cavernous. As such, MSR continues to be a fruitful field for OS investigation. MSR, as conceived in this contribution, broadens its classical conception and simultaneously includes different aspects related to the MS content and process.

Among these aspects are the analysis, selection, and implementation of the so-called Management Practices (MP), which are considered necessary for the achievement of MOS improvement. Typically, the most widely renowned MP in the literature are: Just in Time (JIT), Total Quality Management (TQM), Total Productive Maintenance (TPM), and lean manufacturing. Slack and Lewis (2011) suggest that practitioners tend to confuse MP with OS, which is a mistake, because although they may be considered 'new approaches' to operations management, in the strict sense, they do not constitute an OS. Given that MP have been widely studied in developed countries, it is necessary to examine their applicability in developing countries, so as to test the robustness of existing OS theories (Amoako-Gyampah and Acquaah 2008). Further, it is necessary to test whether practices additional to those typically addressed in the literature are adopted in Latin American countries. This suggests that MP investigation should continue together with the MSR concept.

In order to contribute to this field of study, an empirical investigation was carried out by a group of Colombian industrial companies, located in the coffee region. The investigation strove to respond to the following questions:

- Does MSR affect the achievement of OS objectives?
- What MP are used for performance improvement by companies in the Colombian coffee region?

In order to respond to the first question, survey research was performed in a pilot sample of industrial companies. The second question was answered by combining the survey research (deductive approach) with a case study (inductive approach). As such, the investigation presented is exploratory with a mixed approach.

In accordance with the results obtained, it may be concluded that MSR is a component that should be considered in OS formulation. Concretely, this investigation contributes to the field of study from two perspectives: first, empirical evidence is presented regarding the positive effect of MSR on MOS performance. Second, evidence of the existing relationship between MP implementation levels and the strategic role assigned to MOS is submitted. Thus, the results contribute to research advancement within the OS field, which has few precursors, not only in the Colombian context but also in the Latin American context.

The present chapter was organized as follows: in Sect. 15.2, the review of the literature, from which the hypotheses that guided this investigation were derived, is presented. In Sect. 15.3, the employed methodology is explained. In Sects. 15.4 and 15.5, the results obtained, and discussion are presented. Finally, in Sect. 15.5, the most relevant conclusions and some lines for future investigation are raised.

15.2 Theoretical Background

15.2.1 Operations Strategy

According to Slack and Lewis (2011, p. 7):

> Operations strategy is concerned less with individual processes and more with the total transformation process that is the whole business. It is concerned with how the competitive environment is changing and what the operation has to do in order to meet current and future challenges. It is also concerned with the long-term development of its operations resources and processes so that they can provide the basis for a sustainable advantage.

Operations Strategy (OS) originated with the seminal contributions of Wickham Skinner (1969, 1966), to which a group of later investigations was added, including those of Skinner (1974), Hayes and Schmenner (1978), Wheelwright (1978, 1984), Schmenner (1982), Hayes and Wheelwright (1984), Hill (1985), Swamidass and Newell (1987) and Leong et al. (1990). These authors, among others, helped to consolidate the theoretical bases of this field of study.

These authors initially coined the term “manufacturing strategy”, as its initial interest focused on manufacturing companies. However, the term changed to what is known today as “operations strategy”, for two fundamental reasons: on one hand, there was a possibility to apply its principles to another type of nonmanufacturing organization (Slack and Lewis 2003), and on the other, because of service operations grew in manufacturing companies (Martín-Peña et al. 2017). OS has traditionally been studied in two components: process and content (Boyer et al. 2005; Chatha and Butt 2015; Dangayach and Deshmukh 2001).

The process, which seeks to explain “how” to formulate OS, has been dominated by conceptual models, and investigative contributions in this field are scarce (Jia and Bai 2011). As such, manufacturing strategy formulation continues to be an open problem in the literature, despite having emerged as a field of study decades ago (Platts and Gregory 1990). Years of investigation have allowed for the conclusion that many approaches to OS formulation are possible (Miltenburg 2009), and there is no a predominant approach in the literature (da Silveira 2005).

In content, the “what” of OS is studied. This defines competitive priorities (MOS objectives) and the MOS subsystems intervention. From the OS point of view, competitive priorities represent means of MOS performance, and they also are known as manufacturing outputs (Miltenburg 2005), task of company manufacturing function (Skinner 1969), manufacturing capabilities (Avella et al. 2011), order-winners and qualifiers (Hill 1995), among others. The four most common competitive priorities are quality, cost, flexibility, and delivery (Ibarra and Sarache 2008). That said, environmental protection has positioned itself as a new manufacturing system priority (Avella et al. 2011; Jabbour et al. 2012). Further, MOS subsystems refer to the decision areas that should be modified for performance improvement (Slack and Lewis 2011).

Work on the OS formulation process without having defined content could lead to poor results (Hill 2000). In fact, Brown et al. (2007) found that the highest level plants emphasize process and content simultaneously. Existing OS frameworks, however, have not permitted a deep understanding of how to integrate content and process, in order to deploy a strategy. Hayes and Wheelwright’s (1984) seminal study highlights the importance of defining a strategic role for manufacturing systems, in order to overcome this error. The MSR envisaged in this contribution, then, is not explicitly part of content or process, but rather, of both. This concept, although coined in the 1980s, has scarcely been addressed in research projects that contribute empirical evidence.

15.2.2 Manufacturing’s Strategic Role and Management Practices

In an article entitled “Manufacturing’s strategic role”, Hill (1983, p. 854) expressed that:

> Production managers are too late in the corporate debate. This results in their taking a reactive role, with less likelihood of contributing to strategy alternatives and influencing the eventual decision (…) Part of this problem is the lack of developed language which would

both help express the conceptual insights which need to be considered and also provide the medium to help explain, from a corporate viewpoint, the production issues and perspectives which need to be understood and addressed

The Manufacturing's Strategic Role (MSR) concept is based on the recognition of the importance of MOS in a company's general strategy. Hayes and Wheelwright's (1984) study suggested that MSR could affect the achievement of OS objectives, which would depend on the importance of MOS' role in a company's general strategy. Hayes and Wheelwright (1984) proposed four stages of development for MSR establishment: (1) internally neutral, (2) externally neutral, (3) internally supportive, and (4) externally supportive. Although this model has been widely accepted in the literature, Barnes and Rowbotham (2003, p. 613) affirm that little practical application of the model is reported in the literature.

Despite these early proposals, the present study's literature review confirmed that this concept has been little investigated, and that the few existing advances have focused on expanding Hayes and Wheelwright's proposal. For example, by analyzing four study cases, Swamidass et al. (2001a) took certain proposals in order to analyze the lack of agreement between manufacturing and marketing managers. Also, using the four-stage model of Hayes and Wheelwright, Swamidass et al. (2001b) classified three companies, in accordance with the way in which their manufacturing strategy was developed. Barnes and Rowbotham (2004) operationalized Hayes and Wheelwright's model in a group of variables (customer, quality, technology, and workforce, among others), surveyed 460 managers from different functional areas in the U.K., and found that 263 responses adjusted to the model's four stages. Jain et al. (2013) developed an instrument to measure manufacturing effectiveness, based on Hayes and Wheelwright's model, and proved its consistency in a sample of 28 Indian companies.

Existing MSR contributions are relevant, but the empirical evidence that they have presented offers investigative possibilities, for at least three reasons: first, these studies have been performed in a limited group of countries. Second, the effect of MSR on MOS performance has been little analyzed in the literature. Third, the existing operationalization of the concept may still be considered incomplete, as MSR is not easily categorized as part of content or process and requires measurement of rarely considered variables in the existing OS frameworks.

For example, the commitment of top management is vital for the achievement of substantial system improvements (Miltenburg 2005). Also, the capability of MOS articulation with other functional areas is a relevant OS formulation strength (Hayes and Wheelwright 1984). The study of the MOS environment requires the involvement of different *stakeholders* (Maylor et al. 2015). Furthermore, in the literature review conducted by Lillis and Lane (2007), it was concluded that the methodologies developed to audit operations strategy were limited by their emphasis on the outside-in approach and limited consideration of the inside-out approach. One contribution that addresses some of these variables is that of Gilgeous (2001), which is a study of 295 U.K. companies. Based on Hayes and Wheelwright's model, its goal was to identify relevant dimensions that permit measurement of MOS effectiveness. The study identified critical factors, including the attitude of top managers toward manufacturing, coordination between manufacturing and other functions, and the emphasis on manufacturing strategy formulation. All of these are related to the strategic role of the manufacturing system, and they have often remained suppositions instead of variables that are explicitly measured in existing OS frameworks.

Recently, Vivares et al. (2018) proposed that MSR constituted a key dimension for MOS maturity improvement. Although the said study did not present evidence of the effect of this variable on performance, it suggests that MSR could have a relevant impact, specifically on competitive priorities. According to the literature review that underpins the present investigation, it may be affirmed that MSR has not been sufficiently addressed in academic research. As such, the following investigative hypothesis is proposed:

H_1: *Manufacturing's Strategic Role (MSR)* positively influences *MOS performance.*

One relevant element that emerges from MSR study is the selection and implementation of the so-called Management Practices (MP) for the manufacturing system. In accordance with Slack and Lewis (2011), MP must be understood as management approaches for MOS, and as such, should not be considered an OS replacement. It may, thus, be said that MP are choices that the company can make as part of their OS (Vivares et al. 2017). According to Karim et al. (2008, p. 3583), the identification of effective manufacturing practices has long been considered one of the key elements in manufacturing strategy research.

Slack and Lewis (2011) denominate MP as "substitutes for strategy" because some organizations confuse them with the OS. According to Miltenburg, MP must be understood as "improvement programs in manufacturing" since their goal is to increase the MOS performance. Dangayach and Deshmukh (2001) establish that MP are a set of "best practices" for MOS. In turn, Vivares-Vergara et al. (2014) call them "Management's approaches to manufacturing". A selection of MP that has been frequently studied in the literature is presented in Table 15.1.

Hayes and Wheelwright (1984, p. 399) affirm that companies that pursue a manufacturing-based competitive advantage, "anticipate the potential of new manufacturing practices and technologies and seek to acquire expertise in them long before their implications are fully apparent." However, the purpose is not to implement all the practices. After a number of years of investigation in the global macro project entitled High Performance Manufacturing (HPM), Schroeder and Flynn (2001, p. 4) noted that rejecting new approaches out of hand is a risky business, just as is adopting every new approach that comes along. Therefore, it could be said that the adequate selection and implementation of MP depend on the strategic role that the company assigns to MOS. Thus, the following investigative hypotheses are proposed:

H_2: the companies analyzed *apply MP* which exist in the specialized literature.
H_3: companies that *adopt MP* have *greater MSR strength.*

Table 15.1 Some MP reported in the literature

MP	Authors
Just in time (JIT)	Fullerton and McWatters (2001), Yasin et al. (2003), Swink et al. (2005), Singh and Garg (2011)
Total quality management (TQM)	Phan et al. (2011), Suwandej (2015), Cetindere et al. (2015)
Total productive maintenance (TPM)	Stamatis (2010), Kaur et al. (2013), Piechnicki et al. (2015)
Theory of constraints (TOC)	Vargas et al. (2016), Tsou (2013)
5S	Pheng (2001), Gapp et al. (2008), Chen and Meng (2008), Filip and Marascu-Klein (2015), Kanamori et al. (2015), Jiménez et al. (2015).
Kaizen	Styhre (2001), Bateman and David (2002), Bradley and Willett (2004), Aoki (2008), Farris et al. (2009), Glover et al. (2011), García et al. (2014).
Reengineering	Hammer and Champy (1993), O'Neill and Sohal (1999), Goel and Chen (2008), Radhakrishnan and Balasubramanian (2008)
Lean manufacturing	Womack et al. (1992), Shokri et al. (2016)

Miltenburg (2005, p. 245) calls MP "improvement programs", and argues that their objective is to raise the capabilities of the levers in a production system or a manufacturing network in order to raise the levels of the factory or network outputs. In other words, MP implementation is expected to contribute to improved MOS performance (da Silveira and Sousa 2010; Morita et al. 2001). For this reason, the following investigative hypothesis is proposed:

H_4: companies that *adopt MP* have better *MOS performance* than those that do not.

15.3 Methodology

This study was carried out in two stages, which, respectively, involved survey research and a case study. It may, thus, be considered exploratory and of a mixed perspective. The survey research represents a deductive component, while the case study represents an inductive approach.

15.3.1 Stage 1: Survey Research

In Stage 1, survey research was performed, in order to evaluate hypotheses H_1, H_2, H_3, and H_4. To this end, the companies involved employed at least 50 people, were classified as medium or large in the region in which the investigation was performed (Colombian Coffee Region, Caldas Department). According to Vivares-Vergara et al. (2016, p. 115), this region has based its economy on the commercialization of bulk coffee; however, due to the dropping of international prices and demand volatility, the national economic policy has been geared to promote the strengthening of manufacturing enterprises to preserve the employment level in different sectors.

The resulting population was 48 companies according to the official entities, with a 31.25% response rate (15 companies), which exceeds the 20% minimum suggested by Malhotra and Grover (1998) for survey research in production and operations management. Note that the size of the sample is small, for which it may be considered a pilot sample to support an exploratory investigation. The variables involved in MSR are presented in Table 15.2. A 0.855 Cronbach's alpha coefficient was obtained for said construct.

In order to evaluate MOS performance, five competitive priorities which are accepted in the literature were considered: cost, quality, flexibility, deliveries, and environmental protection. These dimensions obtained Cronbach's alpha coefficients of 0.782, 0.826, 0.673, 0.700, and 0.692, respectively. OS global performance is a simple average of the performance in these priorities. The MSR variables and competitive priority performance were measured with a five-point Likert scale.

The MP involved were JIT, TQM, TPM, TOC, 5S, Kaizen, reengineering, and lean manufacturing. Considering that MP are choices that companies may or may not use, these were measured on a six-point scale, with respect to the level of implementation (0—Not used, 1—Very deficient, 2—Deficient, 3—Medium, 4—High, 5—Very high).

Table 15.2 MSR operationalization

Code	Degree of strength in
MSR01	Commitment of top managers with the MOS improvement
MSR02	Involvement of the different stakeholders interested in the planning processes and the MOS improvement (owners, workers, suppliers, customers, government, or others)
MSR03	Study of the environment (government, society, market, competition, suppliers, institutions, etc.) to support decision-making and the MOS improvement
MSR04	Coordination with other functional areas (human resource management, sales, finances, etc.) to support decision-making and the MOS improvement
MSR05	Implementation of programs and/or MP aimed at the MOS improvement
MSR06	Utilization of strategic planning methodologies for the MOS improvement
MSR07	Evaluation and monitoring of the MOS planning and the goals achievement
MSR08	Actions oriented towards the improvement of strategic decision areas of the MOS

15.3.2 Stage 2: Case Study

In Stage 2, one of the companies with a very low level of the practices addressed in the survey research was studied, in order to identify which practices had been implemented. For confidentiality reasons, in this document, it will be called Company A. This company pertains to the metallurgical sector, has 251 employees, and produces 11 families of products. Exactly 30% of sales come from international markets (in 14 countries).

Top management at Company A agreed to allow the investigator a months-long immersion experience, in order to observe and interact with practitioners. During this time period, MP that said company implemented were detected, and an exercise was carried out with employees, so as to estimate their implementation level (0% to 100%), in accordance with the company's ideal expectation.

15.4 Results

Considering the order in which the methodology was set forth, the survey research investigative results are presented first, followed by those obtained in the case study.

15.4.1 Survey Research

15.4.1.1 Manufacturing's Strategic Role (MSR)

Descriptive statistics related to MSR are presented in Table 15.3. As can be seen, in the analyzed companies, the MSR is suitable. The variables with greater strength were MSR01 (commitment of top managers with the MOS improvement), MSR03 (the study of the environment to support decision making and the MOS improvement), and MSR07 (evaluation and monitoring of the MOS planning and the goals achievement). In contrast, the most lagged variables were MSR05 (implementation of programs and/or MP for the MOS improvement) and MSR06 (Utilization of strategic planning for the MOS improvement).

As can be seen in Fig. 15.1, the MOS' performance for all companies was acceptable. In particular, quality was the best evaluated competitive priority, followed by environmental protection, deliveries, cost, and flexibility.

A regression analysis was performed, with MSR considered an independent variable. The results presented in Table 15.4 corroborate that, in the sample of companies studied, MSR positively affects MOS performance, as the independent variable (MSR) is significant at 0.001 (p-value < 0.001), as in the F test, with which a determination coefficient of 67.3% was obtained. Given that the residuals

Table 15.3 Descriptive statistics for MSR

Variables	Mean	Standard deviation	Percentage of companies (%)			
			1–2	3	4–5	Total
MSR01	4.467	0.516	0.0	0.0	100.0	100
MSR02	4.133	0.743	0.0	20.0	80.0	100
MSR03	4.267	0.704	0.0	13.3	86.7	100
MSR04	4.067	0.799	6.7	6.7	86.6	100
MSR05	3.667	1.234	13.3	13.3	73.4	100
MSR06	3.533	1.457	20.0	13.3	66.7	100
MSR07	4.267	0.594	0.0	6.7	93.3	100
MSR08	4.200	0.775	0.0	20.0	80.0	100
Global MSR	4.075	0.927	5.0	11.7	83.3	100

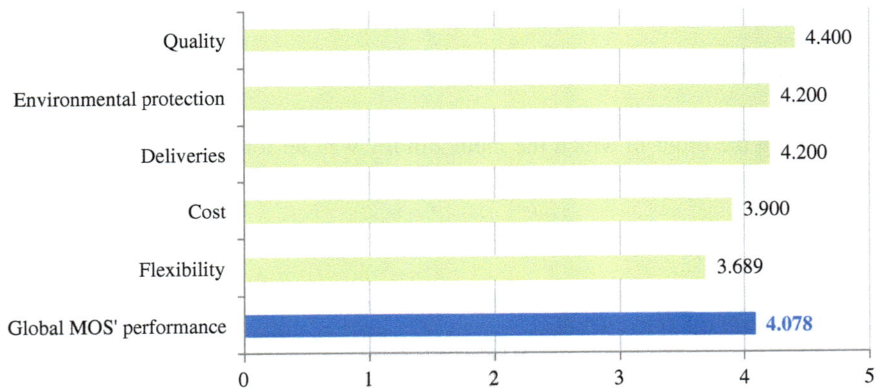

Fig. 15.1 MOS' performance

Table 15.4 Regression analysis

Variables	Value	*p*-value
Constant	1.932	0.000
Independent (MSR)	0.527	0.000
F	26.728	0.000
R^2	0.673	–

had a mean equal to zero and presented a normal distribution (Shapiro–Wilk test: S-W = 0.968; *p*-value = 0.822), model adjustment suitability was corroborated. The individual data for each company is presented in Appendix. These results provide support for Hypothesis 1 (H_1).

15.4.1.2 OS Management Practices

The sample of companies analyzed presented a low level of MP implementation, and on some occasions, a lack of knowledge of the different existing management practices in the OS literature. In Fig. 15.2, observe that JIT principles have not been formally used in 79% of the companies, and those that have applied them have a low-level implementation. In other words, partial evidence was found in the survey research to support Hypothesis 2 (H_2).

Performance of the Mann–Whitney U test (Table 15.5) revealed that, in almost all cases, companies that implemented MP show higher levels of MSR strength. This result supports Hypothesis 3 (H_3).

Table 15.6 contains the Mann–Whitney U test results used to compare MOS performance between companies that report some level of MP implementation and those that report none whatsoever. The results indicate that, of the eight MP included in the study, differences with 95% (p-value < 0.05) and 99% (p-value < 0.01) reliabilities were found in just three MP (JIT, TOC, and lean). In another two (Kaizen and reengineering), differences were observed, but with 90% reliability (p-value < 0.1). In the remaining MP (TQM, TPM, and 5S), significant differences were not found. This evidence is insufficient to support Hypothesis 4 (H_4).

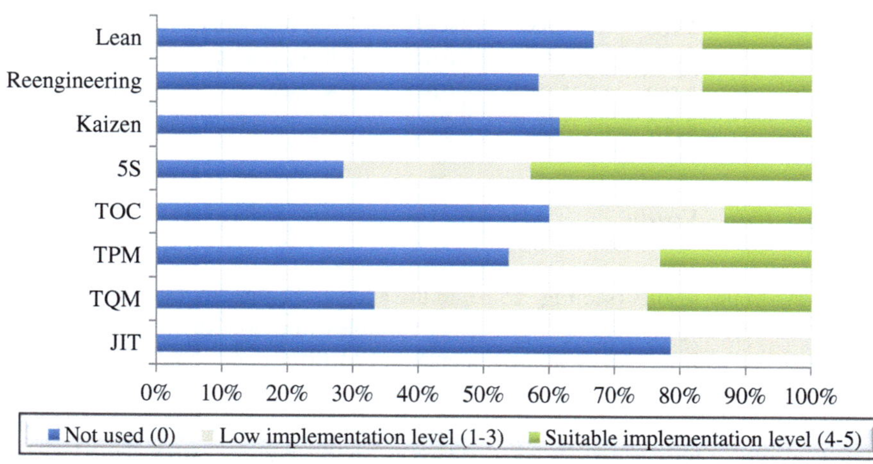

Fig. 15.2 MP implementation levels (evaluated on a six-point scale [0—not used, 1—very deficient, 2—deficient, 3—medium, 4—high, 5—very high])

Table 15.5 MSR differences in accordance with MP implementation

MP	U Mann–Whitney	W Wilcoxon	*p*-value	Interpretation (not used vs. some level of MP implementation)
JIT	0.000	66.000	0.004	Significant differences at 0.01
TQM	15.500	51.500	0.932	Not differences
TPM	6.000	34.000	0.032	Significant differences at 0.05
TOC	6.000	51.000	0.013	Significant differences at 0.05
5S	3.500	13.500	0.019	Significant differences at 0.05
Kaizen	5.000	41.000	0.027	Significant differences at 0.05
Reengineering	0.000	28.000	0.004	Significant differences at 0.01
Lean	0.500	36.500	0.008	Significant differences at 0.01

Table 15.6 Performance differences in accordance with MP implementation

MP	U Mann–Whitney	W Wilcoxon	*p*-value	Interpretation (not used vs. some level of MP implementation)
JIT	0.000	66.000	0.004	Significant differences at 0.01
TQM	14.000	50.000	0.734	Not differences
TPM	9.500	37.500	0.100	Not differences
TOC	9.500	54.500	0.039	Significant differences at 0.05
5S	9.000	19.000	0.119	Not differences
Kaizen	8.500	44.500	0.091	Significant differences at 0.1
Reengineering	5.500	33.500	0.051	Significant differences at 0.1
Lean	0.000	36.000	0.006	Significant differences at 0.01

15.4.2 Case Study

According to the results (see Fig. 15.3), in the analyzed company the MSR showed a strength level of 69%. The best performer variables were MSR01 (commitment of top managers with the MOS improvement) and MSR04 (Coordination with other functional areas to support decision making and the MOS improvement). In turn, weaknesses were observed in MSR05 (implementation of programs and/or MP aimed at the MOS improvement), MSR06 (utilization of strategic planning methodologies for the MOS improvement), and MSR08 (actions oriented towards the improvement of strategic decision areas of the MOS).

The observation process carried out within Company A, from an inductive perspective, proved that it has implemented nine MP (Fig. 15.4). While the 5S form part of the list of recurrent practices in the literature, and the continuous improvement team presents a certain resemblance to the Kaizen, it is different in its conception and implementation. For this reason, together with that found in the survey research, it could be said that there is insufficient evidence to support Hypothesis 2 (H_2).

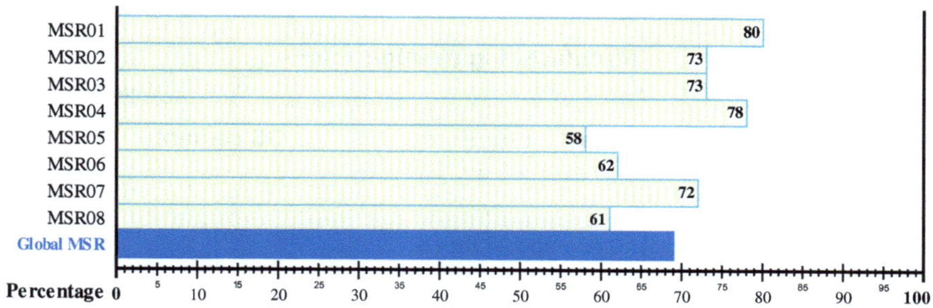

Fig. 15.3 Descriptive statistics for MSR in Company A

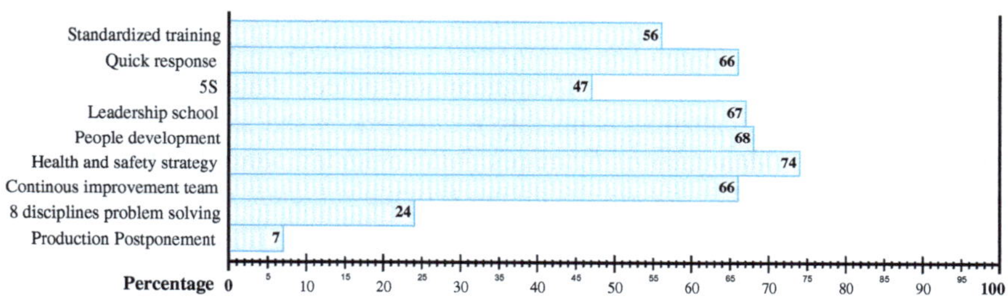

Fig. 15.4 Implementation level of management practices in Company A

15.5 Discussion

In line with the theoretical intentions of this document, in this section, the discussion will be incorporated on two axes: the MSR and MP implemented by the companies studied. With respect to MSR, this investigation provides arguments for its conceptual relevance, as well as empirical evidence to corroborate its importance as a key element to be considered in OS. MSR, to date, has been little studied, as Barnes and Rowbotham (2003) and Vivares et al. (2018) also suggest. Given that evidence was found to support hypothesis H_1, it may be affirmed that MSR, as proposed in this contribution, positively influences MOS performance in the companies analyzed.

The MSR conceptualized in this study operationalizes variables that are transversal to various MOS components, which may be simultaneously associated with OS content and process. For example, the commitment of top management is not a competitive priority, but is vital, in order to achieve substantial improvements to the system (Miltenburg 2005). MOS' joint capability with other functional areas is a relevant strength for OS formulation, although it does not constitute a manufacturing lever in and of itself. The same situation occurs with the study of the MOS environment (internal and external). These variables have often remained suppositions instead of variables that are explicitly measured in existing OS frameworks,

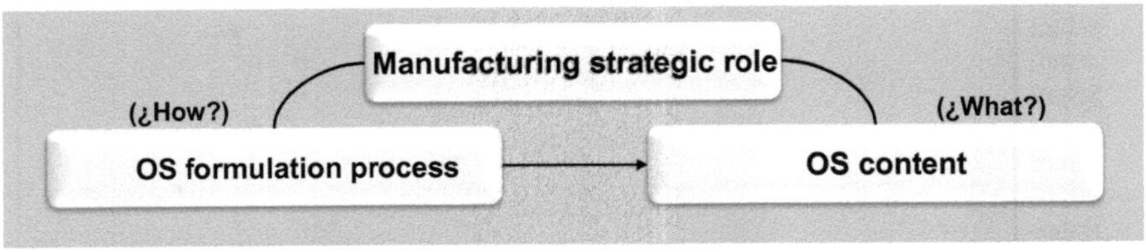

Fig. 15.5 New analytical framework proposed for OS

and which could potentially hinder system performance and decision making. As such, the results suggest the importance of MSR involvement in OS formulation.

This vision is also applicable to services; Roth and Van Der Velde (1991) proposed that operations were a key element for successful OS development. For this reason, a renewed idea of OS is proposed in this study, which adds MSR as a fundamental element to the traditional, content, and process-based view (Fig. 15.5).

One of the MSR challenges with this framework is the study, selection, and implementation of Management Practices (MP) oriented toward MOS improvement. With respect to MP, the results do not provide sufficient evidence to support hypotheses H_2 or H_4.

Regarding hypothesis H_2, the findings indicate that the companies under study scarcely use MP identified in the literature. In the survey research, it was shown that the group of companies in the sample used these practices very little, while in the case study, it was found that, among the nine MP considered, Company A implemented just one, and the rest corresponded to different MP. One new MP that company A decided to implement, during the case study application, was the production postponement strategy; for this reason. It was the MP with the lowest level of application. According to Gómez et al. (2017), the postponement concept is a mass customization strategy used to offer a wide portfolio under uncertain conditions at a lower cost, so as to delay final product assembly and creating work-in-process inventory at a decoupling point.

Reject H_2 may be the result of traditional MP that has been implemented and studied mainly in developing countries. However, the companies studied in this investigation exist in a different context, which prompts them to seek out alternatives. This finding is consistent with conclusions derived from the global macro project entitled High Performance Manufacturing (HPM), in which it is affirmed that "We advance the idea that the management practices leading to global high-performance manufacturing differ by country, industry, and size company, to name just a few contingencies. We take this contingency approach to the adoption of management practices rather than a universal, one-size-fits-all approach" (Schroeder and Flynn 2001, p. 3).

Another interesting finding is that companies that adopt traditional MP show higher levels of MSR strength (H_3). This is consistent with results found in the analyzed literature, as adequate MP selection and implementation in companies depends on their degree of MSR strength. In other words, the more important the role of MOS in a company's competitive strategy, the higher the MP application level.

However, the benefits of MP were limited in the companies studied, as just three of them reported MOS performance (H_4) higher than that of companies that had not implemented them. This situation could be attributed to two reasons: on one hand, there were low MP implementation levels in the majority of the companies analyzed. On the other, MP must be carefully selected, in accordance with the company's strategic context, as well as satisfactorily conceptualized for its successful implementation. Both situations reduce the impact of these practices on MOS performance.

These results coincide with similar weaknesses found in other companies in the Colombian context. Vivares-Vergara et al.'s (2014) investigation found that the majority of MP implemented by a group of companies did not effectively support the MOS performance. In a separate study, Vivares et al. (2017) found that practitioners had flaws in MP conceptualization, which impeded complete exploitation of their potential for MOS improvement. In this study, the additional and even more specific evidence is contributed to the problem resolution.

For example, in Company A, an MP called Eight Disciplines Problem Solving (8Ds) was applied. Although this MP had been implemented in the company several years prior, it had been minimally effective for MOS improvement. During the immersion experience, it was noted that, among other questions, 8Ds is a practice that has not been well adapted for personnel. One concrete reason is that the company does not use eight complete disciplines, but rather seven of them, and so the last discipline on the list was absent: (1) form a team, (2) describe the problem, (3) implement an interim containment action, (4) analyze the root cause, (5) establish permanent corrective actions, (6) implement and validate the permanent actions, (7) prevent the recurrence and the root cause, and (8) recognize/congratulate the team.

The limitations found in 8D implementation, as with other practices reviewed in Company A, provided a better understanding of survey research findings. For example, one of the reasons for which the MP were not contributing to MOS performance was the lack of knowledge and limited use of OS planning methodologies. On occasion, the practices emerged in an improvised fashion, but not as part of a strategic management model. Likewise, these problems reflect certain MSR weaknesses.

15.6 Conclusions

Operations Strategy (OS) seeks to contribute to the consolidation of competitive advantages in companies, through the strategic management of Manufacturing/Operations Systems (MOS). Traditionally, OS study has concentrated on two components: content and process. OS has been consolidated as a field of academic and practical interest that offers successive opportunities for investigation. In this study, it was found that Manufacturing's Strategic Role (MSR) is a missing link variable in the diverse OS *frameworks* studied. Although this concept emerged in the 1980s, the literature review indicates a lack of empirical evidence on the topic.

As such, this investigation provides evidence to enrich the field of study. In particular, it was found that MSR positively influences MOS performance, and that companies that adopt Management Practices (MP) have higher levels of MSR strength. One of the challenges in MSR is the study, selection, and implementation of MP oriented toward MOS improvement. In this sense, various findings may be derived from this investigation: on one hand, the companies studied scarcely used the MP identified in the literature, and some of them applied MP scarcely covered by the academic community. On the other hand, those companies that implemented MP had not achieved improved MOS performance results, partly because of a lack of knowledge, and partly owing to the minimal use of OS planning methodologies, another reflection of MSR weakness.

For this reason, a renewed idea of OS is proposed in this study, which adds MSR as a fundamental element to the traditional content and process-based view. The investigative findings suggest clear implications for practitioners and academics. Practitioners should pay attention to variables which are typically not measured in OS planning processes, such as the commitment of top management, ability of MOS to link with other functional areas, integral study of the MOS environment for decision making, use of models to facilitate OS formulation, and selection and implementation of MP in analytical and orderly form, among other matters that strengthen MSR to effectively contribute to companies' global strategies. Academics should investigate MSR in greater depth, in order to corroborate results and enrich the proposals made in this contribution.

Finally, further investigation is necessary on three fronts: first, the limited size of the sample makes it imperative to replicate the study, so as to verify whether the findings encountered are validated in other regional contexts. Second, MSR conceptualization and operationalization could be improved to better coordinate a general OS *framework*. Third, the investigation should be carried out with a wider variety of MP than those typically reported in the literature.

Acknowledgements This investigation was supported by Universidad Nacional de Colombia (Project code 28741). We also thank Colciencias for the first author's financial funding.

Appendix

List of companies and survey data

	MSR01	MSR02	MSR03	MSR04	MSR05	MSR06	MSR07	MSR08	Total MSR
E01	5	5	5	5	5	5	5	5	5.000
E02	5	4	4	5	4	4	3	5	4.250
E03	4	4	4	4	4	4	4	4	4.000

(continued)

(continued)

	MSR01	MSR02	MSR03	MSR04	MSR05	MSR06	MSR07	MSR08	Total MSR
E04	5	1	1	4	3	3	4	2	2.875
E05	4	4	1	4	4	3	3	3	3.250
E06	4	1	1	4	5	5	4	4	3.500
E07	4	3	3	3	4	4	4	4	3.625
E08	4	4	4	4	4	5	5	4	4.250
E09	4	5	5	5	5	4	5	5	4.750
E10	5	4	4	5	5	4	5	4	4.500
E11	4	4	4	4	3	4	4	4	3.875
E12	4	3	3	4	3	3	4	4	3.500
E13	5	4	5	5	5	5	5	4	4.750
E14	5	4	4	4	4	4	4	4	4.125
E15	5	5	5	4	5	5	5	5	4.875
	JIT	TQM	TPM	TOC	5S	Kaizen	Reengineering	Lean	MOS' performance
E01	3	0	0	0	4	0	3	4	4.400
E02	0	3	2	3	5	4	4	2	4.200
E03	0	1	1	0	0	0	0	0	4.067
E04	0	1	0	0	0	0	0	0	3.167
E05	0	3	0	0	0	0	0	0	4.000
E06	0	0	0	0	0	0	0	0	3.967
E07	0	0	0	0	2	0	0	0	4.133
E08	0	4	4	4	4	4	3	0	3.700
E09	3	4	5	3	5	5	3	4	4.567
E10	0	4	4	4	3	4	5	3	4.367
E11	0	0	0	0	3	0	0	0	3.800
E12	0	2	0	0	2	0	0	0	3.700
E13	3	MD	MD	2	4	4	MD	MD	4.567
E14	0	MD	MD	0	MD	MD	MD	MD	3.833
E15	3	MD	2	2	4	MD	MD	MD	4.700

MD: missing data

References

Amoako-Gyampah K, Acquaah M (2008) Manufacturing strategy, competitive strategy and firm performance: an empirical study in a developing economy environment. Int J Prod Econ 111 (2):575–592. https://doi.org/10.1016/j.ijpe.2007.02.030

Aoki K (2008) Transferring Japanese Kaizen activities to overseas plants in China. Int J Oper Prod Manage 28(6):518–539. https://doi.org/10.1108/01443570810875340

Avella L, Vazquez-Bustelo D, Fernandez E (2011) Cumulative manufacturing capabilities: an extended model and new empirical evidence. Int J Prod Res 49(3):707–729. https://doi.org/10.1080/00207540903460224

Barnes D, Rowbotham F (2003) Developing a questionnaire for the four-stage model of operations strategy. Prod Plan Control 14(7):613–622. https://doi.org/10.1080/09537280310001626205

Barnes D, Rowbotham F (2004) Testing the four-stage model of the strategic role of operations in a UK context. Int J Oper Prod Manage 24(7):701–720. https://doi.org/10.1108/01443570410542000

Bateman N, David A (2002) Process improvement programmes: a model for assessing sustainability. Int J Oper Prod Manage 22(5):515–526. https://doi.org/10.1108/01443570210425156

Boyer KK, Swink M, Rosenzweig ED (2005) Operations strategy research in the POMS journal. Prod Oper Manage 14(4):442–449

Bradley JR, Willett J (2004) Cornell students participate in Lord corporation's Kaizen projects. Interfaces 34(6):451–459. https://doi.org/10.1287/inte.1040.0103

Brown S, Squire B, Blackmon K (2007) The contribution of manufacturing strategy involvement and alignment to world-class manufacturing performance. Int J Oper Prod Manage 27(3):282–302. https://doi.org/10.1108/01443570710725554

Cetindere A, Duran C, Yetisen MS (2015) The effects of total quality management on the business performance: an application in the province of Kütahya. Procedia Econ Finance 23:1376–1382. https://doi.org/10.1016/S2212-5671(15)00366-4

Chatha KA, Butt I (2015) Themes of study in manufacturing strategy literature. Int J Oper Prod Manage 35(4):604–698. https://doi.org/10.1108/IJOPM-07-2013-0328

Chen L, Meng B (2008) How to make 5S as a culture in Chinese enterprises. In: Information management, innovation management and industrial engineering, 2008. ICIII'08. International conference, vol 3, pp 221–224. https://doi.org/10.1109/ICIII.2008.251

da Silveira GJC (2005) Market priorities, manufacturing configuration, and business performance: an empirical analysis of the order-winners framework. J Oper Manage 23(6):662–675. https://doi.org/10.1016/j.jom.2005.01.005

da Silveira GJC, Sousa RS (2010) Paradigms of choice in manufacturing strategy. Exploring performance relationships of fit, best practices, and capability-based approaches. Int J Oper Prod Manage 30(12):1219–1245. https://doi.org/10.1108/01443571011094244

Dangayach GS, Deshmukh SG (2001) Manufacturing strategy: literature review and some issues. Int J Oper Prod Manage 21(7):884–932. https://doi.org/10.1108/01443570110393414

Farris JA, Van Aken EM, Doolen TL, Worley J (2009) Critical success factors for human resource outcomes in Kaizen events: an empirical study. Int J Prod Econ 117(1):42–65. https://doi.org/10.1016/j.ijpe.2008.08.051

Filip FC, Marascu-Klein V (2015) The 5S lean method as a tool of industrial management performances. IOP Conf Ser Mater Sci Eng 95(1):1–6. https://doi.org/10.1088/1757-899X/95/1/012127

Fullerton RR, McWatters CS (2001) The production performance benefits from JIT implementation. J Oper Manage 19(1):81–96. https://doi.org/10.1016/S0272-6963(00)00051-6

Gapp R, Fisher R, Kobayashi K (2008) Implementing 5S within a Japanese context: an integrated management system. Manage Decis 46(4):565–579. https://doi.org/10.1108/00251740810865067

García JL, Maldonado AA, Alvarado A, Rivera DG (2014) Human critical success factors for Kaizen and its impacts in industrial performance. Int J Adv Manuf Technol 70(9–12):2187–2198. https://doi.org/10.1007/s00170-013-5445-4

Gilgeous V (2001) The strategic role of manufacturing. Int J Prod Res 39(6):1257–1287. https://doi.org/10.1080/00207540010022971

Glover W, Farris JA, Van Aken EM, Doolen TL (2011) Critical success factors for the sustainability of Kaizen event human resource outcomes: an empirical study. Int J Prod Econ 132(2):197–213

Goel S, Chen V (2008) Integrating the global enterprise using six sigma: business process reengineering at general electric wind energy. Int J Prod Econ 113(2):914–927. https://doi.org/10.1016/j.ijpe.2007.12.002

Gómez ID, Henao Arango R, Sarache W (2017) Multicriteria methodology for decoupling point placement in manufacturing systems. Ingeniería y Desarrollo 35(1):102–131. https://doi.org/10.14482/inde.35.1.8945

Hammer M, Champy J (1993) Reengineering the corporation: a manifesto for business revolution. Harper Business

Hayes RH, Schmenner RW (1978) How should you organize manufacturing? Harvard Bus Rev 56 (1):105–118

Hayes RH, Wheelwright SC (1984) Restoring our competitive edge: competing through manufacturing. Wiley, New York

Hill TJ (1983) Manufacturing's strategic role. J Oper Res Soc 34(9):853–860

Hill T (1985) Manufacturing's strategy—the strategic management of the manufacturing function. Macmillan, London

Hill T (1995) Manufacturing strategy: text and cases. Macmillan Press, London.

Hill T (2000) Manufacturing strategy: text and cases. McGraw-Hill, Boston

Ibarra S, Sarache WA (2008) Production management: its strategic role in productivity. In: Becerra F (ed) Production management: a conceptual approach. Universidad Nacional de Colombia, Manizales, pp 15–63 (in Spanish)

Jabbour CJC, da Silva ME, Paiva EL, Almada FC (2012) Environmental management in Brazil: is it a completely competitive priority? J Clean Prod 21(1):11–22. https://doi.org/10.1016/j.jclepro.2011.09.003

Jain B, Adil GK, Ananthakumar U (2013) An instrument to measure factors of strategic manufacturing effectiveness based on Hayes and Wheelwright's model. J Manuf Technol Manage 24(6):812–829. https://doi.org/10.1108/JMTM-11-2011-0102

Jia GZ, Bai M (2011) An approach for manufacturing strategy development based on fuzzy-QFD. Comput Ind Eng 60(3):445–454. https://doi.org/10.1016/j.cie.2010.07.003

Jiménez M, Romero L, Domínguez M, del Mar Espinosa M (2015) 5S methodology implementation in the laboratories of an industrial engineering university school. Saf Sci 78:163–172. https://doi.org/10.1016/j.ssci.2015.04.022

Kanamori S, Sow S, Castro MC, Matsuno R, Tsuru A, Jimba M (2015) Implementation of 5S management method for lean healthcare at a health center in Senegal: a qualitative study of staff perception. Glob Health Action 8:1–9. https://doi.org/10.3402/gha.v8.27256

Karim MA, Smith AJR, Halgamuge S (2008) Empirical relationships between some manufacturing practices and performance. Int J Prod Res 46(13):3583–3613. https://doi.org/10.1080/00207540601164201

Kaur M, Singh K, Singh I (2013) An evaluation of the synergic implementation of TQM and TPM paradigms on business performance. Int J Prod Perform Manage 62(1):66–84. https://doi.org/10.1108/17410401311285309

Leong GK, Snyder DL, Ward PT (1990) Research in the process and content of manufacturing strategy. Omega 18(2):109–122. https://doi.org/10.1016/0305-0483(90)90058-H

Lillis B, Lane R (2007) Auditing the strategic role of operations. Int J Manage Rev 9(3):191–210. https://doi.org/10.1111/j.1468-2370.2007.00209.x

Malhotra MK, Grover V (1998) An assessment of survey research in POM: from constructs to theory. J Oper Manage 16(4):407–425. https://doi.org/10.1016/S0272-6963(98)00021-7

Martín-Peña ML, Pinillos MJ, Reyes LE (2017) The intellectual basis of servitization: a bibliometric analysis. J Eng Technol Manage 43(1):83–97. https://doi.org/10.1016/j.jengtecman.2017.01.005

Maylor H, Turner N, Murray-Webster R (2015) It worked for manufacturing…!: operations strategy in project-based operations. Int J Proj Manage 33(1):103–115. https://doi.org/10.1016/j.ijproman.2014.03.009

Miltenburg J (2005) Manufacturing strategy: how to formulate and implement a winning plan, 2nd edn. Productivity Press, New York

Miltenburg J (2008) Setting manufacturing strategy for a factory-within-a-factory. Int J Prod Econ 113(1):307–323. https://doi.org/10.1016/j.ijpe.2007.09.001

Miltenburg J (2009) Setting manufacturing strategy for a company's international manufacturing network. Int J Prod Res 47(22):6179–6203. https://doi.org/10.1080/00207540802126629

Morita N, Flynn J, Milling P (2001) Linking practices to plant performance. In: Schroeder RG, Flynn BB (eds) High performance manufacturing. Global perspectives. Wiley, New York, pp 41–56

O'Neill P, Sohal AS (1999) Business process reengineering a review of recent literature. Technovation 19(9):571–581. https://doi.org/10.1016/S0166-4972(99)00059-0

Phan AC, Abdallah AB, Matsui Y (2011) Quality management practices and competitive performance: empirical evidence from Japanese manufacturing companies. Int J Prod Econ 133 (2):518–529. https://doi.org/10.1016/j.ijpe.2011.01.024

Pheng LS (2001) Towards TQM—integrating Japanese 5-S principles with ISO 9001:2000 requirements. TQM Mag 13(5):334–341. https://doi.org/10.1108/EUM0000000005859

Piechnicki AS, Sola AVH, Trojan F (2015) Decision-making towards achieving world-class total productive maintenance. Int J Oper Prod Manage 35(12):1594–1621. https://doi.org/10.1108/IJOPM-11-2013-0479

Platts KW, Gregory MJ (1990) Manufacturing audit in the process of strategy formulation. Int J Oper Prod Manage 10(9):5–26. https://doi.org/10.1108/EUM0000000001264

Radhakrishnan R, Balasubramanian S (2008) Business process reengineering: text and cases. PHI Learning, New Delhi

Roth A, Van Der Velde M (1991) Operations as marketing: a competitive service strategy. J Oper Manage 10(3):303–328. https://doi.org/10.1016/0272-6963(91)90071-5

Schmenner RW (1982) Multiplant manufacturing strategies among the fortune 500. J Oper Manage 2(2):77–86

Schroeder RG, Flynn BB (2001) High performance manufacturing. Global perspectives. Wiley, New York

Shokri A, Waring TS, Nabhani F (2016) Investigating the readiness of people in manufacturing SMEs to embark on lean six sigma projects: an empirical study in the German manufacturing sector. Int J Oper Prod Manage 36(8):850–878. https://doi.org/10.1108/IJOPM-11-2014-0530

Singh S, Garg D (2011) JIT system: concepts, benefits and motivation in Indian industries. Int J Manag Bus Stud 1(1):26–30

Skinner W (1966) Production under pressure. Harvard Bus Rev 44(6):139–146

Skinner W (1969) Manufacturing-missing link in corporate strategy. Harvard Bus Rev 47(3):136–145

Skinner W (1974) The focused factory. Harvard Bus Rev 52(3):113–121

Slack N, Lewis M (2003) Operations strategy, 1st edn. Prentice Hall, New Jersey

Slack N, Lewis M (2011) Operations strategy, 3rd edn. Pearson Prentice Hall, Edinburgh

Stamatis DH (2010) The OEE primer: understanding overall equipment effectiveness, reliability, and maintainability. CRC Press, New York. https://doi.org/10.1201/EBK1439814062-f

Styhre A (2001) Kaizen, ethics, and care of the operations: management after empowerment. J Manage Stud 38(6):795–810

Suwandej N (2015) Factors influencing total quality management. Procedia Soc Behav Sci 197:2215–2222. https://doi.org/10.1016/j.sbspro.2015.07.361

Swamidass PM, Newell WT (1987) Manufacturing strategy, environmental uncertainty and performance: a path analytic model. Manage Sci 33(4):509–524. https://doi.org/10.1287/mnsc.33.4.509

Swamidass PM, Baines T, Darlow N (2001) The role of manufacturing and marketing managers in strategy development: lessons from three companies. Int J Oper Prod Manage 21(7):933–948. https://doi.org/10.1108/01443570110393423

Swamidass PM, Darlow N, Baines T (2001b) Evolving forms of manufacturing strategy development: evidence and implications. Int J Oper Prod Manage 21(10):1289–1304. https://doi.org/10.1108/EUM0000000005971

Swink M, Narasimhan R, Kim SW (2005) Manufacturing practices and strategy integration: effects on cost efficiency, flexibility, and market-based performance. Decis Sci 36(3):427–457

Tsou CM (2013) On the strategy of supply chain collaboration based on dynamic inventory target level management: a theory of constraint perspective. Appl Math Model 37(7):5204–5214. https://doi.org/10.1016/j.apm.2012.10.031

Vargas JWP, Guerrero MBA, Crespo LLG (2016) Constrains theory applied to manufacturing and service companies. Ingeniare 12:79–86 (in Spanish)

Vivares JA, Sarache W, Naranjo JC (2017) The human management in the strategy of manufacture: an empirical study in the industry caldense. Editorial Universidad Nacional de Colombia, Bogotá (in Spanish)

Vivares JA, Sarache W, Hurtado JE (2018) A maturity assessment model for manufacturing systems. J Manuf Technol Manage 29(5):746–767

Vivares-Vergara JA, Sarache-Castro WA, Naranjo-Valencia JC (2014) The content of manufacturing strategy: a case study in Colombian industries. Dyna 81(183):140–147. https://doi.org/10.15446/dyna.v81n183.37672

Vivares-Vergara JA, Sarache WA, Naranjo-Valencia JC. (2015) Manufacturing strategy: exploring the content and process. Inf Tecnol 26(3):87–98 (in Spanish). https://doi.org/10.4067/S0718-07642015000300013

Vivares-Vergara JA, Sarache-Castro WA, Naranjo-Valencia JC (2016) Impact of human resource management on performance in competitive priorities. Int J Oper Prod Manage 32(2):114–134. https://doi.org/10.1108/IJOPM-11-2013-0484

Wheelwright SC (1978) Reflecting corporate strategy in manufacturing decisions. Bus Horiz 21 (1):57–66

Wheelwright SC (1984) Manufacturing strategy: defining the missing link. Strateg Manage J 5 (1):77–91. https://doi.org/10.1002/smj.4250050106

Womack J, Jone DT, Roos D (1992) The machine that changed the world. McGraw-Hill (in Spanish)

Yasin MM, Small MH, Wafa MA (2003) Organizational modifications to support JIT implementation in manufacturing and service operations. Omega 31(3):213–226. https://doi.org/10.1016/S0305-0483(03)00024-0

Chapter 16
Additive Manufacturing: Fused Deposition Modeling Advances

Julián Israel Aguilar-Duque, Juan Luis Hernández-Arellano, Liliana Avelar-Sosa, Guillermo Amaya-Parra and Ulises Jesús Tamayo-Pérez

Abstract The paradigm of the manufacturing systems was broken in 1980 with the beginning of the Additive Manufacturing (AM). This technology has been considered as the complement of the classic manufacturing technology, where the material is removed from a raw material until getting the final product. The addition of material in layers have been considered the new alternative to face the impact in the environment, the economy of materials and process, and the opportunity to generate new complex shapes limited by the classic manufacturing technology. The present chapter exposes the advances of the Fused Deposition Modeling (FDM), one of the seven technologies of AM which is mostly used during the past three decades. In this field, different adaptations and investigations of the technology have been focused on the increment of the capacity of the production system and improve the quality generated by this technique. The methodology used to determine the advance of AM was to employ a Systematic Literature Review using databases. The search was developed considering the keywords of AM for the construction of specific search syntax of documents associated with this technology. The documents obtained were analyzed to identify the progress in this technology. The results present the advancements of the FDM as a technology that change the industrial processing to customize the process, where the globalization makes possible to have this technology available at each desk.

J. I. Aguilar-Duque (✉) · G. Amaya-Parra · U. J. Tamayo-Pérez
Faculty of Engineering, Architecture and Design, Autonomous University of Baja California, Carretera Transpeninsular Ensenada-Tijuana No. 3917, Colonia Playitas, Ensenada, Baja California, Mexico
e-mail: julian.aguilar@uabc.edu.mx

J. L. Hernández-Arellano
Department of Design, Autonomous University of Ciudad Juarez, Av. Del Charro 450 Norte, Col. Partido Romero, Ciudad Juárez, Chihuahua, Mexico

L. Avelar-Sosa
Department of Industrial Engineering and Manufacturing, Autonomous University of Ciudad Juarez, Av. Del Charro 450 Norte, Col. Partido Romero, Ciudad Juárez, Chihuahua, Mexico

J. L. García Alcaraz et al. (eds.), *Best Practices in Manufacturing Processes*,
https://doi.org/10.1007/978-3-319-99190-0_16

Keywords Additive manufacturing fused deposition modeling · Advantages

16.1 Introduction

Nowadays, the manufacturing systems have developed new strategies to satisfy the needs of markets and organizations that are following a correct and perfect way to face the globalization and to survive. Manufacturing could be divided in two ways: (1) from a technological viewpoint, and (2) from an economic viewpoint. Technologically, manufacturing consists of the application of the physical and chemical process to alter/modify the geometry, the properties, and the appearance of a given starting material to make new parts or products. Manufacturing also includes an assembly process in which the products are made by joining multiple parts into a single entity (Zandin 2001). The process to accomplish manufacturing involves a combination of machinery, tools, power, and manual labor.

As it has been mentioned, the evolution of manufacturing has been a critical factor in wellness and the economic development of many countries. This cadence represents the capacity to count with enough goods and services to satisfy the demand of the population (Degarmo et al. 2003). Altinkemer et al. (2011) declare that, due to important contributions and constant growing, the manufacturing has been classified as a dynamic activity in constant development. This involves the invariable evolution in materials, process, and technologies focused on improving the manufacturing sector (Shunta 1997; Vollman et al. 1997; Schey 2002; Groover 2007; Tauseef 2010; Kalpakjian and Schmid 2014).

According to Srivastava (2010), manufacturing has been classified as subtractive or additive manufacturing due to the development of the process of materials. Before describing the AM technologies, it is necessary to present a resume of the most important developments related to the classic manufacturing. Table 16.1 presents the information.

With its emergence in the 80s, AM, also known as a 3D or rapid prototyping, has been characterized by the process of building parts layer by layer in a tridimensional space where the model comes from a digital design. This type of manufacture has been applied in many industries due to its operation advantages (Scott et al. 2012).

The 3D printing and AM technologies have created high expectations which is a viable option for the future process of manufacture. Formally, AM has been defined as a "process of join materials to build objects coming from 3D data models, usually layer by layer" (ASTM International 2013). AM follows a process that depends basically on four forms of materials: liquid, sheet, filament, and powder. Figure 16.1 shows a representation of the profiles works in additive manufacture.

The AM technologies are classified into seven groups: Stereolithography (SLA), Fused Deposition Modeling (FDM), Laminated Object Manufacturing (LOM), 3D printing (3DP), Selective Laser Sintering (SLS), Laser Engineered Net Shaping (LENSTM), and Electron Beam Melting (EBM) (Bourell et al. 1990; Campbell and Dickens 1994; Beaman et al. 2014; Srivatsan and Sudarshan 2016). The description of these technologies is shown below.

Table 16.1 Historical development of the manufacturing process

Year or period of appearance	Materials	Forming and shaping	Tools and manufacturing systems
4000	Gold, copper	Hammering	Tools of stone, wood, bone
3000	Cooper casting, stone, and metal molds, lost-wax, silver bronze	Stamping, jewelry	Corundum
2000	Bronze casting, gold leaf	Wire by slitting sheet metal	Hoe making, hammered axes, tools for iron making and carpentry
1	Cast iron, cast steel	Stamping of coins	Chisels, saws, woodworking lathes
1 AC	Zinc, steel	Armor, coining, forming, steel swords.	Etching of armor
1000	Type metals, casting of bells, pewter	Wire drawing, gold, and silversmith work	Sandpaper, windmill driven saw
1500	Cast-iron cannon, tinplate	Waterpower for metalworking, rolling mill	Hand lathe for wood
1600	Brass from copper and metallic zinc	Rolling, shape rolling	Boring, turning, drill press
1700	Malleable cast iron, crucible steel	Extrusion, deep drawing, rolling	
1800	Nickel steels, galvanized steel, polyester, styrene, celluloid, rubber extrusion	Steam hammer, steel rolling, seamless tube, steel-rail rolling, continuous rolling, electroplating	Shaping, milling, copying late for gunstocks, turret lathe, universal milling machine, vitrified grinding wheel
1900	Bakelite, borosilicate glass	Tube rolling, hot extrusion	Geared lathe, automatic screw machine, hobbling, high-speed steel tools, aluminum oxide and silicon carbide
1920	Development of plastics, polyvinyl chloride, cellulose acetate, polyethylene glass fibers	Tungsten wire from metal powder	Tungsten carbide, mass production, transfer machine
1940	Acrylics, synthetic rubber, epoxies, photosensitive glass	Extrusion, swaging, powder metals for engineering parts	Phosphate conversion coatings, total quality control
1950	Semiconductors, Acrylonitrile–butadiene–styrene, silicones, fluorocarbons, polyurethane, float glass, glass ceramics	Cold extrusion, explosive forming, thermomechanical processing	Electrical and chemical machining, automatic control

(continued)

Table 16.1 (continued)

Year or period of appearance	Materials	Forming and shaping	Tools and manufacturing systems
1960–1970	Acetals, polycarbonate, cold forming of plastics, reinforced plastics, filament winding	Hydroforming, hydrostatic extrusion, electroforming	Titanium carbide, synthetic diamond, numerical control, the integrated circuit chip
1970–1990	Optical fibers, structural ceramics, ceramic-matrix composites, biodegradable plastics, electrically conducting polymers	Precision forming, isothermal forming, superplastic forming	Coated tools, diamond turning, ultraprecision machining, computer integrated manufacturing, industrial robots, flexible manufacturing systems, machining and turning centers, artificial intelligence, computer simulation and optimization
1990–2010	Nanophase materials, metal foams, diamond like carbon, carbon nanotubes	Rapid prototyping, rapid tooling, environmentally friendly metalworking fluids, digital manufacturing	Micro and nanofabrication, Lithography-electroplating and molding, dry etching, linear motor drives, artificial neural networks, six sigma, three-dimensional computer chips, blue arc machining, soft lithography
2010–2017	Nanomaterials, nanoclay, glass/epoxy composites, composites with nanomagnetic	Matrix friction hot pressing, rapid manufacturing by laser forming	Carbon footprint production systems, Assembly of rigid components, micro milling and micro-drilling by laser

16.1.1 Stereolithography (SLA)

This technique is the first and the most applied process of rapid prototyping. Basically, it is a liquid-based process that works by solidifying a photosensitive polymer. The process of solidification begins with the construction of a model using Computer Assisted Design (CAD). Then, the model is translated to a Standard Triangle Language (STL) file, where the model is transformed into cut pieces or most commonly named as “slices”. Each slice contains the information required for each layer. An ultraviolet laser is applied to the resin indicating solidification in specific locations of each layer. Once the layer is ready, the platform is lowered. Then, the process continues under the algorithm until the piece or component is finished (Hull 1986).

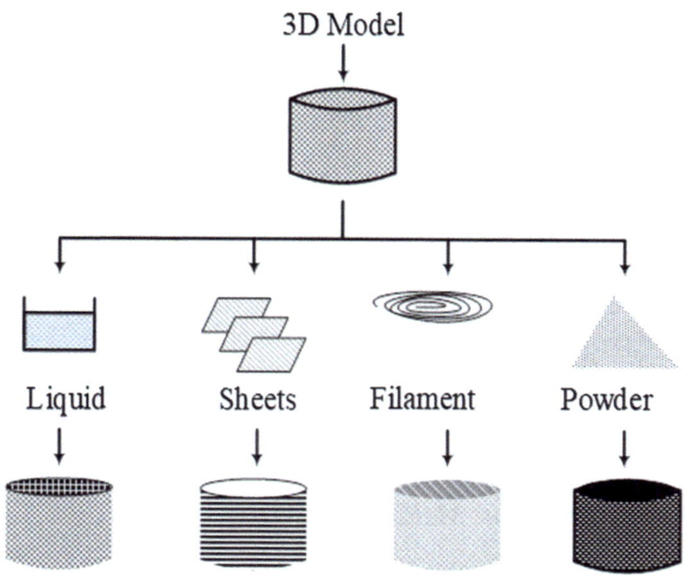

Fig. 16.1 Alternatives of the process to develop a product layer by layer

16.1.2 Laminated Object Manufacturing (LOM)

LOM uses adhesive coated sheet material. The adhesive is precoated onto materials or deposited on the surface immediately. The 3D components are manufactured by a sequence of lamination and coated in sections crossed in two dimensions by laser. The deep of the cut will correspond to the high layer (Feygin and Sik Pak 1999).

16.1.3 Three-Dimensional Printing (3DP)

In this process, a water-based liquid binder is applied in a jet onto a starch-based powder to print the design converted in data from CAD. The powder particles lie on a powder bed and become glued to each other when the binder is applied. Following the sequential application of layers, the unbound powders are carefully removed (Haggerty et al. 1993).

16.1.4 Selective Laser Sintering (SLS)

This process mainly works with high power laser to fuse small particles of the build material. The fabrication powder bed is heated just below the melting point of the material with the primary objective of minimizing thermal distortion and to facilitate fusion to the previous layer. Then, each layer is drawn into the powder bed using a laser to sinter the material (Deckard 1989).

16.1.5 Laser Engineered Net Shaping (LENS)

With this technique, a component is fabricated by focusing a high-powered laser beam onto a substrate. The primary objective is to create a molten pool into which metal powder particles are injected to build each layer gradually. The substrate is moved gently below the laser beam to deposit a thin cross section and thereby create the desired part (Jeantette et al. 2000).

16.1.6 Electron Beam Melting (EBM)

This process is quite similar to SLS; an electron laser beam is used to melt the powder. High voltage powers the laser beam. With this technology, the high-power electron generates the energy needed for high melting capacity and high productivity. The electron beam is managed by electromagnetic coils providing extremely fast and accurate beam control that allows several melt pools to be maintained simultaneously (Yamamoto and Sakai 2005).

16.1.7 Fused Deposition Modeling (FDM)

This technique works extruding material, normally liquid thermoplastic, from an extruder and it is deposited on a hot bed. The material is heated until it reaches the liquid state, in other words, in average one grade above the melting point. This characteristic allows solidifying the material immediately after the extrusion building a layer (Crump and Stratasys 1992).

According to the time to award the patents, it is possible to identify the development of the AM during the past three decades. Figure 16.2 presents a timeline for this technology. It is necessary to specify that due to the variety of process, each category has its characteristics. After this global scenery of AM, it is possible to determinate the importance of FDM in AM.

Up to now, FDM is the most economical technology of AM. As a consequence, this technology has been developed and changed its characteristics quickly. To explain better this technology, it is necessary to describe with more precision what is FDM.

Using FDM, it is possible to fabricate three-dimensional objects from virtual CAD models, sometimes with complex geometrical shapes. When we write

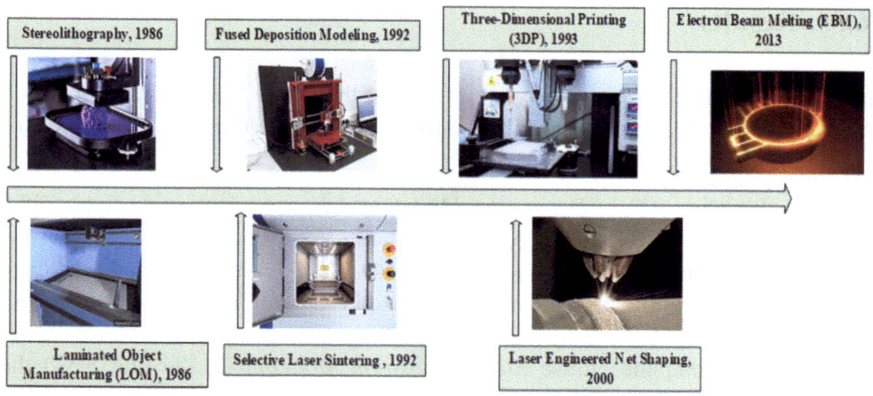

Fig. 16.2 Timeline of additive manufacturing

sometimes, we think about different factors that make possible to create these types of shapes. One of its advantages is the fast way to reduce the product development cycle period (Kovan et al. 2017). Layer-upon-layer, 3D structures are built approximately by hundreds of planar layers without geometric restrictions (Hartke 2011; Wohlers Associates 2011). Another advantage is the low cost, this technology, as it has been mentioned, is cheaper than the other six AM technologies. For example, in the market is possible to find equipment from $100.00 US as Anet-A8, until the most complex equipment with a value of $7000.00–10,000.00 US as the Makerbot Z18. The prices make possible that many people who are interested in creating prototypes or their own parts could manufacture them in a customized way.

Convenient and high material usage efficiency are other characteristics of FDM. Nowadays, the development of materials has given high value to this technology. It is possible to find synthetic polymers as ABS, PLA, PET, NYLON, and so forth; and natural polymer as a special kind of WOOD (Jones et al. 2015).

16.2 Methodology

The methodology used to find out the recent advances in FDM published on database and internet consisted of a Systematic Review designed in two phases:

- *Information resources*: This activity was focused to identify the information associated with FDM advances using different syntaxes. The review of databases considered online was EBSCO, ELSEVIER, EMERALD, IEEE, SCOPUS, and SPRINGER.
- *Classification of information*: This activity consists of generating a timeline using the date of the publications, making convergence of the terms used by the authors and the code generated by the different operators.

16.3 Results

As a first step, the determinations of the syntaxes that allow finding the information associated with the Fused Deposition Advances were determined. The syntaxes obtained are shown below.

- Search for phrase *Fuse Deposition Modeling Advances* in the title field. Title: "Fused Deposition Modeling Advances."
- Search for phrase *Fuse Deposition Modeling Advances* and *Additive Manufacturing* in the body field. Title: "Fused Deposition Modeling Advances" and "Additive Manufacturing."
- Search for either the phrase *Fuse Deposition Modeling* in the title AND the phrase *advances* in the body field or the word *advances* in the title field. Title: "Fused Deposition Modeling" and body "advances."

The second step consisted of using the syntaxes generated in the first step. The syntaxes were typed in each database with the purpose to get the documents associated with the topic of interest. This proceeding considered the number of results obtained and the date of publication. These proceedings considered the number of results obtained and the date of publication. The aim was to determine the number of articles and documents generated by year. The period of search was under the restriction of the history of AM. The first patent was in 1980. In this context, it was possible to define periods of 10 years, until the last that is of 8 years. Table 16.2 presents the results identified with the syntaxes "Fused Deposition Modeling Advances".

Table 16.3 presents the results of the search using the syntaxes "Fused Deposition Modeling Advances" and "Additive Manufacturing".

Table 16.2 Resume of documents found with the syntaxes "fused deposition modeling advances"

Syntaxes	Resource	Period	Documents found
Title: "fused deposition modeling advances"	EBSCO	2010–2018	163
		2000–2009	30
		1990–1999	0
	EMERALD	2010–2018	0
		2000–2009	0
		1990–1999	0
	IEEE	2010–2018	0
		2000–2009	0
		1990–1999	0
	SCOPUS	2010–2018	0
		2000–2009	0
		1990–1999	0
	SPRINGER	2010–2018	0
		2000–2009	0
		1990–1999	0

Table 16.3 Resume of documents found with the syntaxes "fused deposition modeling advances" and "additive manufacturing"

Syntaxes	Resource	Period	Documents found
Title: "fused deposition modeling advances" and "additive manufacturing"	EBSCO	2010–2018	0
		2000–2009	0
		1990–1999	0
	EMERALD	2010–2018	0
		2000–2009	0
		1990–1999	0
	IEEE	2010–2018	231
		2000–2009	2
		1990–1999	0
	SCOPUS	2010–2018	0
		2000–2009	0
		1990–1999	0
	SPRINGER	2010–2018	0
		2000–2009	0
		1990–1999	0

Table 16.4 Resume of documents found with the syntaxes title: "fused deposition modeling advances" and body "advances"

Syntaxes	Resource	Period	Documents found
Title: "fused deposition modeling" and body "advances"	EBSCO	2010–2018	15
		2000–2009	0
		1990–1999	0
	EMERALD	2010–2018	0
		2000–2009	0
		1990–1999	0
	IEEE	2010–2018	0
		2000–2009	0
		1990–1999	0
	SCOPUS	2010–2018	91
		2000–2009	6
		1990–1999	1
	SPRINGER	2010–2018	20
		2000–2009	2
		1990–1999	0

Table 16.4 shows the results obtained from the search using the syntaxes Titled: "Fused Deposition Modeling" and body "advances."

With the research of the syntaxes mentioned above, we identified 561 papers that cover the inclusion and exclusion criteria. This restriction considers the works that include the technology of Fused Deposition Modeling and advances in all areas of engineering, medicine, mechatronics, and so forth.

According to the documents found, a summary of the progress identified was developed. The resume is integrated with the most common topics that were covered by the researchers. The next list describes the topics covered.

16.3.1 Design Methods and Standards

Focused on the advances identified by the concept "Fused Deposition Modeling", the design and manufacturing activities have been increasing their capabilities on the creation of new shapes with unbelievable functions. The evolution of components, assembly, and sub-assembly has demonstrated the utility of the AM in this field. With the AM, designers began to create and explore new forms and functions of their creations, allowing them the facility to reduce the restrictions that classical manufacturing has. The systematic prescription of the shape and the liberty of the designer to create new components has been restricted by characteristics of an artifact to achieve specified objects (Esposito Corcione et al. 2018).

Since the design has been the beginning of the manufacturing process and quality, the quality of some special components have improved significantly from a process point of view, also, the design has been considered a critical activity because it represents as much as 80% of the cost of the product and the success of the product during the production process (Ford and Despeisse 2016). Until this critical percentage, the design in FDM has been changed considering the final function of the component printed.

At the beginning of the FDM, the design of the component only considered the shape, the format of the printer, and the material. Currently, the design is part of the preprocess (Gautam et al. 2018).

This advantage and the development in the FDM represent the integration of the parameters of the design in the specification that manufacture uses.

The phase of design and the advantages associated with FDM have also been associated with the simulation process (Dong et al. 2018). Now, designers could validate their designs using advanced software for design and prototype before sending it to the production process. From this point of view, it is important to declare, that in some cases, FDM is the technology most commonly used due to their low-cost equipment and their economy of operation.

Finally, with the development of FDM, new opportunities have been opened creating new alternatives of design, the economy of process, development of materials, environment protection, and social responsibilities (Raja et al. 2006). These changes allow the technology to create products focused on maximizing the needs and satisfaction of the users and designers.

16.3.2 Process Innovation

Innovation in additive manufacturing processes includes efforts for increasing parts qualities, high productivity rate, high-security requirements, reduction of manufacturing cost, reduction of lead time and, among others. Focused on the satisfaction of the customer needs, usually, the manufacturing processes conditions are established for each application. At this point, it is important to specify the existence of automatized equipment that makes 90% of the adjustment automatically, the other 10% is made by the user. In low-cost equipment, the customer has to make all the adjustments necessary to manufacture the component with acceptable quality.

A general recommendation found in publications declares that the key success of the FDM depends upon the proper selection of process parameters (Jones et al. 2015; Morozov et al. 2016; Vairis et al. 2016; Kovan et al. 2017). The determination of the optimum process has been an important task for users (Mellor et al. 2014; Gardan 2015). As a result of the control of equipment, materials, and environment process, it is possible to assure the quality of the printing parts, improving dimensional precision, avoiding unacceptable wastes, enhancing production rates and reducing production time and cost (Rayna et al. 2015).

Until 2017, the process exhibits much difficulty in determining optimal parameters due to the presence of a large number of conflicting parameters that influence the part quality and material properties (Vairis et al. 2016; Hart and Wetzel 2017; Kovan et al. 2017). At this point, it is possible to declare that there exists a complementary action between materials, designs, components, and the control of the process parameters.

16.3.3 Surface Roughness

This topic has been studied since the technology appeared. This characteristic has been a problem that is faced by the FDM versus classical manufacturing techniques. Until nowadays, the advancement identified has been to modify the dimension of the nozzle hole. However, the time of production has been increased. On the other hand, it is possible to identify that the information developed by researchers has been focused on the orientation of the base considering thickness, road width and speed, raster angle and air gap (Anitha et al. 2001; Nancharaiah et al. 2010; Boschetto et al. 2016). To describe the surface roughness some papers presents a general image of the slices. Figures 16.3 and 16.4 describe the effect of slicing over irregular figures. It is clear that the roughness has been associated to this effect.

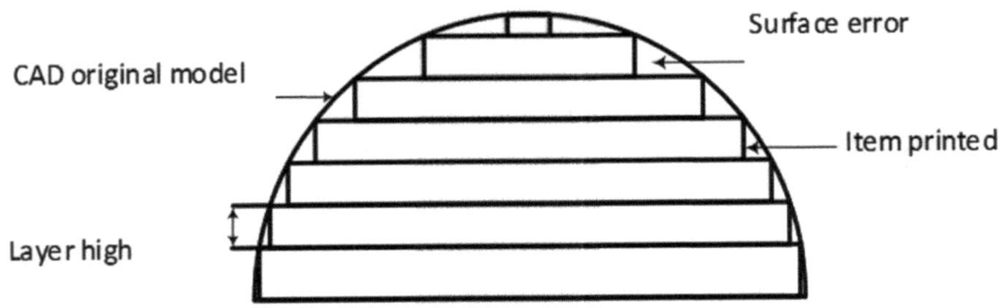

Fig. 16.3 Surface defects over spherical part

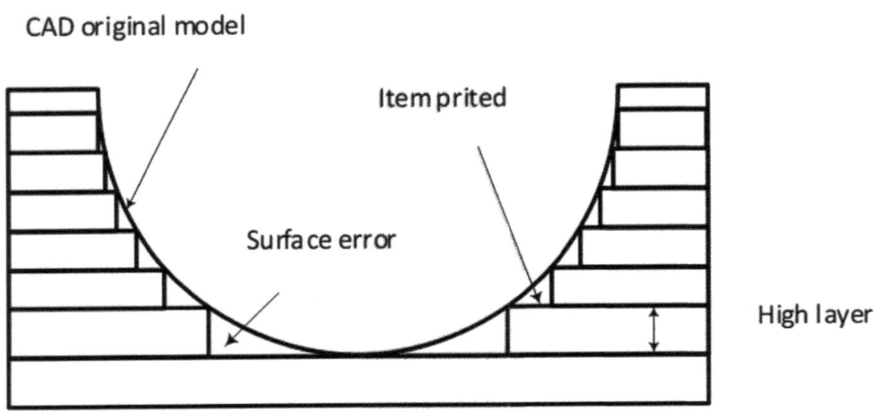

Fig. 16.4 Surface defects in a cavity

16.3.4 Dimensional Accuracy

Because of the deformation of the materials during the process of additive material, the dimensional accuracy depends on build orientation and the position thickness. The advantage is focused on the different deposition patterns. In this case, the advances generated include the influences and control of five processes parameters that include dimensional accuracy, raster angle, air gap, layer thickness, and orientation, these parameters are presented in Fig. 16.5. As a result, in the topic related with the advantage, the literature recommend that the thickness of the fabricated part should be consider a layer thickness of 0.178 mm, a par orientation of 0°, a raster angle of 0°, a road with of 0.4564 mm, and an air gap of 0.008 mm (Raja et al. 2006; Sood et al. 2009; Nancharaiah et al. 2010; Kovan et al. 2017).

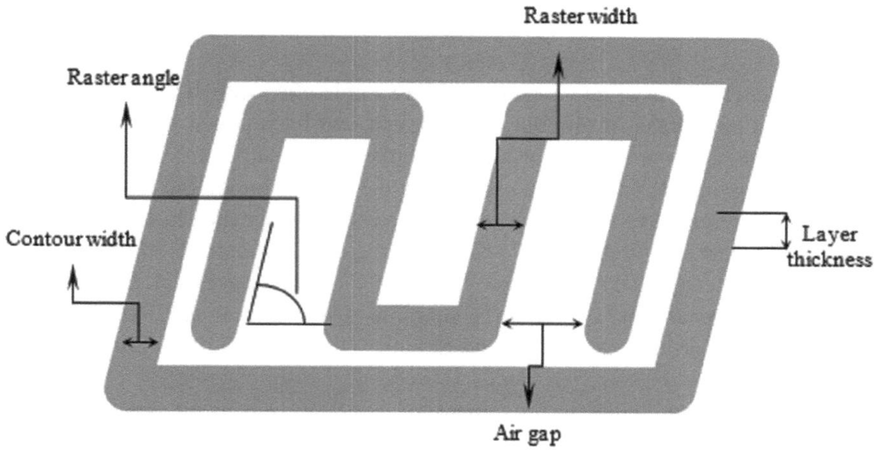

Fig. 16.5 Process parameters for dimensional accuracy

16.3.5 Material Behavior

This part of the process has been investigated considering the inclusion of new materials and additives. In the polymer process, it is natural that the material will develop a behavior in the function of the environment of printing. When the process began, it was usual to find equipment exposed to the environmental factors, which affect the retraction or expansion of the material. Today, under this restriction, most of the high-cost equipment have different complements that allow control or reduce the impact of the environment. In a most specific way, the printing will depend on the materials used during the processes without dismissing the factors associated to the equipment, that are the layer thickness, the raster angle, and the air gap (Peng et al. 2014; Achillas et al. 2015; Boschetto and Bottini 2015).

16.3.6 Build Time

The estimation of the time required to print the component is important. At the beginning of this technology, the time was a forecast estimated in function of the layer thickness. Considering this restriction, the development of new software allows the user to control different factors, principally the idea of the function of the prototype. This means that many users just need to build their ideas without resistance or functionality. Recent results present that the layer thickness and air gap contributed to the 67 and 30% of the build time, respectively. This recommendation is an important advantage because it allows the user to optimize the process time (Beaman et al. 2014).

16.3.7 Mechanical Properties

With the development of new materials, for example, the synthetic polymers, FDM has covered new areas of product development and has opened up new opportunities to create diverse areas of products. Synthetic polymers have been used since this technology appeared in the market, being the most common the ABS and PLA. The new materials have the properties of increase the resistance to tension (Vairis et al. 2016; Hart and Wetzel 2017; Kovan et al. 2017), compression (Vairis et al. 2016; Dickson et al. 2017; Hart and Wetzel 2017; Hinderdael et al. 2017; Kovan et al. 2017; Thomas 2017), torsion (Balderrama-Armendariz et al. 2018; Jiao et al. 2018; Salazar-Martín et al. 2018), and flexion (Dinon et al. 2018; Soriano-Heras et al. 2018).

Finally, the continuous improvement of materials has been converted as a challenge of the AM. The time required by this technology is compensated by the mechanical characteristics that will be modified in a short time.

16.3.8 Economic Implications of Fused Deposition Modeling

According to the literature review, FDM has changed the economic scenario of the manufacturing. Although several companies have used FDM in prototyping for more than 25 years, it was only recently that the techniques gained the attention of the broader public to the point of enthusiastic reports in the mass media. Facing the new manufacturing era, the market for AM, including all products and services worldwide, grew to $3.07 billion with a compound annual growth rate of 34.9%; experts estimate the size of the AM market in 2021 at $10.8 billion (Thiesse et al. 2015). Considering this important forecast of growth, FDM is considering the most useful technology of AM, where the percentage of participation in this growth is projected by at least 50% of the use in FDM.

The current state of technological FDM and skills, define a technology frontier which separates possible production scenarios from fictional devices. For Thiesse et al. (2015), the rise of FDM as a part of AM extends this technology along the flexibility axis and opens opportunities for manufacturing companies in three regards.

- FDM offers the option of generating objects that would have been impossible to make with any other technology. This high level of flexibility refers not only to the actual production outputs but also tools, which can be prepared more efficiently.
- About job shop manufacturing, FDM can be used as an automation technology which substitutes human labor. Though it may seem counterintuitive to the flexibility of 3D printer thus allows for efficiency gains.

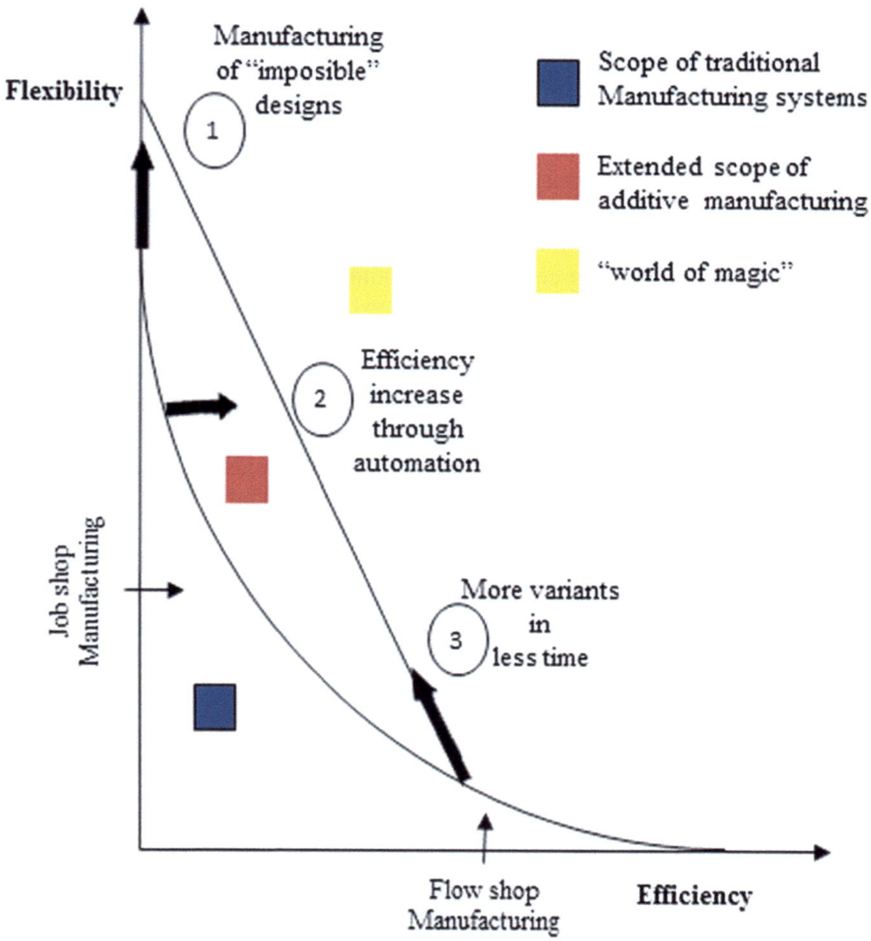

Fig. 16.6 Impacts of 3D printing on manufacturing systems

- Finally, FDM allows for cost-efficient switching from traditional mass production to new areas of mass customization. Here, companies use FDM to offer their customers a broader product range, individualized products, or shorter product life cycles over time.

In Fig. 16.6, it is possible to identify the relationship between the factor flexibility and efficiency. It is essential to mention that the factor flexibility describes design variability, fixed costs, etc., and, the efficiency explains variable costs, lead time, etc.

As it could be seen, manufacturing of "impossible" designs explain a high level of flexibility, at this level designers broke the limit of their imagination. Their designs are developed entirely on function and image. In contrast with flexibility, the variable that opposes the creativity of designers is the efficiency. At this level is where the 3D printing loses the battle of manufacturing systems only for the lead time.

Considering AM as an industrial manufacturing technique it is possible to define its technological potential. The layer-based manufacturing process makes possible production of individual products with high expectations of design, as it has been described in Fig. 16.6. The most important restrictions are associated with the producibility. It is reasonable that AM is limited by the number of pieces generated by each technology. As a consequence, the technological potential is not associated with the number of pieces generated and it is the quality of the pieces that could be considered unique.

One of the characteristics that make profitable AM is the ecological potential. The technology allows increasing the resource efficiency. The aim has applied the material only in those areas where is required. These specific activities allow saving over 30% of the material and weight. Also, reduce the logistics process via digitalization. In other words, the physical flow of materials can be reduced significantly.

Finally, it is possible to resume the economic implications of this technology in just three words, economic, ecological, and customized. The optimization of material favors the economic impact, as a consequence, the ecological impact is measured by the optimal use of the resources and the customized is characterized by the creation of new products focused on the satisfaction of the users.

16.3.9 Workforce in FDM

Technology is rapidly transforming the manufacturing industry but, what will this mean for the future work? The answer is clear, technology demands the development of skills and workforce where the companies and governments have to prepare for a new kind of human capital. According to Soltesz et al. (2016), each job in the advanced manufacturing industry supports another 3.5 jobs in the supply chain.

In Soltesz et al. (2016) research, they found that AM industries account for 13% of all jobs in the U.S., nearly 24 million people have been employed in AM industries, and they are compensated more highly than other workers. In their report, they describe that, on average, a worker makes nearly $95,000, compared to an average of about $73,000 overall manufacturing industries. Workers in traditional manufacturing make only about $57,000.

However, what about FDM, it is clear that their income is equal and is in the range of the workers that develop their skills in AM. Since 2008, Wagner et al. (2008) identifies the necessity of workers open to new technology, people with advanced knowledge and skills in computer programs and design. Nowadays, these skills could be found in 20% of the American houses, where young people begin to identify themselves with entrepreneurial projects. In other words, the economy is facing the new era of business, headed by customized productions.

It is essential to describe that due to the growth of FDM, a range of jobs including engineering, design, software development, material science, and additive manufacturing technicianshas developed (Petch 2018). These are the future for professionals and technicians in FDM.

16.4 Conclusions

With the development and change of the paradigm and manufacturing from subtractive manufacturing to additive manufacturing, it could be possible to declare that the Third Industrial Revolution began with AM technology since 1980. Facing this change is possible to affirm that this technology has increased its development quickly compared with the subtractive manufacturing. Nowadays, it is possible to resume the advantages in three ways.

First, the liberty of design. This point refers to the advantage gained by designers in a way to create components with structures and shapes more complicated than the used ones actually. Restricted by the capacity of the process, the designers had considered this capacity to create their new designs, always limited by internal and external shapes of the component. Using FDM, it is possible to create the prototype and, in some cases, a micro-production of particular components or assembles.

Second, the process has been changed since the apparition of this technology in 1992. Now, it is possible to buy specialized equipment of high cost including all the electronics components that are necessary to create more complex components with high quality in less time than 10 years ago. In another way, it is possible to buy low-cost equipment that works perfectly with the same or better quality than high-cost equipment. If the user knows how to set up correctly the printer, the low-cost equipment will produce high-quality elements. The process will change fast in the next 10 years due to the implementation and use of advanced electronics and technology. These new challenges include equipment, materials, process, social compromise, economy, environment, and satisfaction of user's needs.

Third, materials are the field where FDM have found a remarkable discovery. With the nanoprocess and bioprocess focused on improving the quality of life, the integration of this technology has supported the creation of prosthesis, nanodevices, etc. Materials allow FDM to create a component that could be comparable with components created by a process that use more complex tools, equipment, and personnel.

Finally, future studies would allow FDM not only optimize this technology, materials, and techniques but also to develop effective methods for inspecting their processes and products. With these factors under control, it will be possible to have at least one personnel FDM equipment as a personal computer at home.

References

Achillas C, Aidonis D, Lakovou E et al (2015) A methodological framework for the inclusion of modern additive manufacturing into the production portfolio of a focused factory. J Manuf Syst 1:328–339. https://doi.org/10.1016/j.jmsy.2014.07.014

Altinkemer K, Ozcelik Y, Ozdemir Z (2011) Productivity and performance effects of business process reengineering: a firm-level analysis. J Manage Inf Syst 27:129–162. https://doi.org/10.2753/MIS0742-1222270405

Anitha R, Arunachalam S, Radhakrishnan P (2001) Critical parameters influencing the quality of prototypes in fused deposition modelling. J Mater Process Technol 118:385–388. https://doi.org/10.1016/S0924-0136(01)00980-3

ASTM International (2013) F2792-12a—standard terminology for additive manufacturing technologies. Rapid Manuf Assoc 10–12. https://doi.org/10.1520/f2792-12a.2

Balderrama-Armendariz CO, MacDonald E, Espalin D et al (2018) Torsion analysis of the anisotropic behavior of FDM technology. Int J Adv Manuf Technol 1–11. https://doi.org/10.1007/s00170-018-1602-0

Beaman J, Bourell D, Wallace D (2014) Additive manufacturing and 3d printing. J Manuf Sci Eng

Boschetto A, Bottini L (2015) Roughness prediction in coupled operations of fused deposition modeling and barrel finishing. J Mater Process Technol 219:181–192. https://doi.org/10.1016/j.jmatprotec.2014.12.021

Boschetto A, Bottini L, Veniali F (2016) Finishing of fused deposition modeling parts by CNC machining. Robot Comput Integr Manuf 41:92–101. https://doi.org/10.1016/j.rcim.2016.03.004

Bourell D, Beaman J, Marcus J, Barlow W (1990) Solid freeform fabrication: an advanced manufacturing approach. In: Beaman J, Marcus H, Bourell J, Barlow W (eds) Solid freeform fabrication symposium proceedings, Austin, TX, pp 1–7

Campbell R, Dickens P (1994) Rapid prototyping: a global view. In: Marcus H, Beaman J, Barlow D et al (eds) 5th solid freeform fabrication symposium proceedings, Austin, TX, pp 110–117

Crump S, Stratasys I (1992) Apparatus and method for creating three-dimensional objects

Deckard CR (1989) US patent no 4,863,538. US Patent and Trademark Office, Washington, DC

Degarmo E, Black J, Kohser R (2003) Materials and processes in manufacturing, 9th edn. Wiley International

Dickson AN, Barry JN, McDonnell KA, Dowling DP (2017) Fabrication of continuous carbon, glass and kevlar fibre reinforced polymer composites using additive manufacturing. Addit Manuf 16:146–152. https://doi.org/10.1016/j.addma.2017.06.004

Dinon T, Caimmi M, Chiavenna A, Malosio M, Prini A, Scano A, Molinari Tosatti L, Curró C, Lenzi B, Megale V (2018) DUALarm: an open-source and 3D-printable device for upper limb neurorehabilitation. J Rehabil Assitive Technol Eng 5:1–13. https://doi.org/10.1177/2055668317749989

Dong G, Wijaya G, Tang Y, Zhao YF (2018) Optimizing process parameters of fused deposition modeling by Taguchi method for the fabrication of lattice structures. Addit Manuf 19:62–72. https://doi.org/10.1016/j.addma.2017.11.004

Esposito Corcione C, Palumbo E, Masciullo A et al (2018) Fused deposition modeling (FDM): an innovative technique aimed at reusing Lecce stone waste for industrial design and building applications. Constr Build Mater 158:276–284. https://doi.org/10.1016/j.conbuildmat.2017.10.011

Feygin M, Sik Pak S (1999) US patent no 5,876,550. US Patent and Trademark Office, Washington, DC

Ford S, Despeisse M (2016) Additive manufacturing and sustainability: an exploratory study of the advantages and challenges. J Clean Prod 137:1573–1587. https://doi.org/10.1016/j.jclepro.2016.04.150

Gardan J (2015) Additive manufacturing technologies: state of the art and trends. Int J Prod Res 54:3118–3132. https://doi.org/10.1080/00207543.2015.1115909

Gautam R, Idapalapati S, Feih S (2018) Printing and characterisation of Kagome lattice structures by fused deposition modelling. Mater Des 137:266–275. https://doi.org/10.1016/j.matdes.2017.10.022

Groover M (2007) Fundamentals of modern manufacturing: materials, processes and systems, 1st edn. Pearson Prentice-Hall, Mexico DF (in Spanish)

Haggerty S, Michael J, Williams PA (1993) Three-dimensional printing techniques

Hart KR, Wetzel ED (2017) Fracture behavior of additively manufactured acrylonitrile butadiene styrene (ABS) materials. Eng Fract Mech 177:1–13. https://doi.org/10.1016/j.engfracmech.2017.03.028

Hartke K (2011) Manufacturing technology support. OH

Hinderdael M, Jardon Z, Lison M et al (2017) Proof of concept of integrated load measurement in 3D printed structures. Sensors 17:328. https://doi.org/10.3390/s17020328

Hull CW (1986) Apparatus for production of three-dimensional objects by stereolithography, pp 1–16

Jeantette FP, Keicher DM, Romero JA, Schanwald LP (2000) Method and system for producing complex-shape objects. US006046426A

Jiao Z, Li F, Xie L et al (2018) Experimental research of drop-on-demand droplet jetting 3D printing with molten polymer. J Appl Polym Sci 135:1–12. https://doi.org/10.1002/app.45933

Jones JB, Wimpenny DI, Gibbons GJ (2015) Additive manufacturing under pressure. Rapid Prototyp J 21:89–97. https://doi.org/10.1108/RPJ-02-2013-0016

Kalpakjian S, Schmid S (2014) Manufacturing engineering and technology, 7th edn. Pearson, New Jersey

Kovan V, Altan G, Topal ES (2017) Effect of layer thickness and print orientation on strength of 3D printed and adhesively bonded single lap joints. J Mech Sci Technol 31:2197–2201. https://doi.org/10.1007/s12206-017-0415-7

Mellor S, Hao L, Zhang D (2014) Additive manufacturing: a framework for implementation. Int J Prod Econ 149:194–201. https://doi.org/10.1016/j.ijpe.2013.07.008

Morozov EV, Novikov MM, Bouznik VM (2016) MRI monitoring and non-destructive quality measurement of polymeric patterns manufactured via stereolithography. INFONA 12:16–24

Nancharaiah T, Raju DR, Raju VR (2010) An experimental investigation on surface quality and dimensional accuracy of FDM components. Int J Emerg Technol 1:106–111

Peng A, Xiao X, Yue R (2014) Process parameter optimization for fused deposition modeling using response surface methodology combined with fuzzy inference system. Int J Adv Manuf Technol 73:87–100. https://doi.org/10.1007/s00170-014-5796-5

Petch M (2018) 3D printing industry jobs board launches. In: 3D Print Ind. https://3dprintingindustry.com/news/3d-printing-industry-jobs-board-launches-130004/. Accessed 7 Mar 2018

Raja V, Zhang S, Garside J et al (2006) Rapid and cost-effective manufacturing of high-integrity aerospace components. Int J Adv Manuf Technol 27:759–773. https://doi.org/10.1007/s00170-004-2251-z

Rayna T, Striukova L, Darlington J (2015) Co-creation and user innovation: the role of online 3D printing platforms. J Eng Technol Manag 37:90–102. https://doi.org/10.1016/j.jengtecman.2015.07.002

Salazar-Martín AG, Pérez MA, García-Granada A-A et al (2018) A study of creep in polycarbonate fused deposition modelling parts. Mater Des 141:414–425. https://doi.org/10.1016/j.matdes.2018.01.008

Schey J (2002) Manufacturing process (Procesos de Manufactura). MCGRAW-HILL/INTERAMERICANA DE MEXICO, Mexico, DF

Scott J, Gupta N, Weber C, Newsome S (2012) Additive manufacturing: status and opportunities

Shunta J (1997) Achieving world class manufacturing through process control, 1st edn. Prentice Hall, Saddle River

Soltesz JP, Rutkofsky M, Kerr K, Annunziata M (2016) The workforce of the future—advanced manufacturing's impact on the global economy. GE Digit 1–27

Sood AK, Ohdar RK, Mahapatra SS (2009) Improving dimensional accuracy of fused deposition modelling processed part using grey Taguchi method. Mater Des 30:4243–4252. https://doi.org/10.1016/j.matdes.2009.04.030

Soriano-Heras E, Blaya-Haro F, Molino C, de Agustín del Burgo JM (2018) Rapid prototyping prosthetic hand acting by a low-cost shape-memory-alloy actuator. J Artif Organs. https://doi.org/10.1007/s10047-017-1014-1

Srivastava S (2010) Process modeling and simulation. Wipro Applying Thought

Srivatsan TS, Sudarshan TS (2016) Additive manufacturing, innovations, advances, and aplications. CRC Press, Boca Raton

Tauseef A (2010) Flexible manufacturing system : hardware components. J Future Manuf Syst. https://doi.org/10.5772/10237

Thiesse, Morar D, Lasi H et al (2015) Economic implications of additive manufacturing and the contribution of MIS. 57:139–148. https://doi.org/10.1007/s12599-015-0374-4

Thomas D (2017) Developing enhanced carbon nanotube reinforced composites for full-scale 3D printed components. Reinf Plast 0:4–7. https://doi.org/10.1016/j.repl.2017.06.089

Vairis A, Petousis M, Vidakis N, Savvakis K (2016) On the strain rate sensitivity of abs and abs plus fused deposition modeling parts. J Mater Eng Perform 25:3558–3565. https://doi.org/10.1007/s11665-016-2198-x

Vollman T, Berry W, Whybark D (1997) Manufacturing planning and control systems, 4th edn. APICS

Wagner H, Dainty A, Hague R et al (2008) The effects of new technology adoption on employee skills in the prosthetics profession. Int J Prod Res 46:6461–6478. https://doi.org/10.1080/00207540701432623

Wohlers Associates Inc (2011) Additive manufacturing technology roadmap for Australia, Australia. http://3dprintingexpo.org/wp-content/uploads/Additive-Manufacturing-Technology-Roadmap-CSIRO-2011.pdf

Yamamoto N, Sakai K (2005) US patent no 6,858,059 B2. US Patent and Trademark Office, Washington, DC

Zandin KB (ed) (2001) Maynard's industrial engineering handbook, 5th edn. MCGRAW-HILL, New York

Chapter 17
Analysis of the Productivity of a Shoe Production Line—Application of Queueing Theory and Lean Manufacturing

Salvador Hernández-González, Ricardo Ramírez-Tapia and José Alfredo Jiménez-García

Abstract Footwear production is a complex process. Producers often do not know-how to quantify the performance of their production lines in addition to their total lack of knowledge about tools that would enable them to make the right decisions about future events. We analyze a shoe production line of a factory located in the city of León, state of Guanajuato. This analysis is done by combining two approaches: on the one hand, we apply the Factory Physics approach, to understand how the materials flow within the production line; on the other hand, the proposals for improvements are made under the guidance of a lean manufacturing approach, in order to focus attention on very specific factors in the system. We determined the number of additional workers needed to obtain the required production. The results indicate that a saving of 16.73% would be obtained in the operation costs of the line.

Keywords Shoe production · Management · Productivity · Queueing theory Lean

17.1 Introduction

Shoe production in Mexico has been going on for 400 years; the Mexican shoe industry is a recognized economic activity that generated around 112,000 jobs, which represented 2.4% of the people employed in the manufacturing industry (INEGI 2014).

S. Hernández-González (✉) · R. Ramírez-Tapia · J. A. Jiménez-García
Department of Industrial Engineering, Tecnológico Nacional de México en Celaya, Antonio Garcia Cubas Pte. No. 600 esq. Av. Tecnológico, Col. Alfredo V. Bonfil, C.P. 38010 Celaya, Guanajuato, Mexico
e-mail: salvador.hernandez@itcelaya.edu.mx

J. L. García Alcaraz et al. (eds.), *Best Practices in Manufacturing Processes*,
https://doi.org/10.1007/978-3-319-99190-0_17

Four entities currently concentrate 94% of the value of footwear production: Guanajuato (70%), Jalisco (15%), State of Mexico (5%), and Mexico City (3%) (SE 2015). The leather-footwear sector represents 11% of the state's gross domestic product, in 2016 exceeded 27 billion pesos, in other words, 5% more than in 2015 and created jobs for more than 150 thousand people (Noticias Guanajuato 2017) (Table 17.1). The price of footwear is the focal point of competition in the sector, because the strategy of companies is to lower costs, with manpower being the most significant element (Duana Ávila and González Pedraza 2014).

Leather-footwear concentrates the majority of the economic units, people in employment, and gross production in the state of Guanajuato (INEGI 2014).

Nowadays Mexican producers are experiencing growing competition from imports from Brazil, China, Spain, and Italy. And, although the advantage of their access to raw materials has been adequately exploited, the market has contracted over the past decade (Duana Ávila and González Pedraza 2014).

The footwear industry has reached an intermediate level of technology based on foreign techniques, which makes it evident that Mexico needs to develop its own technology (Fig. 17.1). One feature of this industry is its intensive use of manpower

Table 17.1 Economic data for shoe manufacturing in Mexico (INEGI 2014)

Type	Economic units	People employed	Gross production
Leather and hide cutting	62.1	73.3	78
Fabric cutting	3.2	5.4	6.1
Plastic footwear	13.8	12.8	10.9
Rubber footwear	3.2	3.2	2.5
Sandals and other types of materials	17.7	5.3	2.5

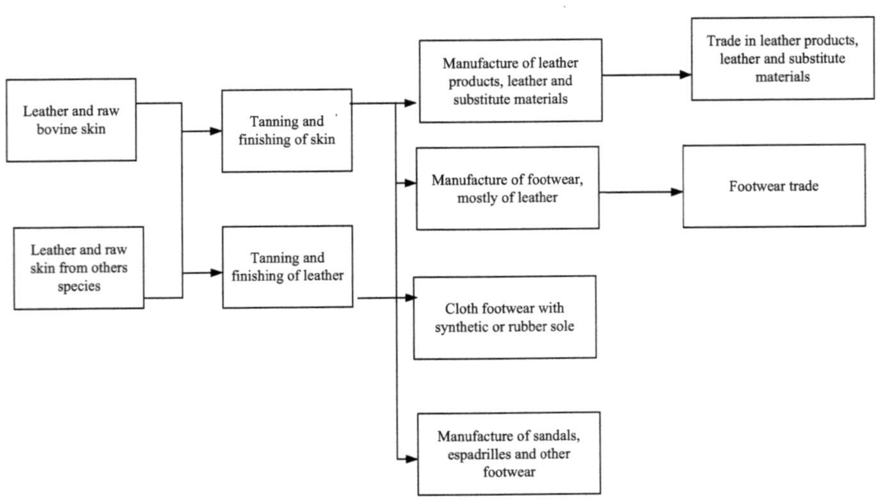

Fig. 17.1 Production chain of leather and footwear products in Mexico (INAES 2017)

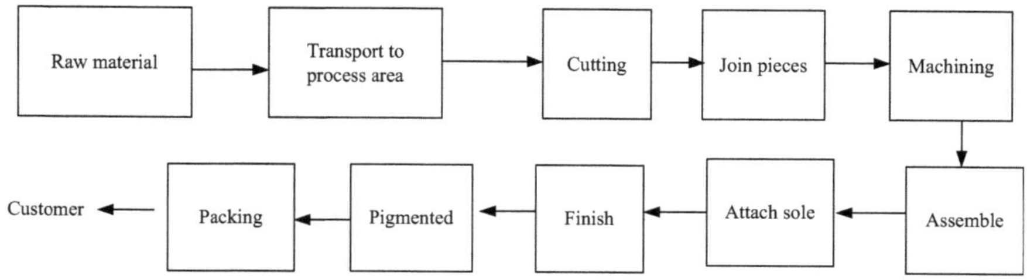

Fig. 17.2 Shoe manufacturing process (INAES 2017)

and it is common to find that companies are using obsolete production systems with very low productivity levels. In general consumables, machinery and manpower concentrate the majority of this industry's problems in respect of efficiency (Torres Noyola 2015).

Among the most important costs to produce footwear are the inputs from the activity of tanning and leather finishing; next are the costs of rubber products, resins, and synthetic rubbers, then fibers and finally the cost of materials from the footwear industry itself (Fig. 17.2) (INEGI 2014).

Moreover, the choice of production equipment is a decision that is important for the operations of a company in view of the high outgoings and limited financing options and can mean the success or failure of a company. As regards the management of resources, the administrators also have to make decisions about the resources needed to produce a particular model. How do you manage an industry of this nature? (INAES 2017).

The above implies that the people in charge of decision-making use tools and models that represent the system in order to analyze the different options available in the management of a production line. A model that represents a system allows an administrator to understand how each one of the system's variables are related to each other; minimizing the uncertainty associated with proposing changes to the system's operating conditions.

In this paper, we analyze a shoe production line of a factory located in the city of León, state of Guanajuato. This analysis is done by combining two approaches: on the one hand, we apply the Factory Physics approach, using queueing theory to understand how the materials flow within the production line and taking into account the fact that demand and the production time in each stage are random processes; on the other hand, the proposals for improvements are made under the guidance of a lean manufacturing approach, in order to focus attention on very specific factors in the system.

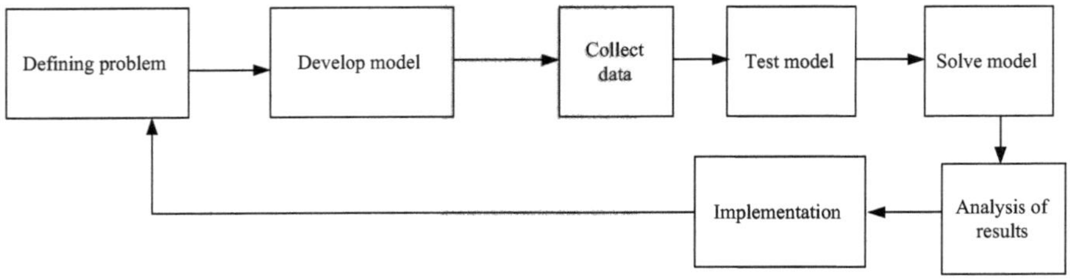

Fig. 17.3 A scientific approach to decision-making

17.1.1 Quantitative Analysis

Quantitative analysis refers to the scientific approach to decision-making in systems (Fig. 17.3). The starting point for this approach is the availability of data, which must be manipulated and processed to extract valuable information that will then be used to achieve a company's objectives (Render et al. 2009).

Once the problem has been defined, we have reached the stage of developing the model, which can be either analytical or simulation. Mathematical models are tools that help us understand the behavior of a system or subsystem. It is worth clarifying that mathematical models do not only consist of equations, but also have a set of suppositions and constraints that delimit their field of application.

There are mainly two different types of models: Deterministic and Stochastic. In a deterministic model, each variable and each parameter can be adjusted to a value or series of fixed values for a given series of conditions. On the other hand, the analysis of stochastic models is based on the probability of occurrence (Ramírez-Tapia 2017).

17.1.2 Models for Decision-Making

A decision model is a set of expressions, where the variables represent decisions that should or could be taken (Eppen et al. 2000). The models offer a congruent and logical reference for analysis and we can list the following advantages, among others:

- They help us to have and define the objectives.
- They help us to identify all the interactions between the decisions as well as their advantages and disadvantages.
- They enable us to find the relevant variables in the system.
- They invite us to consider which data are suitable for quantifying said variables and identifying their interactions.
- They enable us to communicate ideas and know-how that facilitate teamwork.

17.2 State of the Art

As with other branches of manufacturing, the shoe industry has benefitted from the use of tools that reduce uncertainty in analysis and decision-making.

Staikos and Rahimifard (2007) develop a tool for identifying the best strategy for the recovery, recycling, and reuse of shoes discarded by their users. Facchin and Sellitto (2012) develop a simulation model based on Petri networks for representing the behavior of a production line. Eryilmaz et al. (2012) built a simulation model for analyzing the performance of a production line faced with high volatility in demand for the product. Lin et al. (2015) develop an analytical procedure in three stages, considering failures in the equipment and apply it to measure the performance of a shoe production line. This is the only one of the papers consulted where the stochastic part is modeled using analytical expressions.

Sayid Mia et al. (2017) apply lean manufacturing tools to improve the productivity of a production line and use simulation to assess the proposals for improvement.

Latin American shoe manufacturing companies have opted for the application of tools to improve their processes and be competitive; however, there are still very few articles and case studies reported in high impact operations management and engineering journals.

Pérez-Gallardo et al. (2014) developed a model to integrate a supply chain of small companies and artisan companies, Ortiz-Triana and Caicedo-Tolón (2014) develop a mathematical programming model for calculating the number of pairs of shoes to produce, assuming that the process is deterministic. Reyes-Vasquez et al. (2015) present a heuristic approach to evaluate the production capacity of footwear companies.

Selitto et al. (2015) develop a SCOR model for performance measurement in the footwear industry of Brazil.

The manufacturing processes are subject to random factors (demand, process times, and failures), and in this case, the preferred tool is a simulation; Mejía Ospina et al. (2016) construct a dynamic model of systems to simulate the operations of a footwear company. Furthermore, Silva de Lima et al. (2016) apply Value Stream Mapping and discrete event simulation to propose improvements to the shoe manufacturing process

Ramírez-Betancourt et al. (2017) developed an analytical expression of operation cost of the production line and applied it to evaluate the benefits of improving the quality of the shoe.

17.3 Queueing Theory

Waiting is a phenomenon where a set of customers arrive at a system in search of a service where they have to wait if the service is not fast. Then, when the customers are served, they leave the system. Figure 17.4 shows a basic queuing system. Queues have the following components (Hall 1991):

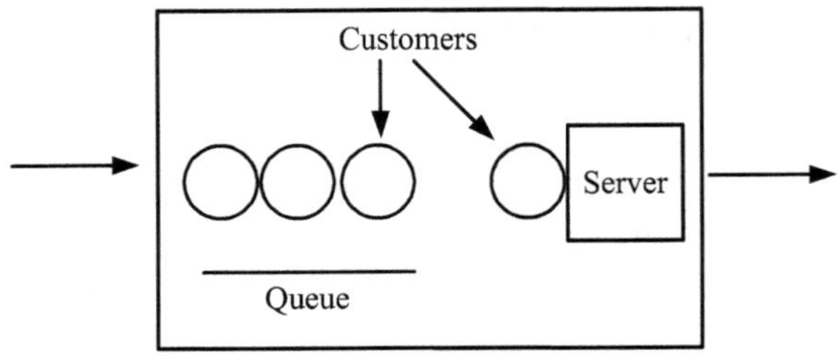

Fig. 17.4 Queueing systems with a single station

Customer Arrival Rate (λ): Arrivals to the system are normally stochastic, in other words, the pattern of arrival is given by a random variable. In this case, we need to know the probability distribution between 2 successive customer arrivals. At the same time whether the customers arrive independently or simultaneously (in other words if they arrive in batches) must be taken into account and the probability distribution would have to be defined for these cases (Gross et al. 2008).

Service Rate (t_s): The servers can have a variable service time; when this happens we need to use a probability function in order to be able to define it. The pattern of service can be both batch or individual.

Queueing discipline: It is the order in which customers are selected to be served. A common policy is to serve in accordance with the order of arrival, and that is known as first-in first-out (FIFO). Other strategies are last-in first-out (LIFO) and according to the importance of the customers (priorities) (Gross et al. 2008)

Number of servers (c): This is the number of customers that can be served in the system at the same time.

Service Stages: The queueing system can be both single stage and multi-stage. In multi-stage systems, the customer can pass through a number of stages that is higher than 1. In some multi-stage queueing systems, there can be reverse or recycled stages.

Capacity for the number of customers in the System (K): It is the constraint on the number of customers that can wait in the queue to be served; the system is also called a finite queueing system (Gross et al. 2008).

17.3.1 Performance Measurements in a Queueing System

The queueing theory results generate mathematical expressions that enable us to measure the performance and make decisions about the system. We will now show the general expressions that link the main variables that make up a queuing system.

17.3.1.1 Little's Law

In a queueing system in equilibrium with a mean arrival rate (λ), a mean queue size (WIP) and a mean residence time in the system or cycle time (CT), the work in process varies as a direct proportion of arrival rate and cycle time. This is known as Little's formula (17.1) (Hillier and Lieberman 2010).

$$\text{WIP} = \lambda \text{CT} \quad (17.1)$$

17.3.1.2 Waiting Time in the Queue (CT_q) and Waiting Time in the System (CT_S)

The cycle time in the queue or simply CT_q, is the time that the customer spends in the queue before being served. On the other hand, the cycle time in the system (CT_S), is the time the customer takes from its arrival in the system until it leaves. This time may indicate the capacity or efficiency of the server (Curry and Feldman 2009; Gross et al. 2008). Assuming that the mean service time is l/μ then the cycle time in the system is expressed as follows (Fig. 17.5):

$$\text{CT}_S = \text{CT}_q + t_S \quad (17.2)$$

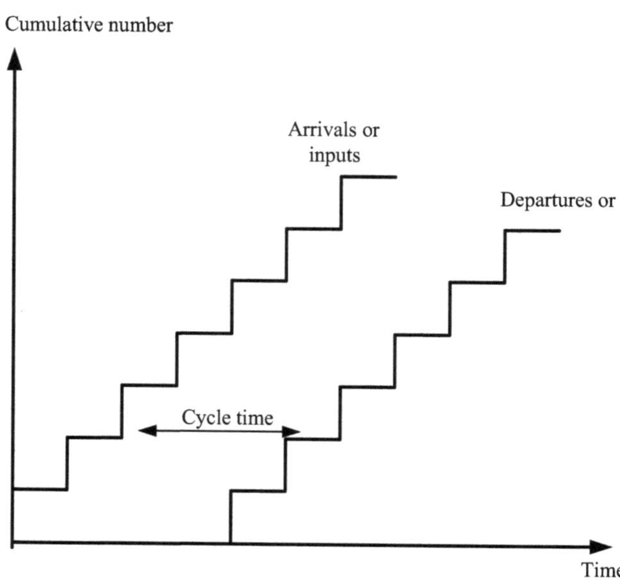

Fig. 17.5 Cycle time in a station

17.3.1.3 Number of Customers in the System (WIP_S)

The number of customers in the system includes queueing customers as well as those being served. This quantity is also connected to the service's efficiency and capacity, as a large number of customers in the system causes congestion, generating customer dissatisfaction. The number of customers in the system is also known as work in process and represented by WIP (Work in Process) (Curry and Feldman 2009).

17.3.1.4 Length of the Queue (WIP_q)

This is the number of customers to be found queueing in the system waiting for a service. Depending on its length or size, WIP_q reflects any of the following situations: A queue with few customers is consequence either of an efficient service or an excess of capacity. A long queue reflects either the server is not very efficient, or the capacity of the system needs to be increased.

17.3.1.5 Utilization of the Servers

Utilization is the level of occupation of each one of the servers. Usually, we want to maintain a high level of utilization in order for the operation to be reasonably profitable (Krajewski et al. 2013). The server's utilization factor ρ is given in the following equation (Hillier and Lieberman 2010):

$$\rho = \lambda t_S < 1 \tag{17.3}$$

It is a necessary condition for the capacity to be higher than the service demand, otherwise customers enter the system too fast and pile up, making the queue in front of the server endlessly grow; however, when the system has a finite number of customers that can stay within the system, this condition does not need to be fulfilled, as we will show further on.

17.3.1.6 Throughput

Furthermore, the throughput is the number of finished customers or jobs that leave the system per unit of time. The average throughput rate is known as Th. (Curry and Feldman 2009).

A / B / C / D

Fig. 17.6 Kendall's notation

17.3.1.7 Process Path

This is the sequence the customer or job follows until it leaves the system. It is said that when a job follows the same path as another, then it is the same type of job. Therefore, we conclude that different jobs follow different paths (Curry and Feldman 2009).

17.3.1.8 Kendall's Notation and Terminology

The queues can be described by a notation introduced by Kendall, which is shown in Fig. 17.6.

Where: A corresponds to the probability distribution of inter-arrival time, B is the probability distribution of service time, C is the number of servers that can work in parallel, and D refers to the system's capacity.

The simplest system, where both the inter-arrival time and service time are distributed exponentially, with one server and infinite capacity, is M/M/1 where the M represents the Markov system (Taha 2011). Table 17.2 gives the types of distributions most commonly found in queues.

17.3.2 Topology of Queues: Queues in Series and Queues with a Network Arrangement

Topology in queueing theory refers to the type of arrangement that a queueing system has; these arrangements can be series or network. Network systems can, in general terms, be defined as a group of nodes that represent the installation of a service with *c* number of servers. In this case, the customers can enter through any node in the system and move through the network according to the customer's needs, finally leaving through any of the nodes (Gross et al. 2008).

Table 17.2 Symbols for probability distributions in Kendall's notation

Probability distribution	Symbol
Exponential	M (Markovian)
Deterministic	D
Erlang	E
General	G

17.3.2.1 Open Network Systems (Jackson Networks)

Jackson networks are systems that comply with the following assumptions: there is only one type of job or customer in the network; and the number of jobs in the network is limitless. At each node, customers may or may not arrive. It is not necessary for the inter-arrival times or service times to be exponentially distributed and the customers can leave the system through any node (Jackson 1957, 1963).

The service discipline at the nodes is FIFO, for every stage i, a customer is moved to another stage with probability r_{ij}, and outside the system with probability r_{i0}, there is no limit on capacity in any queue. Figure 17.7 shows an open Jackson network.

17.3.2.2 Network System and Finite Queueing Capacity

It is generally assumed that there is enough space for the customers to line up in front of a station ($K = \infty$); however, in reality, this is not always the case: sometimes there are a finite number of places for the customers.

Assuming that the inter-arrival time and the service time are random variables that follow an exponential distribution, the classification of these types of systems is M/M/c/K, according to Kendall's notation. Figure 17.8 illustrates a system with n stations in a series arrangement and where, between each station, there is limited space for the work in process.

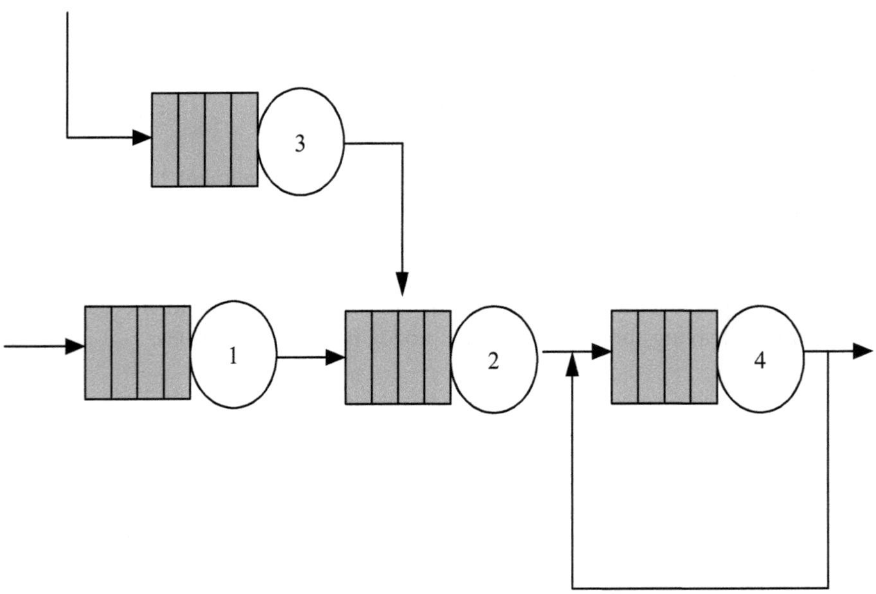

Fig. 17.7 Open Jackson network

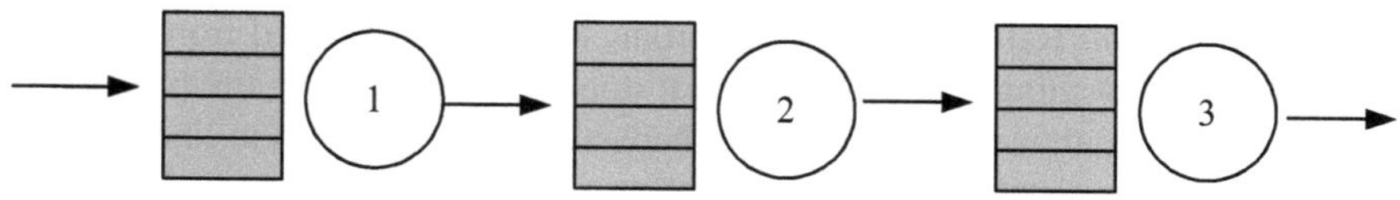

Fig. 17.8 System in series with a buffer between stages

The capacity of each station will be the size of the buffer (b) plus the number of parallel servers; if stations have one server then: $K = b + 1$ (Balsamo et al. 2001).

Owing to the restriction on the size of the queue, the system will have all the spaces full for part of the time; then the probability of there being K customers in the system will be

$$p_K = \frac{1 - \rho}{1 - \rho^{K+1}} \tag{17.4}$$

Given that not all the customers enter the system; the output will be less than inflow to the system; then the system's production is

$$\lambda_{\text{Out}} = \lambda_{\text{in}}(1 - p_K) \tag{17.5}$$

The work in process is calculated using the following expression:

$$\text{WIP}_{\text{S}} = \frac{\rho[1 - (K+1)\rho^K + K\rho^{K+1}]}{(1-\rho)(1-\rho^{K+1})}, \quad \rho \neq 1 \tag{17.6}$$

If $\rho = 1$, then

$$\text{WIP}_{\text{S}} = \frac{K}{2} \tag{17.7}$$

On the other hand, the cycle time is obtained as follows:

$$\text{CT}_{\text{S}} = \frac{\text{WIP}_{\text{S}}}{\lambda_{\text{in}}(1 - p_K)} \tag{17.8}$$

17.3.3 Competitive Edge Obtained Using the Lean Manufacturing Approach

The Lean Manufacturing approach seeks to supply the required products "exactly" when the customer needs them. The Lean Manufacturing approach focuses on a set of elements known as the 7 wastes, which should be considered when this philosophy is implemented in a company's resource management.

- *Overproduction*: Refers to the excess production that is obtained from the line.
- *Queues*: Any waiting implies a time that does not add value to the product.
- *Transport*: The movement of material between plants, production lines and, in general, between stations implies a waste.
- *Inventory*: The excess work in process, raw materials, and, in general, any stored material.
- *Movement*: Unnecessary movements to perform an operation.
- *Reprocesses*: Any additional operation that does not add value to the product.
- *Defectives, reprocesses,* and *returns* are considered as waste.

The lean approach works under a paradigm of pull production, in other words, the product is manufactured as the customer asks for it. Some authors consider that the lean approach is the antithesis of push systems (or mass production systems) such as the case set forth in this chapter (Heizer et al. 2017); however, this is not in any way an impediment to applying the approach and guiding the improvement process.

17.4 Case of Application in a Footwear Company

The footwear manufacturing company has a demand for 2100 pairs of lady's ballet flat shoes a day, every shoe is assembled by means of a seam (also known as backstitch) with four pieces per foot; one pair of shoes requires eight assembled parts.

The backstitching process and preliminary processes, where the pair of shoes is assembled consist of the following stages (Fig. 17.9):

- Sew together the parts that form the lining del shoe
- Join the front part of the shoe (vamp) to the heel
- Join the parts that make up the shoe's heel
- Join the lining to the leather
- Fold the seam made in stage 4
- Spread glue on the lining and turn it over to stick it onto the leather
- Remove the left-over thread with scissors or by singing the thread after the seams are made.

Every operation requires a service time and, moreover, every station has only enough space for a certain amount of work in process (buffer). At the present time, the company has to monitor the operations of the production line, for which it needs to apply the proper tools for the management and control of the resources, so an analytical model was developed for the analysis of the line.

Samples are taken of the operation times at each station, and then a goodness-of-fit test is performed to determine what probability distribution is the one that best fits the data. Table 17.3 gives the results of the sampling of the operation times and the capacity of each station.

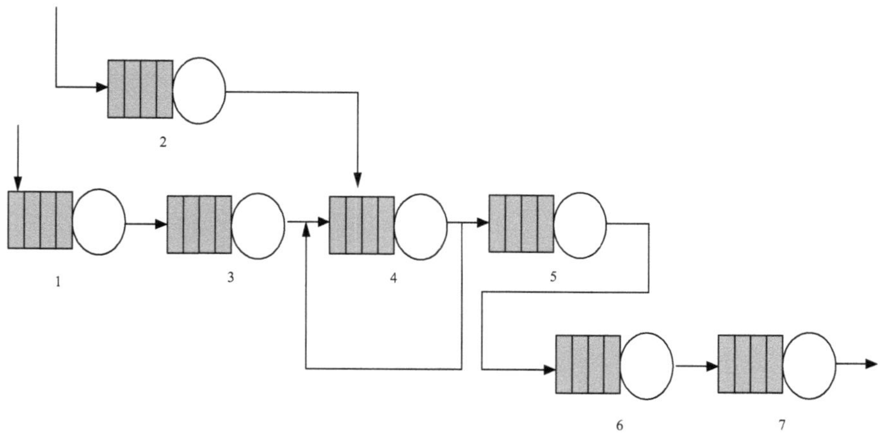

Fig. 17.9 Process flow diagram for the selected shoe model

Table 17.3 Data of the stations

Operation	Number of servers	Service time (s)	Standard deviation (s)	CV
1.	2	31.59	3.07	0.10
2.	2	22.02	2.29	0.10
3.	2	19.81	1.51	0.08
4.	4	55.04	3.39	0.06
5.	2	18.93	1.19	0.06
6.	2	29.28	1.27	0.04
7.	2	28.62	1.52	0.05

Given that an approximation method was used to get the line's performance measurement that assumes that the service times follow an exponential distribution, it was necessary to carry out a goodness-of-fit test to check this assumption. The results are given below identified by the operation number (Fig. 17.10).

As can be observed, the service time of station 4 fails to comply with the assumption, so there will be a deviation in the estimations. Even so, the calculations are still done because, as we will show further on, the analytical results were validated by simulation. In the following stage, supplements were added to the service times to get a value that includes skill, conditions, effort, and consistency. For reasons of confidentiality, these are not given in the paper, but they can be obtained from the authors of the document.

The daily demand for shoes is 2100 pairs. We must point out that operations 1 and 2 assemble 2 pieces per shoe (4 pieces per pair), so the conversion is as follows: (2100 pairs $\times$ 2 shoes $\times$ 2 pieces)/(8 h) = 1050 pieces. Station 4 is where the 8 remaining pieces are joined to form the shoe. Table 17.4 shows the process path in

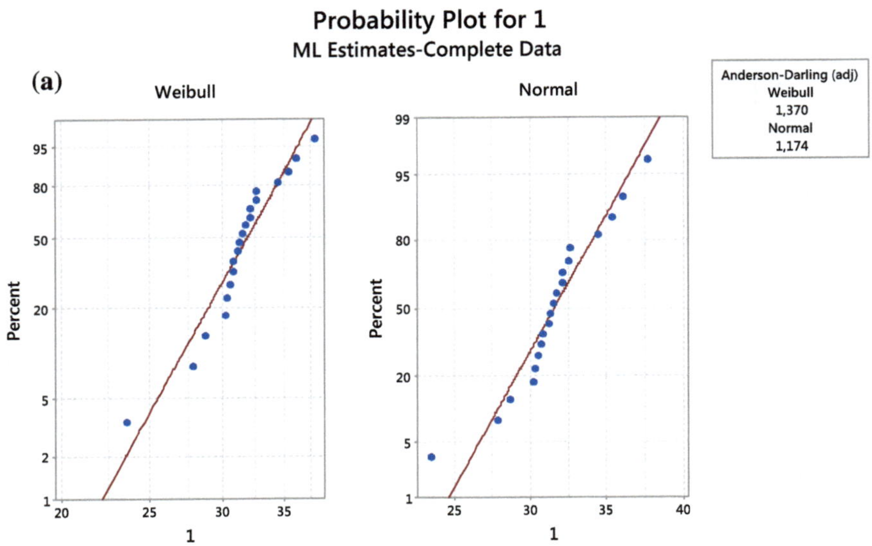

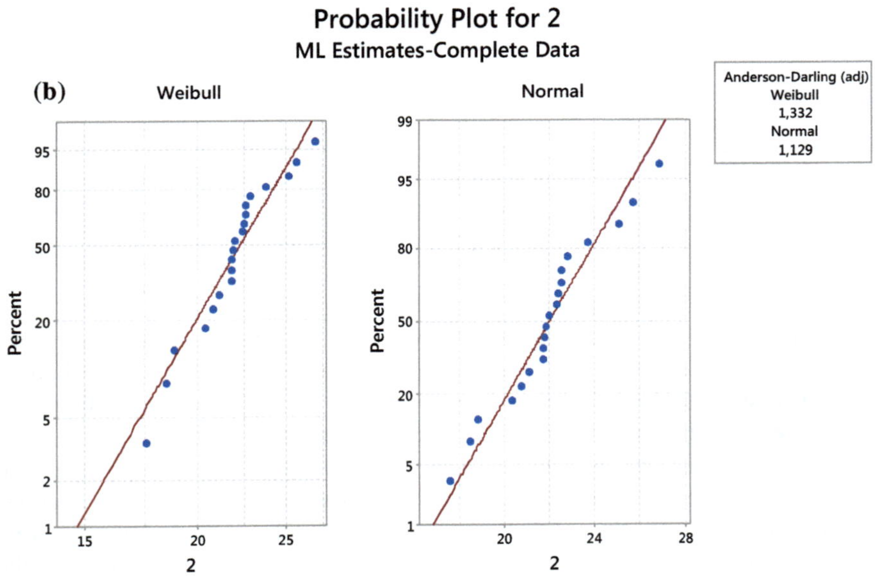

Fig. 17.10 Results of the goodness-of-fit test for the stations. **a** Station 1, **b** station 2, **c** station 3, **d** station 4, **e** station 5, **f** station 6, and **g** station 7

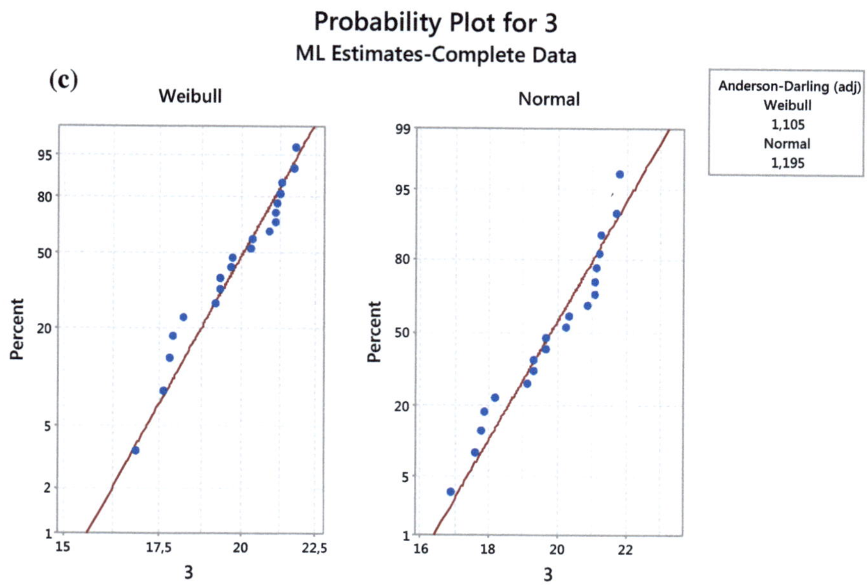

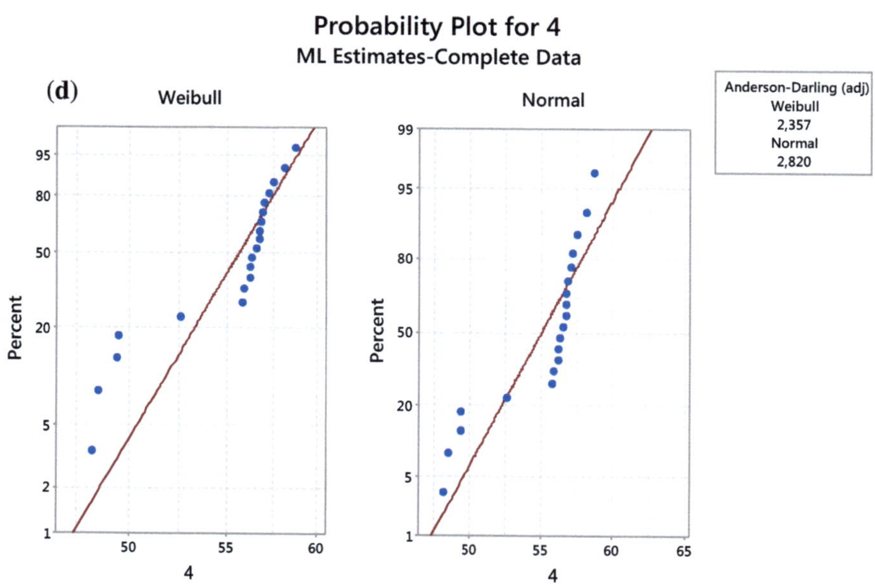

Fig. 17.10 (continued)

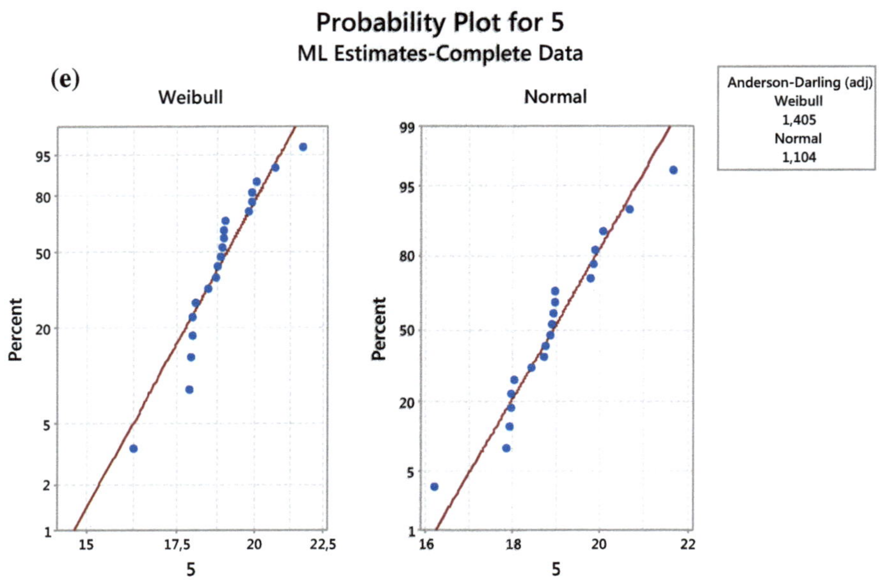

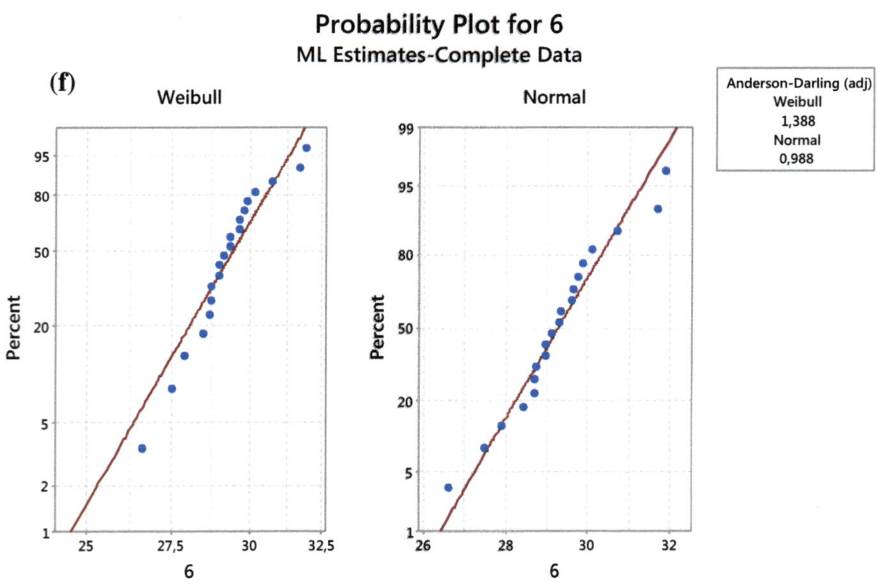

Fig. 17.10 (continued)

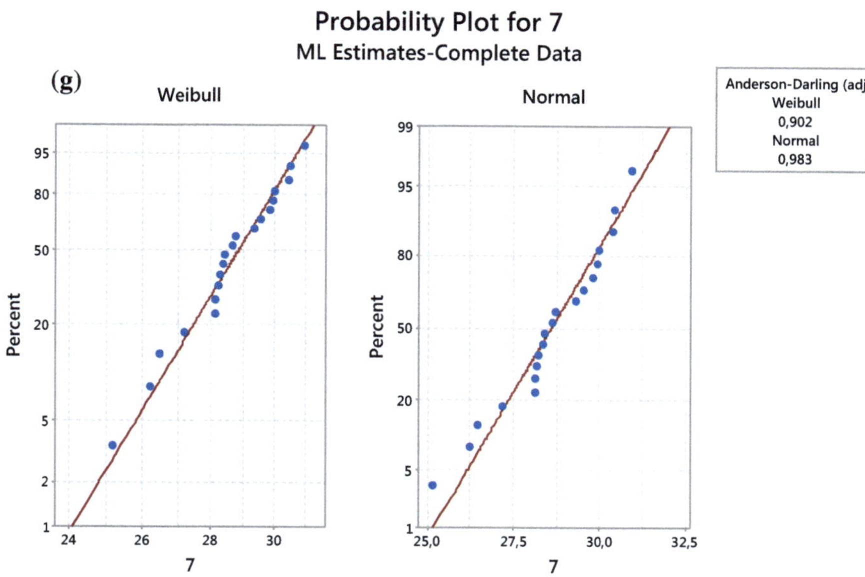

Fig. 17.10 (continued)

Table 17.4 Process path

	1	2	3	4	5	6	7
1	0	0	1	0	0	0	0
2	0	0	0	1	0	0	0
3	0	0	0	1	0	0	0
4	0	0	0	0.1	0.9	0	0
5	0	0	0	0	0	1	0
6	0	0	0	0	0	0	1
7	0	0	0	0	0	0	0

a matrix arrangement. During the analysis of the process, we found that station 4 has to reprocess 10% of the end product because of quality errors.

We use Takahashi et al. (1980) approximation algorithm based on the system decomposition process; encoded in Scilab language (Scilab Enterprises 2018) and run on a computer with an i3 processor, 3.3 GHz and 4 Gb RAM, the program takes less than 1 s to finish. To validate the results, a simulation model is constructed in the ARENA package, where 5 simulations of 8 h each were carried out, thus ensuring a confidence level of 95%.

17.4.1 Diagnosis of the Production Line

Table 17.5 shows the analytical results for the average queue size and the cycle time in each station divided into two sections: analytical and simulated. The lean approach was applied to identify waste, focusing on the work in process and cycle time. We can see that in station 2 (attaching the front part of the shoe) we have the highest cycle time of all the production line followed by stations 6 (spreading glue) and 7 (removing left-over bits). The highest number of jobs in the queue (in pairs of shoes) is found at station 2.

The production obtained through the last station is 252 pairs of shoes. The total work in process of the entire line is the equivalent of 4.6 pairs and the overall cycle time on the line is approximately 65 s (Table 17.6).

An analysis of the operation costs of the line is given below; only the cost of the work in process and the station's operation cost were considered. For the cost of the work in process, the costs of the materials (synthetic materials and threads) were used; from station 4 to 7 it is the same cost because no new components are added to the pair of shoes. The manpower costs were determined using the table of Minimum Wages in Force (CONASAMI 2017), which shows that a worker in a footwear workshop earns \$99.93 Mexican pesos/8 h = \$12.5 an hour (1 dollar ≈ \$19.21 Mexican pesos) (Table 17.7). The cost of the materials in the line is around \$118.24 an hour or \$946 per 8 h working day.

Table 17.5 Results of the analytical model

Operation	Analytic		Simulated	
	Work in process (pairs)	Cycle time (s/pair)	Work in process (pairs)	Cycle time (s/pair)
1	0.29	35.78	0.32	37.48
2	2.38	168.88	3.06	207.74
3	0.23	30.82	0.23	30.45
4	0.22	74.04	0.39	79.2
5	0.083	26.46	0.06	24.49
6	0.63	102	0.50	90.43
7	0.54	87.55	0.33	67.39

Table 17.6 Performance of the production line

Parameter	Analytic	Simulated	Diff (%)
Total production (pairs)	252.3	263.5	−4.24
WIP global (pairs)	4.6	4.7	−1.64
Cycle time (s/pair)	65.6	64	2.72

Table 17.7 Cost data and cost per station (Mexican pesos, MXN)

Operation	1	2	3	4	5	6	7
Manpower cost ($)	12.5	12.5	12.5	12.5	12.5	12.5	12.5
Raw material cost ($)	15	22.5	22.5	37.5	37.5	37.5	37.5
Operation cost ($)	24.5	24.5	24.5	50	24.5	24.5	24.5
Waiting cost ($)	4.4	53.5	5.2	8.2	3.1	23.7	20.3

The operation cost of the line is $318.14 an hour; a working day has 8 h, then operation costs are $2545.15; a month has 30 days, then costs are $76,354.5 Mexican pesos a month.

17.4.2 Study for Improvement and Assessment of Costs

In this case, the objective parameter is the quantity of shoe pairs to be produced that, on this line, is less than the demand; we observe that stations 2, 6, and 7 do not have enough capacity to produce the number of pairs required (262.5 pairs an hour or 2100 pairs a day).

The economic impact of increasing the number of workers at said stations was analyzed, in the first place, to achieve the desired production and, second, to estimate by how much the reduction in the work in process and the saving in cycle time, mainly at station 2, would be. In the process of discriminating between possible actions, for the moment we consider increasing capacity to be primordial for immediately reaching the desired production level.

In each of the aforementioned stations, we increased the number of workers until the desired output was reached. In the program, only the number of parallel stations (c) was modified in each case. As can be appreciated in Table 17.8, in this case, an increase of one worker in each one of the three stages is enough. The total cost is $264.92/h or $63,581/month. The saving in costs in respect of the original conditions is 86,354 − 63,581 = $12,773; so, the result suggests that increasing capacity is a decision that will benefit the system not only in terms of production volume, but also in economic terms (Table 17.8).

Table 17.8 Economic assessment of increasing the capacity of stations 2, 6, and 7

Operation	1	2	3	4	5	6	7
Work in process (pairs)	0.29	0.18	0.23	0.22	0.08	0.06	0.02
Waiting cost ($)	4.35	4.11	5.16	8.22	2.94	2.11	0.7
Servers	2	3	2	4	2	3	3
Cost ($)	24.98	37.47	24.98	49.97	24.98	37.47	37.47

17.5 Conclusions

Leather-footwear currently concentrates the majority of the economic units, people in employment and gross production in the state of Guanajuato; a large percentage of these workers are to be found working in small companies.

Owing to growing competition, businessmen must implement the necessary improvements not only in their production lines but also in the way they use the available resources.

There are not always enough financial resources to invest in tools such as simulation packages but, on the other hand, analytical models are within everyone's reach, and moreover, they enable us to make estimations faster and they help us to understand how the system's variables relate to each other.

This paper presented the development of a tool for analyzing the production of a shoe factory and this tool is based on the expressions of queueing theory.

The production line is a network of stations with reprocessing, which is a manual operation. The operation times in each stage were determined by sampling and we assumed that the service times are exponentially distributed random variables. This was verified by means of a goodness-of-fit test; we also assumed that the demand is a Poisson type; the system can be classified as an M/M/c/K network of stations.

Using an approximation algorithm, we determined the line's cycle time and work in process; and from the results, we deduced that three stations are not producing the required quantity of shoes.

Applying the queueing theory model, we determined the number of additional workers needed to obtain the required production in stations 2, 4, and 6, while the criterion for determining the benefit was cost. The results indicate that a saving of 16.73% would be obtained in the operation costs of the line.

References

Balsamo S, Nitto-Persone V, Onvural R (2001) Analysis of queueing networks with blocking. Springer, New York

CONASAMI (2017) Minimum wages in force. Retrieved 22/01/2018 from https://www.gob.mx/conasami

Curry GL, Feldman RM (2009) Manufacturing systems modeling and analysis, 2nd edn. Springer, Berlin

Duana Ávila D, González Pedraza F (2014) Commercial performance of the footwear industry in Mexico. Boletín Científico de las ciencias Económico—Administrativas 3(5). Retrieved 02/10/2017 from https://www.uaeh.edu.mx/scige/boletin/icea/n5/e4.html (in Spanish)

Eppen GD, Gould FJ, Schmidt CP, Moore JH, Weatherford LR (2000) Introductory management science. Prentice Hall, New Jersey

Eryilmaz MS, Kusacki AO, Gavraanovic H, Findik F (2012) Analysis of shoe manufacturing factory by simulation of production processes. Southeats J Soft Comput 1(1):120–127

Facchin T, Sellitto MA (2012) Measurement of work-in-process and manufacturing lead-time by petri nets modeling and throughput diagram. Retrieved 04/12/2017 from http://cdn.intechopen.

com/pdfs/38503/InTech-Measurment_of_work_in_process_and_manufacturing_lead_time_by_petri_nets_modeling_and_throughput_diagram.pdf
Gross D, Shortle JF, Thompson JM, Harris CM (2008) Fundamentals of queueing theory. Wiley, Hoboken, NJ
Hall RW (1991) Queueing methods: for services and manufacturing. Prentice Hall, Englewood Cliffs, NJ
Heizer J, Render B, Munson C (2017) Operations management. Sustainability and supply chain management, 12th edn. Pearson, Singapore
Hillier FS, Lieberman GJ (2010) Introduction to operations research, 8th edn. McGraw-Hill, New York, NY
INAES (2017) Leather footwear. Retrieved 04/12/2017 from http://www.inaes.gob.mx/doctos/pdf/guia_empresarial/calzado_de_cuero.pdf (in Spanish)
INEGI (2014) Statistics on footwear. National Institute of Statistics Geography and Informatics. Retrieved 2/10/2017 from http://internet.contenidos.inegi.org.mx/contenidos/productos/prod_serv/contenidos/espanol/bvinegi/productos/nueva_estruc/702825068332.pdf (in Spanish)
Jackson JR (1957) Networks of waiting lines. Oper Res 5(4):518–521
Jackson JR (1963) Job-shop like queueing systems. Manage Sci 10(1):131–142
Krajewski LJ, Ritzman LP, Malhotra MK (2013) Operations management: processes and supply chains, 10th edn. Pearson, New Jersey
Lin YK, Chang PC, Chen J (2015) Performance evaluation for a footwear manufacturing system with multiple production lines and different station failure rates. Int J Prod Res 51(5):1603–1617. https://doi.org/10.1080/00207543.2012.718451
Mejía Ospina LA, Arroyave Contreras O, González Bernal RA (2016) Analysis of the critical variables of a productive system through the system dynamics. Scientia Et Technica 21 (2):135–143 (in Spanish)
Noticias Guanajato (2017) The footwear leather sector: spearhead at the national level in exports, innovation, fashion and design. Retrieved 2/10/2017 from http://noticias.guanajuato.gob.mx/2017/03/el-sector-cuero-calzado-punta-de-lanza-a-nivel-nacional-en-exportaciones-innovacion-moda-y-diseno-mmm/ (in Spanish)
Ortiz-Triana VK, Caicedo-Rolón AJ (2014) Optimal production scheduling in a small shoe business in Colombia. Revista de Ingeniería Ind 35(2):114–130
Pérez-Gallardo JR, Hernández-Vera B, Moras-Sánchez CG, Posada-Gómez R, Alor-Hernández G (2014) Methodology for supply chain integration: a case study in the artisan industry of footwear. Math Probl Eng. Retrieved 2/10/2017 from https://www.hindawi.com/journals/mpe/2014/508314/
Ramírez-Betancourt FC, Vitery-Moya JR, García-Rodríguez EG, Carrión-Palacios VM (2017) Optimal value of management efficiency. Shoe process case. Ingeniería Ind XXXVI(2):163–174 (in Spanish)
Ramírez-Tapia R (2017) Calculation of performance averages for Markov systems in buffer lines and in network. Master's thesis, Tecnológico Nacional de México in Celaya (in Spanish)
Render B, Stair RM, Hanna ME. (2009) Quantitative analysis and management. Pearson—Prentice Hall, Upper Saddle River
Reyes-Vasquez JP, Aldas-Salazar DS, Morales-Perrazo LA, García-Carrillo MG (2015) Evaluation of the capacity for assembly in the footwear manufacturing industry. Ingeniería Ind XXXVII(2):14–23 (in Spanish)
Sayid Mia MA, Nur-E-Alam MD, Uddin MK (2017) Curt shoe production line: improvement of process cycle efficiency by using lean tools. Leather Footwear J 17(3):135–146
Scilab Enterprises (2018) Scilab: open software for numerical computation. http://www.scilab.org/en
SE (2015) The footwear industry in Mexico. Retrieved 2/10/2017 from https://www.gob.mx/se/articulos/la-industria-del-calzado-en-mexico (in Spanish)
Selitto MA, Pereira GM, Borchardt M, da-Silva RI, Viegas CV (2015) SCOR-based model for supply chain performance measurement: application in the footwear industry. Int J Prod Res 53 (16):4917–4926. Doi:https://doi.org/10.1080/00207543.2015.1005251

Silva de Lima DF, Franca Alcantara PG, Costa Santos L, Freitas de Silva LM, Moreira da Silva R (2016) Value stream mapping and simulation for implementation of lean manufacturing practices in a foot wear company. Revista Produção Online 16(1):366–392. Doi:https://doi.org/10.14488/1676-1901.v16i1.2183

Staikos T, Rahimifard S (2007) A decision-making model for waste management in the footwear industry. Int J Prod Res 45(18–19):4403–4442. https://doi.org/10.1080/00207540701450187

Taha HA (2011) Operations research: an introduction, 9th edn. Prentice Hall, Upper Saddle River, NJ

Takahashi Y, Miyahara H, Hasegawa T (1980) An approximation method for open restricted queueing networks. Oper Res 28(3):591–602

Torres Noyola F (2015) Footwear. Retrieved 04/12/2017 from http://www.contactopyme.gob.mx/estudios/docs/calzado_edomex.PDF (in Spanish)

Chapter 18
Performance Evaluation of a Commercial 3D Printer that Uses Fused Filament Deposition Technology

Secundino Ramos-Lozano, Javier Molina-Salazar, Lázaro Rico-Pérez and David Atayde-Campos

Abstract Since 3D printer home and industrial applications have increased in the last years, the need for a reliable tool to evaluate 3D printer capabilities has become necessary. A DOE was performed to determine the optimum parameters for dimensional accuracy and finished surface on printed pieces on a commercial 3D printer that uses fused filament deposition technique (FDM) with polylactic acid (PLA) as print material. A 3D digital model with geometric internal and external features was generated with a CAD software; this model was used to print two sets of physical samples, one set with the 3D printer adjusted according to DOE optimal settings results and the other set with the manufacturer recommended setup. Measurement system analysis bias was applied to evaluate the dimensional and geometrical performance of the 3D printer for each set. Samples were measured and compared against dimensional specifications on the drawing. No evidence of works related to the finished surface or geometrical analysis was found in the literature reviewed about 3D printed models. Finished surface was evaluated, and it was found that roughness depends mainly on the layer thickness of lateral walls of the piece, while on the upper face, the infill density has a major influence on the finished part. Most geometries and dimensions were rejected according to bias criteria, and no significant difference was found between both evaluated setups, so the printed models should be used only where dimensional accuracy is not critical. MSA bias can be used as an alternative method to make a dimensional and geometrical evaluation of printed models on 3D printers.

Keywords 3D printer · CAD · FDM · PLA

S. Ramos-Lozano (✉) · J. Molina-Salazar · L. Rico-Pérez · D. Atayde-Campos
Department of Industrial Engineering and Manufacturing, Universidad Autónoma de Ciudad Juárez, Av. Del Charro 450 Norte. Col. Partido Romero, Juárez, Chihuahua, Mexico
e-mail: secundino.ramos@uacj.mx

J. L. García Alcaraz et al. (eds.), *Best Practices in Manufacturing Processes*,
https://doi.org/10.1007/978-3-319-99190-0_18

18.1 Introduction

3D printing is a technology which converts 3D computer-aided design (CAD) data into a physical prototype (Dawoud et al. 2016). The beginning of the additive manufacturing dates back to 1976, with the invention of the inkjet printer. In 1984, designers modified technology and created the first 3D printer with some adaptations and improvements on the concept of the inkjet printer. Now, this print technology is able to work with different materials. Figure 18.1 shows different 3D printing techniques.

Figure 18.1a shows the first method created by Charles Hull, who introduced the first 3D printer using stereolithographic (SLA) technique; SLA is a photopolymerization process where a build tray is submerged in a basin of photosensitive liquid material. The basin depth varies based on laser strength, material, or desired tolerance. A UV laser (not lamp) solidifies one slice of the part onto the build tray. Then, the tray is submerged, and the laser solidifies the next slice of the part. The layer thickness affects the quality of the print and its tolerances. The laser travels the entire path of the part's cross section as it builds up each layer, so the speed becomes an important consideration. When the part is complete, the printer drains the resin excess, which is reusable. The operator washes the formed parts to remove resin excess and the support structures are physically removed (Stanbury 2016).

Figure 18.1b shows the laser printing technology (SLS) developed in 1986. SLS process uses a layer of powdered material carefully laid down by a leveler or roller on the build tray. Then, a laser sinters the part's cross section, and the tray goes down to repeat the process. Similar to SLA, layer thickness varies based on laser strength, material, or desired tolerance.

Figure 18.1c shows a widely used 3D printing technique called fused deposition modeling (FDM), introduced by the end of the 80s. The printer melts the filament of material in a heated nozzle and leaves it on a platform, once the printer finishes the

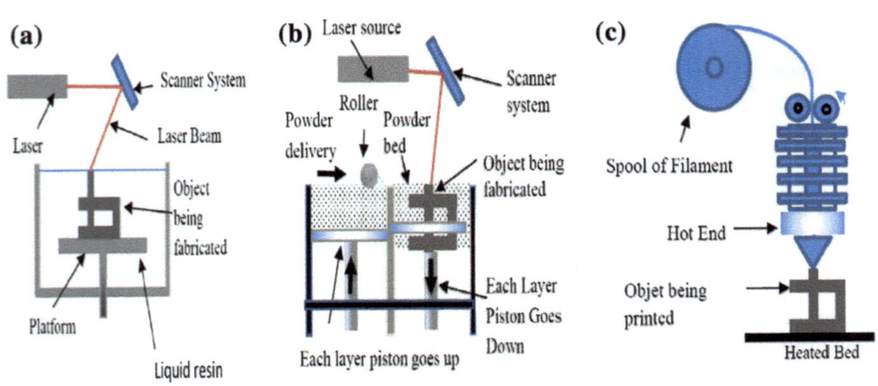

Fig. 18.1 3D printer technologies

layer, the platform lowers a layer thick on the *Z* axis and the next layer deposition begins. The printer may leave a secondary sacrificial material to support the construction of overhanging geometries.

The choice of printing material depends on the type of application and desired properties. Commonly applied materials include polylactic acid (PLA) as a stiff and environmentally friendly material, nylon, for soft applications, high-density polyethylene (HDPE) for the production of food compatible parts, and acrylonitrile butadiene styrene (ABS) a solution for tough parts with acceptable strength (Dawoud et al. 2016).

There are different techniques for creating layers on 3D printing, from jetting a binder into a polymeric powder, using an ultraviolet laser to harden a photosensitive polymer (stereolithographic), to using a laser to selectively melt metal or polymeric powder (laser sintering) (Campbell et al. 2011).

In the last decades, several industries developed a great variety of applications of 3D printing technology. Additive manufacturing processes make three-dimensional objects of almost any form, based on digital models through successive layers of material placed on a print platform under numerical control (Satyyanarayana and Jaya 2015) and generate less waste than traditional subtractive production methods.

Digital models are usually created using computer-aided design (CAD) software. In addition, 3D scanners automatically generate digital models from physical objects (just like 2D scanners are used to digitize photos, drawings or documents) (Rayna and Striukova 2016).

The adoption of additive manufacturing and other manufacturing technologies predicts a future in which chains value are shorter, smaller, more localized, more collaborative, and able to deliver substantial benefits of sustainability (Gebler et al. 2014). Figure 18.2 shows a generalized 3D printing process starting with the creation of a digital model and finishing with the part printed.

The accessibility of 3D printers for industrial and general public applications has grown dramatically in the past decade (Stanbury 2016), due mainly to the price

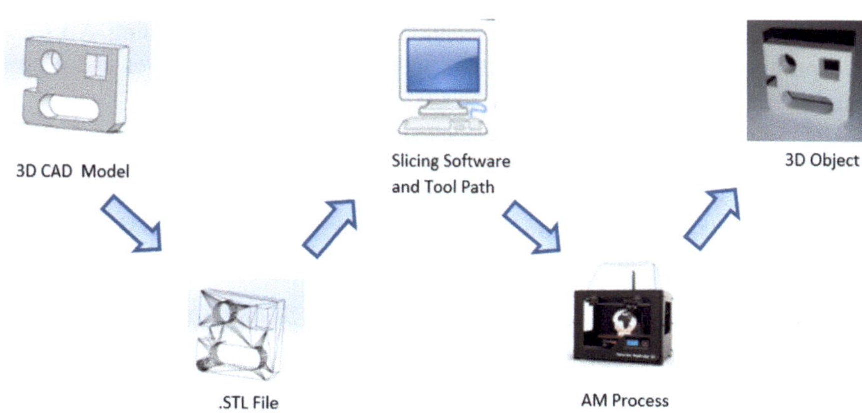

Fig. 18.2 Generalized additive manufacturing process

drop and to the vast quantity of applications found on 3D printers. Currently, this technology has a major impact on objects that could cost up to $30,000 dollars in the traditional industry; 3D printers give almost the same quality for less than $2000 (Evans 2012); however, the results do not allow their use in finished products, leaving their use mainly relegated to prototypes and pre-series (Cajal et al. 2013). The increase in the number of low-cost 3D printers on the market and a wide range of manufacturers makes the evaluation of the capabilities of different 3D systems paramount to the proliferation of low-cost additive manufacturing in terms of consumer confidence in this technology.

Since 3D printer home applications have increased substantially, a need to establish a tangible metric to evaluate the units' accuracy in an easier way is necessary. For the time being, we have been unable to identify a tool to evaluate the consumer worthiness of an additive manufacturing unit (Roberson et al. 2013).

Manufacturers of commercial 3D printers provide little or none information in terms of the quality of the manufactured product, such as reproducibility in extruder positioning, dimensional and geometric precision, surface texture, and crucial data for established elements (Nunez et al. 2015). Dimensional and geometric accuracy can be improved by reducing print speed but increasing the overall printing time (Galantucci et al. 2015).

This work evaluates the geometrical and dimensional performance of a commercial 3D printer (Flashforge Creator Pro) analyzing printed samples under vendor recommended specifications in order to determine the performance and accuracy of the printer using PLA material.

18.2 Materials and Methods

18.2.1 3D Printer

The Flashforge Creator Pro (FlasforgeUSA™) is a 3D printer with a dual extruder that uses fused deposition filament technology, capable of processing ABS and PLA printing material with a resolution up to 100 μm for each layer and a print space of 22.5 cm × 14.5 cm × 15.0 cm (4893.75 cm^3).

The print material selected for the analysis is a 1.75 mm diameter PLA filament fused at 180–220 °C during the printing process. This 3D printer is compatible with .STL and .OBJ files on Windows, Mac, and Linux operating systems. The extruder moves through *X*–*Y* planes putting fused PLA material on the print bed that moves down on the *Z* plane allowing the extrusion to place a new layer until a physical model is formed.

SolidWorks 2016 (Dassault Systemes 2016) is the software used to generate digital model saved as .STL file (stereolithography). This type of file created by the company 3D systems is compatible with several commercial software, and it is widely used for rapid prototyping and computer-aided manufacturing.

A .STL model is transferred to a Flashforge Creator Pro printer (FlashfogeUSA 2017) that uses a software called Replicator G (Hoeken et al. 2012) which makes the slicing operation, calculates the extruder path, and generates the G code to communicate the printer commands in order to build the 3D object.

18.2.2 DOE for Printer Settings Selection

The 2^k designs (k corresponds to the number of factors or printer settings to be analyzed and 2 refers to the number of levels for each factor) are particularly useful when there are many factors to be investigated, since they provide the smallest number of runs using k factors that can be studied in a complete factorial design. Because there are only two levels of each factor, we assume that the response is approximately linear over the range of the factor levels chosen (Montgomery 2001). The levels are "low" and "high".

Five factors could influence the dimensional performance of the Flashforge Creator Pro 3D printer. These are: Extruder temperature, the levels selected are within the range of melting temperature to print with PLA filament; infill density that refers to the material inside the contour of the piece printed; infill percentage could have a value of 0% (no material inside the printed part) to 100% where no empty space left; layer thickness is the height of each material layer stacked during the printing process; and number of shells refers to the perimeter thickness. Infill density and number of shells levels chosen give good mechanical properties, and print speed refers to the printer speed. Faster 3D print speed generally means a lower quality of the printed part and problems, as the filament tends to slip at higher speeds. Levels of layer thickness and print speed selected range result in a good quality print and a reliable printing process. Table 18.1 shows the low and high DOE levels factors and the recommended manufacturer settings.

In this study, we performed a 2^5 DOE full factorial plan in order to determine the main effects that affect the dimensional performance of the 3D printer with one replication for a total of 32 experiments. Table 18.2 shows the full factor array with all the possible combinations of the five factor levels performed in the experiment.

Table 18.1 Selected factors and its corresponding values

Factor	Low level (1)	High level (2)	Manufacturer recommended settings
Extruder temperature (°C)	190	220	220
Infill density (%)	20	60	10
Layer thickness (mm)	0.178	0.270	0.200
Print speed (mm/s)	40	70	80
No. of shells	1	3	1

Table 18.2 Full factorial 2^5 array

Std. order	Extrude temp. (°C)	Infill density (%)	Layer thickness (mm)	Print speed (mm/s)	Number of shells
1	190	20	0.178	40	1
2	220	20	0.178	40	1
3	190	60	0.178	40	1
4	220	60	0.178	40	1
5	190	20	0.270	40	1
6	220	20	0.270	40	1
7	190	60	0.270	40	1
8	220	60	0.270	40	1
9	190	20	0.178	70	1
10	220	20	0.178	70	1
11	190	60	0.178	70	1
12	220	60	0.178	70	1
13	190	20	0.270	70	1
14	220	20	0.270	70	1
15	190	60	0.270	70	1
16	220	60	0.270	70	1
17	190	20	0.178	40	3
18	220	20	0.178	40	3
19	190	60	0.178	40	3
20	220	60	0.178	40	3
21	190	20	0.270	40	3
22	220	20	0.270	40	3
23	190	60	0.270	40	3
24	220	60	0.270	40	3
25	190	20	0.178	70	3
26	220	20	0.178	70	3
27	190	60	0.178	70	3
28	220	60	0.178	70	3
29	190	20	0.270	70	3
30	220	20	0.270	70	3
31	190	60	0.270	70	3
32	220	60	0.270	70	3

18.2.3 Model Design

Figure 18.3 shows a designed rectangular workpiece of $L_1 = 22$ mm, $L_2 =$ 24 mm, $L_3 = 12$ mm, with an internal cubic feature of $L_4 = 10$ mm, $L_5 = 8$ mm, $L_3 = 12$ mm, in order to perform volume and roughness analysis. The calculated error

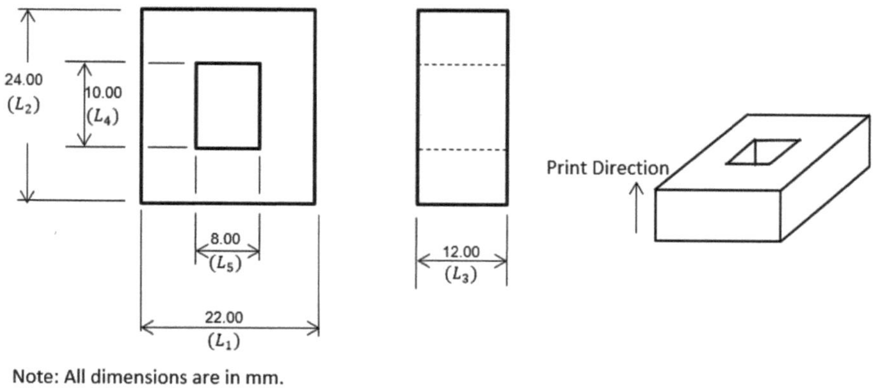

Fig. 18.3 Sample design with an internal square feature used for DOE

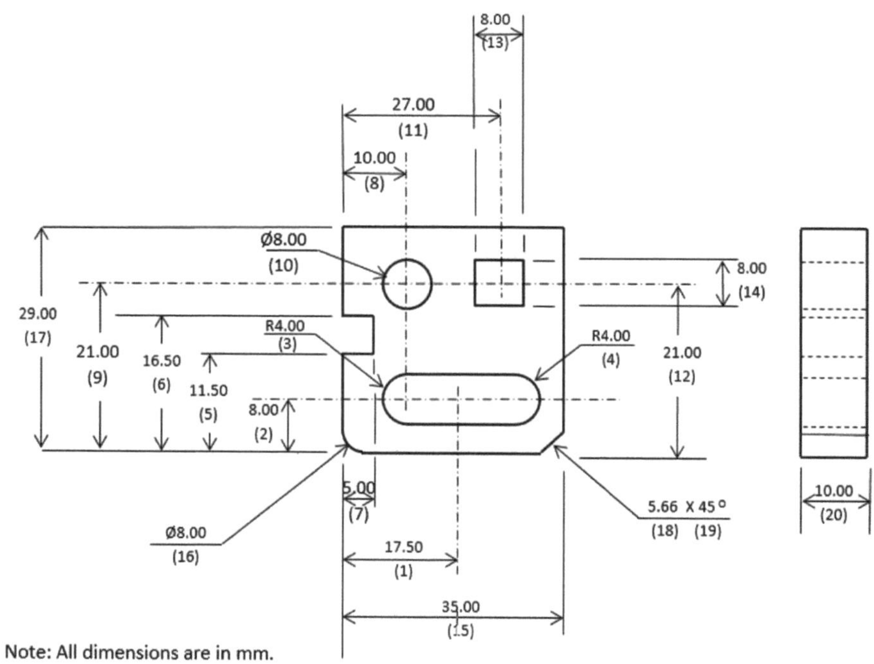

Fig. 18.4 Model drawing

is the difference between the measured and the nominal value of the workpiece volume, used as a response in order to determine the factor that influences the dimensional accuracy.

Figure 18.4 shows the sample model of 35 mm length, 29 mm width, and 10 mm thickness designed for dimensional and geometric analysis to facilitate the dimensional measurements including labeled dimensions as radio, chamfer, and basic internal features. We adjusted the Flashforge Creator Pro 3D printer according to the DOE's factor levels to obtain the optimal printer settings in order to optimize dimensional and roughness parameters using PLA printing material.

18.2.4 Dimensional and Roughness Measurement

We measured all printed samples using a PH14-A Mitutoyo profile projector; which uses a horizontal optical system with a resolution of 0.001 mm, and a 10X projection lens. The projector uses a QM DATA 200 coordinate system alignment function to align the workpiece and the axes. We measured roughness using an SJ210 Mitutoyo Surftest high resolution, with a measuring range of 17.5 mm on the *X* axis, and speed of 0.25, 0.5, 0.75 mm/s. We measured the roughness at vertical (lateral surface) and the horizontal plane (upper surface).

18.2.5 Dimensional Analysis

The measurement system analysis (MSA) workgroup manual contains an important method used to analyze the system variation called bias. The bias is a systematic error component of the measurement system that evaluates the difference between the observed measurements average and the reference value (accepted value of an artifact).

Figure 18.5 shows a graphical description of bias concept. In general, the bias or linearity error of a system is acceptable if it is not statistically significantly different from zero when compared to repeatability. Consequently, the repeatability must be acceptable when compared to the process variation in order to be useful (Automotive Industry Action Group 2010). In this work, the bias determines the location variation of the features measured to make the geometric study.

We performed the following steps in order to make this research:

1. For each reading, Eq. (18.1) calculates the difference between the measurements average and the reference value.

$$\text{bias}_i = x_i - \text{reference value} \tag{18.1}$$

2. Equation (18.2) calculates the average bias of the n readings.

$$\text{Average bias} = \bar{x} - \text{reference value} \tag{18.2}$$

3. Equation (18.3) computes the repeatability standard deviation (σ_r).

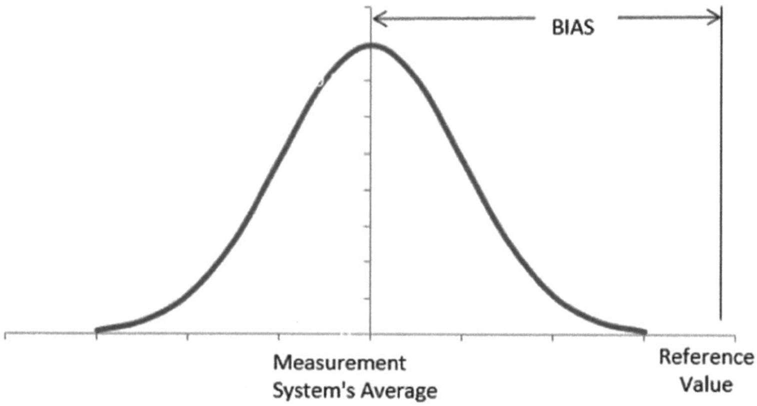

Fig. 18.5 BIAS. *Source* MSA

$$\sigma_r = \frac{\mathrm{Max}(x_i) - \mathrm{Min}(x_i)}{d_2^*} \tag{18.3}$$

$\mathrm{Max}(x_i)$ is the maximum measurement reading, $\mathrm{Min}(x_i)$ is the minimum measurement reading and d_2^* is obtained with $g = 1$ and $m = n$ of "Values Associated with the Distribution of Average Range" table located on Append C of the MSA manual.

4. Equation (18.4) determines the t statistics for the bias.

$$t_{\text{statistic}} = t_{\text{bias}} = \frac{\text{average bias}}{\sigma_r/\sqrt{n}} \tag{18.4}$$

5. Equation (18.5) gives the uncertainty for the bias σ_b.

$$\sigma_b = \sigma_r/\sqrt{n} \tag{18.5}$$

6. Equation (18.6) establishes the criteria: bias is acceptable at the $\propto$ level if zero falls between $1-\propto$ confidence bounds based on the bias value.

$$\text{Bias} - \left[\frac{d_2 x\,\sigma_b}{d_2^*}\left(t_{v,1-\propto/2}\right)\right] \le 0 \le \text{Bias} + \left[\frac{d_2 x\,\sigma_b}{d_2^*}\left(t_{v,1-\propto/2}\right)\right] \tag{18.6}$$

Standard t tables use $v = n - 1$ and $t_{v,1-\propto/2}$ (Walpole et al. 1999).

18.2.6 Geometric Analysis

Geometric tolerances specify the maximum variation allowed on form, orientation, and location of an element workpiece. A positional geometric tolerance is the width of the diameter or a tolerance zone within some surface axis of a hole or cylinder must remain in order to meet the functional part requirements or appropriate interchangeability (Warren and Duff 1994). True position tolerance increases the permissible tolerance in all directions; the real position takes into account all the relationships that must be kept in the assembly of interchangeable parts. We performed the geometric analysis locating the center of geometric figures and computing the difference against the drawing specification. The drawing specifies geometric tolerances in accordance with functional requirements although manufacturing and inspection requirements can also influence the geometric tolerance (ISO 1101 2012). Equation (18.7) calculates the deviation from the center of the internal features on the sample respect the center specified on the drawing.

$$\oplus = 2 * \sqrt[2]{(x_2 - x_1)^2 + (y_2 - y_1)^2} \tag{18.7}$$

The coordinates (x_1, y_1) are the values specified on the drawing and (x_2, y_2) are values measured on the profile projector.

18.2.7 Surface Analysis

The aim of this study is to find the optimal printer parameters to obtain the best quality surface by evaluating the factors selected for the dimensional performance. We used the same printed parts as in the DOE for dimensional and geometrical analysis because parts dimensions allow enough sampling surface for the roughness analysis. We used the same procedure and factor levels considered for dimensional accuracy to perform the surface analysis. Since there is a considerable roughness difference between lateral and upper surfaces, we performed the analysis in separate events.

The arithmetic mean surface roughness R_a is one of several different parameters that describe the deviation of a surface from an ideal level, and it is defined according to the international standard ISO 4287-1997 (ISO4287 1997); Eq. (18.8) defines the arithmetic mean surface roughness R_a and describes the deviation of the surface from a theoretical centerline R_{mean}. Figure 18.6 shows both R_a, and R_{mean}. Over a measurement length L_m (Stahl et al. 2011).

$$R_a = \frac{1}{L_m} * \int_0^{L_m} |y| * dx \tag{18.8}$$

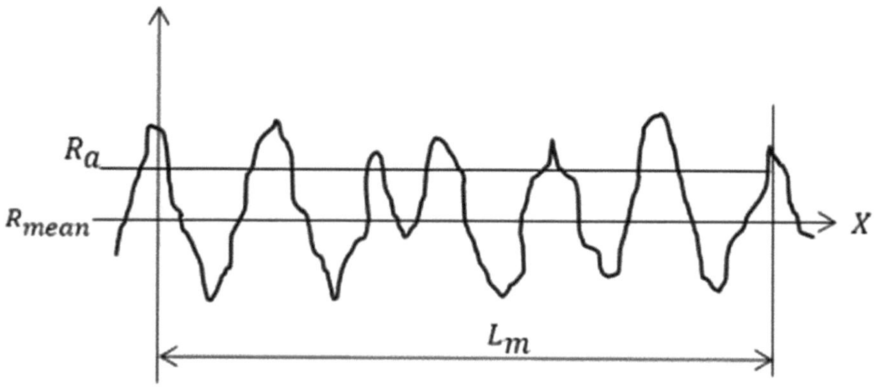

Fig. 18.6 Surface profile showing R_a and R_{mean}

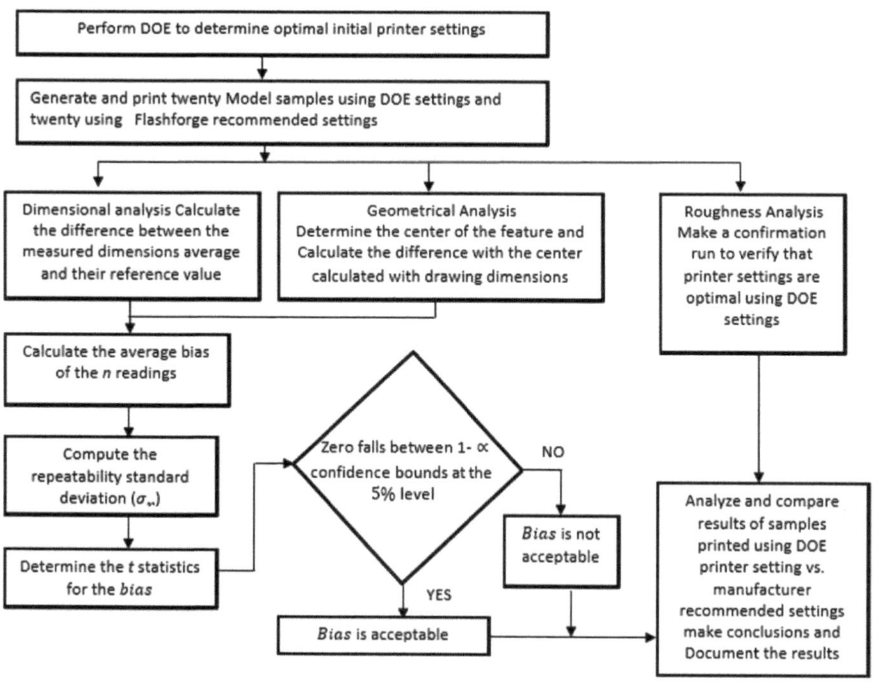

Fig. 18.7 Method diagram

We measured roughness at three locations along the lateral surface and the average of these measurements is the response to the factorial analysis, we applied the same procedure to the upper surface area.

Figure 18.7 shows all the steps taken to analyze Flashforge 3D printer dimensional and finished performance, beginning with a DOE to establish the initial 3D printer settings that give a better dimensional and roughness performance using PLA as print material. We printed 20 samples for the analysis using DOE optimal settings and 20 samples using Flashforge recommended settings. For roughness analysis, we performed a confirmation run to verify if DOE results are better than samples printed using the manufacturer recommended setup.

MSA methodology analysis determines the dimensional and geometrical performance of the Flashforge printer with a confidence level of 95%. We labeled each dimension in the drawing with a consecutive number to make easier the comparative process and keep traceability of the dimensions. We compared the average of the 20 dimensions against drawing specification. If zero falls between confidence bounds, the deviation of the reference value (bias) is acceptable.

18.3 Results

Table 18.3 shows the measured value of length, width, and height of the printed samples. Column seven shows the calculated volume $L_1 * L_2 * L_3$ and column eight gives the volume of the internal square feature $L_4 * L_5 * L_3$. Column nine shows the measured roughness for the lateral side of the part and column ten the upper surface roughness measurement expressed in μm. These values are the response in the factorial design to define the significant factors for dimensional and roughness analysis.

18.3.1 External Volume Response Analysis

Figure 18.8 shows the Pareto chart of the effects for the external volume. We used the hypothesis testing to verify if the factors and their interactions have a significant effect on the response. Analyzing the external dimensional test, the main effect (*D*) and the five terms interaction (ABCDE) have a significant effect over the external dimensions and consequently, they affect the volume involved with these dimensions.

When an interaction is significant, it has priority over main individual effects to select the optimal factor levels. Since five terms interaction is significant no main terms or interactions can be eliminated, so all factors should be controlled to optimize the response. Figure 18.9 shows the results obtained from Minitab17 response optimizer to determine the optimum level of each factor. Since all factors are significant for external dimensions, the factor levels should be adjusted according to the values listed in order to get the optimal dimensional performance.

Table 18.3 Nominal and measured values

Run	L_1 (mm) (X axis)	L_2 (mm) (Y axis)	L_3 (mm) (Z axis)	L_4 (mm) (Y axis)	L_5 (mm) (X axis)	Externa volume (mm^3)	Internal volume (mm^3)	R_a lateral (μm)	R_a upper (μm)
Nominal	22.00	24.00	12.00	10.00	8.00	6336.00	960.00		
1	21.95	24.02	11.88	9.92	8.02	6263.60	945.15	11.436	3.663
2	21.96	24.05	11.83	9.96	8.03	6247.87	946.15	12.333	6.921
3	21.96	24.05	11.85	9.95	7.97	6258.44	939.72	12.033	4.303
4	22.05	24.08	11.88	9.84	7.97	6307.85	931.69	12.164	4.266
5	21.95	24.09	11.80	9.94	7.94	6239.55	931.30	19.607	7.703
6	22.06	24.11	11.84	9.90	8.00	6297.30	937.73	18.941	5.748
7	21.95	24.15	11.79	9.87	7.90	6249.79	919.30	19.204	7.157
8	22.08	24.16	11.78	9.70	7.72	6284.07	882.13	17.679	6.213
9	21.94	24.07	11.89	9.98	8.03	6279.06	952.86	11.945	5.843
10	22.09	24.13	11.86	10.08	8.05	6321.76	962.37	12.993	8.746
11	22.01	24.10	11.90	9.98	8.02	6312.25	952.47	11.027	4.846
12	22.17	24.17	11.83	9.97	7.90	6339.09	931.77	11.899	6.708
13	21.98	24.22	11.88	9.91	7.89	6324.38	928.90	18.110	6.977
14	22.12	24.19	11.78	9.99	8.06	6303.28	948.52	18.155	7.557
15	22.10	24.18	11.82	9.95	7.76	6316.35	912.65	17.767	7.652
16	22.23	24.26	11.81	9.84	7.85	6369.13	912.25	18.404	9.659
17	21.91	24.01	11.84	10.06	8.03	6228.54	956.46	19.201	4.979
18	21.97	24.09	11.87	9.99	8.06	6282.28	955.77	11.522	7.062
19	21.94	24.04	11.86	9.98	7.97	6255.41	943.35	14.005	4.156
20	21.97	24.04	11.81	10.06	7.99	6237.56	949.28	11.892	5.468
21	22.03	24.15	11.88	9.79	7.80	6320.45	907.18	21.158	11.443
22	22.09	24.16	11.75	9.81	7.94	6270.91	915.22	19.394	11.406
23	21.99	24.15	11.80	9.82	7.92	6266.49	917.74	19.593	9.163
24	22.07	24.11	11.74	9.67	7.69	6246.94	873.01	17.923	9.119
25	21.96	24.05	11.91	9.95	7.97	6290.12	944.48	10.911	7.601
26	22.07	24.12	11.84	9.90	7.85	6302.77	920.15	13.215	12.460
27	21.99	24.05	11.85	10.00	7.90	6266.99	936.15	12.285	6.168
28	22.10	24.11	11.84	10.03	8.05	6308.72	955.98	11.324	10.449
29	21.99	24.07	11.87	10.06	8.08	6282.78	964.85	11.270	7.532
30	22.08	24.19	11.74	9.98	8.02	6270.51	939.66	17.866	7.828
31	21.98	24.12	11.90	9.87	7.88	6308.88	925.53	18.083	10.436
32	22.14	24.08	11.80	9.94	8.00	6290.95	938.34	18.322	13.417

18.3.2 Internal Volume Response Analysis

Figure 18.10a shows the Pareto chart before we eliminated all nonsignificant effects, and Fig. 18.10b shows the model with all the nonsignificant terms removed. The selections begin with higher order interactions until it is not possible to remove

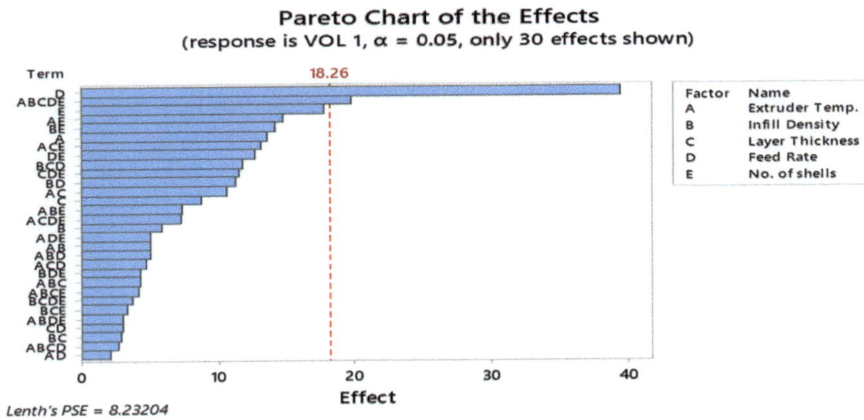

Fig. 18.8 Pareto chart of the factors effects of external volume

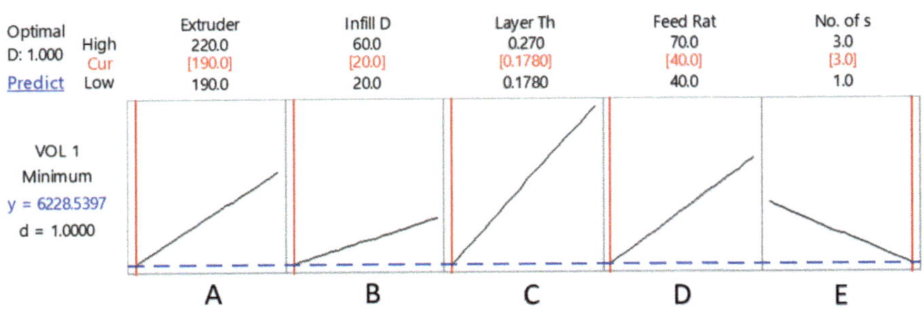

Fig. 18.9 Optimal values for external volume using optimizer Minitab tool

the terms since they are included in the significant effects. Figure 18.10b helps to determine which factors should be controlled since they have a direct impact on the response. Figure 18.11 shows the results obtained to optimize the response for internal dimensions. For the extruder temperature (A), there is almost no difference between low and high levels for this factor. Thus, this is the only nonsignificant term for the internal dimensions and it may be low or high.

18.3.3 Lateral Surface Roughness Response Analysis

Figure 18.12a shows the Pareto chart before the elimination of all nonsignificant effects for the lateral surface roughness. Figure 18.12b shows the results of the analysis using Minitab after all the nonsignificant terms have been removed for lateral surface roughness test, two main effects (C and D) and one interaction

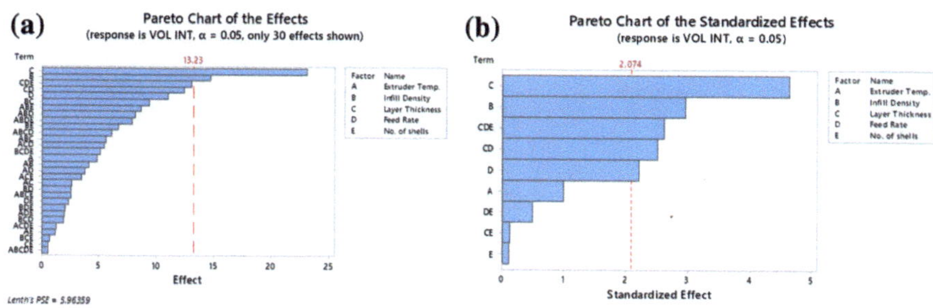

Fig. 18.10 Pareto chart of the factor effects of internal volume

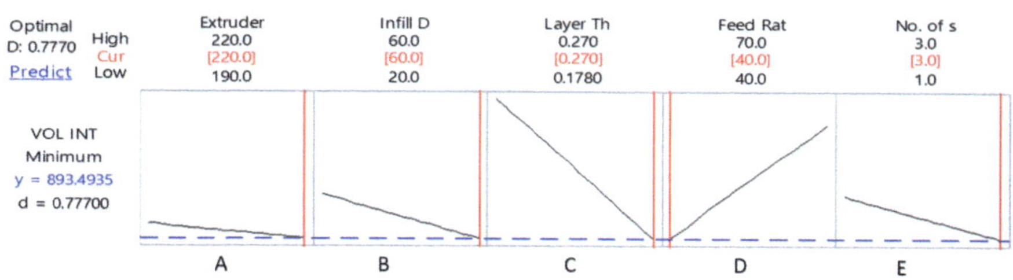

Fig. 18.11 Optimal values using optimizer Minitab tool for internal dimensions

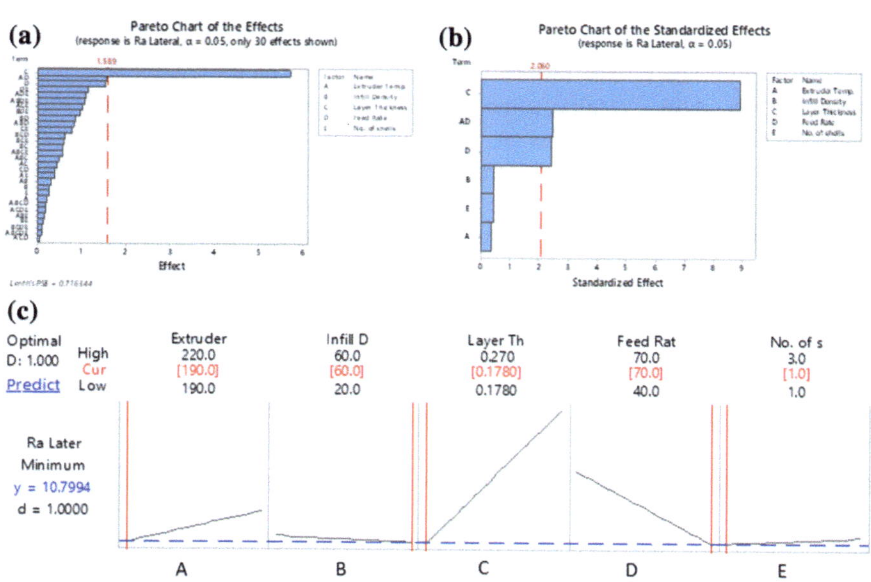

Fig. 18.12 Pareto charts and optimum levels for lateral surface roughness using Minitab optimizer

Table 18.4 Optimal factor values for volume and roughness

Factor	External volume	Internal volume	Ra lateral	Ra upper	Selected levels
Extruder temperature (°C)	190	220[a]	190[a]	190	190
Infill density (%)	20	60	60[a]	60	60
Layer thickness (mm)	0.178	0.270	0.178	0.178	0.178
Print speed (mm/s)	40	40	70	40	40
No. of shells	3	3	1[a]	3	3

[a]Nonsignificant

(AD) have a significant effect over the lateral surface roughness. According to Fig. 18.12b, significant and nonsignificant terms affect the lateral surface roughness response.

Layer thickness, extruder temperature, and print speed are the factors that should be controlled while the infill density and number of shells do not have a significant impact on lateral roughness response, as it can be seen in Fig. 18.12c where there is no significant difference between low and high level for the factors B and E. Table 18.4 shows recommended settings for optimal values to minimize lateral surface roughness.

18.3.4 Upper Surface Roughness Response Analysis

Figure 18.13a shows the results of the upper surface roughness analysis. Figure 18.13b shows results without nonsignificant effects for upper surface roughness test. Four main effects (A, C, D, and E), five two-term interactions (CD, AC, BC, AD, and BD), and two three terms interactions (BCD and CDE) have a significant effect over the response. Figure 18.13c shows the results obtained to optimize upper surface roughness. Since all factors appear at least one time on the interactions, all of them should be controlled according to the optimizer results. Table 18.4 shows the settings of each factor to optimize the desired response, numbers marked with an asterisk corresponds to nonsignificant factors adjusted to low or high levels without any effect on the response.

18.3.5 Dimensional Analysis

We measured all samples to analyze the dimensional and geometric accuracy of the printed pieces. We found an important dimensional variation in relation to drawing specifications that could affect the part functionality. Table 18.5 shows the

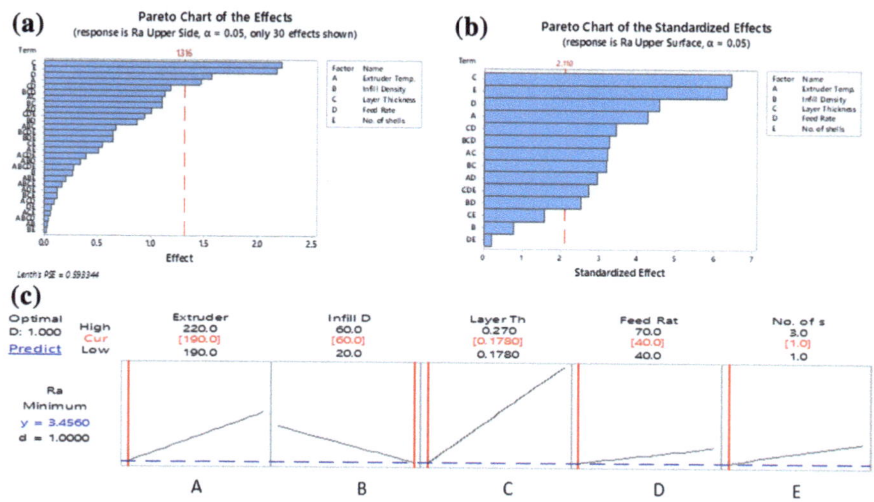

Fig. 18.13 Optimum levels for upper surface roughness using Minitab optimizer

dimensional results measured on the profile projector using DOE printer settings (columns 3–7), and Flashforge Creator Pro settings (columns 8–12). The first column shows the dimension label according to the drawing, the next column shows the dimension expressed in mm (reference value), the third column gives the average of all measurements, and the fourth column shows the average bias, while the fifth and sixth columns include the upper and lower limits of the acceptance criteria, and the seventh column is the decision-maker. Table 18.5 shows the same information for Flashforge settings on the last five columns.

Table 18.6 shows the variation expressed in the average percentage of measures for each dimension, and the significant deviations between the ideal dimension and the real one, on DOE and Flashforge Creator Pro settings. Most dimensions are below the drawing specifications, due mainly to PLA material properties that shrink during the cooling process from filament fusion temperature (180–220 °C) to room temperature. The effect of drastic temperature changes during the printing process could be reduced by heating the building print bed; however, this affects only the first layers of the printed piece.

18.3.6 Geometrical Analysis

Table 18.7 shows the results obtained from the measurements of the geometric dimensions of the model. The first column shows the feature shape, next column identifies the measurements used to calculate deviation in relation to the center point of the feature, third to sixth columns show to the measurements of the printed pieces under DOE adjustments, and the last four columns refer to the manufacturer recommended settings.

Table 18.5 Results of dimensional analysis

Dim.	Ref. value (mm)	DOE settings					Flashforge settings				
		Ave. reading (mm)	Ave. bias (mm)	Lower limit (mm)	Upper limit (mm)	Bias is accepted	Ave. reading (mm)	Ave. bias (mm)	Lower limit (mm)	Upper limit (mm)	Bias is accepted
1	17.50	17.523	0.0235	0.0006	0.040	NO	17.569	0.069	0.034	0.104	NO
2	8.00	8.087	0.087	0.027	0.146	NO	8.064	0.064	0.037	0.090	NO
3	4.00	3.869	−0.267	0.130	−0.059	NO	3.843	−0.156	−0.191	−0.121	NO
4	4.00	3.826	−0.173	−0.199	−0.147	NO	3.884	−0.116	−0.166	−0.065	NO
5	11.50	11.752	0.252	0.213	0.290	NO	11.643	0.143	0.105	0.182	NO
6	16.50	16.465	−0.034	−0.074	−0.005	YES	16.362	−0.137	−0.166	−0.109	NO
7	5.00	4.975	−0.025	−0.046	−0.003	NO	4.945	−0.055	−0.079	−0.030	NO
8	10.00	10.092	0.092	0.072	0.111	NO	10.034	0.034	−0.001	0.068	YES
9	21.00	21.015	0.015	−0.028	0.058	YES	21.031	0.031	0.128	0.050	NO
10	8.00	7.644	−0.356	−0.385	−0.326	NO	7.671	−0.328	−0.375	−0.281	NO
11	27.00	26.948	−0.051	−0.076	−0.026	NO	26.945	−0.055	−0.087	−0.022	NO
12	21.00	20.987	−0.013	−0.058	0.032	YES	21.058	0.058	0.028	0.087	NO
13	8.00	7.675	−0.324	−0.348	−0.300	NO	7.759	−0.240	−0.264	−0.216	NO
14	8.00	7.623	−0.377	−0.399	−0.354	NO	7.704	−0.296	−0.329	−0.262	NO
15	35.00	35.094	0.094	0.055	0.132	NO	35.112	0.112	0.077	0.147	NO
16	8.00	8.203	0.203	0.040	0.365	NO	8.352	0.352	0.124	0.579	NO
17	29.00	29.140	0.140	0.093	0.187	NO	28.635	−0.346	−1.446	0.753	YES
18	5.66	6.008	0.348	0.288	0.407	NO	5.777	0.117	0.048	0.185	NO
19	45.00^{O}	45.269^{O}	0.269	−0.130	0.551	YES	45.743^{O}	0.743	0.447	1.038	NO
20	10	9.921	−0.078	−0.104	−0.052	NO	9.901	−0.099	−0.141	−0.056	NO

Table 18.6 Percentage of variation in relation to reference value

Dim.	Ref value (mm)	Variation DOE settings (%)	Variation Flashforge settings (%)	Dim.	Ref value (mm)	Variation DOE settings (%)	Variation Flashforge settings (%)
1	17.50	0.13	0.40	11	27.00	−0.19	−0.20
2	8.00	1.09	0.80	12	21.00	−0.06	−0.28
3	4.00	−6.68	−3.90	13	8.00	−4.05	−3.00
4	4.00	−4.33	−2.90	14	8.00	−4.71	−3.70
5	11.50	2.19	1.24	15	35.00	0.27	0.32
6	16.50	−0.21	−0.83	16	8.00	2.54	4.40
7	5.00	−0.50	−1.10	17	29.00	0.48	−1.19
8	10.00	0.92	0.34	18	5.66	6.15	2.07
9	21.00	0.07	0.15	19	45.00^{O}	0.60	1.65
10	8.00	−4.45	−4.10	20	10	−0.78	−0.99

Third and seventh columns show the average actual position expressed in mm. In this table, we can observe a greater deviation on the internal features compared to the external feature identified by (5, 6) position deviation is only 0.02 mm compared with internal features where deviation exceeds the 0.2 mm. Columns four and five show the bias upper and lower limits. Column six shows the decisions for the DOE settings, while Flashforge settings are shown in the last three columns.

18.4 Conclusions

We concluded that the MSA bias concept could be used as an alternative, and an easy to use, method to evaluate the accuracy of commercial 3D printers that use fused filament deposition technique in order to determine the dimensional and geometric performance of the printer with specific materials, since this information is not given by the manufacturer in most cases.

We did not find a significant difference between the DOE optimal settings printer adjustments and manufacturer recommended settings for dimensional and geometrical analysis. Bias acceptance criteria for dimensional analysis rejected most of the evaluated dimensions. However, 4 out of 20 parts printed using DOE settings meet the acceptance criteria while only two dimensions of printed pieces using manufacturer recommended setup meet the acceptance bias criteria. We found comparing the average bias for each setup that about half of the dimensions of the DOE setup are smaller than the manufacturer recommended setup. The same situation is present in the geometric analysis, where only one out of four features evaluated meets the bias acceptance criteria using manufacturer recommended setup and none using DOE parameters setup, the average bias presents same behavior

Table 18.7 Geometrical analysis

Feature	Dim. (x_1, y_1)	DOE settings				Flashforge settings			
		⌖ Average (mm)	Lower limit (mm)	Upper limit (mm)	Bias is accepted	⌖ Average (mm)	Lower limit (mm)	Upper limit (mm)	Bias is accepted
	(1, 2)	0.244	0.162	0.328	NO	0.230	0.178	0.282	NO
	(5, 6)	−0.108	−0.143	−0.073	NO	0.003	−0.028	0.022	SI
	(8, 9)	0.248	0.156	0.191	NO	0.167	0.126	0.208	NO
	(11, 12)	0.310	0.217	0.306	NO	0.215	0.167	0.263	NO

than the dimensional analysis, so the conclusion is that there is no difference between the DOE and the manufacturer recommended setup.

3D printers are able to process different types of materials and produce at low cost with enough quality for prototyping parts; 3D printers have become a useful tool for companies that have capabilities to handle this technology since it facilitates and encourages innovation. For this reason, there is a high interest of companies to evaluate the accuracy of low-cost 3D printers in order to improve their processes.

Surface roughness depends on the print direction of the parts; therefore, it is important which side of the part is the upper face and which are the lateral sides. For lateral roughness, layer thickness, extruder temperature, and print speed are the factors that should be controlled while infill density and number of shells are nonsignificant factors. Nevertheless, all five factors evaluated in DOE should be controlled to get the optimal roughness on the upper surface of the printed piece.

Further analysis is required to determine causes of variation, in order to take actions to correct them and improve the quality of printed parts. The α level can also be modified but it depends on the level of sensitivity associated with the loss of function of the printed part. One source of variation is the drastic temperature change during the print process when the material (PLA) changes from fused to room temperature, the cooling process shrinking causes this variation. However, dimensional variation does not follow a consistent pattern, and percentage of variation in some dimensions could be too high for some applications where a dimensional performance is important for the part functionality.

We recommend performing studies with different filament materials in order to find which filament material gives a better dimensional performance, based on the different thermal properties.

References

Automotive Industry Action Group (2010) MSA-4 Measurement systems analysis. http://www.aiag.org/store/publications/details?ProductCode=MSA-4

Cajal C, Santoralia J, Velazquez J, Aguado S, Abajez J (2013) Volumetric error compensation technique for 3D printers. Proc Eng 63:126–132. https://www.sciencedirect.com/science/article/pii/S1877705813014896

Campbell T, Williams C, Ivanova O, Garett B (2011) Could 3D printing change the world?. Atlantic council, strategic foresight report, Oct 2011, pp 1–14

Dassault Systems (2016) SolidWorks corporation. Available at: http://Solidworks.es/sw date accessed: 14 06 2016

Dawoud M, Taha I, Ebeid S (2016) Mechanical behaviour of ABS: an experimental study using FDM and injection moulding techniques. J Manuf Process 21:39–45. https://www.sciencedirect.com/science/article/pii/S1526612515001395

Evans B (2012) Technology in action practical 3D printers: the science and art of 3D printing. Apress Media, New York, USA

FlashfogeUSA (2017) FlashforgeUSA. Available at: http://www.flashforge-usa.com date accessed: 16 05 2017

Galantucci L, Bodi I, Kacani J, Laveccia F (2015) Analysis of dimensional performance for a 3D open source printer based on fused deposition modeling technique. Proc CIRP 28:82–87. https://doi.org/10.1016/j.procir.2015.04.014

Gebler M, Schoot-Uiterkamp A, Visser C (2014) A global sustainability perspective on 3D printing technologies. Energy Policies 74:215–243. https://doi.org/10.1016/j.enpol.2014.08.033

Hoeken Z, Kintel M, Mayer A, Mets M (2012) ReplicatorG. Lowering the barrier to 3D printing at http://replicat.org date accessed: 18 04 2017

ISO 1101 (2012) Geometrical product specifications (GPS)—geometrical tolerancing—tolerances of form, orientation, location and run-out. Available at: https://www.iso.org/obp/ui/es/#iso:std:iso:1101:ed-4:v1:en date accessed: 19 06 2017

ISO4287 (1997) Geometrical product specifications (GPS)–Surface texture: profile method—terms, definitions and surface texture parameters. Available at: https://www.iso.org/standard/10132.html date accessed: 18 04 2017. (in Switzerland)

Montgomery DC (2001) Design and analysis of experiments. Willey, New York

Nunez P, Rivas A, Garcia EBE, Sanz A (2015) Dimensional and surface texture characterization in fused deposition modelling (FDM) with ABS plus. Proc Eng 132:856–863. https://doi.org/10.1016/j.proeng.2015.12.570

Rayna T, Striukova L (2016) From rapid prototyping to home fabrication: how 3D printing is changing business model innovation. Technol Forecast Soc Change 102:214–224. https://doi.org/10.1016/j.techfore.2015.07.023

Roberson D, Espalin D, Wicker R (2013) 3D printer selection: a decision-making evaluation and ranking model. Virtual Phys Prototyping 8(3):201–212. https://doi.org/10.1080/17452759.2013.830939

Satyanarayana B, Jaya K (2015) Component replication using 3D Printing technology. Proc Mater Sci 10:263–269. https://doi.org/10.1016/j.mspro.2015.06.049

Stahl JE, Schultheiss F, Hagglund S (2011) Analytical and experimental determination of the Ra surface roughness during turning. Proc Eng 19:349–356. https://doi.org/10.1016/j.proeng.2011.11.124

Stanbury JIM (2016) 3D printing with polymers: challenges among expanding options and opportunities. Dent Mater 32:54–64. https://doi.org/10.1016/j.dental.2015.09.018

Walpole R, Myers R, Myers S (1999) Probability and statistics for engineers. Prentice Hall Hispanoamericana S.A. Mexico (in Spanish)

Warren J, Duff J (1994) Fundamentals of drawing in engineering. Prentice Hall Inc, Mexico (in Spanish)

Chapter 19
Organizational Systems Convergence with the Industry 4.0 Challenge

Magdiel Pérez-Lara, Jania Astrid Saucedo-Martínez, José Antonio Marmolejo-Saucedo and Tomás Eloy Salais-Fierro

Abstract The concept of Industry 4.0 consists of the digital technologies introduction in companies. It is the way to call the phenomenon of digital transformation applied to the production industry. The problem faced by organizations is that they assimilate the global idea of Industry 4.0 and the technical aspects associated with it. Industry 4.0 is highly linked to technologies such as the Internet of things, which favors predictive analysis, based on the information collected through these solutions, making it easier for companies to anticipate consumer requests. It is necessary that companies can readjust and plan their operations quickly and accurately to respond to the new demands of consumers. The products proliferation will continue to increase in the coming years, in a context that will allow greater customization. Therefore, companies must take advantage of market analysis to improve their supply chains and meet the demands of their users, since the customization of products requires a clear connection with production capabilities. The objective of the research is to describe the properties of Industry 4.0, to propose a measurement system according to the principles of the supply chain, by means of the descriptive and comparative research method.

Keywords Organizational systems · Supply chain · Industry 4.0
Business model · Maturity level

M. Pérez-Lara (✉) · J. A. Saucedo-Martínez · T. E. Salais-Fierro
Department of Logistic and Supply Chain, Universidad Autónoma de Nuevo León, Av. Pedro de Alba s/n. Col. Ciudad Universitaria, San Nicolás de Los Garza, Nuevo León, Mexico
e-mail: magdiel.perezlr@uanl.edu.mx

J. A. Marmolejo-Saucedo
Facultad de Ingeniería, Universidad Panamericana, Augusto Rodin 498, México, 03920 Ciudad de México, Mexico

J. L. García Alcaraz et al. (eds.), *Best Practices in Manufacturing Processes*,
https://doi.org/10.1007/978-3-319-99190-0_19

19.1 Introduction

The production systems transformation has been presented at different rates depending on market needs and available tools. Within the organizational and business area we can conceive all activities as operations set linked to computer systems and network information management, allowing efficient flow. In management, it is necessarily new techniques application that provides value generation and market stability with trends to increase. The Industry 4.0 development is framed in three main scenarios:

- *Operations*: The rethinking of these is not intended to mark a radical change in the organization operations, but to adapt the technological and administrative systems to a new level, in which the tools enhance the value of the activities to the product or service creation.
- *Organization*: It is a significant criterion in the company operation; it is involved in structural innovations in which the departments' synergy influences the performance of the human capital; therefore, the company vision must be global and open to the low change concepts of integration; through organizational structures contributing to the value creation, it requires excellent information management and teamwork.
- *Customers*: With the availability of multiple technological tools, the customers demand is increasing gradually, in terms of quality and time, which triggers an operations series in organizations throughout its value creation network to fulfill its market.

The perspective of the evolution of the industry and the incorporation of technologies has generated a compilation of new challenges. These challenges must be fulfilled and addressed to achieve stability and permanence in the market, from the point of view of world economies, with a high emphasis on production, as well as on the main aspects on which the basis of intelligent manufacturing rests, with the approach of strategies and platforms management.

19.2 Methodology

The methodology used in the development of the research is of the descriptive type, and subsequently, the information obtained was used in a comparison as validation of criteria.

Table 19.1 shows the sequence of the methodology.

Table 19.1 Sequence description

Step	Description
Literature review	Data collection—Sampling strategy—Material revision/evaluation
Proposal design	The design of measurement levels of industry level 4.0 around the integration systems
Comparative	Presents a comparison of the study carried out from the point of view of supply chain, with existing models of measurement

19.3 Literature Review

19.3.1 Industry 4.0

Industry 4.0 refers to a new model of organization and control of the value chain through the life cycle of the product and throughout the manufacturing systems supported and made possible by information technologies (Schlechtendahl et al. 2014; Simionis et al. 2016; Botthof and Hartmann 2015; Wolter et al. 2015).

The term Industry 4.0 is widely used in Europe, it was coined in Germany. It is common to refer to this concept with terms such as "Smart Factory" or "Industrial Internet". In short, it is the application to the industry of the Internet of things (IoT) model. All these terms have in common the recognition that manufacturing processes are in a process of digital transformation, an "industrial revolution" produced by the advancement of information technologies and, particularly, of information technology and software (Saucedo-Martínez et al. 2017; Kang et al. 2016; Toro et al. 2015; Stojkić et al. 2016; Anderl 2014). Figure 19.1 shows the industry evolution.

Next, Fig. 19.2 shows the prominent technologies in Industry 4.0.

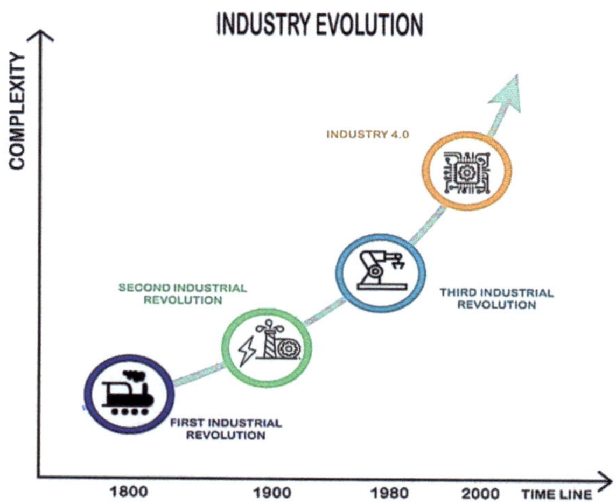

Fig. 19.1 The industry evolution

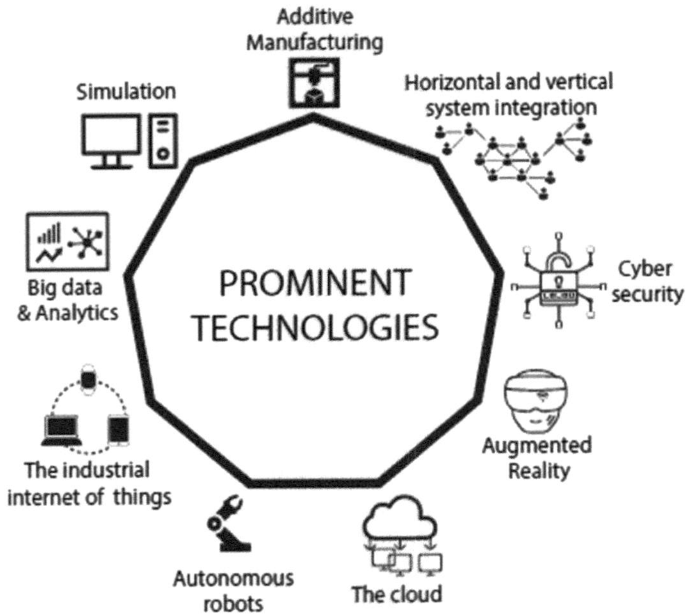

Fig. 19.2 Group technologies in Industry 4.0

- *Big Data and Analytics*: Analytics based on large data sets has only recently emerged in the world of manufacturing, where it optimizes production quality, saves energy and improves equipment service (Chen et al. 2015; Esmaeilian et al. 2016; Kagermann 2015; Matties 2016; Schuh et al. 2014).
- *Autonomous Robots*: Manufacturers in many industries have long used robots to tackle complex tasks, but robots are evolving for even greater utility (Li 2016; Gilchrist 2016). They are becoming more autonomous, flexible, and cooperative (Stock and Seliger 2016; Matties 2016; Thames and Schaefer 2016; Fedorov et al. 2015; Brettel et al. 2016a, b; Bagheri et al. 2015; Schlott 2013; Botthof and Hartmann 2015; Kolberg and Zühlke 2015).
- *Simulation*: In the engineering phase, three-dimensional simulations of products, materials, and production processes are already used, but in the future the simulations will be used more widely in the operations of the plant (Chen 2016; Abduaziz et al. 2015; Lim et al. 2014; Wang et al. 2016; Avci and Selim 2016; Mosterman and Zander 2016; Moreno et al. 2016; Li 2016; Syntetos et al. 2016; Ojha et al. 2016; Laghari and Niazi 2016).
- *Horizontal and Vertical System Integration*: Companies, suppliers, and customers are rarely closely linked. With Industry 4.0, companies, departments, functions and capabilities will become much more coherent, as data integration networks between companies and universals will evolve and allow truly automated value chains (Möller 2016; Kang et al. 2016; Saucedo-Martínez et al. 2017; Igor et al. 2016; Shafiq et al. 2015, Mazark and Huemer 2015; Stock and Seliger 2016; Prause and Weigand 2016; Senter and Flynn 1999).

- *The Industrial Internet of Things*: With the industrial Internet of things, more devices—even unfinished products—will be enriched with integrated computing and connected using standard technologies (Saucedo-Martínez et al. 2017; Rizzardi et al. 2016; Igor et al. 2016; Mashal et al. 2016; Moosavi et al. 2016; Seo et al. 2016; Khan and Turowski 2016; Macaulay 2017; Lin et al. 2016; Shafiq et al. 2015; Lu et al. 2015; Decker et al. 2008).
- *Cybersecurity*: With the increase in connectivity and the use of standard communications protocols that come with Industry 4.0, the need to protect critical industrial systems and manufacturing lines from cybersecurity threats increases dramatically (Möller 2016; Wildgoose 2016; Thames and Schaefer 2016; Weber and Studer 2016).
- *The Cloud*: Companies are already using cloud-based software for some business and analytics applications, but with Industry 4.0, more production-related companies will require more data exchange between sites and business boundaries (Albers et al. 2016; Igor et al. 2016; Chen 2016; Salcito et al. 2014; Ahn 2016; Shafiq et al. 2015; Stock and Seliger 2016; Li et al. 2017; Atanasov et al. 2015; Colom et al. 2017; Raza et al. 2017; Alberti et al. 2017; Khan and Turowski 2016; Macaulay 2017; Bian et al. 2016).
- *Additive Manufacturing*: The additive manufacturing methods will be widely used to produce small batches of customized products that offer construction advantages, such as complex and lightweight designs (Yao and Lin 2016; Vogel-Heuser et al. 2017; Scholz et al. 2016; Wahlster and Beste 2016; Stock and Seliger 2016; Brettel et al. 2016a, b). High performance decentralized additive manufacturing systems will reduce transportation distances and available stocks (Vogel-Heuser et al. 2017; Scholz et al. 2016).
- *Augmented Reality*: The role of data is in this scheme as a basis for the generation of knowledge for decision-making, taking as a reference the data generated involved in the development of activities, including it in decision-making is essential, the big data are closely related to data mining, and share information management elements to generate knowledge (Gilchrist 2016; Kang et al. 2016; Kans et al. 2016; Kolberg and Zühlke 2015; Vyas et al. 2016; Shafiq et al. 2015).

If decisions are made without analysis, the response time would be prolonged and most cases inefficient. While all the abovementioned technologies are important, it should be considered in supply chain context.

19.3.2 Industry 4.0 and IoT

IoT dates to 1999 and was coined by British technology pioneer Kevin Ashton. It defines how physical objects become "intelligent" by connecting them to the Internet using ubiquitous sensors. We could define it as the interconnection between manufacturing and distribution.

Its purpose is to be a means for machines, robots, sensors and human beings to connect independently of their geographical location. Currently, the environment is completely interconnected, fast and accessible; we will see that the IoT offers various intelligent business systems that take advantage of the use of data in the cloud to deliver more and better solutions in manufacturing and logistics processes. An important point to consider for the IoT, are has adequate analytical tools, like Software as a Service (SaaS) applications, Industry 4.0 will undoubtedly increase its productivity more and more, optimizing the weak points in each of the processes.

With the use of mobile applications, it is possible to monitor the status of assets, inventory management and activities of people throughout the value chain, and most importantly, in real time. With IoT, it is possible to measure the performance of these assets and make changes in the processes that are currently carried out and that will be done in the future.

Among the benefits of implementing these services in the cloud are those that allow automating business processes to eliminate manual interventions and avoid possible errors, improve quality and visibility, and reduce costs.

The incorporation of Industry 4.0 technologies to the company, allows obtaining and processing a large volume and information, obtained throughout the life cycle of the product, for decision-making and the development of new business models.

19.3.3 Business Model

The incorporation of Industry 4.0 technologies in the company not only results in an improvement in the optimization and efficiency of industrial processes and in the quality of products but also offers new business opportunities.

A business model, also known as business design, is the planning that a company makes with respect to the income and benefits it seeks to obtain. The guidelines are established to follow to attract customers, define product offers and implement advertising strategies, among many other issues related to the configuration of the company's resources.

When establishing the business model, it is important that the person in question analyze the company in depth and answer a series of questions, since based on the answers, one or the other type of business model can be implemented. In this case, it is important that you establish whether you have competition in that service or product that you have, what makes you different from the rest of business rivals, how you are going to get customers, how growth will occur and how you will win.

The logistics and distribution of industrial products are one of the processes that most changes and that more business opportunities will be presented soon new models in delivery services using drones or the sale of personalized products made on demand through electronic commerce.

19.3.4 Smart Supply Chains

Smart supply chains will be highly automated and integrated and, again, made possible through the integration of software and communications in the industry.

The basis for these networks in production environments and engineering platforms connected in a network together with interfaces between companies. Also, in this aspect, the base is the computer science and the software will be a decisive and good sample of it is the leadership of SAP in the impulse of the Industry 4.0. The connected supply chain is a central piece in any Industry 4.0 strategy. To manage the increasing complexity of supply chains, physical flows are replicated on digital platforms (Ji et al. 2016; Fedorov et al. 2015).

Throughout the supply chain, cyber-physical system (CPS) generates real-time data on their position and status. This digitalization allows automating the processes of the Supply Chain and identifying the product throughout the production process allowing the manufacturer to be more sensitive to changes in orders. The visibility of the movements of the supply network provides transparency. It allows recognizing inefficiencies and risks and increases the robustness and the ability to respond to incidents, increase reliability, and reduce costs.

Production could be organized according to a supply–demand model where the capacity of systems is supply and demand arises from the orders that must be met. Each CPPS could decide its production schedule (based on its processing time, delivery dates or profit or sustainability objectives).

19.4 Discussion and Analysis

19.4.1 An Assessment Proposal from the Systems Integration

Vertical and horizontal integration and the relationship of these with the business model (Chen et al. 2015; Tripathy et al. 2016; Neeliah and Seetanah 2016; Thomé et al. 2014; Dombrowski and Wagne 2014; Moreno et al. 2016), observe Tables 19.2 and 19.3, can be identified the relationship between supply chain processes and their relationship with horizontal and vertical integration.

The environment in which the organizations work and the lack of control they experience, sooner or later make it indispensable to put in place strategies that allow them to face these needs, especially when the dependence of third parties and intermediaries is very strong throughout the supply chain or when the competition represents serious threats (Saucedo-Martínez et al. 2017). Either to reduce the risks associated with the operation of the organization, ensure a competitive position or have a more dominant position in the market (Sherwin 2016; Lim et al. 2014; Wende and Kiradjiev 2014; Gurcaylilar-Yenidogan and Windsperger 2014; Abramowicz et al. 2016).

Table 19.2 Relationship between supply chain processes and horizontal integration

Horizontal integration		
Supplier	Supply chain	Customers
Business processes and operations	Time reduction	Inventory management
Purchase-sale orders	Short- and long-term programming	Manufacturing by orders
Buying locations	Organization by models and families	Assemblies by orders
Transaction management	Moves and logistics	Keep inventories
Business processes and operations	Simulations	Warehouses management

Table 19.3 Value creation modules

Vertical integration	
Socio-technical system + Value creation modules	Characteristics
Human operating system	Scales of learning, culture, ability to adapt
Organizational system	Integration by transactions, shared information, strategic collaboration
Technological system	Physical systems, cyber infrastructure
Product	Variants, size, components
Processes	Lots, quantities, automation

19.4.1.1 Vertical Integration

Vertical integration occurs when the same organization takes over activities that it has traditionally delegated to third parties. For this, it can create or acquire other companies through which it can be self-supplying regarding the supply of materials and supplies, and/or take over some or all the tasks related to the distribution of the goods it produces (such as it can be having its own warehouses or distribution centers, transporting the merchandise or selling it to the final consumer).

The company performance lies in the level of synergy it possesses, for it must be considered the crucial elements involved in the creation, development, and manufacture of the product as well as its administration, the vertical integration mapping, considering the value creation modules, see Fig. 19.3.

19.4.1.2 Horizontal Integration

Horizontal integration strategy occurs when a company acquires, merges, or creates another or other companies that perform the same activity; that is, they produce goods of the same type or that can even substitute, generally to cover other market segments and increase their participation and power within it, Fig. 19.4.

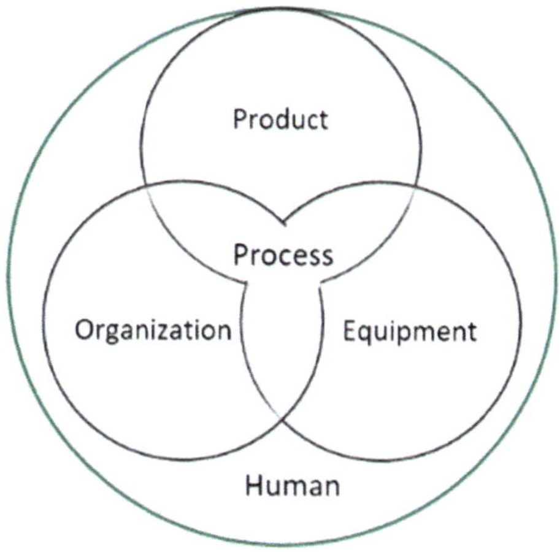

Fig. 19.3 Value creation modules

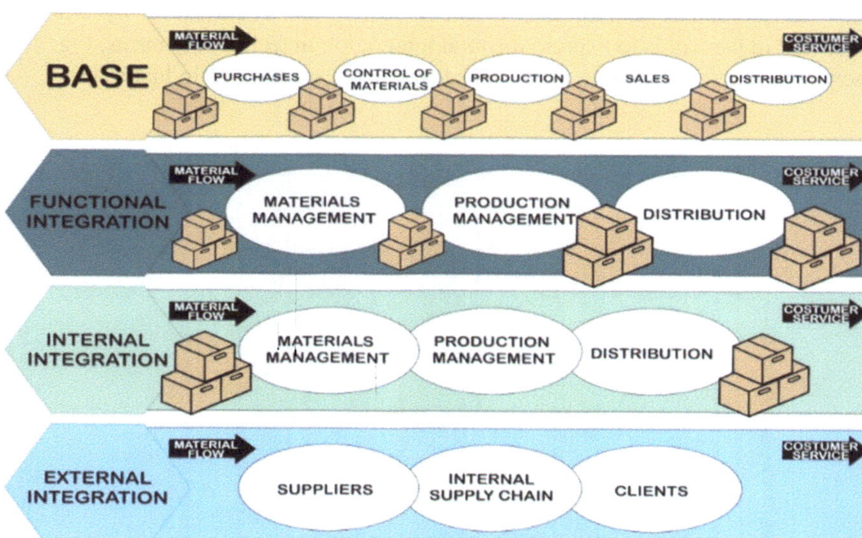

Fig. 19.4 Supply chain integration

Main objectives that it pursues are as follows: In addition to the ones just mentioned, companies can achieve some of the following objectives through a strategy of this type:

- Take advantage of the positioning that a brand enjoys a segment by complementing the current portfolio of products (for example, instead of creating a new brand from scratch).

- Take advantage of access to new distribution channels to which the acquisition or merger with another company can give access.
- Achieve greater bargaining power with suppliers.
- Help create economies of scale and, as in vertical integration, allow greater efficiency by sharing human and technological resources.

19.4.2 Industry 4.0 Proposal Levels

In order to determine the level of Industry 4.0 of the company and place the potential of development of new technologies, the previously described criteria are evaluated, where vertical integration is measured in per percentage, according to the development of the value creation modules, horizontal integration, places the company in one of the four existing levels according to its integration with suppliers and customers, and the last criterion is measured as a percentage and is a diagonal line that measures the implementation of the different technologies available in what We know as prominent technologies of the Industry 4.0. The relation by which is measured in percent is to express the total integration in technology terms, Fig. 19.5 shows the marker according to the four levels of integration of Industry 4.0.

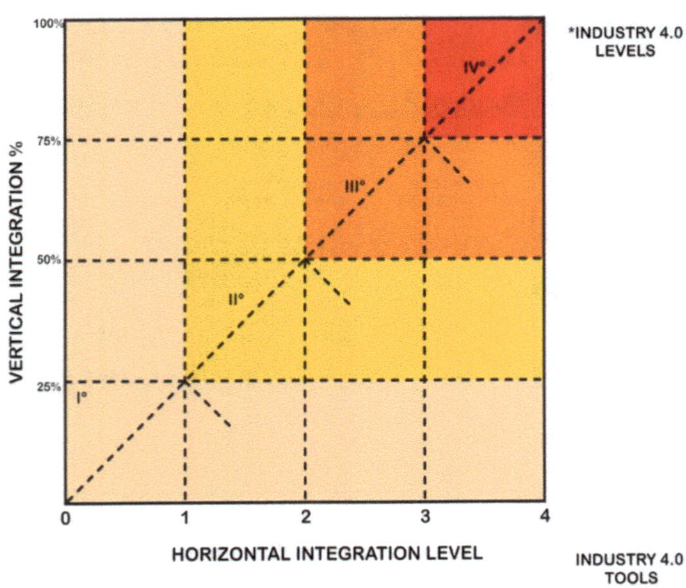

Fig. 19.5 Level scoreboard of Industry 4.0

To achieve this, they can either seek to assume a more active role and less dependent on third parties in the supply chain (vertical integration) or on the other hand merge or acquire other companies that operate within their own market (horizontal integration).

19.4.3 Comparisons of Models

We reviewed six models that measure the level of maturity in the framework of Industry 4.0; Table 19.4 shows the comparison of these models.

Once the Industry 4.0 evaluation models were analyzed, an in-depth analytical study was conducted of those that fulfilled characteristics like the instrument designed.

Figure 19.6 shows the selected models, they are based on the structure that is proposed and they are also measurement instruments that are organized in a systematic way and are available on the Internet, so that access to them allowed obtaining specific information about the approach of reagents that allowed the exploration of the instrument from the direction of vertical and horizontal integration of systems.

Table 19.4 Evaluation models in Industry 4.0

Model	Author	Description
The Connected Enterprise Maturity Model	Rockwell-Automation Rockwell (2014)	It is a maturity model that consists of five stages to measure the level of Industry 4.0 with four dimensions (model published)
Fitness-Check for Industrie 4.0 Entwickelt	FH-Obersterreich Sciences (2015)	Maturity evaluation in three dimensions and 13 reagents for positioning in 10 levels (without specific detail)
IMPULS Industrie 4.0 Readiness	VDMA RWTH Aachen IW Consult IMPULS (2015)	Provides an evaluation in six dimensions, including 18 reagents to indicate performance in five levels (electronic model)
HADA Advanced Digital Self-Diagnostic Tool	Ministerio de economía, industria y competitividad Competitividad (2016)	Evaluates five key dimensions, divided into 16 levers in the performance of six levels (electronic model)
Empowered and Implementation Strategy for Industry 4.0	Lanza et al. (2016)	Maturity evaluation of Industry 4.0 as a quick check and part of a process model for its realization (without details of reagents and levels)
The Industry 4.0/ Digital Operations Self-Assessment	PWC PricewaterhouseCoopers (2016)	Evaluation in six dimensions, focused on four levels of maturity (electronic model access to three of six dimensions)

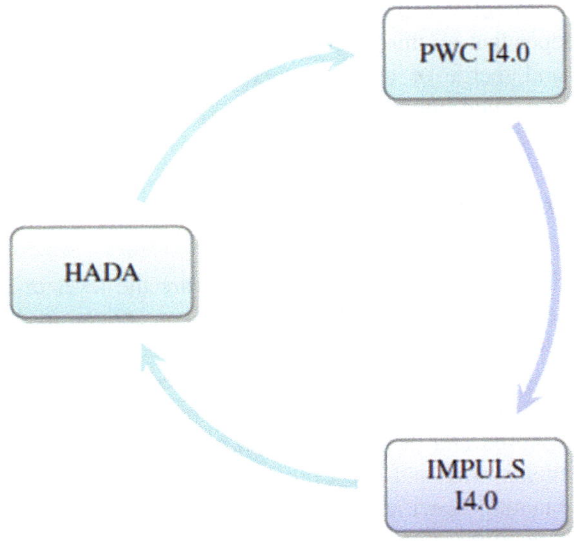

Fig. 19.6 Selected models on Industry 4.0

From these models, findings of elements and criteria were analyzed to understand new technologies; it allows more efficient production chain virtually simulate the manufacture of a product, prevent and repair damage to remote and flexible production, all without firing operating costs. The hyper-connectivity and interoperability, the emergence of technologies such as cloud or Internet of things, data analytics, cybersecurity, and IoT, among others, offer great new possibilities for the industry and above all pose new challenges to solve. The fourth industrial revolution is to make the binomial humans and machines will be more productive. And this goes through digitization in production processes.

The use of new technologies IoT and big data allow hybridization between the physical and the digital. Industry 4.0 must link the physical world to the virtual to make an intelligent industry. The "digital enablers" are the set of technologies that enable this new industrial revolution is carried out. With the use of these technologies, it is possible to obtain and analyze information on production costs and the behavior of the production process in real time. The fourth industrial revolution will be not only in the factories, in the outlets or logistics processes. The next industrial revolution will cover the entire value chain. One consequence of the onset of Industry 4.0 has to do with security. In a world where more and more machines and other devices connected to the network, the risk of a potential cyberattack is also higher.

The data obtained in the production process and the conclusions obtained from them will be as or more valuable than the machines in companies. The loss of data obtained during production, or theft, would be insurmountable for many companies' brokenness. Therefore, it is essential also to innovate in cybersecurity, to develop safe protection systems and measures anti—attacks that are scalable and effective. Ultimately, the Industry 4.0 allows connecting with consumers.

The digital channels will connect you with the end customer. What is new business models will be deeply influenced by the connectivity and intelligence together, open great possibilities of evolution, and development (Kang et al. 2016).

19.5 Application of the 9 Tools in the Supply Chain

19.5.1 Big Data and Analytics

With an increasing number of products (CPS) and intelligent systems (CPPS) in the factories and the market, the amount of data available to manufacturers will multiply. Their analysis will identify patterns and interdependencies, analyze processes and discover inefficiencies, and even predict future events. This will open new opportunities, not only improve efficiency but also discover services for the client, which will be much better known.

19.5.2 Autonomous Robots

The improvement of artificial intelligence along with a new sensor has allowed creating robots increasingly autonomous, flexible and cooperative. Over time, they will interact with each other and work safely with human beings, learning from them by offering a range of capabilities far superior to those used in today's manufacturing.

19.5.3 Simulation

It refers to the production of three-dimensional objects from virtual models. Although its industrial use is limited, additive manufacturing will eliminate the disadvantages in the production efficiency of personalized products. It will allow the rapid creation of prototypes and a highly decentralized manufacturing: the model of the product could be sent to the "print" site closest to the client eliminating intermediate steps.

19.5.4 Horizontal and Vertical System Integration

19.5.4.1 Vertical Integration

It happens when the same company takes over activities that it has traditionally delegated to third parties. To this end, the organization can create or acquire other companies through which it can be self-sufficient in terms of the supply of materials and supplies, and/or take over some or all of the tasks related to the distribution of the goods it produces (as it can be to have their own warehouses or distribution centers, carry out the transport of the merchandise or the sale to the final consumer).

Depending on the activities that the company wants to take over, this integration can be made backward (when it decides to become its own supplier in terms of inputs or raw materials), forward (when it wishes to take over tasks related to distribution and sale of products, for example, to the final consumer or wholesale level if it is the case), or compensated; When the parent company has subsidiaries that exercise the role of suppliers and also has other that allows it to deal with the distribution and/or sale to the final consumer.

Main objectives pursued: Regardless of whether it is backward or forward when talking about vertical integration in general, companies seek greater control over the competition and about the tasks and activities carried out by the intermediaries within the supply chain (while reducing their dependence on these).

But to be more exact, some things that companies can achieve through a strategy of this type are as follows:

- Significantly reduce production costs from the primary sector and achieve economies of scale in each subsidiary by eliminating the margins normally left to third parties.
- Ensure a continuity of supply and quality of products that allow maintaining a competitive advantage and a differentiated offer.
- Have a greater capacity for negotiation (a group of companies that belong to the same company or owners can get better negotiations by sharing different resources).
- To be able to do a better planning of the work as well as to have a better control over the processes that allow focussing in a greater generation of value for the client when reducing the dependency of third parties.
- By sharing human and material resources, a group of companies can also achieve greater efficiency by sharing different activities and creating synergies.
- Can set prices for the company itself and exercise greater control over them.
- Serve customers directly and at the same time obtain valuable market information by being closer to the final consumer.

19.5.4.2 Horizontal Integration

Unlike vertical integration, horizontal integration strategy occurs when a company acquires, merges, or creates another or other companies that perform the same activity; that is, they produce goods of the same type or that can even substitute, generally to cover other market segments and increase their participation and power within it.

Main objectives pursued: In addition to the ones just mentioned, companies can achieve some of the following objectives through a strategy of this type:

- Take brand positioning advantage brand within a segment by complementing the current portfolio of products (for example, instead of creating a new brand from scratch).
- Take access advantage to new distribution channels to which the acquisition or merger with another company can give access.
- Achieve greater bargaining power with suppliers.
- Help create economies of scale and, as in vertical integration, allow greater efficiency by sharing human and technological resources.
- Thanks to this strategy, a company can strengthen its position in the market as well as reduce rivalry in the industry.

Despite its advantages, integration strategies can have their cons. Although there may be several advantages that can be represented by implementing an integration strategy, be it vertical or horizontal, they can also bring some inconveniences to the companies, which can either directly affect them or, in other cases, generate negative effects for consumers and the market.

For example, while controlling several organizations with different activities for the management of a business group, it can become highly complex and somewhat difficult to maneuver in the face of sudden changes in a vertical integration strategy, in the case horizontal integration consequences to seek control over the competition and to have greater market power is without doubt that the organization tends to become a monopoly. For that reason, beyond the advantages that both strategies can represent to any company, we must consider that they can also lead them to lose the domain they claim to have on different issues when they may just be in search of the opposite: having the control over different aspects related to the business and the market.

19.5.5 The Industrial Internet of Things

M2M communication is the basic technology of the "Internet of things" (IoT), it refers to technologies that allow the exchange of information between the intelligent products and systems that constitute the Industry 4.0 environment, and with this information it is possible to build a virtual physical factory, which will simulate not only products but complete manufacturing processes. The most obvious form of communication use M2M will be in the connection of intracompany systems will also be a key factor in intercompany collaboration.

19.5.6 Cybersecurity

With the increase in connectivity and the use of standard communication protocols, the need to protect critical industrial systems and manufacturing lines from security threats increases dramatically. As a result, secure and reliable communications as well as sophisticated identity and access management of machines and users will be necessary.

A large amount of data that is generated in the environment of an Industry 4.0 through devices such as sensors and the fact that part of this data is shared between companies to achieve greater efficiency in the supply chain make a vital factor importance in this field is that of security.

19.5.7 The Cloud

The cloud includes applications and infrastructures offered as a service through public or private networks, often on a pay-per-use basis. Smart products and systems (CPS and CPPS) will generate huge amounts of data to store and process that must be accessible online from anywhere. The cloud allows this flow of data without borders and eliminates the need for investment in infrastructure to increase capacity, allowing unprecedented flexibility.

19.5.8 Additive Manufacturing

It refers to the production of three-dimensional objects from virtual models. Although its industrial use is limited, additive manufacturing will eliminate the disadvantages in the production efficiency of personalized products. It will allow the rapid creation of prototypes and a highly decentralized manufacturing: the model of the product could be sent to the “print” site closest to the client eliminating intermediate steps.

19.6 Conclusions

Industry 4.0 is presented imminently in the global market, radically transforming the industry, business, systems, society and holistically the way of life. The following statements can be cited as conclusions of the research developed:

- This revolution is considered integrative from the point of view that includes technologies, tools, skills, and available knowledge, to provide autonomy to the systems, increase efficiency levels, customer service, and sustainability.
- The technologies and the Internet play a preponderant role in the new era; de facto, its adoption in companies is essential.
- Industry 4.0 seeks to interrelate autonomously to provide excellence in service through system integration.

- The inclusion of companies in the new way of operating requires dissemination of information that allows the understanding and practice of the new modality of carrying out productive activities.
- Companies must have a pre-application preparation to operate in Industry 4.0 schemes, to ensure their development in the new system.
- The person must have technical knowledge for both administrative and operational areas, as well as the willingness to acquire new knowledge and skills that allow them to carry out their activities efficiently.
- The process of research and development of new services and products must be analyzed from a technological and sustainable point of view.
- The supplier's management and consumers must operate under integrated and synchronized chains.
- Operations management in the new system integrates companies regardless of their size and turns, must adapt according to their immediate needs.
- The importance of evaluating falls on the impulse that the companies have of knowing the current state and determining the slopes that can lead to the digital transformation and the use of data by means of the identification of specific activities.
- Follow up on evaluated processes of technological development, it allows seeing the progress through time and the impact on the flow of income in the company, as well as the relationships with customers and suppliers in the management of the supply chain.
- The identification in the research process of the need to train human sources with profiles of technical analysis for operations and decision-making in administration, as a reference for the development of new educational plans.
- The training and acquisition of knowledge of technological tools of analysis and management in response to emerging needs in the labor market.
- In industrial terms, this revolution has an impact on all aspects of production processes, without leaving room for expectation; therefore, its adoption requires assertiveness and pragmatism.

References

Abduaziz O, Cheng JK, Tahar RM, Varma R (2015) A hybrid simulation model for green logistics assessment in automotive industry. Proc Eng 100:960–969. https://doi.org/10.1016/j.proeng.2015.01.455

Abramowicz W, Alt R, Franczyk B (2016) Business information systems. Springer, Cham, Leipzig, Germany. https://doi.org/10.1007/978-3-319-39426-8

Ahn S (2016) Becoming a network beyond boundaries: brain-machine interfaces (BMIs) as the actor-networks after the internet of things. Technol Soc 47:49–59. https://doi.org/10.1016/j.techsoc.2016.08.003

Albers A, Gladysz B, Pinner T, Butenko V, Stürmlinger T (2016) Procedure for defining the system of objectives in the initial phase of an Industry 4.0 project focusing on intelligent quality control systems. Proc CIRP 52:262–267. https://doi.org/10.1016/j.procir.2016.07.067

Alberti AM, Mazzer D, Bontempo M, de Oliveira LH, Righi Rd, Sodr AC (2017) Cognitive radio in the context of internet of things using a novel future internet architecture called NovaGenesis. J Comput Electr Eng 57(C):147–161. https://doi.org/10.1016/j.compeleceng.2016.07.008

Anderl R (2014) Industrie 4.0—advanced engineering of smart products and smart production. In: Proceedings 19th international seminar on high technology technological innovations in the product development, pp 1–14

Atanasov I, Nikolov A, Pencheva E, Dimova R, Ivanov M (2015) An approach to data annotation for internet of things. In: Alkhatib GI (ed) Int J Inf Technol Web Eng (IJITWE). IGI Glogal. https://doi.org/10.4018/IJITWE.2015100101

Avci MG, Selim H (2016) A multi-agent system model for supply chains with lateral preventive transshipments: application in a multi-national automotive supply chain. Comput Ind 82:28–39. https://doi.org/10.1016/j.compind.2016.05.005

Bagheri B, Yang S, Kao HA, Lee J (2015) Cyber-physical systems architecture for self-aware machines in Industry 4.0 environment. IFAC-PapersOnLine 48(3):1622–1627. https://doi.org/10.1016/j.ifacol.2015.06.318

Bian J, Yoshigoe K, Hicks A, Yuan J, He Z, Xie M, Modave F (2016) Mining twitter to assess the public perception of the "Internet of Things". PLoS ONE 11(7):1–14. https://doi.org/10.1371/journal.pone.0158450

Botthof A, Hartmann EA (2015) Zukunft der Arbeit in Industrie 4.0. Springer. https://doi.org/10.1007/978-3-662-45915-7

Brettel M, Fischer FG, Bendig D, Weber AR, Wolff B (2016a) Enablers for Self-optimizing production systems in the context of Industrie 4.0. Proc CIRP 41:93–98. https://doi.org/10.1016/j.procir.2015.12.065

Brettel M, Klein M, Friederichsen N (2016b) The relevance of manufacturing flexibility in the context of Industrie 4.0. Proc CIRP 41:105–110. https://doi.org/10.1016/j.procir.2015.12.047

Chen Y (2016) Analyzing and visual programming internet of things and autonomous decentralized systems. Simul Model Pract Theory 65:1–10. https://doi.org/10.1016/j.simpat.2016.05.002

Chen F, Deng P, Wan J, Zhang D, Vasilakos AV, Rong X (2015) Data mining for the internet of things: literature review and challenges. Int J Distrib Sens Netw 11(8):1–14. https://doi.org/10.1155/2015/431047

Colom JF, Mora H, Gil D, Signes-Pont MT (2017) Collaborative building of behavioural models based on internet of things. Comput Electr Eng 58(C):385–396. https://doi.org/10.1016/j.compeleceng.2016.08.019

Competitividad Md (2016) Industria Conectada 4.0. Retrieved from https://hada.industriaconectada40.gob.es/hada/

Decker C, Berchtold M, Chaves LW, Beigl M, Roehr D, Riedel T, Herzig D (2008) Cost-benefit model for smart items in the supply chain. The internet of things. Lecture Notes in Computer Science vol 4952, pp 155–172. Springer, Berlin, Heidelberg. https://doi.org/10.1007/978-3-540-78731-0_10

Dombrowski U, Wagne T (2014) Mental strain as field of action in the 4th industrial revolution. Proc. CIRP 17:100–105. https://doi.org/10.1016/j.procir.2014.01.077

Esmaeilian B, Behdad S, Wang B (2016) The evolution and future of manufacturing: a review. J Manuf Syst 39:79–100. https://doi.org/10.1016/j.jmsy.2016.03.001

Fedorov A, Goloschchapov E, Ipatov O, Potekhin V, Shkodyrev V, Zobnin S (2015) Aspects of smart manufacturing via agent-based approach. Proc Eng 100:1572–1581. https://doi.org/10.1016/j.proeng.2015.01.530

Gilchrist A (2016) Industry 4.0: the industrial internet of things. (Apress, ed) Springer. https://doi.org/10.1007/978-1-4842-2047-4

Gurcaylilar-Yenidogan T, Windsperger J (2014) Inter-organizational performance in the automotive supply networks: the role of environmental uncertainty, specific investments and formal contracts. Proc Soc Behav Sci 150:813–822. https://doi.org/10.1016/j.sbspro.2014.09.090

Igor H, Bohuslava J, Martin J (2016) Proposal of communication standardization of industrial networks in Industry 4.0. In: IEEE 20th Jubilee international conference on intelligent engineering systems (INES). Budapest, Hungary: IEEE, pp 119–124. https://doi.org/10.1109/INES.2016.7555105

IMPULS (2015) IMPULS foundation of the German engineering federation (VDMA). Retrieved from https://www.industrie40-readiness.de/?lang=en

Ji C, Shao Q, Sun J, Shijun L, Pan L, Wu L, Yang C (2016) Device data ingestion for industrial big data platforms with a case study. Sensors 16(3):279–294. https://doi.org/10.3390/s16030279

Kagermann H (2015) Change through digitization—value creation in the age of Industry 4.0. In: Albach H, Meffert H, Pinkwart A, Reichwald R (eds) Management of permanent change. Springer. Obtenido de, pp 23–45. http://www.springer.com/gp/book/9783658050139

Kang HS, Lee JY, Choi S, Kim H, Park JH, Son JY, Kim BH (2016) Smart manufacturing: Past research, present findings, and future directions. Int J Precis Eng Manuf Green Technol 3 (1):111–128. https://doi.org/10.1007/s40684-016-0015-5

Kans M, Galar D, Thaduri A (2016) Maintenance 4.0 in railway transportation industry. In: Proceedings of the 10th world congress on engineering asset management, Lectures notes in mechanical engineering, pp 317–331. https://doi.org/10.1007/978-3-319-27064-7_30

Khan A, Turowski K (2016) A survey of current challenges in manufacturing industry and preparation for Industry 4.0. In: Abraham A, Kovalev S, Tarassov V, Snášel V (eds) Advances in intelligent systems and computing, vol 1, pp 15–26. Springer, Russia. https://doi.org/10.1007/978-3-319-33609-1

Kolberg D, Zühlke D (2015) Lean Automation enabled by Industry 4.0 technologies. IFAC-PapersOnLine 48(3):1870–1875. https://doi.org/10.1016/j.ifacol.2015.06.359

Laghari S, Niazi MA (2016) Modeling the internet of things, self-organizing and other complex adaptive communication networks: a cognitive agent-based computing approach. PLoS ONE 11(1):e0146760. https://doi.org/10.1371/journal.pone.0146760

Lanza G, Nyhuis P, Ansari SM, Kuprat T, Liebrecht C (2016) Empowerment and implementation strategies for Industry 4.0. ZWF Zeitschrift für wirtschaftlichen Fabrikbetrieb 111(2):76–79. https://doi.org/10.3139/104.111462

Li D (2016) Perspective for smart factory in petrochemical industry. Comput Chem Eng 91:136–148. https://doi.org/10.1016/j.compchemeng.2016.03.006

Li Y, Chen X, Wang X, Xu Y, Chen PH (2017) A review of studies on green building assessment methods by comparative analysis. Energy Build 146:152–159. https://doi.org/10.1016/j.enbuild.2017.04.076

Lim LL, Alpan G, Penz B (2014) Reconciling sales and operations management with distant suppliers in the automotive industry: a simulation approach. Int J Prod Econ 151:20–36. https://doi.org/10.1016/j.ijpe.2014.01.011

Lin K, Wang W, Bi Y, Qiu M, Hassan MM (2016) Human localization based on inertial sensors and fingerprints in the industrial internet of things. Comput Netw 101:113–126. https://doi.org/10.1016/j.comnet.2015.11.012

Lu X, Qu Z, Li Q, Hui P (2015) Privacy information security classification for internet of things based on internet data. Int J Distrib Sensor Netw (Special issue on Big Data and Knowledge Extraction for Cyber-Physical Systems) 23:23. https://doi.org/10.1155/2015/932941

Macaulay T (2017) Safety requirements in the internet of the things. In: Macaulay T (ed) RIoT Control: understanding and managing risks and the internet of things, 3rd edn. Elseiver, United States, pp 113–123. https://doi.org/10.1016/B978-0-12-419971-2.00006-6

Mashal I, Alsaryrah O, Chung TD (2016) Performance evaluation of recommendation algorithms on internet of things services. Phys A 451:646–656. https://doi.org/10.1016/j.physa.2016.01.051

Matties B (2016) SMT007 magazine. Obtenido de http://smt.iconnect007.com/landing/smt/smt-magazine?skin=smt

Mazark A, Huemer C (2015) A standards framework for value networks in the context of Industry 4.0. In: IEEE international conference on industrial engineering and engineering management (IEEM). IEEE, Singapore, pp 1342–1346. https://doi.org/10.1109/IEEM.2015.7385866

Möller DP (2016) Guide to computing fundamentals in cyber-physical systems: concepts, design methods, and applications. In: Editor S (ed) Springer International Publishing Switzerland, Swindon, Switzerland, UK. https://doi.org/10.1007/978-3-319-25178-3

Moosavi SR, Gia TN, Nigussie E, Rahmani AM, Virtanen S, Tenhunen H, Isoaho J (2016) End-to-end security scheme for mobility enabled healthcare internet of things. Future Gener Comput Syst 64:108–124. https://doi.org/10.1016/j.future.2016.02.020

Moreno A, Velez G, Ardanza A, Barandiaran I, de Infante ÁR, Chopitea R (2016) Virtualisation process of a sheet metal punching machine within the Industry 4.0 vision. Int J Interact Des Manuf 11:365–373. https://doi.org/10.1007/s12008-016-0319-2

Mosterman PJ, Zander J (2016) Industry 4.0 as a cyber-physical system study. Softw Syst Model 15(1):17–29. https://doi.org/10.1007/s10270-015-0493-x

Neeliah H, Seetanah B (2016) Does human capital contribute to economic growth in Mauritius? Eur J Training Dev 40(4):248–261. https://doi.org/10.1108/EJTD-02-2014-0019

Ojha D, Shockley J, Acharya C (2016) Supply chain organizational infrastructure for promoting entrepreneurial emphasis and innovativeness: the role of trust and learning. Int J Prod Econ 179:212–227. https://doi.org/10.1016/j.ijpe.2016.06.011

Prause M, Weigand J (2016) Industry 4.0 and object-oriented development: incremental and architectural change. J Technol Manage Innov 11(2):104–111. https://doi.org/10.4067/S0718-27242016000200010

PricewaterhouseCoopers (2016) Industry 4.0 Self Assessment. Retrieved from PWC: https://i40-self-assessment.pwc.de/i40/landing/

Raza S, Misra P, He Z, Voigt T (2017) Building the internet of things with bluetooth smart. Ad Hoc Netw 57:19–31. https://doi.org/10.1016/j.adhoc.2016.08.012

Rizzardi A, Sicari S, Miorandi D, Coen-Porisini A (2016) AUPS: an open source authenticated publish/subscribe system for the internet of things. Inf Syst 62:29–41. https://doi.org/10.1016/j.is.2016.05.004

Rockwell (2014) Rockwell automation. Retrieved from http://literature.rockwellautomation.com/idc/groups/literature/documents/wp/cie-wp002_enp.pdf

Salcito K, Singer BH, Krieger GR, Weiss MG, Wielga M, Utzinger J (2014) Assessing corporate project impacts in changeable contexts: a human rights perspective. Environ Impact Assess Rev 47:36–46. https://doi.org/10.1016/j.eiar.2014.03.004

Saucedo-Martínez J, Pérez-Lara M, Marmolejo-Saucedo J, Salais-Fierro T, Vasant P (2017) Industry 4.0 framework for management and operations: a review. J Ambient Intell Humanized Comput 1–13. https://doi.org/10.1007/s12652-017-0533-1

Schlechtendahl J, Keinert M, Kretschmer F, Lechler A, Verl A (2014) Making existing production systems Industry 4.0-ready: holistic approach to the integration of existing production systems in Industry 4.0 environments. Prod Eng Res Devel 9(1):143–148. https://doi.org/10.1007/s11740-014-0586-3

Schlott S (2013) Industrie 4.0 - Abschied von der Automatisierungspyramide. ATZ - Automobiltechnische Zeitschrift 115(4):306–307. https://doi.org/10.1007/s35148-013-0098-z

Scholz S, Mueller T, Plasch M, Limbeck H, Adamietz R, Iseringhausen T, Woegerer C (2016) A modular flexible scalable and reconfigurable system for manufacturing of microsystems based on additive manufacturing and e-printing. Rob Comput Integr Manuf 40:14–23. https://doi.org/10.1016/j.rcim.2015.12.006

Schuh G, Potente T, Varandani R, Schmitz T (2014) Global footprint design based on genetic algorithms—an "Industry 4.0" perspective. CIRP Ann Manuf Technol 63(1):433–436. https://doi.org/10.1016/j.cirp.2014.03.121

Sciences UO (2015) FH - Obersterreich. Retrieved from FITNESS-CHECK FÜR INDUSTRIE 4.0 ENTWICKELT: https://www.fh-ooe.at/facebook/fitness-checkfuer-

Senter RJ, Flynn MS (1999) Changing interorganizational patterns in the North American automotive supply chain. Appl Behav Sci Rev 7(1):59–80. https://doi.org/10.1016/S1068-8595(99)80122-6

Seo SM, Kim SW, Jeon JW, Kim JH, Kim HS, Cho JH, Paek SH (2016) Food contamination monitoring via internet of things, exemplified by using pocket-sized immunosensor as terminal unit. Sens Actuators B: Chem 233:148–156. https://doi.org/10.1016/j.snb.2016.04.061
Shafiq SI, Sanin C, Szczerbicki E, Toro C (2015) Virtual engineering object/virtual engineering process: a specialized form of cyber physical system for Industrie 4.0. Proc Comput Sci 60:1146–1155. https://doi.org/10.1016/j.procs.2015.08.166
Sherwin P (2016) Industry 4.0 + IOT = Smart industrial ovens & furnaces. Process instrumentation, pp 12–18. Retrieved from https://www.eurotherm.de/smart-industrial-ovens
Simionis K, Glory YS, Gries T (2016) INDUSTRIE 4.0—automation in weft knitting technology. IOP Conf Ser Mater Sci Eng 141(1):1–11. https://doi.org/10.1088/1757-899X/141/1/012014
Stock T, Seliger G (2016) Opportunities of sustainable manufacturing in Industry 4.0. Proc CIRP 40:536–541. https://doi.org/10.1016/j.procir.2016.01.129
Stojkić Ž, Veža I, Bošnjak I (2016) A concept of information system implementation (CRM AND ERP) within Industry 4.0. In: Proceedings of the 26th DAAAM international, pp 912–919. https://doi.org/10.2507/26th.daaam.proceedings.127
Syntetos A, Kolassa B, Kolassa Z, Kolassa B, Kolassa J, Konstantinos S, Konstantinos N (2016) Supply chain forecasting: theory, practice, their gap and the future. Eur J Oper Res 252(1):1–26. https://doi.org/10.1016/j.ejor.2015.11.010
Thames L, Schaefer D (2016) Software-defined cloud manufacturing for Industry 4.0. Proc CIRP 52:12–17. https://doi.org/10.1016/j.procir.2016.07.041
Thomé AM, Scavarda LF, Pires SR, Ceryno P, Klingebiel K (2014) A multi-tier study on supply chain flexibility in the automotive industry. Int J Prod Econ 158:91–105. https://doi.org/10.1016/j.ijpe.2014.07.024
Toro C, Barandiaran I, Posada J (2015) A perspective on knowledge based and intelligent systems implementation in Industrie 4.0. Proc Comput Sci 60:362–370. https://doi.org/10.1016/j.procs.2015.08.143
Tripathy S, Aich S, Chakraborty A, Lee GM (2016) Information technology is an enabling factor affecting supply chain performance in Indian SMEs: a structural equation modelling approach. J Model Manage 11(1):269–287. https://doi.org/10.1108/JM2-01-2014-0004
Vogel-Heuser B, Bauernhansl T, ten Hompel M (2017) Handbuch Industrie 4.0 Bd.1. Deutchsland: Springer. https://doi.org/10.1007/978-3-662-45537-1
Vyas D, Kröner A, Nijholt A (2016) From mundane to smart: exploring interactions with 'smart' design objects. Int J Mob Hum Comput Interact 8(1):63–88. https://doi.org/10.4018/IJMHCI.2016010103
Wahlster W, Beste D (2016) HERMES AWARD—Internationaler Technologiepreis der HANNOVER MESSE. Springer Vieweg, Wiesbaden, Germany. https://doi.org/10.1007/978-3-658-12834-0
Wang S, Wan J, Zhang D, Li D, Zhang C (2016) Towards smart factory for industry 4.0. Int J Comput Telecommun Networking 101(C):158–168. https://doi.org/10.1016/j.comnet.2015.12.017
Weber RH, Studer E (2016) Cybersecurity in the internet of things: legal aspects. Comput Law Secur Rev 32(5):715–728. https://doi.org/10.1016/j.clsr.2016.07.002
Wende J, Kiradjiev P (2014) Eine Implementierung von Losgröße 1. Elektrotechnik & Informationstechnik 131(7):202–206. https://doi.org/10.1007/s00502-014-0222-0
Wildgoose N (2016) Supply chain risk management. In: Green PE (ed) Enterprise risk management: a common framework for the entire organization. Elseiver, Tokyo, pp 75–87. doi:https://doi.org/10.1016/B978-0-12-800633-7.00006-7
Wolter MI, Mönnig A, Hummel M, Schneemann C, Weber E, Zika G, Neuber-Pohl C (2015). Industrie 4.0 und die Folgen für Arbeitsmarkt und Wirtschaft, 1 edn. IAB Forschungsbericht, Germany. Retrieved from http://doku.iab.de/forschungsbericht/2015/fb0815.pdf
Yao X, Lin Y (2016) Emerging manufacturing paradigm shifts for the incoming industrial revolution. Int J Adv Manuf Technol 85(5–8):1665–1676. https://doi.org/10.1007/s00170-015-8076-0

Chapter 20
Modeling by Finite Element of a Turning Process with Chip Detachment

Baldomero Lucero-Velázquez, Juan José Delfín-Vázquez, José Efrén Ruelas-Ruíz, Eusebio Jiménez-López and Mario Acosta-Flores

Abstract In the present chapter, we have developed mathematical modeling and analysis by finite element (FEA) process of turning with chip detachment; the study incorporated the dynamic and structural cutting orthogonal and oblique mathematical models in three dimensions (3D) and with incorporating thermal mesh structural static fluid for your simulation in FEA software. Validation is performed on an experimental basis by record temperatures and mediated the release of chip in the process of turning with octagonal and oblique cut. With the simulation are determined variables important to optimize; these are orthogonal and oblique cutting: cutting speed, cutting depth, and angle cutting, presenting a variation of 4–8% with respect to the experimental values. In this process, it was determined that cutting temperature presents a value means of 132 °C and a maximum of 147 °C and a minimum of 115–14 °C cutting angle and a forward speed of a 0.1466 speed m/s temperature-optimized cutting. In the process of chip detachment are decisive values for temperature, speed, depth, and angle of cut according to a previous simulation incorporating the models mentioned above.

Keywords Analysis by finite element · Turning · Cutting temperature Orthogonal · Oblique

B. Lucero-Velázquez (✉) · J. J. Delfín-Vázquez · J. E. Ruelas-Ruíz
Department of Mechanical Engineering, Instituto Tecnológico Superior de Cajeme, Carretera Internacional a Nogales km-2, Col. Amaneceres 2, PO Box 754 Ciudad Obregón, Sonora, Mexico
e-mail: blv@itesca.edu.mx

E. Jiménez-López
Department of Research and Development, Technological University of Southern Sonora-ULSA NOROESTE-IIMM, Doctor Norman E. Borlaug Street, km 14, Ciudad Obregón, Sonora, Mexico

M. Acosta-Flores
Faculty of Chemical Sciences and Engineering, Autonomous University of the State of Morelos, Avenida Universidad 1001, Col. Chamilpa, Cuernavaca, Morelos, Mexico

J. L. García Alcaraz et al. (eds.), *Best Practices in Manufacturing Processes*,
https://doi.org/10.1007/978-3-319-99190-0_20

20.1 Introduction

20.1.1 Mathematical Modeling and Computational

Simulation through finite element analysis contributes significantly in the determination of physical parameter of the manufacturing process or operation of industrial components; in this case of the detachment of chip, different mathematical models have been developed through time.

20.1.2 The First Theories for the Prediction of the Results

Merchant et al. (1945a, b) published the first predictive model focused on the material start operation. This model is still today one of the most important in the academic field, Merchant focused on the orthogonal cut, so the model is based on an analysis under the hypothesis of material flow in the plane parallel to the cutting edge.

The working material was considered perfect plastic and the contact between the chip and the tool defined it as totally sliding under a Coulomb friction ratio, with a constant coefficient of friction.

Molinary and Moufki et al. (2008) proposed a modified Merchant model that improves the definition of the kinematics of the problem. The authors include a criterion of supplemental stability for the morphology of the chip. Specifically, this new model takes into account a softened shape of the free surface at the base of the chip that allows correcting the angle of shear.

There are other models of orthogonal cutting to predict the cutting forces as the models of Lee et al. (1951) and Oxley et al. (1989). This model was later based in Kudo et al. (1965), and then also by Dewhurst et al. (1978), thereby improving the prediction of the curvature of the chip.

They considered the material of the piece as rigid-perfectly plastic. The main contribution of this model is that it considers a plastic contact at the interface.

Oxley et al. (1989) developed a thermomechanical model in which he considered thermo-visco-plastic behavior for the material, in addition to defining a non-infinitesimal thickness for the primary shear zone. Thus, he defined a non-plastic contact at the interface where the material does not slide on the tool. This hypothesis has the consequence of the appearance of the secondary shear zone.

It is important to note that this orthogonal cutting model considers friction as a phenomenon of shearing of material within the chip along the contact area.

20.1.3 The Contact in the Chip/Tool Interface

Takeyama et al. (1958) published a study on the variation of shear forces with respect to contact surfaces. For this, the authors used a tool designed to keep the contact surface constant.

The main conclusion of this work is that the friction stress is proportional to the contact surface, that is, the mean friction stress at the interface is constant. In addition, the authors demonstrated that this medium friction stress is almost equal to the elastic shear strength of the material.

Kattwinkel et al. (1957), Andreev et al. (1958) and Usui (1960) were the first to experimentally observe the distribution of efforts at the interface. To achieve this, these authors used a photoelastic tool to machine lead at low cutting speeds (between 0.02 and 0.08 m/s).

Usui and Takeyama observed in their research that normal stress presented a significant peak of effort at the cutting edge, and then stabilized in a small part of the interface before falling back sharply in the second part of the contact zone. The main observation of Usui (1960) was that the friction stress presented a plateau along almost the entire interface before falling sharply at the end of the contact.

The same tendencies of distribution of tensions were published by Zorev et al. (1963), who proposed the model presented in 1963. In this model, the normal stress σ decreases strongly and not linearly from the cutting edge until the end of the contact, following a power-type law.

Wallace et al. (1964) and Usui et al. (1960) highlight the difficulty in extrapolating the stress distribution at the interface for several materials from the cutting forces (Al–O, Al–H, Cu, Pb, Pb–Sn, and Zn) with several advances and detachment angles; they did not consider the measurements near the cutting edge important. These authors demonstrated that the distribution of the stresses at the interface was related to the type of material observed, both in the form and in the voltage, levels measured.

Childs et al. (1998) considered of interest not only the tip of the cutting tool but also the chip–tool interface, so he experimentally obtained for a mild steel a distribution of friction stress similar to that of Kato et al. (1972).

20.1.4 Mechanisms of Friction Between the Tool and the Chip

The simultaneous existence of an adhesions zone and a gliding zone shows the discontinuous and complex character of the contact at the interface (Usui et al. 1960; Zorev et al. 1963; Childs et al. 1998). To illustrate, the model presented by Zorev et al. (1963), for example, describes two relations for the friction stress, depending on whether it is located in the glued or sliding part of the contact.

Shaw et al. (1960) explains that friction has a different behavior according to whether the pressure at the interface is relatively high or relatively low: for relatively low pressures, the friction stress ratio between normal stresses has a constant value for a pair of materials in contact regardless of the applied pressure (as long as it is low).

Lim et al. (1989) observed that the friction obtained depends to a large extent on the roughness at the interface for cutting speeds lower than 1 m/s. For higher cutting speeds, the authors showed that the friction depends on both the friction and the normal pressure at the interface.

Finnie et al. (1956) related the physics of the phenomenon of friction, for high pressures, with Moore's effect (Moore et al. 1948). According to the excessive plasticization of one of the two surfaces in contact, the ratio *R* between the real contact surfaces and the apparent contact surface increases logarithmically to reach values greater than 1, from a certain pressure level. The behavior of the friction between two surfaces can then be modeled with an exponential type law, where three friction regimes are distinguished according to the level of normal stress applied between the two surfaces (Childs et al. 1989).

The large pressure levels observed near the cutting edge were the main argument of several authors, such as Boothroyd et al. (2006) and Bailey et al. (1975) suggest the existence of a contact area stuck located near the cutting edge.

Shirakashi et al. (1973) relied on the work of Finnie (1956) to propose a model, which continuously relates friction stress to normal stress in cases of low and high pressure at the interface.

20.1.5 Explicit Finite Element Numeric Models

In the Lagrangian case, attention is focused on analyzing how the velocity of a particular element varies with respect to time, Dixit, Uday et al. (2008). Examples are the three-dimensional cutting process analysis with different cutting velocities (Lin et al. 1996); others are elastoplastic, orthogonal, and hardened steel; the parameters used are as follows: effort–deformation, cutting forces, cutting speed, residual stresses, and 3D.

A study of an oblique cutting model is reported by Lin et al. (1999) considering elastoplastic, oblique, cast steel, and P20 tool; the parameters used are as follows: cutting forces, depth of cut, specific cutting power, residual stresses, deformation, cutting speed, distribution of efforts, and 3D model.

Simulation of sliding wear with finite element method is reported by Anderson et al. (1999) using FEA (contact), circular, and steel; the parameters used are as follows: wear, temperature, forces, and 3D and 2D models.

Turning simulations using a three-dimensional FEM code are reported by Ceretti et al. (2000) using thermo-elastoplastic, orthogonal and oblique, low carbon steel, tool steel, was cutting parameters temperature, cutting forces, stresses and deformation, and 3D and 2D models.

An analytical finite element model for predicting three-dimensional tool forces and chip flow by Strenkowski et al. (2002) using thermo-elastoplastic, orthogonal cutting model, AISI 1020 steel, carbide insert, cutting angles, shear stress, friction angle, nose radius the tool, advance, 3D models.

Advances in machining process modeling are reported by Chigurupati et al. (2004) using thermo-elastoplastic, orthogonal cut, 1015 carbon steel, and carbide insert tungsten (uncoated); the parameters used are as follows: cutting speed, cutting angle, cutting time, temperature, feed, wear, and 3D and 2D models.

The temperature on a machined surface was estimated using an inverse approach by Battaglia et al. (2005) with thermo-elastoplastic, orthogonal, machine hard steel, and CBN insert; the parameters used are as follows: temperature, heat flow, cutting speed, feed rate, gradient thermal, and 3D and 2D models.

A three-dimensional inverse problem in predicting the heat fluxes distribution in the cutting tools was reported by Hung et al. (2005) using thermoplastic, orthogonal, heat flow, *h*, and 3D models.

The effect of cutting tool material and edge geometry on tool life and workpiece surface integrity was studied by Hughes et al. (2005) using thermoplastic, orthogonal cut, Ti–6Al–4V, and uncoated tungsten carbide (10 mm diameter); the parameters used are as follows: roughness, Vc, f, depth of cut, radius of the tip of the tool, residual stress, pressure, wear, and 3D and 2D models.

Arbitrary Lagrangian–Eulerian analysis on cutting with a honed tool was reported by Kshawy et al. (2007), using thermo-elastoplastic, orthogonal cut, AISI 4140 steel, and cemented carbide; the used parameters are as follows: radius of the tip of the tool, f, V_c, cutting angles, contact forces, T, deformation, F_c, von Mises stresses, and 2D models.

Determinate of optimal cutting conditions in orthogonal metal cutting using LS-DYNA with the design of experiments approach by Masillamani et al. (2009), using thermo-elastoplastic, orthogonal, and aluminum 6061-T6, is simulated as a rigid body; the parameters used are as follows: cutting speed, angle of incidence, depth of cut and temperature, and 3D numerical models.

However, the previous studies suggest requiring the incorporation of friction and shear failure parameters in the oblique and orthogonal cut simultaneously in 3D, so in this chapter, the incorporation of the previous models is developed and validated.

20.1.6 Physical Model of the Orthogonal Cut, Process Parameters, and Geometry

The orthogonal cutting model can be used to approximate turning and some machining operations with a tip. In this way, the majority of the cut will take place in the direction of the advance. Coming up next the conversion of one cutting situation to the other is indicated. In a few words, the conversion of the cut into the lathe is made as if a brushing of a plate were made.

Table 20.1 Conversion: operation of turning against orthogonal cutting

Turning operation	Orthogonal cutting model
Advance f	Thickness of the chip before cutting t_0
Depth d	Cutting width w
Curing force F_c	Cutting force F_c
Cutting speed v	Cutting speed v
Breakthrough force F_f	Pushing force F_t

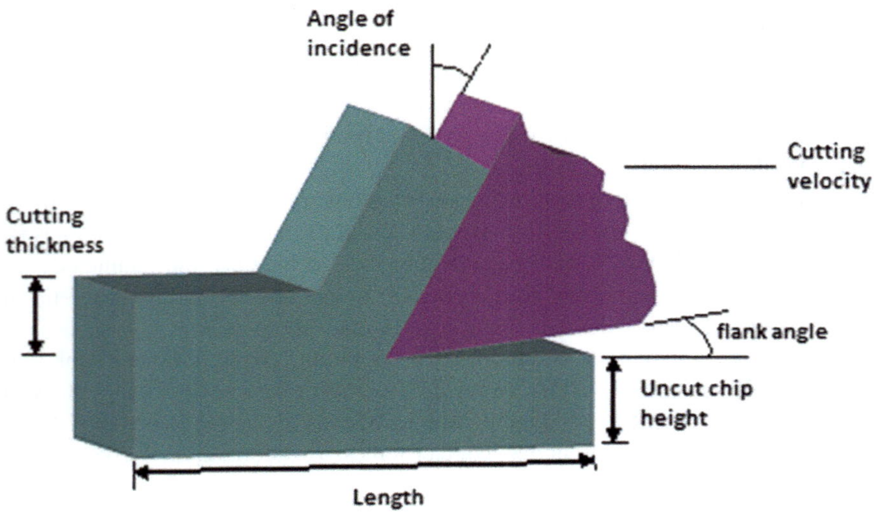

Fig. 20.1 Physical model of the process of orthogonal cut in three dimensions

The interpretation of the cutting conditions is different in the two cases. In the orthogonal cut, the thickness of the chip before cutting (t_0) corresponds to the advance (f) in turning, and the width of cut (w) corresponds to the cutting depth in turning (see Table 20.1). In addition to the pushing force, (F_t) in the orthogonal cutting model corresponds to the breakthrough force (F_f) in the turning. Table 20.1 summarizes the conversions.

The geometry and dimensions of the CAD model, as shown in Fig. 20.1, represent the real physical model that later is checked in the experimental tests.

The physical model of the cutting process is based on the orthogonal cut; this model is shown in Fig. 20.1, where the workpiece and the cutting tool can be seen. The said model aims to consider the cutting process in three-dimensional form, and the cutting width is five times higher in some cases than the depth of cut.

Such situation is obtained under the following conditions: the piece has a length $L = 16$ mm, $h = 4$ mm, and a width of cut of 5 mm, and the angle of clear is $\alpha = 10°$, which will remain constant. The effective normal angle of incidence (γ), the depth of cut (t_c), and the cutting speed (V_c) will be variable parameters in the

Table 20.2 Results of the maximum shear stresses and maximum cutting temperatures at different cutting conditions

Maximum shear stress ($T_{máx}$)	Main parameters in finite element models with ANSYS/ LS-DYNA
640.966 MPa	$C = 940$, $P = 3.5$, $\beta = 0$
950.13 MPa	$C = 220$, $P = 5$, $\beta = 0.5$
1340.78 MPa	$C = 940$, $P = 3.5$, $\beta = 1$
$\Delta t_{máx}$	Main parameters in finite element models with ANSYS/ LS-DYNA
(188-20) °C	$C = 220$, $P = 5$, $\beta = 0$
(262-20) °C	$C = 220$, $P = 5$, $\beta = 0.5$
(357-20) °C	$C = 220$, $P = 5$, $\beta = 1$

determination of the stresses, deformations, and temperature distribution in the process of formation of the chip. For the case of the numerical simulation of this study, we will work with a cutting speed of 2 m/s, the material of the workpiece will be steel 1018 and a cobalt steel tool will be used.

In the cutting process, the lower surfaces and the left side will fix the workpiece. The tool will move with a constant speed, the displacement of the edge of the tool will be horizontal.

The conditions of the cutting process for the model presented in this research are presented in Table 20.2; this model is representative for the study of boundary conditions, for the model presented are follows: cutting model 1, $\gamma = 14°$, $t_c = 1.2$ mm, $V_C = 2$ m/sec, Cuting Width = 5 mm.

20.1.7 Behavior of the Material and Its Properties

Steel 1018 is used, because it is a commercial steel easy to get; in addition, the properties of the material are also common. The material of the workpiece will be a low carbon steel AISI 1018, and the material of the cutting tool will be cobalt steel, whose mechanical properties are given below (LSTC et al. 2007).

- Material properties of the Steel 1018:
  - $\sigma_y = 310$ MPa,
  - $E = 200$ GPa,
  - $\sigma_{ult} = 780.93$ MPa,
  - $v = 0.29$,
  - $\eta = 0.21$,
  - $\rho = 7865$ kg/m^3, and
  - $C = 500$ joules/(kg-°C).

- Material properties of the cobalt steel:
  - E = 200 GPa,
  - ν = 0.3,
  - ρ = 7865 kg/m^3, and
  - C = 456 joules/(kg-°C).

Note: In the numerical part, the cutting tool of cobalt steel will be simulated as a rigid body.

The yield stress (σ) of the material was obtained by Shikarashi et al. (1983), in compression tests at a temperature of 700 °C, and a deformation velocity of 2 × 103s-1.

The empirical equation of Shikarashi and Maekawa includes the effect of temperature, the rate of deformation speed and has the following form:

$$\sigma = \sigma_0(T, \dot{\varepsilon}) \left[\int_{T, \dot{\varepsilon} \equiv h(\varepsilon)} e^{\frac{k}{n}T} \dot{\varepsilon}^{-\frac{m}{n}} d\varepsilon \right]^n \tag{20.1}$$

where σ is yield stress, σ_0 is initial stress, T is temperature, $\dot{\varepsilon}$ is strain speed, $e^{\frac{k}{n}T}$ is antilogarithm of binomial exponent with respect to time in its application elastic load, and $\dot{\varepsilon}^{-\frac{m}{n}}$ binomial is exponent of the strain velocity.

Solving Eq. (20.1), the Shirakashi equation takes the form

$$\sigma = A_0(T, \dot{\varepsilon}) \left(\frac{\dot{\varepsilon}}{1000} \right)^{0.0195} . \varepsilon^{0.21} \tag{20.2}$$

where A_0 is initial area.

The fluency of the material (ε^{el}) begins with a stress value of ($\sigma_y = 310$ MPa), and with a modulus of elasticity of ($E = 200$ GPa), so the elastic component of the deformations will be:

$$\varepsilon^{\text{el}} = \frac{\sigma_y}{E} \tag{20.3}$$

$$\varepsilon^{\text{el}} = \frac{310}{200,000} = 0.00155 \tag{20.4}$$

To determine the behavior of the curve of the material in the plastic zone, we start from the principle that the component of the plastic deformation (ε^p), for a value of the stress equal to the yield stress, that is to say ($\sigma_y = 310$ MPa), the plastic deformation will be considered as ($\varepsilon^p = 0$).

20.2 Analytical Considerations for a Lagrangian Model

20.2.1 Basic Equations

A description of the Lagrangian is the equations of motion, which are given by

$$\rho \dot{W}i = \frac{\partial \sigma_{ij}}{\partial x_j}, \quad Wi = \dot{x}_i \tag{20.5}$$

where ρ is the density of mass, and $\dot{W}i$ is the acceleration.

The point represents the derivative in time along the pattern of the cut surface, and $\dot{W}i$ are the components of the velocity (Stefanov et al. 1997).

To ensure continuity, the following relationship is proposed:

$$\frac{\dot{\vartheta}}{\vartheta} = \dot{\varepsilon}_{\mathrm{kk}} \tag{20.6}$$

where ϑ is the relative volume, $\dot{\vartheta}$ is the volumetric deformation velocity, and $\dot{\varepsilon}_{\mathrm{kk}}$ is the variation of the volumetric strain.

The energy equation is

$$\rho \dot{E} = \sigma_{ij} \dot{\varepsilon}_{ij} \tag{20.7}$$

where $\dot{E}$ is the variation of Young's modulus.

For any strain velocity, we have

$$\dot{\varepsilon}_{ij} = \frac{1}{2}\left(\frac{\partial W_i}{\partial x_j} + \frac{\partial W_j}{\partial x_i}\right) \tag{20.8}$$

20.2.2 Domain, Governing Equations, and Mechanical Boundary Conditions

In the simulation by finite element in the metal cutting process, it is of great interest to define the initial and border conditions and to study the effects in the orthogonal metal cutting process on the distribution of stress in three dimensions. The sequential proposal in coupled thermomechanical analysis involves two steps: performing an explicit transient analysis, followed by a thermo-elastoplastic structural analysis. This numerical simulation is the validation of the experimental results shown later (Stefanov et al. 1997).

20.2.2.1 Domain

In the present formulation, the problem is decoupled. When mention is made of the decoupled term means that first, the structural part of the physical phenomenon is considered. Later the thermal part is coupled, and in that way, we already have the physical phenomenon under study complete, the equations below are presented, are the structural part first, so in this case, the elastic strain is small compared to the plastic strain, and the working material is elastoplastic. The transitory term in the equation of movement is taken into account. In addition, body forces are negligible as initial conditions.

20.2.2.2 Government Equations

For the uncoupled problem, the workpiece is elastoplastic. The body forces are neglected; the velocity fields are represented, the fields of the strain velocities with $\dot{\varepsilon}$, the field of hydrostatic stress (or pressure) with p, and the field of deviating stresses σ'; and the control volume is governed by the following equations:

1. Strain speed ratio,
2. Strain velocity,
3. Movement equations, and
4. Incompressibility constraint.

For the purpose of the finite element formulation, the aforementioned equations need to be expressed in form of components. So, the problem is three-dimensional, the velocity vector has three components that can be different from zero, any component is considered independent.

In terms of components with respect to the system of coordinate axes according to Fig. 20.2.

The equations of government are as follows:

In terms of components with respect to the system of coordinate axes according to Fig. 20.2.

20.3 State of the Problem to Analyze

If the machined material is considered as a connected region according to Fig. 20.2, then the formation of the chip could be considered and referred to as a "flow in motion". The continuity of the material is interrupted along the operation pattern only while a space is opened over the local region in front of the tip of the tool. Thus, in this way, the surface is continuously transformed into the chip and where the chip–tool slip occurs, which occurs along the cutting tip.

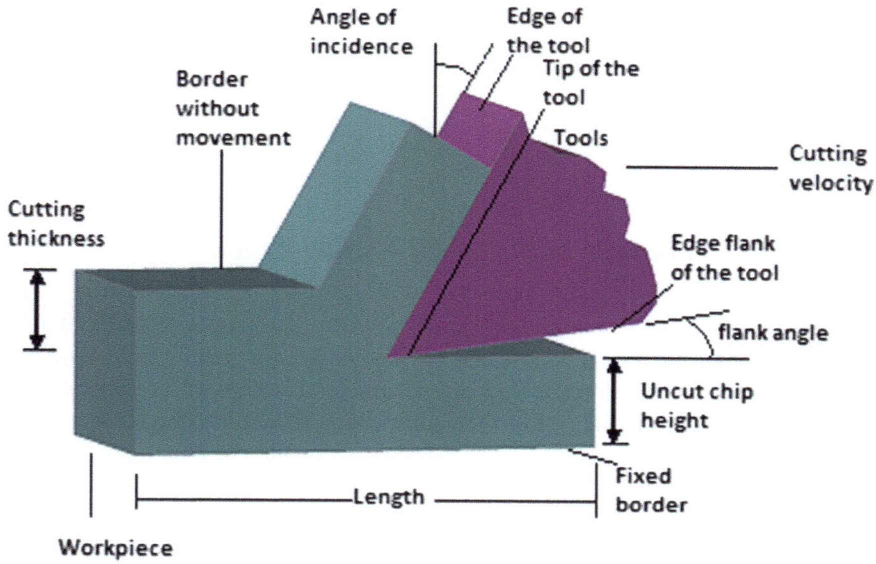

Fig. 20.2 Orthogonal 3D cutting process where the tool and the workpiece are shown

Referring to the coordinates *x*, *y*, *z*, of Fig. 20.2, the workpiece is rectangular and is fixed at the boundary B. The tool has a horizontal movement with constant cutting speed (V_c). That corresponds to an orthogonal cut, with an effective angle of incidence (γ) (Shirakashi et al. 1983).

The initial conditions are

$$t = 0, \ \sigma, v_x, v_y, v_z, \ E = 0 \quad (20.9)$$

The border conditions are

The left lateral part of the workpiece is a fixed surface, and the base of the workpiece is too fixed by what we now have

$$v_x, v_y, v_z = 0, \text{ for } (x, y, z) \in B, \text{for } (x, y, z) \in \Gamma_1 \quad (20.10)$$

$$\sigma_{i,j,k} = 0, \text{ for } (x, y, z) \in \Gamma \quad (20.11)$$

where B is the lateral border of the piece, Γ_1 is the front face preceding, and Γ is the upper face (see Fig. 20.2).

The other surfaces are free.

In the tool:

$$u_x = \boldsymbol{v} \quad (20.12)$$

In the contact region or slip zone workpiece–tool.

$$u_x = \nu v_x; u_y = \nu v_y; u_z = \nu v_z \tag{20.13}$$

where u_x, u_y, u_z are speeds in x, y, z.

20.3.1 Heat Transfer in the Metal Cutting Process

The generation of heat that brings as consequence the distribution of temperature in the metal cutting process is determined by the contact for friction chip–tool, and workpiece–tool mainly. In metallurgical and structural studies of the metal cutting process, the transfer of energy from the cut metal is represented by a surface as a heat source or by the generation of energy that gives the cut metal with a distribution of initial temperature.

In general, the temperature field should be taken as thermodynamically coupled to the mechanical field.

The governing equation of heat flow follows the first law of thermodynamics (conservation of energy). This state law says that the rate of change of the internal energy (ρCdt) and the conduction ($\nabla.q$) should be in equilibrium with the production of heat (Q) and the power of elasticity and plasticity strains ($\dot{\varepsilon}^e$) and ($\dot{\varepsilon}^p$), respectively (Stefanov et al. 1997).

$$\rho C \frac{\partial T}{\partial t} + \nabla.q = Q - \frac{E\alpha T}{1 - 2v} \dot{\varepsilon}^e + \xi s \dot{\varepsilon}^{\mathrm{ep}} \tag{20.14}$$

where ρ is density, C is specific heat, $\nabla.q$ is heat conduction, Q is the heat generated, E is the Young's modulus, α thermal expansion coefficient, T is temperature, v is the Poisson relation, s is the relative density, $\dot{\varepsilon}^e$ is elastic deformation speed, and $\dot{\varepsilon}^{\mathrm{ep}}$ is elastoplastic deformation speed.

The parameter ξ takes a value of 1, if all the dissipation of the energy is inelastic ($\dot{\varepsilon}^p$) and is converted into heat. The term of mechanical coupling in Eq. (20.14) is in many cases not considered, because its influence on the temperature field is very small (Shirakashi et al. 1983). Therefore, in this way, it is possible to divide the thermomechanical analysis into a metal cutting process into two main parts: the analysis of the mechanical field and the analysis of the temperature field.

20.3.2 Constitutive Equation for Structural and Thermal Models, Kinematic Plastic Equation of Cowper–Symonds (Manual of ANSYS Version 11, 2011)

In the simulation by finite element in the structural and thermal part, a kinematic elastoplastic material model was used, which can be utilized with isotropic and kinematic hardness, or a combination of both, with dependence on the strain speed and failure. The combination of kinematic and isotropic can be varied by adjusting the parameter of hardness (β); its value is between 0 (only kinematic hardness) and 1 (only isotropic hardness). The strain speed is supported by the Cowper–Symonds model, which scales the yield stress by the deformation rate, a factor that is dependent as shown in Eq. (20.15).

$$\sigma_y = \left[1 + \left[\frac{\dot{\varepsilon}}{C}\right]^{\frac{1}{p}}\right]\left(\sigma_o + \beta E_{\mathrm{p}} \epsilon_{\mathrm{p}}^{\mathrm{eff}}\right) \tag{20.15}$$

where $\dot{\varepsilon}$ is strain speed; Cyp are the parameters of strain rate Cowper–Symonds, which is the parameter of deformation speed Cowper–Symonds; $\epsilon_{\mathrm{p}}^{\mathrm{eff}}$ is the effective plastic strain; σ_o is initial yield stress; and E_{p} is plastic hardness module (Groover et al. 2004).

With Eq. (20.16), you can get the module E_{p}.

$$E_{\mathrm{p}} = \frac{E_{\mathrm{tan}} E}{E - E_{\mathrm{tan}}} \tag{20.16}$$

20.3.3 Modeling of the Contact and of Friction

The friction in the orthogonal cutting process occurs at the contact surface of the tool and the workpiece.

Near the tip of the cutting tool, friction exists in the contact area of the tool's impact face and the chip. On the face of the tool, sliding friction occurs in the remaining area. In the present work, the Coloumb friction law is used for the friction model described by Eq. (20.17) (Manual of ANSYS version 11 et al. 2011).

$$\mu_{\mathrm{c}} = \mathrm{FD} + (\mathrm{FS} - \mathrm{FD})\left(e^{(-DC * V_{\mathrm{rel}})}\right) \tag{20.17}$$

where FS is the static friction coefficient; FD is the dynamic friction coefficient; DC is the exponential decay coefficient; and V_{rel} is the relative speed.

In the cutting process, the part of the material of the piece that is converted into chips slides on the surface of attack of the tool, causing a contact surface with high friction between both surfaces. In this work, modulations were worked with a coefficient of static friction equal to the dynamic of 0.4; with these values in Eq. (20.16) is obtained (μ_c) the dependent on the relative velocity of the contact surface.

The viscous friction coefficient V_c is necessary to define the limit of maximum friction. So, a limit force (F_{lim}) is calculated numerically.

$$F_{\text{lim}} = V_c(Acont) \tag{20.18}$$

where *Acont* is the area of the segment contacted.

The suggested value for the yield stress in the cut is

$$V_c \approx \frac{\sigma_y}{\sqrt{3}} \quad \sigma_y \approx \sqrt{3}\mathrm{V} \tag{20.19}$$

Note: Eq. (20.19) is an empirical equation.

20.3.4 Methodology for Modeling by Finite Element Analysis

For the modeling of the orthogonal cutting process in low carbon steel, with the use of coolant by means of finite element analysis the process is:

1. The physical model of the orthogonal or oblique cutting process is defined, by establishing the geometry of its components, physical magnitudes, use of materials and their physical–mechanical properties, and the parameters of the cutting regime.
2. Based on the physical model, the finite element model is established, with the definition of the appropriate boundary conditions, so that the model approaches the physical model of orthogonal cut or oblique cut.
3. When establishing the boundary conditions, they are discretized by mean elements.
4. Known the behavior of the physical system, which is nonlinear in nature from the material and geometric point of view, the type of solution is determined. This behavior is due to the presence of large plastic deformations, high deformation rates, large displacements with rotations, fracture of the material, and generation of heat with a considerable increase in temperature and contact surfaces with high friction value during the formation of the chip and the machined surface.
5. It is used in the FE model, the analysis of orthogonal cut or oblique cut in AISI 1018 steel, with the use of cooling fluid, as cooling medium in the process.

20.4 Results of Numerical Simulation with Reference to Example

20.4.1 Explicit Structural Models

The results come from the case presented, and it is a representative model of one of the simulations performed. For this, several models were simulated to make the comparison of parameters or input variables in the numerical simulation, of which the results of only two combinations are shown. The idea is to look for the most appropriate combination and that will be as close as possible to results similar to each other. When comparing both combinations shown in Fig. 20.5, according to Eq. (20.15) (kinematic plastic equation known as the Cowper–Symonds model). The combination of results refers to the comparing results to adjust the input variables of the constitutive equation used in the models by analysis by finite element, and which more resemble each other according to the input parameters and those are the most suitable for the study that was carried out.

Figure 20.3 shows the variation of the maximum shear stress (T), for the constants of Cowper–Symonds. Also present the coefficients of hardness of the material (β), for the two combinations found, ($C = 220\ \mathrm{s}^{-1}$, $P = 5$) and ($C = 940\ \mathrm{s}^{-1}$, $P = 5$), with a deformation due to material failure of 0.75, for a friction coefficient of the material of $\mu = 0.4$, for steel 1018.

Both series of simulations are intended to be compared, with the results of the maximum cutting temperatures in the metal, in order to verify the experimental part.

The intention to perform the tests with the parameters and constants mentioned above is, to present a study to calibrate the finite element models, cutting with orthogonal turning. The parameters and constants were adjusted during several tests

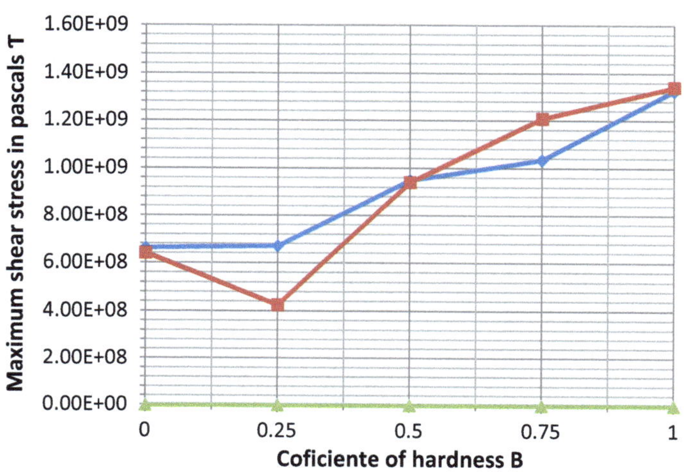

Fig. 20.3 The influence of hardness coefficient (β) with the maximum shear stress (T)

of numerical simulation, in the process of removal of the chip, these parameters are as follows: (a) the hardness coefficient of the material, deformation by (b) fault and the constants of Cowper–Symonds.

In Fig. 20.3, we can observe that the results are very similar in three points of the graph, where the parameters represent the beginning of the graph or the lowest values. The intermediate part and the final part in the simulations, the similarity in the results were given for the values ($C = 220\ \mathrm{s}^{-1}$, $P = 5$) and ($C = 940\ \mathrm{s}^{-1}$, $P = 3.5$), with values of hardness by deformation of ($\beta = 0, 0.25, 0.5, 0.75$, and 1), the deformation value for material failure was kept constant and it was 0.75.

In Fig. 20.3, the results of the maximum shearing stress (τ) are shown for values where the results obtained are similar according to the graphs shown above, for the parameters ($C = 220\ \mathrm{s}^{-1}$, $P = 5$) and ($C = 940\ \mathrm{s}^{-1}$, $P = 3.5$), with values of hardness by deformation of ($\beta = 0, 0.5$, and 1). The value of deformation by failure of the material was kept constant at 0.75; the coefficient of friction was maintained at $\mu = 0.4$.

Table 20.2 shows the results of the maximum shearing stress (τ_{max}), where we can observe the similarity in the results, according to the graph of Fig. 20.3.

The maximum shear stresses are shown below; at the end of the simulation, stage by finite element analysis, said maximum shear stress is presented in the tool–chip interface, due to the fracture of the material in the primary cutting zone.

20.4.2 Structural Thermal Models

The results of the temperature distribution were simulated for the same case (Table 20.2), by natural convection (air static) and heat conduction, without considering the heat radiation; this is because the results that we want to analyze. In this case, the isotherms or temperature distribution and the higher cutting temperature were considered for the study. Also, because we have an open system, we do not consider the radiation important, since as time passes, the cutting temperature will decrease, until it becomes equivalent with respect to the environment temperature, and that process of energy transfer is not important at the moment, for the purposes of this article.

20.4.3 Thermal Properties of the Workpiece and the Tool

- Workpiece (Steel 1018):
  - Specific heat capacity $C = 500$ joules/(kg-°C),
  - Thermal conductivity $K = 51.82$ W/m-K (for 20 °C), and
  - The convection coefficient for natural air $h_f = 15\ \mathrm{W/(m^2\text{-}°C)}$ at room temperature.

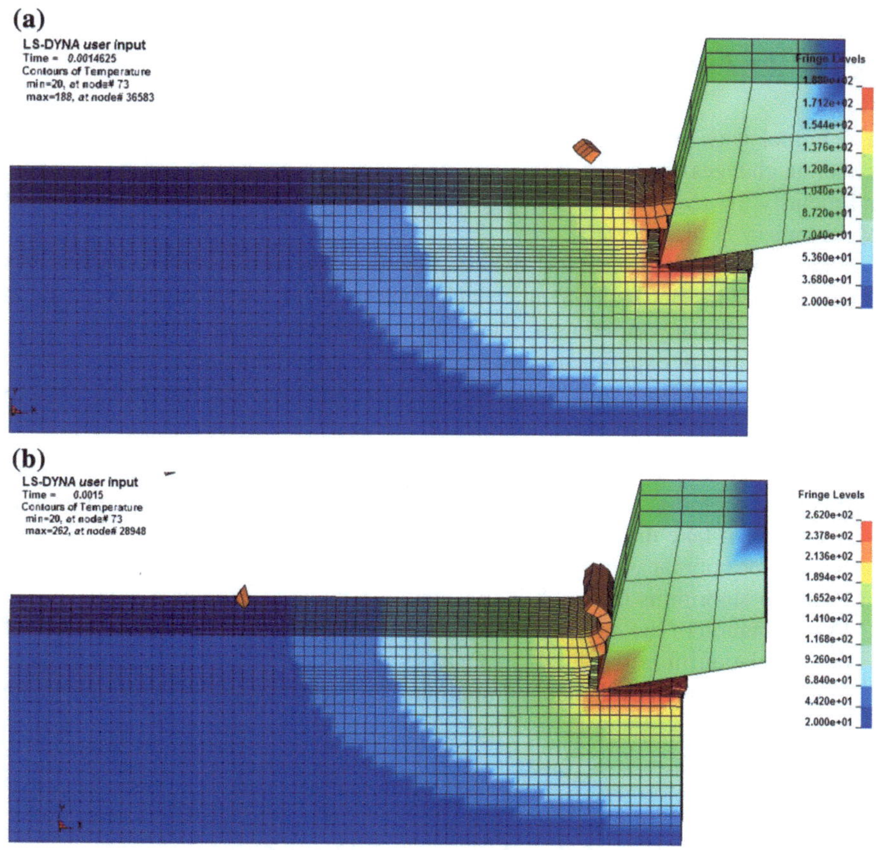

Fig. 20.4 Temperature distribution to different hardnesses: **a** temperature distribution, to $C = 220$, $P = 5$, $\beta = 0$ and **b** temperature distribution, to $C = 220$, $P = 5$, $\beta = 0.5$

- Work tool (Cobalt steel)
  - Specific heat capacity $C = 456$ joules/(Kg-°C) and
  - Convection coefficient for natural air $h_f = 15$ W/(m^2-°C).

Coming up next you can see Fig. 20.4, where the results of the temperature distribution or isotherms are shown, at the end of the simulation stage by finite element analysis, the maximum temperature in the cut of the material; it is presented in the tool–chip interface, due to the fracture of the material, in the primary cutting zone.

20.4.4 Design of Experiments Applied in Numerical Models by Finite Element

In a formal way, an experiment is defined as a test or series of tests in which deliberate changes are made in the input variables of a process in order to observe and identify the reasons for the changes that could be seen in the response of exit (Montgomery et al. 2009).

When working with several factors that affect a process, the correct approach is to perform a factorial experiment, which consists of an experimental strategy in which the factors are varied as a whole, in order to analyze their behavior.

So, the design of experiments can be of various types, depending on the specific purposes that have been raised. Among the most common and important, the following can be mentioned:

1. Comparison of the means and variances of the alternatives analyzed. They are commonly referred to as the design of simple factor experiments: to do this, it is necessary to take into account the size of the sample and the initial conditions.
2. Search for optimal values of a set of variables: Generally, it is necessary to use heuristic algorithms (a search is invented and found to provide a clue).
3. Determination of the importance and effect of different variables in the results of the simulation: it is based on the analysis of the variance and regression techniques as means to evaluate the importance and effect of several variables in the operating results of a system.

20.4.5 Description of the Orthogonal Metal Cutting Problem

In this work, the validation of the models by finite elements was presented, where the margin of error compared to the experimental part was approximately 4% to 8%, considering the maximum cutoff temperature as the main comparison parameter. The average of the data of the maximum temperature of cut was 137.80 °C, for the experimental part, and 150.40 °C, for the modeling by finite elements.

In addition to the model by finite elements, it was necessary to check the impact and the subsequent fracture of the material, which is something important that was achieved in the present research work and later, the plastic thermoelastic analysis of the material.

Thus, in this way, the finite element models were validated and accepted, for which the decision was made to work with three parameters (factors) in the metal cutting, which directly affect the maximum temperature of the cut.

- Cutting speed (V_c),
- Cut depth (t_c), and

- Effective normal incidence angle (ɣ).

In addition, an output variable:

- The maximum temperature that occurs in the cutting process (It is an average in the measurement zone) (*T*).

Therefore, a 2 k factorial experiment of three factors was designed, so that eight runs were made, with the combinations of the maximum and minimum levels, such as those shown below. The following levels were taken for the cutting parameters:

- For the cutting speed, it was taken between 1 and 2 m/s.
- For the depth of cut was taken between 1 and 1.2 mm in order to analyze the response data based on this variability.
- For the effective normal incidence angle, 2 data were taken; 100 and 140 which are common data used in practice (for the purposes of this work only the mentioned angles were considered.

20.4.6 The Factorial Design 2^3

There are three factors in numerical models per finite element that need to be analyzed, which we will name: *A* (cutting speed), *B* (depth of cut), and *C* (effective normal angle of incidence), either as in our case in two levels. The design is called 2^3 (factorial design) and eight combinations in the treatment, with the use of the notation "+" and "−", which represent the high and low level of the factors; you can show the eight runs for the design 2^3.

This is sometimes called the design of the matrix, in which you can write the notation with the combination of the treatment in standard order such as (1), A, B, C, AB, AC, BC, and ABC. These symbols also represent the total of all (*n*) observations, in this particular combination of treatment.

In the design of experiments used, there are seven degrees of freedom among eight combinations treated in the factorial design 2^3. The three degrees of freedom are associated with the main effects of A, B, and C. Four degrees of freedom are associated with the interactions of the factors: AB, AC, BC, and ABC.

20.4.7 Results of the Design of Experiments for the Factorial Analysis 2^3 in the Orthogonal Metal Cut

Model comparison by finite element with an effective normal angle of incidence of ɣ = 10° and 14°; depth of cut 1 and 1.2 mm; cutting speed 1 and 2 m/s, relative to

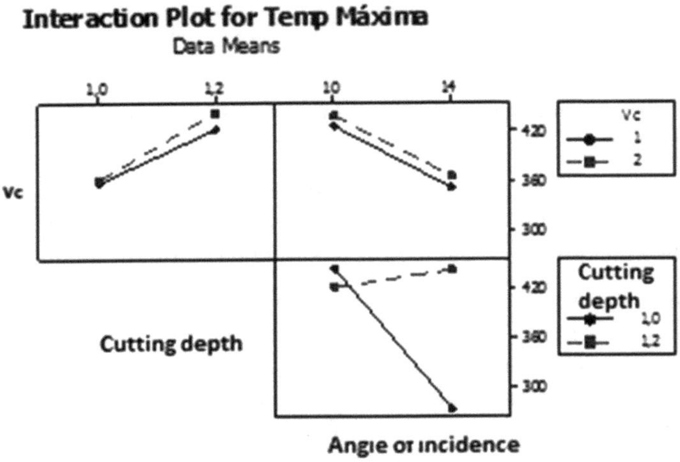

Fig. 20.5 Interaction of ABC factors, in relation to the maximum temperature in the cut

the maximum temperature as response variable, in the orthogonal metal cut of STEEL 1018, with the use of cobalt steel tool.

In the ANOVA, you can observe high values of standard deviation (SS) and high values in the standard deviation of the means (MS), especially for the effective normal angle of incidence that is the most significant value, because it has the highest value. The second result with significance is that of the factor of the depth of cut because it has a high value for both the standard deviation of the means (MS) and the standard deviation (SS).

In Fig. 20.5, in the interaction of factors in relation to the maximum metal cutoff temperature as a response variable, we have the factors interaction (*B*) Depth of cut and (*C*) effective normal angle of incidence. The slopes are cross over what is here where greater significance is observed in the interaction of factors. In the interaction of factors (*A*) V_c and (*B*) depth of cut is observed at the beginning of the slopes of the straight lines interaction at a point but without crossing of the slopes. So, there is significance, but it is not as illustrative as the case previously described, with the factors (*A*) V_c and (*C*) Effective angle of normal incidence there is no significance or interaction because the slopes remain parallel.

20.5 Experimental Test

Input parameters for cutting, instrumentation, and equipment are shown below:

1. To perform the experimental tests, in order to obtain the maximum temperature that the working material undergoes due to the cutting process, as an output parameter.

2. The depth of cut that was worked was of 5 mm for the experimental part, with a speed of advance in the cut of $V_c = 0.1466$ m/s. The piece of work that was used was a tube of steel 1018 of 25.4 mm of outer diameter and an internal diameter of 15.4 mm with a thickness of 5 mm, and the working tool that was used, is cobalt steel with an effective normal incidence angle of $\gamma = 14°$ and a clear angle of $\alpha = 10°$.
3. Several experimental tests were made according to the data shown.

20.5.1 Statistical Analysis of the Sample for the Experimental Tests

We can observe that the sample data in the experimental tests, which were 22, this data were first subjected to a correlation analysis, to verify the linearity in the sample and, as we can see, the data have a linear trend. The average in this case as maximum temperature in the orthogonal metal cut was 132 °C; according to the graph Fig. 20.6, these data were subjected to analysis in a computer statistical package of engineering which is called "mini tab".

In this last graph, it can be observed that the sample was 22 tests, in which we show an average cut temperature of 132 °C in the experimental tests.

The data sampled were subjected to a statistical analysis in the "mini tab" computational package, in order to verify the trend toward a normal distribution. As a result, it was found that if there is a normal tendency, there is no well-defined data due to the size of the sample according to the population. But it is acceptable because if it is achieved in a certain way the normal distribution of data within the graph. In this analysis, the average of the sampled data was 132 °C, the lower limit

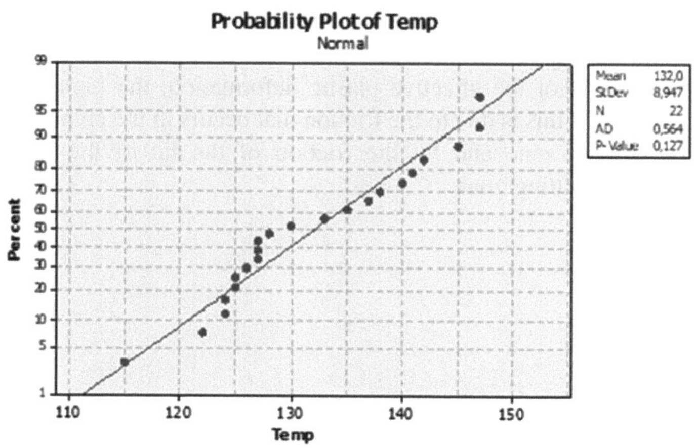

Fig. 20.6 Graph of correlation of data of maximum temperature of cutting of experimental tests

(left side of the graph) resulted from 115 °C, and the upper right limit of the graph (right side of the graph) was 147 °C, according to the data sampled in the experimental tests.

An analysis of the sample was carried out, to verify the linear trend of the data, with the purpose of estimating the value of the average of the data if the whole population was taken and the results were that the average of the temperature data maximum of the metal cut would be 136.55 °C. This result was compared with the finite element modeling; the difference with respect to the sample is that the average was 132.5 °C, so the values are very similar, and the difference is about 4 °C.

20.5.2 Results of the Numerical Models by Finite Element, for the Case of Orthogonal Cut, by Convection with Static Water, with a Lateral Incidence Angle of $\gamma = 10^{\circ}$, $V_C = 2$ m/s, Depth of Cut = 1.2 Mm

In the distribution of von Mises stresses, the maximum values are in the primary cutting zone (incidence of the tool with the workpiece), this is due to the friction of the tip of the tool with the workpiece. Too are maximum in the secondary cutting area (interface chip cutting tool), and this is due to the friction between the chip–tool interface. The highest value in the results of the stresses of von Mises is presented in the element 27,413 that is located in the area Primary cutting or flat cutting chip, and it is $2.82385e^9$ Pa.

The distribution of maximum shear stresses is very similar; in terms of the distribution of stresses; the highest value is in element 27,413 of the primary cutting area, and it is $1.62978e^9$ Pa.

Distribution of the displacements in the workpiece; the highest values are found in the chip, this is due to the friction phenomenon in the chip–tool interface or secondary cutting zone, during the cutting process, the highest value is 0.00110057 m and is given in node 32,582, which It is located in the chip.

In the distribution of the effective plastic deformation, the highest values are given in the chip, and this is due to the friction that occurs in the chip–tool interface or secondary cutting zone and by the friction of the tip of the tool with the workpiece primary cutting zone.

20.5.3 Properties for Structural Thermal Models

20.5.3.1 Thermal Properties

- Workpiece Steel (1018):
  - $C = 500$ joules/(kg-°C),
  - $K = 51{,}082$ W/(m-°K) for 20 °C, and
  - $h_f = 293.15$ W/(m^2-°K).
- Work Tool (cobalt steel)
  - $C = 456$ joules/(kg-°C) and
  - $h_f = 293.15$ W/(m^2-°K).

where C is Specific heat capacity; k is thermal conductivity; and h_f is natural convection coefficient for water.

20.5.4 Result Thermal Model for Orthogonal Cutting, by Natural Convection with Static Water

References for thermal simulations are the isotherms or the temperature distribution. The highest temperature in the metal cutting is presented in the interface of the workpiece–tool tip (primary cutting zone) and in the chip–tool interface (secondary cutting zone). This is due to the friction that occurs both in the primary cutting zone, and in the secondary zone (chip), the maximum temperature in the metal cut is 90.5 °C for this case.

20.5.5 Numerical Model of Finite Elements by Natural Convection with Static Water for Oblique Cutting

Before analyzing the results of numerical models by finite element for oblique cutting in metal cutting, we will give some important definitions that are related to the subject.

All metal cutting operations can be similar to the orthogonal cutting process or oblique cut as seen in Fig. 20.6, where the tool is wedge-shaped, has a straight edge, and its movement is restricted with respect to the workpiece, such a way that a layer of material is removed in the form of chip.

Oblique cutting: Most cutting operations involve three-dimensional tool shapes. In this type of cutting, in this type of cut, the edge of the tool forms an angle called

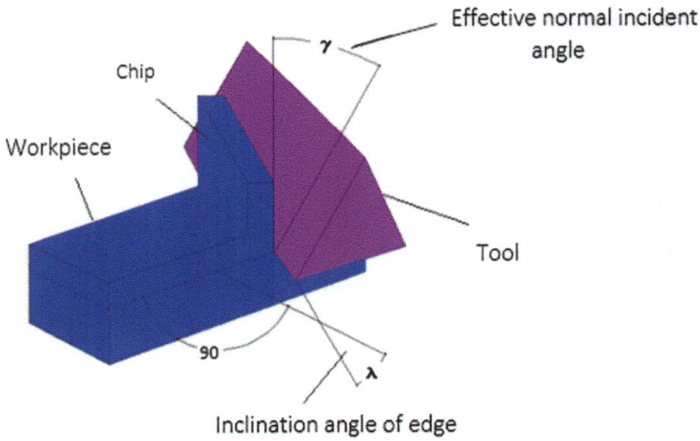

Fig. 20.7 Oblique cut

the angle of inclination. The lateral direction of the chip movement looks like that of a motor-shaper machine with an oblique blade, which pushes the earth to the side as shown in Fig. 20.7 (Boothroyd et al. 2006).

It was decided to consider the cutting tool as a rigid and symmetrical body, in order to avoid possible interactions in the results due to the fine meshing of the workpiece.

The finite element model of oblique cut takes the same restrictions discussed below.

The constraints imposed on the finite element model for oblique cutting are as follows: (a) it is restricted from any movement in the left side of the workpiece; (b) it is restricted too in the bottom or base of the workpiece; and (c) the cutting tool moves in the x direction at a constant cutting speed horizontally.

20.5.6 Results of the Numerical Models by Finite Element, for the Case of Oblique Cutting with a Lateral Incidence Angle of $\gamma = 14^{\circ}$, $V_C = 2$ m/s, $Depth of Cut = 1$ mm, an Angle of Inclination of the Edge $\gamma = 10^{\circ}$ and an Anhel of Clear $\alpha = 10^{\circ}$

The highest values of the distribution of von Mises stresses were found in the workpiece and in the chip; this was due to the friction between the chip–tool interface and the friction of the tip of the tool with the workpiece. The highest value of the stress was found in the element 3854 that is located in the output of the chip, and it is $2.65916e^{9}$ Pa.

In relation to the distribution of maximum shear stresses was observed a similar trend to one of the simulations analyzed above. In relation to the distribution of stresses; the highest value was in element 3854 and $1.52796e^9$ Pa, and it is located at the exit of the chip and is due to the chip–tool interface in the secondary cutting area.

20.6 General Conclusions

The general conclusions related to this research work are as follows:

1. A three-dimensional model of metal cutting was obtained using the finite element technique, the model includes orthogonal cutting; with heat transfer by natural convection to still air or natural convection to water at rest.
2. For the realization of the model of finite elements, we started from a physical model of orthogonal cut, where the input variables of the process were defined; geometry of the cutting tool, material properties of the workpiece and the tool. This physical model contains the fundamental elements of any cutting process by turning.
3. The verification in the development of the generation and formation of the chip was achieved in this simulation by finite element, from the point of view of Lagrange; this was obtained because the developed models focused their attention on analyzing how the speed of an element in particular with respect to time (Δt). This verification was accomplished when performing the metal cut at a cutting speed of 1, 2, and 0.15 m/s for numerical validation with the experimental part.
4. Regarding the experience that was achieved, in relation to the use of this computer package, it is of great advantage to have practical knowledge in the use of the software, since results are obtained dynamically that can be compared with results analytical. So it can be simulated with a time differential; this helps us to know in different stages the cutting process by means of the dynamic simulation by finite element.
5. The realization of experimentally tests with data according to the numerical simulations by finite element is important, considering for this the mechanical boundary conditions (structural and thermal) established in the analytical part.
6. The solution of the thermal part is obtained on the same finite element mesh generated in the mechanical–structural part, determining the temperature distribution in the workpiece, chip, and cutting tool, by means of the conduction and the convection. The results were compared with experiments carried out in the mechanical workshop.
7. The results obtained by the model of finite elements and the experimental ones, in the orthogonal cutting process, showed a difference of 4%, which is in a very similar range in the finite element models developed at present.

8. It is concluded that when comparing results, for example, of the orthogonal cut simulated by natural convection with static air, according to the parameters, $V_C = 2\,\text{m/sec}$, depth of cut 1.2 mm, and an effective angle of incidence of $\gamma = 10°$, with the model of orthogonal cut by natural convection with static water the results differ. This difference is in terms of the maximum cut temperature obtained, by natural convection with static air resulting from 434.44 ° C, and the simulation by finite element by natural convection with static water resulted in 90.5 °C.
9. The results are therefore very different, because the heat capacity of the fluids used (air and water) is not equal, so the water cools more than the air.
10. But as for the comparison of the models of oblique cut, by natural convection and orthogonal cut by natural convection both with static water, there was very little difference in terms of the results, of the maximum cutting temperatures obtained. This is most likely due to the angle of the inclination of the cutting edge in the oblique cut; in the oblique cut, it was 86.05 °C and in the orthogonal cut of 90.5 °C.
11. The fact to perform experimental tests, with data according to the numerical simulations by finite element and considering the mechanical (structural) and thermal boundary conditions, established in the analytical part of this work, finally led us to find the results that were sought.
12. Performing a statistical analysis of the maximum temperature data in the orthogonal metal cut gives us more certainty based on the sample used in the experimental tests. Based also on sample measurement which was 132.05 °C, because the estimated average of the data of the population was 136.5 °C, the means are very similar, and this gave us a margin of error of approximately 3% in the experimental tests, with respect to the means.
13. In the analysis of the results, by means of the statistical technique of design of experiments, it is concluded that this tool provides us with a lot of help and support, to discern and better understand the results obtained, in this case of the models by finite element.
14. So, when reviewing, the results are obtained with the combinations of the factorial design described in this chapter, of the eight runs for the design 2^3 as shown in the results, made with the statistical software MINITAB, with three factors, which they named: A (cutting speed), B (depth of cut), and C (effective normal angle of incidence). The results are similar to our two levels case.
15. We can say that of the ANOVA gives us results that support and reaffirm those obtained data in the interactions of the factors. In these, we can observe high values of standard deviation (SS) and high values in the standard deviation of the means (MS), especially for the effective normal angle of incidence that is the most significant value, for having the data of the highest value. The second result with significance is the factor of the depth of cut, for having a high value for both the standard deviation of the means (MS), like that of (SS).

16. In graph 5, in the interaction of factors in relation to the maximum metal cutoff temperature as a response variable, we have the factors interaction (*B*) depth of cut and (*C*) effective normal angle of incidence, the slopes are cross what is here where there is greater significance in the interaction of factors.
17. This way it is verified, in the analysis of the results of this statistical study of this work that the factors that most affect the orthogonal cut in the steel 1018, for the parameters given in this chapter, are effective normal angle of incidence and the one of the depth of cut, in that order. According to the ANOVA table (analysis of the variance), for the results found and that are supported with the interactions of Fig. 20.5.
18. As for the simulations by finite element, the investment of computation time for each of the runs was quite since each model took approximately 8 h, between the structural, thermal, and coupling. Therefore, if we add that there were eight combinations, meaning that the convergence time of all the models in this Chap. 4 was 64 h, plus the time spent on the design.

References

Andreev GS (1958) Photoelastic study of stress in a cutting tool by means of cinematography. Vestnik Mashines Troeniya 5:54–57. https://doi.org/10.1016/0020-7357(65)90015-6

Bailey A (1975) Friction in metal machining—mechanical aspects. Wear 31:243–275

Battaglia JI, Puigsegur CO (2005) Estimated temperature on a machined surface using an inverse approach Manual de ANSYS version 11. Available at: https://www.ems.psu.edu/~radovic/ChemKinTutorials_PaSR.pdf. Accessed 21 June

Battaglia JL, Puigsegur L, Cahuc O Estimated temperature on a machined surface using an inverse approach. Exp Heat Transf 18:13–32

Beck JV, Blackwell S, Clair CR (1985) Inverse heat conduction Ill-posed problems. Wiley, New York, USA

Boothroyd G, Knight WA (2006) Fundamentals of machining and machine tools, 3rd edn. CRC Press editorial, New York, USA

Ceretti E, Lazzaroni C, Menegardo L, Altan T (2000) Turning simulations with a three-dimensional FEM code. Mater Process Technol Mag 98(1):99–103. https://doi.org/10.1016/S0924-0136(99)00310-6

Chigurupati P, Jinn JT, Oh JY, Yin Y, Zhang H, Wu WTS (2004) Advances in machining process modeling. In: AIP conference proceedings, p 712. https://doi.org/10.1063/1.1766718

Childs HC, Mahdi MJ, Barrow G (1989) Metal machining: theory and applications. CIRP Ann Manuf Technol 38:55–58

Childs HC, Dirikolu MH, Maekawa K (1998) Modelling of friction in the simulation of metal machining. In: 24th leeds—lyon symposium on tribology. Tribol Ser 34:337–346

Dewhurst P (1978) On the non-uniqueness of the machining process, Proc Royal Soc London A360:587–610. http://rspa.royalsocietypublishing.org/content/360/1703/587

Dixit PM, Dixit US (2008) Modeling of metal forming and machining processes by FE and soft computing methods. Springer, London, UK

Finnie J, Shaw MC (1956) The friction process in metal cutting. ASME J Eng. Ind 79B:1649–1657. http://manufacturingscience.asmedigitalcollection.asme.org/article.aspx?articleid=1437560

Groover MP (2004) Modern manufacturing process, 6th edn. McGraw Hill editorial, Mexico

Kato S, Yamagachi Y, Yamada M (1972) Stress distribution at the interface between chip and Tool in machining. ASME J. Engineering for industry 94B:683–689

Kattwinkel UW (1957) A schnerden spanender workzeuge mit hilfe der spannungsoptik. Industrie-Anzeiger 36:29–36

Kishawy HA, Ibrahim D (2006) An analysis of machining with honed tools using ale finite element model: Ploughing force and minimum chip thickness. Trans North Am Manuf Res Inst SME 34:277–284

Kudo H (1965) Some new slip-line solutions for two-dimensional steady-state machining. Inter J Mech Sci 7:43–45. https://doi.org/10.1016/0020-7403(65)90084-6

Lee HEH, Shaffer BE (1951) The theory of plasticity applied to a problem of machining. J Appl Mech 3(1):47–53. https://doi.org/10.1016/0022-5096(54)90038-1

Lim SC, Ashby MF, Brunton JH (1989) The effects of sliding conditions on the dry friction of metals. Acta Metallica 37(3):767–772. https://doi.org/10.1016/0001-6160(89)90003-5

Lin ZC, Yarng YD (1996) Three-dimensional cutting process analysis with different Cutting velocities. J Mater Process Technol 70(1–3):22–33. https://doi.org/10.1016/S0924-0136(97)00036-8

Livermore software technology corporation (LSTC) (2007) LS-DYNA keyword user's manual, vol 1, version 971, May 2007

Masillamani DP, Chesa J (2009) Determination of optimal cutting conditions in orthogonal metal cutting using LS-DYNA with design of experiments approach. In: 8th international LS-DYNA users conference, pp 9–27. Available at: https://www.dynalook.com/international-conf-2004/09-3.pdf/at_download/file

Merchant ME (1945a) Mechanics of the metal cutting process I. Orthogonal cutting. J Appl Phys 16:267–275. https://doi.org/10.1063/1.1707586

Merchant ME (1945b) Mechanics of the metal cutting process II, plasticity conditions in orthogonal cutting. J Appl Phys 16:318–324. https://doi.org/10.1063/1.1707586

Molinary A, Moufki A (2008) The Merchant's model of orthogonal cutting revisited: a new insight into the modeling of chip formation. Int J Mech Sci 50(2):124–131

Montgomery D (2009) Design of experiments, 3rd edn. Mc. Graw Hill, México

Moore AJW (1948) Deformation of metals in static and sliding contact. Proc Royal Soc London A195:231–243. Available at: http://rspa.royalsocietypublishing.org/content/195/1041/231

Oxley PLB (1989) Mechanics of machining. Ellis Hoopwood, Chichester, UK. https://trove.nla.gov.au/version/45461553

Shaw MC, Berand AM (1960) Friction characteristic of sliding surfaces undergoing subsurface plastic flow. J Basic Eng 82:342–346 https://link.springer.com/book/10.1007%2F978-1-4613-9942-1

Shirakashi MTK (1983) Flow stress of low carbon steel at high temperature and strain rate. J Jpn Soc Precis Eng 17:166–172. https://doi.org/10.2493/jjspe1933.44.1393

Shirakashi T, Usui E (1973) Friction characteristic on tool face in metal machining. J Jpn Soc Precis Eng 39:966–972. https://doi.org/10.1016/0043-1648(75)90161-1

Stefanov YP, Makarov PV, Burkov PV, Matveev VS (1997) Dynamic simulation of generation and formation in metal cutting. Theoret Appl Fract Mech 28(2):117–124. https://doi.org/10.1016/S0167-8442(97)00036-0

Strenkowski JS, Shih AJ, Jin JC (2002) An analytical model of finite elements to predict forces of three-dimensional tools and flux of chip. Int J Mach Tools Manuf 42(6):723–731. https://doi.org/10.1016/S0890-6955(01)00162-6

Takeyama H, Usui E(1958) The effect of tool chip contact area in metal cutting. ASME J Eng Ind 80:1089–1096. http://manufacturingscience.asmedigitalcollection.asme.org/article.aspx?articleid=1437544

Usui E, Takeyama H (1960) A photoelastic analysis of machining stresses. ASME J Eng Ind 82:303. http://manufacturingscience.asmedigitalcollection.asme.org/article.aspx?articleid=1437547

Wallace PW, Boothroyd G (1964) Tool forces and tool-chip friction in orthogonal machining. J Mech Eng Sci 6:74–87

Zorev NN (1963) Interrelationship between shear process occurring along tool face and on shear Plane in metal cutting. In: Proceedings of the thirteenth international machine tool design and research conference. Palgrave, London, pp 42–49

Part IV
Human Factors

Chapter 21
The Role of Knowledge Transfer in Supply Chain Flexibility and Performance

José Roberto Díaz-Reza, Valeria Martínez-Loya, Jorge Luis García Alcaraz and Ismael Canales-Valdiviezo

Abstract This chapter proposes a structural equation model to analyze four latent variables: internal knowledge transfer, external knowledge transfer, supply chain flexibility, and supply chain performance. These variables are interrelated through three research hypotheses. The model is validated with data obtained through a questionnaire that was administered to 269 Mexican manufacturing companies. Then, the hypotheses are tested at a 95% confidence level using partial least squares. The results indicate that internal knowledge transfer has the most important direct effect on supply chain flexibility and the highest indirect effect on supply chain performance.

Keywords Knowledge transfer · Structural equation modeling
Supply chain

21.1 Introduction

In today's volatile environments, knowledge has become an essential constructive and active organizational element within companies, due to its power and influence at an operational and administrative level (Kathiravelu et al. 2014). For this reason, companies are constantly taking advantage of the maximum level of knowledge they have, or they seek to acquire and/or generate new, so they can compete and survive in global markets (Samuel et al. 2011). Knowledge transfer is an important

J. R. Díaz-Reza (✉) · I. Canales-Valdiviezo
Department of Electrical Engineering and Computational, Universidad Autónoma de Ciudad Juárez, Av. Del Charro 450 Norte. Col. Partido Romero, Juárez, Chihuahua, Mexico
e-mail: al164440@alumnos.uacj.mx

V. Martínez-Loya · J. L. García Alcaraz
Department of Industrial Engineering and Manufacturing, Universidad Autónoma de Ciudad Juárez, Av. Del Charro 450 Norte. Col. Partido Romero, Juárez, Chihuahua, Mexico

J. L. García Alcaraz et al. (eds.), *Best Practices in Manufacturing Processes*,
https://doi.org/10.1007/978-3-319-99190-0_21

business tool since it contributes to success (Kaszás et al. 2016); however, inappropriate knowledge management can be a cause of failure.

In corporate contexts, new knowledge can be obtained in two ways: through internal activities and through external sources, such as strategic alliances with universities, research institutions, and other companies, among others (Wu and Lee 2012). Despite the fact that knowledge is a highly important factor for expansion and corporate memory (Kaszás et al. 2016), both knowledge transfer and knowledge exchange are difficult to achieve, not only internally, but also externally, since the purposes and practices are not always shared by all members of a productive chain (Samuel et al. 2011).

In an era of globalized production systems, knowledge transfer and acquisition play crucial roles in supply chains (SC), which are cultivated through trust, commitment, interdependence, shared meaning, and balanced power among SC participants (Qile et al. 2011). In this sense, knowledge has its greatest impact on Supply Chain Flexibility (Clemons and Slotnick 2016), defined as the ability to adapt quickly to changes or problematic situations (Sreedevi and Saranga 2017). Flexibility is usually discussed with respect to risk and uncertainty. If companies do not have enough knowledge to adapt to changes, they are more likely to face problems and perish (Shakerian et al. 2016). Therefore, the goal of this work is to analyze the impact of knowledge transfer on the performance and flexibility of the Mexican manufacturing supply chain.

21.2 Literature Review and Hypotheses Statement

21.2.1 Knowledge Transfer in the Supply Chain

The supply chain (SC) can be considered as the cradle of knowledge whose multiple members are autonomous and have different cultures, managerial backgrounds, and supply chain management practices (Sachin and Ravi 2014). Knowledge exchange adds value and increases efficiency (Sachin and Ravi 2014); therefore, companies should focus more on knowledge transfer than on cost savings. Namely, knowledge transfer is the source to solve problems.

Multiple works have highlighted and reported the impact of knowledge transfer on both *Supply Chain Flexibility* and *Supply Chain Performance*. Similarly, it has been stated that the exchange of knowledge in manufacturing supply chains can come from both internal and external sources. According to Blome et al. (2014), internal knowledge exchange is the ability or capability of a company to share information among its departments, production, and administrative areas; in other words, companies are able to internally relate their functions. Also, internal knowledge is related to production and operations and covers all the main functional areas, such as purchasing, procurement, sales, and marketing (Ryoo and Kim 2015).

To measure *Internal Knowledge Transfer*, some authors recommend verifying the following points (De Silva et al. 2018; Vlajčić et al. 2018; Wehn and Montalvo 2018; Liu 2018):

- *If the supply chain department has a relationship and a common approach with other departments*, otherwise it would be necessary to execute a campaign to understand the objectives of that department.
- *Verify whether there is an adequate exchange of information relevant to the supply chain department among the remaining departments*. This increases communication and good understanding of product quality standards, not to mention supply chain visibility.
- *Check whether there is an intense pursue of cross-functional new product development*. Ideas from all the departments must be gathered to work collectively in the potential design to be manufactured. Sometimes it is possible to design products, but their specifications cannot be met by using the installed technology.

On the other hand, external knowledge exchange is the ability of a company to use external experience or capabilities to benefit its products and processes (suppliers and customers) (Ben Arfi et al. 2017). External partners are important because internal knowledge needs to be supplemented with the external knowledge to ensure the correct execution of supply chain plans (Ryoo and Kim 2015). In that sense, companies need to establish solid relationships that ensure the exchange of necessary interorganizational knowledge (Kim et al. 2012), For instance, machinery and equipment suppliers must provide adequate training to both operators and maintenance staff. Likewise, manufacturers should develop product manuals that provide customers with the necessary information, while customer needs and preferences should be monitored on an ongoing basis (Roper et al. 2017).

To assess *Internal Knowledge Transfer* in manufacturing supply chains, experts recommend evaluating the following points (Ben Arfi et al. 2017; Segarra-Ciprés et al. 2014; Najafi-Tavani et al. 2012; Monteiro 2015):

- *If suppliers can share their experience and knowledge about modern technologies*. This strategy allows maintaining the technological levels required in production processes and quality standards. For instance, advanced manufacturing technologies are useless if workers do not know how to operate or maintain them. In this sense, supplier support is a key element.
- *Whether supplier technical knowledge is integrated into new products and processes*. This strategy ensures supply chain communication in both directions (buyer–supplier). Suppliers first recommend alternatives for raw materials with higher quality and lower costs; then, customers quickly indicate their needs and the required product features.
- *Whether meetings with suppliers and buyers are held to support knowledge development,* which must be mainly oriented toward new product design generation, problem-solving in current products, or the creation of collaborative research and product development strategies.

Knowledge transfer among SC members is related to SC management efficiency and performance (Lu et al. 2012). Some benefits of knowledge transfer can be listed as follows:

- Knowledge from customers and/or suppliers can improve overall SC performance, internal decision-making processes, and operational performance (Kim et al. 2012), since knowledge facilitates communication through multiple channels.
- Knowledge transfer allows understanding of quality system requirements. It enables companies to know the extent to which products meet their requirements (Clemons and Slotnick 2016).
- Knowledge transfer can benefit production process factors, employee performance, organizational culture, and even corporate reputation (Shakerian et al. 2016).
- Knowledge transfer helps suppliers improve processes and products (Clemons and Slotnick 2016).

Despite these benefits, multiple constraints prevent knowledge from flowing along the entire supply chain. According to Bhosale et al. (2016), the most significant include: lack of managerial commitment, cultural differences, inappropriate incentives, technological incompatibilities, knowledge leakage, and lack of negotiation strategies among companies.

21.2.2 Supply Chain Flexibility

Supply chain flexibility is a complex and multidimensional variable that cannot be easily measured or valued. However, multiple concepts have been proposed to integrate this construct, which, for instance, is defined as the ability to solve aspects associated with changes in partner needs, which in turn are related to supply chain risk (Sushil 2017). Similarly, *Supply Chain Flexibility* must be discussed in terms of efficiency parameters, such as delivery time, final product costs, and product modifications (Hu and Cardin 2015), which might be compromised as companies embrace greater flexibility.

Supply Chain Flexibility has several sources. In their work, authors Gothwal and Raj (2017) identified and listed a vast array of flexibility sources, including External and Internal Knowledge Transfer and knowledge transfer speed (Seebacher and Winkler 2015; Gong 2008). In order to measure flexibility in supply chains, experts recommend taking into account the following aspects (Chryssolouris et al. 2012; Teich and Claus 2017; Baharmand et al. 2017; Asad et al. 2016):

- Adjust deliveries to customer changes
- Adjust manufacturing process capabilities
- Reduce manufacturing lead time
- Reduce development cycle times
- Short-term adjust supplier order of goods and services

At this stage, it is worth wondering how internal and *Internal Knowledge Transfer* influence *Supply Chain Flexibility* and performance. On the one hand, Blome et al. (2014) studied the impact of these two variables within complex supply chains, since knowledge is easier to manage and transmit in flat supply chain systems than in those that are vertically complex (Hwang et al. 2018). On the other hand, authors Vlajčić et al. (2018) stated that companies having subsidiaries abroad must consider cultural aspects to prevent information flow blockages, and thus increase *Supply Chain Flexibility*.

Another important aspect of knowledge transfer is product innovation. As Kamasak et al. (2016) claim, information exchange between manufacturers and customers is essential in understanding what the latter expect or need. Likewise, Wehn and Montalvo (2018) reported how interactions between both innovation and knowledge transfer actors and customers reduce production and product cycle times. Moreover, since these interactions guarantee an early entry of the product into the market, it is necessary to generate knowledge networks specialized in that product (Xie et al. 2016).

However, knowledge transfer can also come from suppliers (before the production process) and provide flexibility to the company. Balboni et al. (2017) studied this phenomenon of *Internal Knowledge Transfer* and analyzed the importance of external information reliability in improvement changes. Additionally, Liu et al. (2017) conducted a study among 225 cases of business–supplier pairs and analyzed how suppliers helped improve the quality of the manufacturer and the credibility of its relationships. In turn, this allowed manufacturing companies to reduce raw material delivery times, cycle times, and final product delivery times. To contribute to this discussion on the impact of *Internal Knowledge Transfer* on *Supply Chain Flexibility*, the first hypothesis of this research is proposed as follows:

H_1: External Knowledge Transfer has a positive direct effect on Supply Chain Flexibility.

As regards the role of *Internal Knowledge Transfer* in supply chain flexibility, Machikita et al. (2016) found that kaizen or continuous improvement events are an opportunity to transmit the knowledge and experiences acquired over the years within the company. This strategy, in turn, increases the number of documented problem-solving strategies available for employees, which consequently increase production process flexibility. According to Clemons and Slotnick (2016), corporate maturity in terms of *Internal Knowledge Transfer* is tested when quality problems arise in the production processes. Companies must rely on their experience and human resources to establish proper communication among departments to address the quality deviations previously detected.

In this context, Ai and Tan (2017) analyzed the *Internal Knowledge Transfer* phenomenon among Chinese companies that hold commercial relationships with European companies and found that knowledge transfer strategies must consider cultural aspects to ensure the flow of products and services while ensuring flexibility in the two companies. In order to determine the relationship between *Internal Knowledge Transfer* and supply chain flexibility, the following hypothesis is proposed:

H_2: Internal Knowledge Transfer has a positive direct effect on Supply Chain Flexibility.

21.2.3 Supply Chain Performance

In a world of high competition and constant changes, supply chains are one of the most valuable assets of businesses (Shakerian et al. 2016). The benefits of evaluating Supply Chain Performance are many and cover aspects that allow controlling productive processes, highlighting achievements, improving key processes, identifying potential problems, and providing information about possible improvement actions (Lima-Junior and Carpinetti 2017). To quantify aspects such as efficiency and effectiveness, metrics known as Supply Chain Performance measurement systems (SCPMSs) are used. These systems integrate organizational functions, partner companies, and SC activities (Maestrini et al. 2017).

However, there are not strict rules to evaluate supply chain. For instance, Mathiyalagan et al. (2014) advise focusing on finances (sales, net returns, manufacturing cost reduction, inventories, etc.), customers (lead time, defects, timely deliveries, and demand forecasts), the internal business process (productivity, flow of cash, manufacturing cycle times, etc.) and knowledge (new product development, complaints, training).

On the other hand, Maestrini et al. (2017) highlight the importance of managing relationships with customers and suppliers in order to ensure the flow of information, raw materials, and finished products, as well as financial resources. Additionally, Butzer et al. (2017) remark five approaches to measuring SC performance: flexibility, innovation, processes, SC partners (customers and supplier), and the government. For a complete review of *Supply Chain Performance* indices, we recommend the work of Balfaqih et al. (2016), who consider the following elements as performance indicators:

- Customers satisfaction, no complaints
- Cycle time from supplier to customer delivery
- Full product deliveries on time
- A cost-reduction focus
- Supply chain performance improvements
- Product customization levels
- Cash flow rates
- Supply chain synergy
- Supply chain visibility

Multiple research works have found a direct relationship between *Supply Chain Flexibility* resulting from knowledge transfer and *Supply Chain Performance*. Researchers Lima-Junior and Carpinetti (2017) analyzed a set of quantitative *Supply Chain Performance* evaluation models and found that *Supply Chain*

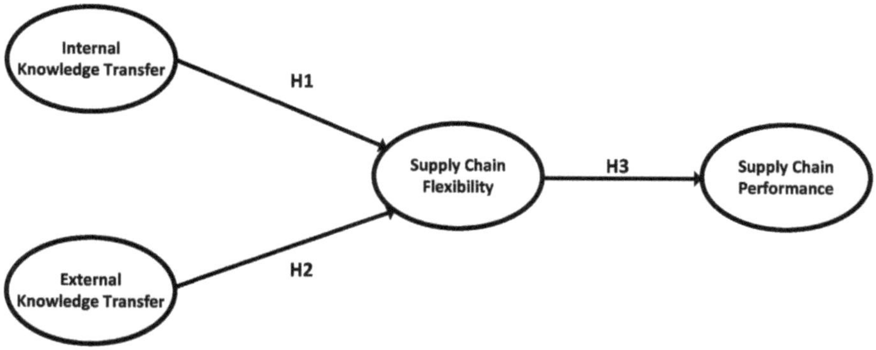

Fig. 21.1 Proposed model

Flexibility allows companies to make appropriate production adjustments prior to customer demand changes without compromising delivery times and by minimizing risks. Similarly, Avelar-Sosa et al. (2014) claimed that SC risks might be the cause of delivery delays due to a lack of flexibility in cycle times.

Moreover, according to Vlachos (2014) and Lozano et al. (2017), information technologies along with lean manufacturing practices, such as just in time, can mitigate SC risks while simultaneously increasing product customization rates. Therefore, to determine the relationship between SC flexibility and SC performance, the following hypothesis is proposed:

H_3: Supply Chain Flexibility has a positive direct effect on Supply Chain Performance.

Figure 21.1 introduces the graphical representation of the three hypotheses.

21.3 Methodology

The following sections describe the methodology followed to validate the research hypotheses depicted in Fig. 21.1.

21.3.1 Survey Design

The data used to validate the model were gathered, thanks to a survey developed for and administered in the Mexican manufacturing industry. The survey took as grounds the research work of Blome et al. (2014), whose questionnaire includes five latent variables: flexibility, *Internal Knowledge Transfer*, *Internal Knowledge Transfer*, supply chain complexity, and product complexity.

Table 21.1 Survey scale

1	2	3	4	5
Never	Rarely	Often	Frequently	Always
Very poor	Poor	Fair	Good	Excellent

Our research, however, does not take into account the last two constructs. Instead, we replaced them with two aspects: *Supply Chain Performance* (nine items) and employee performance (five items). These categories were carefully developed after conducting a literature review. Likewise, our survey includes a demographic section to gather data such as employee seniority (in years), gender, work position, and company working sector. Finally, for the purpose of this research, only those latent variables depicted in Fig. 21.1 are studied, whereas the remaining ones are proposed as future work.

21.3.2 Survey Administration

The questionnaire was administered in the form of face-to-face interviews with employees that are directly involved in the supply chain departments of manufacturing companies located in northern Mexico. The sample varied from managers to operators. Each item was answered using a five-point Likert scale as illustrated in Table 21.1, where the lowest value (1) implied that an activity or benefit was never performed, whereas the highest value (5) indicated that an activity was always performed, or a benefit was always obtained.

For the survey administration process, we consulted a list of supply chain managers provided by AMAC (Asociación de Maquiladoras, A.C.), a Mexican manufacturing association headquartered in Ciudad Juárez, Chihuahua that provides support to export-oriented manufacturing companies in Mexico. To ensure the sample, we first relied on stratified sampling, since we organized meetings with managers to invite the companies to participate in the research.

In this sense, only export-oriented manufacturing companies with solid supply chains were taken into account. Then, after the initial interview with managers, we relied on the snowball sampling technique by interviewing the managers' colleagues and subordinates with similar responsibilities.

21.3.3 Data Capture and Screening

Once the questionnaires were collected, we constructed a database using program SPSS 21®, where the columns represented the survey items or observed variables, and each row represented an answered questionnaire or case. To screen the data, we first estimated the standard deviation of each of case, and surveys with a standard deviation value lower than 0.5 were discarded. Then, missing values (unanswered questions) were identified, thereby removing cases with 10% or more of missing values, and thus increasing data reliability (Nunnally and Bernstein 1994).

On the other hand, in cases with less than 10% missing values, these unanswered items were replaced by the median value of the scale, since the data collected were ordinal. Finally, extreme values were identified by standardizing each item or observed variable. Extreme or atypical values have a standardized absolute value greater than 4 (Kohler et al. 2015), which was replaced by the median.

21.3.4 Survey Validation

Once the database was screened, the latent variables and their corresponding observed variables were statistically validated. To this end, the following six indices or latent variable coefficients recommended by Kock (2015) were estimated:

- *R-Squared (R^2) and Adjusted R-Squared*: they are indicators of survey predictive validity from a parametric perspective. Only values greater than 0.2 are acceptable.
- Q-Squared *(Q^2)*: it is an indicator of nonparametric predictive validity. Acceptable values must be higher than 0 and similar to their corresponding R^2 values.
- *Cronbach's alpha index* (*Cronbach* 1951) *and composite reliability index*: they are used to measure internal reliability in latent variables, which can be estimated according to either the variance or the correlation indices between the items of a latent variable (Adamson and Prion 2013). Acceptable values must be greater than 0.7 (Nunnally and Bernstein 1994; Fornell and Larcker 1981).
- *Average Variance Extracted (AVE)*: it is a measure of the convergent validity of latent variable elements. Acceptable values must be higher than 0.5 (Fornell and Larcker 1981; Kock 2015).
- *Variance inflation factors (VIF) and Average full collinearity VIF (AFVIF)*: they are estimated to measure collinearity in items or observed variables. Acceptable values must be lower than 3.3 (Cenfetelli and Bassellier 2009; Petter et al. 2007).

The aforementioned coefficients have been successfully employed in multiple supply chain research works. For instance, Elbaz et al. (2018) reported the critical competencies affecting the performance of Egyptian travel agents while assessing the negative influence of nepotism on those competencies. Likewise, Díaz-Reza et al. (2017) analyzed the relationships between SMED activities among manufacturing companies located in Ciudad Juárez, Mexico, whereas García-Alcaraz et al. (2014) explored key Just In Time (JIT) success factors in the Mexican manufacturing industry.

21.3.5 Structural Equation Modeling

Structural Equation Model (SEM) is a multivariate analysis technique that determines the relationships between manifested or observed variables through latent or unobserved variables (Samee and Pongpeng 2016). SEM relies on multiple models to represent relationships between the observed variables. Its goal is to provide a quantitative test of a hypothetical theoretical model proposed by the researcher, as that illustrated in Fig. 21.1. Specifically, several theoretical models tested in SEM can hypothesize how sets of variables define constructs and how they relate to each other (Schumacker and Lomax 2010).

The objective of the SEM analysis is to determine how sample data support theoretical models. If a given sample dataset supports a given theoretical model, hypotheses about more complex theoretical models can be formulated; however, if the sample dataset is not compatible with the theoretical model, the original model can be modified and tested, or different theoretical models should be developed and tested (Schumacker and Lomax 2010). Consequently, SEM tests theoretical models using the scientific method of hypothesis testing to understand complex relationships between constructs (2010). In addition, SEM is a reliable technique even when using non-normal data, ordinal data, or small samples (Jenatabadi and Ismail 2014).

The four latent variables studied in this research were integrated into a model that interrelates them through three research hypotheses, as illustrated in Fig. 21.1. The model was tested using partial least squares (PLS), widely accepted in multiple research areas (Evermann and Tate 2016). Then, the hypotheses were validated using PLS-based software WarpPLS 5.0®, widely recommended by Ekrot et al. (2016). Next, the model was tested at a 95% confidence level, thereby implying that the p values of the hypotheses had to be less lower than 0.05. Finally, before interpreting the model, six efficiency indices proposed by Kock (2015) were estimated:

- *Average Path Coefficient (APC)*: it statistically validates the hypotheses in a generalized way. The hypotheses are tested with a significance level of 0.05, therefore, the p values must be lower than 0.05 to be statistically significant.
- *Average R-Squared (ARS) and Average adjusted R-Squared (AARS)*: they measure the model's predictive validity. Acceptable p values for ARS and AARS must be lower than 0.05. Moreover, two hypotheses are proposed for each relationship between latent variables: the null hypothesis, where APC = 0 and ARS = 0, and the alternative hypothesis, where APC $\neq$ 0 and ARS $\neq$ 0.
- *Average Variance Inflation Factor (AVIF) and Average Full collinearity VIF (AFVIF)*: they are a measure of collinearity between latent variables. Acceptable values must be lower than 3.3.
- *Tenenhaus Goodness of Fit index (GoF)*: it measures the explanatory power of models. Recommended values must be greater than 0.36.

21.4 Results

The survey was administered from May to July 2017 among different manufacturing industries in northern Mexico. In total, 290 questionnaires were collected, yet 21 of them were removed from the analysis due to missing values. Hence, 269 surveys were considered as valid, and thus were used to validate the model. The following subsections discuss the descriptive analysis of the collected data.

21.4.1 Sample Description

According to Table 21.2, the automotive industry represents 44.2% of the total sample since. In fact, this industry is one of the most representative in northern Mexico. Similarly, the electronics industry ranked second with 15.6% of collected surveys, whereas the electrical manufacturing industry shows the lowest representation, with only 2% of surveys.

As for the surveyed job positions listed in Table 21.3, almost half of the sample, namely 42.75%, is represented by technicians.

Table 21.2 Surveyed industries

Sector	Frequency
Automotive	119
Electronics	42
Machining	27
Aeronautics	25
Nonanswered	22
Medical	15
Other	10
Logistics	7
Electrical	2
Total	269

Table 21.3 Surveyed job positions

Position	Frequency
Technician	115
Operator	76
Engineer	37
Supervisor	25
Manager	8
Nonanswered	8
Total	269

21.4.2 Descriptive Analysis of the Items

Following, Table 21.4 summarizes the results from the descriptive analysis of the survey items. The results can be discussed with respect to median values and interquartile range values. The four latent variables and their corresponding items are listed in descending order, according to their median values.

Notice that none of the observed variables has a median value higher than 4, but they are all close to it. Such results imply that the listed activities are frequently performed in the surveyed companies and the listed benefits are also frequently obtained.

Table 21.4 Descriptive analysis of the items

Latent variable/item	Percentile			IR
	25	50	75	
Internal knowledge transfer				
We effectively exchange supply-chain-related information with other departments	2.92	3.73	4.54	1.62
We intensely pursue cross-functional development of new products	2.98	3.73	4.51	1.53
The supply chain department understands its importance	2.95	3.67	4.45	1.5
External knowledge transfer				
Suppliers share their expertise in new technology with us	2.72	3.63	4.47	1.75
Suppliers share technical know-how to support our new products and processes	2.76	3.62	4.47	1.71
We hold frequent meetings with suppliers to develop new knowledge	2.59	3.57	4.44	1.85
Supply chain flexibility				
We adjust deliveries to customer changes	3.12	3.84	4.57	1.45
We can adjust our manufacturing process capabilities	3.01	3.74	4.53	1.52
We reduce manufacturing lead times	2.88	3.69	4.51	1.63
We reduce development cycle times	2.88	3.67	4.47	1.59
We make short-term adjustments in supplier orders	2.80	3.56	4.34	1.54
Supply chain performance				
Final product deliveries are full and timely	3.18	3.94	4.68	1.5
We focus on cost reduction	3.08	3.88	4.65	1.57
We have high product customization levels	3.06	3.87	4.66	1.6
SC performance allows for cash flow	3.09	3.83	4.60	1.51
We rely on supply chain synergy	3.08	3.83	4.61	1.53
Customers are completely satisfied; no complaints are filed	3.07	3.80	4.58	1.51
The supply chain is visible	3.00	3.80	4.60	1.60
Supply chain performance is continuously improved	3.01	3.79	4.57	1.56
Product cycle times are long	2.82	3.63	4.47	1.65

As regards to *Internal Knowledge Transfer*, the first two observed variables have the same median value, and thus hold the first place together. Such results indicate that companies are frequently able to effectively exchange information among departments, which in turn allows them to intensely pursue cross-functional development of new products and services. On the other hand, the value of the last item indicates that companies must work harder to ensure an appropriate flow of materials, information, and resources inside of their facilities.

As for *External Knowledge Transfer*, the most important elements are suppliers sharing their experience with existing technologies. In fact, their knowledge directly impacts on their relationship with manufacturers and might be related to new process development, raw materials, and machine/tools efficiency, among others. Conversely, it seems that frequent meetings with suppliers take place less often. Unfortunately, this might have an impact on project development, information sharing, and thus SC integration.

Flexibility in the supply chain is something that managers seek but not always achieved. In this research, it is observed that the most important thing for manufacturing companies is to adjust to demand changes. To this end, production capacity must be adjusted, setup times must be minimized, and machine breakdowns must be prevented. In this sense, companies must improve two aspects: product cycle times (from design to delivery) and delivery times, since the two items hold the last two positions. Finally, as regards *Supply Chain Performance*, the best-scored item is associated with complete and on-time product deliveries, performed under a cost-reduction approach.

As de Jong et al. (2017) claim, a large percentage of production costs is associated with logistics aspects; thus, companies should give priority to this aspect that does not add value to the product. On the other hand, the lowest score in this latent variable indicates that cycle times in the surveyed companies are too long, which indicates a lack of integration.

In this sense, supply chains can be continuously improved by implementing new performance indices and organizing meetings between partners to identify failures. In fact, as Shou et al. (2017) and Qi et al. (2017) argue, SC integration is essential, otherwise operational performance indices cannot be guaranteed. However, according to Zhao et al. (2015), SC integration levels are not always associated with greater benefits.

21.4.3 Survey Statistical Validation

Table 21.5 reports the results from the latent variable validation process. According to such results, it is possible to highlight the following conclusions:

- According to the *R*-Squared and Adj. *R*-Squared values, the survey has enough predictive validity from a parametric perspective.

Table 21.5 Latent variable coefficients

Coefficient	Internal knowledge transfer	Supply chain flexibility	Supply chain performance	External knowledge transfer
R-squared		0.489	0.485	
Adj. *R*-squared		0.485	0.483	
Composite reliability	0.851	0.911	0.930	0.901
Cronbach's alpha	0.738	0.877	0.915	0.835
AVE	0.656	0.671	0.595	0.752
Full collin. VIF	2.184	2.157	2.243	1.731
Q-squared		0.488	0.487	

- According to the composite reliability index and the Cronbach's alpha, the instrument has enough internal reliability. Likewise, AVE and Full Collinearity VIF values are lower than 3.3; thus, collinearity problems are discarded.
- The *Q*-Squared values—all of them higher than 0—indicate that the survey has enough predictive validity from a nonparametric perspective.

In conclusion, the latent variables passed the validation process and can be used to test the model.

21.4.4 Structural Equation Modeling

Five model fit and quality indices were estimated to test the model as a construct. According to the values of APC, ARS, and AARS, the model has predictive validity. Similarly, the values of AVIF and AFVIF discard collinearity problems between latent variables, whereas the value of the GoF index indicates that the model has sufficient explanatory power. In conclusion, the model is valid and can be interpreted accordingly.

- Average Path Coefficient (APC) = 0.489, $P < 0.001$
- Average *R*-Squared (ARS) = 0.487, $P < 0.001$
- Average Adjusted *R*-Squared (AARS) = 0.484, $P < 0.001$
- Average block VIF (AVIF) = 1.468, acceptable if $\leq$ 5, ideally $\leq$ 3.3
- Average Full collinearity VIF (AFVIF) = 2.079, acceptable if $\leq$ 5, ideally $\leq$ 3.3
- Tenenhaus GoF (GoF) = 0.570, small $\geq$ 0.1, medium $\geq$ 0.25, large $\geq$ 0.36.

21.4.5 Direct Effects

Figure 21.2 depicts the model with its corresponding effects. Direct effects can be appreciated as arrows directly connecting two latent variables: one independent latent variable (where the effect originates) with one dependent latent variable (where the effect falls). Notice that the three direct effects correspond to the three research hypotheses and have p values lower than 0.05. Therefore, they are statistically significant at a 95% confidence level. In this sense, the following conclusions can be highlighted:

- H_1 *Internal Knowledge Transfer* has a positive direct effect on *Supply Chain Flexibility*, since when the first latent variable increases by one standard deviation, the second latent variable increases by 0.549 standard deviations.
- H_2 *External Knowledge Transfer* has a positive direct effect on *Supply Chain Flexibility*, since when the first latent variable increases by one standard deviation, the second latent variable increases by 0.222 standard deviations.
- H_3 *Supply Chain Flexibility* has a positive direct effect on *Supply Chain Performance*, since when the first latent variable increases by one standard deviation, the second latent variable increases by 0.696 standard deviations.

Notice that *Supply Chain Flexibility* shows $R^2 = 0.488$, thereby implying that both *Internal Knowledge Transfer* and *External Knowledge Transfer* can explain 48.8% of its variability. However, the decomposition of the R^2 value indicates that *Internal Knowledge Transfer* explains 37%, whereas *External Knowledge Transfer* explains 11.8%.

In conclusion, companies must prioritize appropriate knowledge transfer inside their facilities and not so much outside. In other words, companies should invest in appropriate employee training and education management systems.

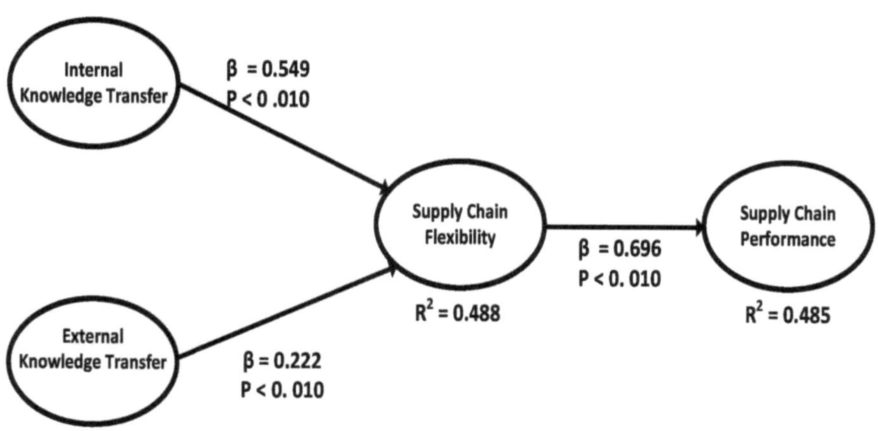

Fig. 21.2 Evaluated model

21.4.6 *Indirect Effects*

Indirect effects occur between two latent variables through mediating variables. In this model, two indirect effects were identified. The first occurs from *Internal Knowledge Transfer* on *Supply Chain Performance* thanks to *Supply Chain Flexibility*. The relationship is statistically significant since it has a p value lower than 0.05.

Moreover, the effect size (ES) is equal to 0.244. The second indirect effect occurs from *External Knowledge Transfer* on *Supply Chain Performance* thanks to *Supply Chain Flexibility*. This relationship is also significant according to the *p* value, and the effect size is equal to 0.094.

21.4.7 *Total Effects*

Total effects are the sum of direct and indirect effects in a relationship. Table 21.6 summarizes the total effects estimated for each relationship. As in previous cases, all the total effects are statistically significant at a 95% confidence level. The greatest total effects occur in the relationship between *Supply Chain Flexibility* and *Supply Chain Performance*, with a value of 0.696 and ES = 0.485. The second greatest effects were found in the relationship between *Internal Knowledge Transfer* and *Supply Chain Flexibility*, with a value of 0.549 and ES = 0.373.

21.5 Conclusions and Industrial Implications

This research proposes a structural equation model to explore and measure the relationships among four latent variables—*Internal Knowledge Transfer*, *Internal Knowledge Transfer*, *Supply Chain Flexibility*, and *Supply Chain Performance*—through three research hypotheses. According to our findings, the following conclusions can be proposed:

Table 21.6 Total effects

Dependent variable	Independent variable		
	Internal knowledge transfer	Supply chain flexibility	External knowledge transfer
Supply chain flexibility	0.549 $P < 0.001$ ES = 0.370	–	0.222 $P < 0.001$ ES = 0.118
Supply chain performance	0.382 $P < 0.001$ ES = 0.244	0.696 $P < 0.001$ ES = 0.485	0.155 $P < 0.001$ ES = 0.094

- Internal and *Internal Knowledge Transfer* are both sources of supply chain flexibility and risk mitigation strategies. However, when analyzing the beta value of the coefficients, we concluded that *Internal Knowledge Transfer* has a higher impact on flexibility since it has the highest value with an intensity of change of 0.549 units.
- Since *Internal Knowledge Transfer* is strongly related to SC flexibility, SC managers and administrators must ensure adequate training programs to manage knowledge. As Kiran (2017) claims, continuous improvement groups and kaizen teams should always be encouraged. Similar findings were reported by Machikita et al. (2016), who point out that kaizen as a lean manufacturing tool is an effective knowledge transfer mechanism. Likewise, our findings on the relationship between internal and *Internal Knowledge Transfer* and *Supply Chain Flexibility* are consistent with those reported by Blome et al. (2014).
- Cultural aspects are important factors to be further studied in the manufacturing industry. Since most of the export-oriented manufacturing companies located in northern Mexico are foreign-owned, cultural practices and preferences can have an impact on knowledge sharing practices and approaches, and thus might be sources of problems, as Vlajčić et al. (2018) argue.
- As for the role of *Internal Knowledge Transfer*, it is important for companies to establish long-term relationships with suppliers and make efforts to achieve higher integration levels to promote and facilitate innovation. According to Kamasak et al. (2016), external knowledge does not necessarily have to be generated from scratch, it can simply be transferred by using new materials or processes.
- Since SC integration plays an important role in effective knowledge transfer, an instrument is required to determine both internal and external integration levels in Mexican manufacturing companies. As Chen et al. (2014) claim, in supply chain systems not only materials and technologies must flow, but also the knowledge associated with them.

References

Adamson KA, Prion S (2013) Reliability: measuring internal consistency using cronbach's α. Clin Simul Nurs 9(2)

Ai Q, Tan H (2017) The intra-firm knowledge transfer in the outward M&A of EMNCs: evidence from Chinese manufacturing firms. Asia Pac J Manag. https://doi.org/10.1007/s10490-017-9518-z

Asad MM, Mohammadi V, Shirani M (2016) Modeling flexibility capabilities of IT-based supply chain, using a grey-based DEMATEL method. Procedia Econ Finance 36:220–231. https://doi.org/10.1016/S2212-5671(16)30033-8

Avelar-Sosa L, García-Alcaraz JL, Castrellón-Torres JP (2014) The effects of some risk factors in the supply chains performance: a case of study. J Appl Res Technol 12(5):958–968. https://doi.org/10.1016/S1665-6423(14)70602-9

Baharmand H, Comes T, Lauras M (2017) Defining and measuring the network flexibility of humanitarian supply chains: insights from the 2015 Nepal earthquake. Ann Oper Res. https://doi.org/10.1007/s10479-017-2713-y

Balboni B, Marchi G, Vignola M (2017) Knowledge transfer in the context of buyer–supplier relationship: an analysis of a supplier's customer portfolio. J Bus Res 80:277–287. https://doi.org/10.1016/j.jbusres.2017.06.022

Balfaqih H, Nopiah ZM, Saibani N, Al-Nory MT (2016) Review of supply chain performance measurement systems: 1998–2015. Comput Ind 82:135–150. https://doi.org/10.1016/j.compind.2016.07.002

Ben Arfi W, Hikkerova L, Sahut J-M (2017) External knowledge sources, green innovation and performance. Technol Forecast Soc Chang. https://doi.org/10.1016/j.techfore.2017.09.017

Bhosale VA, Kant R, Bhandarkar BM identifying causal relations of critical supply chain knowledge flow barriers for an Indian case organization. In: 2016 IEEE international conference on management of innovation and technology (ICMIT), 19–22 Sept 2016, pp 83–88. https://doi.org/10.1109/icmit.2016.7605012

Blome C, Schoenherr T, Eckstein D (2014) The impact of knowledge transfer and complexity on supply chain flexibility: a knowledge-based view. Int J Prod Econ 147:307–316

Butzer S, Schötz S, Petroschke M, Steinhilper R (2017) Development of a performance measurement system for international reverse supply chains. Procedia CIRP 61:251–256. https://doi.org/10.1016/j.procir.2016.11.264

Cenfetelli RT, Bassellier G (2009) Interpretation of formative measurement in information systems research. Mis Q 19

Chen C-J, Hsiao Y-C, Chu M-A (2014) Transfer mechanisms and knowledge transfer: the cooperative competency perspective. J Bus Res 67(12):2531–2541. https://doi.org/10.1016/j.jbusres.2014.03.011

Chryssolouris G, Georgoulias K, Michalos G (2012) Production systems flexibility: theory and practice. IFAC Proc Vol 45(6):15–21. https://doi.org/10.3182/20120523-3-RO-2023.00442

Clemons R, Slotnick SA (2016) The effect of supply-chain disruption, quality and knowledge transfer on firm strategy. Int J Prod Econ 178:169–186. https://doi.org/10.1016/j.ijpe.2016.05.012

Cronbach LJ (1951) Coefficient alpha and the internal structure of tests. Psychometrika 16(3):8

de Jong S, Hoefnagels R, Wetterlund E, Pettersson K, Faaij A, Junginger M (2017) Cost optimization of biofuel production—the impact of scale, integration, transport and supply chain configurations. Appl Energy 195:1055–1070. https://doi.org/10.1016/j.apenergy.2017.03.109

De Silva M, Howells J, Meyer M (2018) Innovation intermediaries and collaboration: knowledge-based practices and internal value creation. Res Policy 47(1):70–87. https://doi.org/10.1016/j.respol.2017.09.011

Díaz-Reza JR, García-Alcaraz JL, Mendoza-Fong JR, Martínez-Loya V, Jiménez Macías E, Blanco-Fernández J (2017) Interrelations among SMED stages: a causal model. Complexity 2017:10

Ekrot B, Kock A, Gemünden HG (2016) Retaining project management competence—antecedents and consequences. Int J Project Manage 34(2):145–157. https://doi.org/10.1016/j.ijproman.2015.10.010

Elbaz AM, Haddoud MY, Shehawy YM (2018) Nepotism, employees' competencies and firm performance in the tourism sector: a dual multivariate and qualitative comparative analysis approach. Tour Manag 67:3–16. https://doi.org/10.1016/j.tourman.2018.01.002

Evermann J, Tate M (2016) Assessing the predictive performance of structural equation model estimators. J Bus Res 69(10):4565–4582. https://doi.org/10.1016/j.jbusres.2016.03.050

Fornell C, Larcker DF (1981) Evaluating structural equation models with unobservable variables and measurement error. J Mark Res 18 (1)

García-Alcaraz JL, Maldonado AA, Iniesta AA, Robles GC, Hernández GA (2014) A systematic review/survey for JIT implementation: Mexican maquiladoras as case study. Comput Ind 65 (4):761–773. https://doi.org/10.1016/j.compind.2014.02.013

Gong Z (2008) An economic evaluation model of supply chain flexibility. Eur J Oper Res 184 (2):745–758. https://doi.org/10.1016/j.ejor.2006.11.013

Gothwal S, Raj T (2017) Analyzing the factors affecting the flexibility in FMS using weighted interpretive structural modeling (WISM) approach. Int J Syst Assur Eng Manag 8(2):408–422. https://doi.org/10.1007/s13198-016-0443-1

Hu J, Cardin M-A (2015) Generating flexibility in the design of engineering systems to enable better sustainability and lifecycle performance. Res Eng Design 26(2):121–143. https://doi.org/10.1007/s00163-015-0189-9

Hwang Y, Lin H, Shin D (2018) Knowledge system commitment and knowledge sharing intention: the role of personal information management motivation. Int J Inf Manag 39:220–227. https://doi.org/10.1016/j.ijinfomgt.2017.12.009

Jenatabadi HS, Ismail NA (2014) Application of structural equation modelling for estimating airline performance. J Air Transp Manag 40:25–33. https://doi.org/10.1016/j.jairtraman.2014.05.005

Kamasak R, Yavuz M, Karagulle AO, Agca T (2016) Importance of strategic flexibility on the knowledge and innovation relationship: an emerging market study. Procedia Soc Behav Sci 229:126–132. https://doi.org/10.1016/j.sbspro.2016.07.121

Kaszás N, Keller K, Birkner Z (2016) The role of transferring knowledge in case of non-business sector projects. Procedia Soc Behav Sci 221:226–235. https://doi.org/10.1016/j.sbspro.2016.05.110

Kathiravelu SR, Mansor NNA, Ramayah T, Idris N (2014) Why organisational culture drives knowledge sharing? Procedia Soc Behav Sci 129:119–126. https://doi.org/10.1016/j.sbspro.2014.03.656

Kim KK, Umanath NS, Kim JY, Ahrens F, Kim B (2012) Knowledge complementarity and knowledge exchange in supply channel relationships. Int J Inf Manag 32(1):35–49. https://doi.org/10.1016/j.ijinfomgt.2011.05.002

Kiran DR (2017) Kaizen and continuous improvement (Chap. 22). In: Kiran DR (ed) Total quality management. Butterworth-Heinemann, pp 313–332. https://doi.org/10.1016/B978-0-12-811035-5.00022-2

Kock N (2015) WarpPLS 5.0 User Manual. ScriptWarp Systems, Laredo, Texas, USA

Kohler M, Müller F, Walk H (2015) Estimation of a regression function corresponding to latent variables. J Stat Plann Inference 162:22

Lima-Junior FR, Carpinetti LCR (2017) Quantitative models for supply chain performance evaluation: a literature review. Comput Ind Eng 113:333–346. https://doi.org/10.1016/j.cie.2017.09.022

Liu C-HS (2018) Examining social capital, organizational learning and knowledge transfer in cultural and creative industries of practice. Tour Manag 64:258–270. https://doi.org/10.1016/j.tourman.2017.09.001

Liu Y, Li Y, Shi LH, Liu T (2017) Knowledge transfer in buyer-supplier relationships: the role of transactional and relational governance mechanisms. J Bus Res 78:285–293. https://doi.org/10.1016/j.jbusres.2016.12.024

Lozano J, Saenz-Díez JC, Martínez E, Jiménez E, Blanco J (2017) Methodology to improve machine changeover performance on food industry based on SMED. Int J Adv Manuf Technol 90(9):3607–3618. https://doi.org/10.1007/s00170-016-9686-x

Lu C, Yanfei F, Yongcai Y Stackelberg game analysis of knowledge transfer between supply chain partners. In: 2012 international conference on information management, innovation management and industrial engineering, 20–21 Oct 2012, pp 163–166. https://doi.org/10.1109/iciii.2012.6339803

Machikita T, Tsuji M, Ueki Y (2016) Does Kaizen create backward knowledge transfer to Southeast Asian firms? J Bus Res 69(5):1556–1561. https://doi.org/10.1016/j.jbusres.2015.10.016

Maestrini V, Luzzini D, Maccarrone P, Caniato F (2017) Supply chain performance measurement systems: a systematic review and research agenda. Int J Prod Econ 183:299–315. https://doi.org/10.1016/j.ijpe.2016.11.005

Mathiyalagan P, Tamil Mannan K, Parthiban P (2014) Performance evaluation in supply chain using balanced scorecard. Int J Adv Mech Automob Eng (IJAMAE) 1 (2014). doi:http://dx.doi.org/10.15242/IJAMAE.E1113024

Monteiro LF (2015) Selective attention and the initiation of the global knowledge-sourcing process in multinational corporations. J Int Bus Stud 46(5):505–527. https://doi.org/10.1057/jibs.2014.72

Najafi-Tavani Z, Giroud A, Sinkovics RR (2012) Mediating effects in reverse knowledge transfer processes. Manag Int Rev 52(3):461–488. https://doi.org/10.1007/s11575-011-0097-9

Nunnally JC, Bernstein I (1994) Psychometric theory. McGraw-Hill, New York

Petter S, Straub D, Rai A (2007) Specifying formative constructs in information systems research. MIS Q 31(4):623–656. https://doi.org/10.2307/25148814

Qi Y, Huo B, Wang Z, Yeung HYJ (2017) The impact of operations and supply chain strategies on integration and performance. Int J Prod Econ 185:162–174. https://doi.org/10.1016/j.ijpe.2016.12.028

Qile H, Gallear D, Ghobadian A (2011) Knowledge transfer: the facilitating attributes in supply-chain partnerships. Inf Syst Manag 28(1):57–70. https://doi.org/10.1080/10580530.2011.536114

Roper S, Love JH, Bonner K (2017) Firms' knowledge search and local knowledge externalities in innovation performance. Res Policy 46(1):43–56. https://doi.org/10.1016/j.respol.2016.10.004

Ryoo SY, Kim KK (2015) The impact of knowledge complementarities on supply chain performance through knowledge exchange. Expert Syst Appl 42(6):3029–3040. https://doi.org/10.1016/j.eswa.2014.11.055

Sachin KP, Ravi K (2014) Knowledge management adoption in supply chain: identifying critical success factors using fuzzy DEMATEL approach. J Model Manag 9(2):160–178. https://doi.org/10.1108/JM2-08-2012-0025

Samee K, Pongpeng J (2016) Structural equation model for construction equipment selection and contractor competitive advantages. KSCE J Civil Eng 20(1):77–89. https://doi.org/10.1007/s12205-015-0632-5

Samuel KE, Goury M-L, Gunasekaran A, Spalanzani A (2011) Knowledge management in supply chain: an empirical study from France. J Strateg Inf Syst 20(3):283–306. https://doi.org/10.1016/j.jsis.2010.11.001

Schumacker RE, Lomax RG (2010) A beginner's guide to structural equation modeling, 3rd edn. Taylor and Francis, NY

Seebacher G, Winkler H (2015) A capability approach to evaluate supply chain flexibility. Int J Prod Econ 167:177–186. https://doi.org/10.1016/j.ijpe.2015.05.035

Segarra-Ciprés M, Roca-Puig V, Bou-Llusar JC (2014) External knowledge acquisition and innovation output: an analysis of the moderating effect of internal knowledge transfer. Knowl Manag Res Pract 12(2):203–214. https://doi.org/10.1057/kmrp.2012.55

Shakerian H, Dehnavi HD, Shateri F (2016) A framework for the implementation of knowledge management in supply chain management. Procedia Soc Behav Sci 230:176–183. https://doi.org/10.1016/j.sbspro.2016.09.022

Shou Y, Li Y, Park Y, Kang M (2017) Supply chain integration and operational performance: the contingency effects of production systems. J Purch Supply Manag. https://doi.org/10.1016/j.pursup.2017.11.004

Sreedevi R, Saranga H (2017) Uncertainty and supply chain risk: the moderating role of supply chain flexibility in risk mitigation. Int J Prod Econ 193:332–342. https://doi.org/10.1016/j.ijpe.2017.07.024

Sushil (2017) Does flexibility mitigate or enhance risk? Global J Flex Syst Manag 18(3):169–171. https://doi.org/10.1007/s40171-017-0156-6

Teich E, Claus T (2017) Measurement of load and capacity flexibility in manufacturing. Global J Flex Syst Manag 18(4):291–302. https://doi.org/10.1007/s40171-017-0165-5

Vlachos IP (2014) A hierarchical model of the impact of RFID practices on retail supply chain performance. Expert Syst Appl 41(1):5–15. https://doi.org/10.1016/j.eswa.2013.07.006

Vlajčić D, Caputo A, Marzi G, Dabić M (2018) Expatriates managers' cultural intelligence as promoter of knowledge transfer in multination companies. J Bus Res. https://doi.org/10.1016/j.jbusres.2018.01.033

Wehn U, Montalvo C (2018) Knowledge transfer dynamics and innovation: behaviour, interactions and aggregated outcomes. J Clean Prod 171:S56–S68. https://doi.org/10.1016/j.jclepro.2016.09.198

Wu WL, Lee YC (2012) Exploring the nonlinear relationships between learning traits and knowledge transfer in outsourcing alliances. In: 2012 proceedings of PICMET '12: technology management for emerging technologies, 29 July 2012–2 Aug 2012, pp 1692–1699

Xie X, Fang L, Zeng S (2016) Collaborative innovation network and knowledge transfer performance: A fsQCA approach. J Bus Res 69(11):5210–5215. https://doi.org/10.1016/j.jbusres.2016.04.114

Zhao G, Feng T, Wang D (2015) Is more supply chain integration always beneficial to financial performance? Ind Mark Manage 45:162–172. https://doi.org/10.1016/j.indmarman.2015.02.015

Chapter 22
Magnitude of Low Back Pain, Occupation, Education, and Economic Level in Mexican Workers

Lilia Roselia Prado-León and Rosa Amelia Rosales-Cinco

Abstract The objective of this investigation was to describe the magnitude of low back pain in workers with lumbar spondyloarthrosis and determine occupational and socio-demographic factors related to the disorder. A cross-sectional, comparative survey was made of 231 (77 cases and 154 controls) Mexican workers affiliated with the Mexican Institute of Social Security. Demographic data, workplace, and the section of the Nordic Questionnaire corresponding to back problems were collected. The cases demonstrated that the disorder had presented itself at a young age and in most of them, the LBP was intermittent and chronic. The lumbar symptoms had developed over a long period of time up to 20 years, including accumulated periods of disability totaling up to 5 years. The prevalence of pain in these cases was 83% (64). The occupations which presented the greatest differences in distribution between cases and controls were those of workers handling merchandise and materials, drivers, and members of the nursing profession. Scholarship and salary were protective factors (odds ratio (OR) = 0.17, IC95% 0.03–0.09, and OR 0.36, IC95% 0.1–0.8). Low back pain is a problem with serious repercussions in Mexico at personal and socioeconomic levels. The results suggest that the higher education and socioeconomic level a person has, the low the probability will be of developing lumbar spondyloarthrosis. Occupations which implicitly include heavy lifting and fixed sitting with vibration were those which demonstrated the most relevant difference between the two groups.

Keywords Low back pain · Epidemiology · Occupational health
Occupational diseases · Risk factors

L. R. Prado-León (✉) · R. A. Rosales-Cinco
University of Guadalajara, Calzada Independencia Norte no. 5075.
Huentitán El Bajo S.H. C.P., 44250 Guadalajara, Jalisco, Mexico
e-mail: aililpleon@gmail.com

J. L. García Alcaraz et al. (eds.), *Best Practices in Manufacturing Processes*,
https://doi.org/10.1007/978-3-319-99190-0_22

22.1 Introduction

The spinal column is a supporting structure that can be affected, often negatively, by posture, by exercise, or by standing. The principal symptom is pain—pain that can be located in different parts of the spine. When this pain originates in the lumbar region, it results in a syndrome which is considered common. This syndrome manifests itself as pain in the lower back, often accompanied by neurogenic pain in the pelvic extremities (Elders and Burdof 2001). That is why we can find the terms Lumbar Pain, Lumbago, or Low Back Pain in much of the literature.

According to Seguí-Díaz and Gérvas (2002), low back pain is defined as a painful sensation in the area of the lumbar spine that impedes its normal mobility.

Low Back Pain (LBP) has been associated with certain occupational activities, poor spinal hygiene habits, trauma, and degenerative factors associated with aging (Ordoñez et al. 2012).

According to Sobrino (2003), the occupational factors are as follows:

- Repetitiveness
- Force
- Awkward postures
- Static overload
- Chronic cumulative pathology due to repetitive microtrauma
- Vibrations
- Low temperatures
- Duration and exposure
- Lack of rest and/or recovery
- Psychosocial factors

The factors of repetitiveness, force, poor posture, and mechanical stress are often implicit in Manual Materials Handling (MMH). For this reason, MMH has been linked to LBP in several studies (Elders and Burdof 2001; Plouvier et al. 2008).

Another determinant, less frequently mentioned, is the one that Katz (2006) points out: socioeconomic factors. Socioeconomic status refers to the amount of resources an individual possesses to meet his or her needs. The three main indicators for measuring socioeconomic status are income, occupation, and educational level.

The ergonomic approach, which is the focus of this study, presupposes that many low back problems may be due to cumulative overuse, it refers to the Cumulative Trauma Disorder Model. This model is based on the hypothesis that all people do things that potentially damage their backs, but if these activities are carried out repetitively, a process sets in where such damage begins to accumulate over weeks, months, or years. This cumulative damage exceeds the body's ability to recuperate, producing degenerative deterioration of the lumbar-sacral region of the spinal column (Putz-Anderson 1988; Konz and Johnson 2008). These degenerative disorders are shared among workers in various occupations, occupations which implicate the repetitive and prolonged execution of tasks which prove hazardous in relation to the development of LBP.

Fifty percent of occupational illness in Europe is due to musculoskeletal disorders (Eurofound 2007 cited by Norasteh 2012).

LBP is a very common ailment in Western societies (Bassols et al. 2003). Devo and Weinstein (in: Bassols et al. 2003) mention that two-thirds of the adult population suffer from back pain at one time or another. Seguí Díaz and Gérvas (2008) suggest that 80% experience the condition at least once, over the course of their lifetimes.

Garro-Vargas (2012), for her part, mentions that low back pain is the most frequent cause of work-related accident claims, occurring among 80–90% of the adult population at some point in their lives, and usually recurrent. According to the World Health Organization (WHO), LBP is the leading cause of medical consultations worldwide (70%); only 4% of these cases require surgery (Garro-Vargas 2012).

Because of this, LBP represents a considerable public health problem with important socioeconomic repercussions: It generates numerous consultations with professionals, an intensified use of available health services, and significant absenteeism in the labor force with a substantial loss of working days; in addition to the condition's individual and industrial impact, it affects family, social, and public systems (Dionne et al. 2006; Thelin et al. 2008).

In the United States, the cost of occupational injuries is more than $179 billion annually (Occupational Safety and Health Administration 2004). It is estimated that the recurrence of episodes over a 1-year period ranges from 24 to 80% (Hoy et al. 2010).

One of the disorders that fall under the category of LBP is lumbar spondylarthrosis, described as chronic spinal disc degeneration with reactive alterations in neighboring vertebrae (Cailliet 1990). Another, more detailed definition, is presented by Wipf and Deyo (1995). It refers to the narrowing of disc spaces and to arthritic changes in the joint facets. The intradiscal pressure causes ligaments to be pulled out of their normal places (inserted in the vertebral bodies), and this material forms a soft spur. Figure 22.1 shows the described alterations.

Mexico lacks precise data on the impact of this problem, but according to information provided by the Mexican Institute of Social Security (MISS, a national healthcare system, which ensures and treats a large number of workers in Mexico), in 2016 disorders of the back and spine were the third cause of disability nationwide (Mexican Institute of Social Security 2017).

Covarrubias-Gómez (2010) considers lumbago a public health problem in Mexico. He states that approximately 30% of working people who suffer low back pain in Mexico require disability, a significant economic loss considering that it amounts to approximately 1.2 billion Mexican pesos per nonworking day.

Ordoñez et al. (2012), affirm that in Tacuba General Hospital, low back pain takes seventh place among all causes in terms of both medical attention and disability leave.

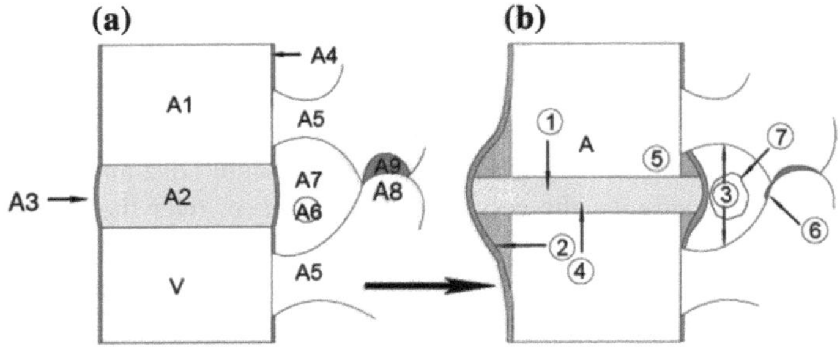

Fig. 22.1 Sequence of degeneration by spondylosis. **a** Normal functional unit: A1 = vertebral body; A2 = intervertebral disc; A3 = anterior common vertebral ligament; A4 = posterior common vertebral ligament; A5 = pedicle; A6 = nerve; A7 = conjunctive hole; A8 = facet and A9 = cartilage. **b** The disc (1) degenerates; the common vertebral ligament (2) loosens and separates from the vertebral body; the conjunction hole (3) narrows; the cartilaginous terminal plates (4) show sclerotic changes; posterior or osteophytic spurs (5) begin to form; cationic degenerative changes (6) occur in the facets; conjuctive hole stenosis appears (7) leading to nerve root compression

Based on the above, the objectives of the present study were as follows:

- Describe the magnitude of lumbar symptoms in workers with lumbar spondylarthrosis, who are insured (by MISS) in the Metropolitan Area of Guadalajara.
- Identify the most common occupations of workers with lumbar spondylarthrosis, in comparison to occupations of workers who do not manifest the disorder.
- Identify socio-demographic risk factors in workers with lumbar spondylarthrosis.

22.2 Materials and Methods

22.2.1 Study Population

The participants were 231 patients enrolled in MISS, 77 cases (patients with lumbar spondylarthrosis), and 154 controls ("healthy" people), both sexes, between 18 and 55 years old, grouped in pairs of equal age, ±2 years.

22.2.1.1 Exclusion Criteria for Cases

All patients whose infirmity included not only the lumbar spine, but also other parts of the body were excluded from the study. Those who had musculoskeletal diseases whose etiology was traumatic (contact), infectious, hereditary, neoplastic, metabolic, and visceral, congenital and idiopathic, or nonspecific, in addition to their lumbar spondylarthrosis, were also excluded, as well as persons of less than 1.45 m in stature, pregnant women, and those persons whose files stated that they had mental or psychological disorders of such a nature as to preclude reliable responses to the interview.

22.2.1.2 Exclusion Criteria for Controls

In addition to the conditions mentioned in the previous paragraph, potential control subjects were excluded from consideration under the following criteria: if the subject's most recent doctor visit was due to hypertension, heart disease (heart failure, heart attacks, enlarged heart, etc.), or serious lung diseases such as asthma, pulmonary emphysema, and tuberculosis. Among the possible control subjects, those who, when interviewed, may have reported that they had experienced LBP at some point in their lives, were also excluded from the study.

22.2.2 Procedure

Both groups were interviewed on the basis of a 23-question survey, with the prior personal authorization of each participant by means of informed consent. The questions in the first part of the survey were designed to obtain height, weight, age, scholarship, job, and salary data. In order to characterize the symptoms of LBP, the second part of the survey implemented the Nordic Questionnaire (Kuorinka et al. 1987), which consists of 11 questions that collect information on pain, fatigue, or discomfort in different body areas, such as the neck, shoulder, dorsal or lumbar spine, elbow or forearm, and wrist or hand.

The participants' body weight was used to obtain the Quetelet index (degree of obesity). To ensure standardization, cases and controls were compared by measuring elapsed time during each interview, finding minor differences between cases (25.0 $\pm$ 9.8) and controls (20.0 $\pm$ 8.4).

Descriptive data were obtained in the SPSS V20 program and estimates of OR (Odds Ratio) for socio-demographic risk factors were obtained in the EGRET V2.0.3 program, applying conditional logistic regression.

22.3 Results

22.3.1 Magnitude of Lumbar Symptoms in the Case Group

Responses to the questions focused on lumbar symptoms (Tables 22.1 and 22.2) show that a higher proportion of subjects with lumbar spondylarthrosis suffer from intermittent pain, as opposed to continuous pain. These people have been afflicted

Table 22.1 Characteristics of lumbar symptoms in the case group

	Frequency	%
Type of pain		
Intermittent	51	66.2
Continuous	26	33.8
Amount of time with low back pain		
6 months–4 years	40	51.9
>4–20 years	37	48.1
Number of days hospitalized		
1–22 days	46	59.7
>22–99 days	31	40.3

Table 22.2 Characteristics of lumbar symptoms in the case group

	Frequency	%
Total disability time taken from work		
2 months–1 year	41	53.2
>1–5 years	36	46.8
Hospitalization	54	70.1
Outpatient consultation	23	29.9
No specific medical attention	19	24.7
Accumulated time with low back pain within the past year		
None	13	16.9
2–6 months	29	37.6
6 months–1 year	35	45.5
Low back pain within the past week		
Yes	64	83.1

with back pain for periods ranging from 6 months to 20 years, and their suffering has caused all of the interviewed to require official disability leave from their work, authorized by the MISS, for cumulative periods from 2 months to 5 years in total.

Likewise, their impairment has been so extreme that more than half of the subjects have had to be hospitalized, with a hospitalization range between 1 and 3 months. Sixty-four of the respondents reported pain in the last week and in the last year (83%).

22.3.2 Occupation and Lumbar Spondylarthrosis

Table 22.3 shows occupational distribution among the case and control groups, utilizing the subgroups found in the International Classification of Occupations. The occupations that presented the greatest differences in their percentages between

Table 22.3 Distribution of occupation data between cases and controls

CIO1	Position	Frequency, cases	%	Frequency, controls	%
4	Merchants and salespeople	8	10.4	27	17.0
0–1	Physical and chemical science professionals and technicians	1	1.3	3	2.0
0–24	Mechanical engineers	2	2.6	1	0.6
0-61	Medical doctors and surgeons	1	1.3	5	3.2
0–71	Nurses	4	5.2	2	1.5
1–1 and 3	Accountants and administrative personnel	8	10.4	29	19.0
5	Service workers	9	13.0	21	14.0
7–99	Tailors, dressmakers, furriers, upholsterers and garment trade workers	1	1.3	0	0
8–02	Shoe manufacturers	3	3.9	5	3.2
0–35 and 0–34	Mechanical technicians and electrical and electronics workers	8	10.4	14	9.0
7	Industrial workers	5	6.5	14	9.0
9–5	Construction workers	3	3.9	5	3.2
9–7	Manual laborers and material (including earth) movers	9	10.4	3	2.0
9–8	Drivers of transport vehicles	15	18.2	9	6.0
1–3	Teaching professionals	0	0	7	4.5
1–6 and 6	Sculptors, photographers, painters, artists	0	0	2	1.5
7–2	Agriculture and forestry workers, metalworkers	0	0	6	4.0

International classification of occupations, International Labor Office

cases and controls were: drivers of transport vehicles with 18.2 and 6.0% (15 and 9), manual laborers and material (including earth) movers, 10.4 and 2.0% (9 and 3), and nurses, 5.2 and 1.5% (4 and 2).

22.3.3 Socio-demographic Aspects

Table 22.4 shows that the sex distribution is very similar between the groups of cases and controls, even though this variable was not considered for matching between groups. With regard to age, the similarity between cases and controls is explained by the subjects having been paired according to age.

Table 22.4 Description of socio-demographic data in cases and controls

Variable	Cases		Controls	
	Number	%	Number	%
Sex				
Male	57	74.0	106	68.8
Female	20	26.0	48	31.2
Age (years)				
18–41	19	24.7	44	28.6
>41–46	19	24.7	34	22.1
>46–51	19	24.7	39	25.3
>51–55	20	26.0	37	24.0
Obesity index				
Not obese	28	36.4	73	47.4
Mild obesity	38	49.4	64	41.6
Moderate obesity	11	14.3	17	11.0
Educational level				
Illiterate	4	5.2	4	2.6
Elementary	40	51.9	66	42.9
Middle school	20	26.0	48	31.2
High school	7	9.1	14	9.1
College and Postgraduate	6	7.8	22	14.3
Wages and salaries (dollars)				
\$300.00–\$2,000.00	11	14.3	37	24.0
>\$160.00–\$300.00	17	22.1	47	30.5
>\$80.00–\$100.00	32	41.6	54	35.1
>\$70.00–\$80.00	17	22.1	16	10.4

With respect to scholarship, it was observed that a greater percentage of the control group had attained a higher level of education in comparison to the case group, by a statistically significant figure: OR = 0.17, 95% CI 0.03–0.9.

Obesity data revealed a slight difference in distribution between cases and controls, with minor risk but without statistical significance. With respect to salary, in the case group, there is a greater proportion of subjects with low wages, and a lesser proportion with higher wages. This category (wages and salaries) did present statistical significance, with an OR = 0.3 95% CI 0.1–0.8 (see Table 22.5).

Table 22.5 OR (odds ratio) and confidence intervals for socio-demographic variables

Variable	OR	95% CI
Obesity index		
Not obese	1.0	
Mild obesity	1.6	0.9–3.1
Moderate obesity	1.7	0.7–4.0
Educational level		
Illiterate	1.0	
Elementary	0.5	0.1–2.2
Middle school	0.3	0.07–1.4
High school	0.3	0.06–2.0
College and postgraduate	0.2	0.03–0.9
Wages and salaries (dollars per month)		
$70.00–$80.00	1.0	
>$80.00–$160.00	0.9	0.4–1.8
>$160.00–$300.00	0.5	0.2–1.0
>$300.00–$2000.00	0.3	0.1–0.9

22.4 Discussion

22.4.1 Magnitude of Lumbar Symptoms in the Cases Group

Observation of the data obtained during the interviews revealed that this condition begins to present itself at an early age, a fact which contradicts any assumption that the disease is a normal degenerative process due to aging. The data that have been provided would instead support an explanation derived from an ergonomic approach, an explanation which emphasizes the microtrauma caused by ordinary tasks implicit in the subjects' occupations (Sobrino 2003; Ordoñez et al. 2012).

Likewise, the large proportion of days lost to the subjects' suffering of low back pain, disability leave, and hospitalization, provides evidence of the personal, social and economic impact of this problem.

22.4.2 Occupation and Lumbar Spondylarthrosis

Occupations that showed the greatest difference between cases and controls involve either heavy physical labor (mechanical engineers, manual laborers and material and nurses) or a fixed seated posture plus vibration (drivers of transport vehicles). All of these types of activity have been reported in the literature as risk factors for LBP (Nahit et al. 2001; Smith et al. 2006; Mitchell et al. 2008; Okunribido et al. 2007; Alperovitch-Najenson et al. 2010).

By the same token, the majority of these occupations implicate MMH tasks (lift, drop, push, pull, carry; see Fig. 22.2).

Fig. 22.2 Activities which involve manual materials handling

With regard to the nursing occupation, it is worth nothing that the study realized by Duque-Vera et al. (2011), demonstrated through their results that LBP is a common condition in this population. Even when male nurses in the study had more patients in their care than the female nurses, the men did not reveal a higher incidence of low back pain. This could be explained by the greater strength that is inherent in the male sex.

However, overall results, regardless of gender, indicate that the greater the frequency of movement of patients, the higher the prevalence of LBP, suggesting a direct relationship.

Moving onward to the theme of exposure to vibrations, considered in Spain to be one of the most important factors in the etiology of occupationally derived illness (Sobrino 2003): Drivers of transport vehicles are exposed to vibration; many, to MMH as well.

Okunribido et al. (2007) conducted a cross-sectional study of transport drivers, in order to determine their exposure to posture demands, MMH, and whole-body vibration as risks for LBP. The results showed that city bus drivers work under conditions of inadequate postural support, perform occasional light MMH, and experience discomforting shock/jerking vibration events. The prevalence of light LBP among the group suggested a need for ergonomic evaluation of the drivers' seats.

Motor vehicle mechanics often assume awkward working postures and MMH (Kant et al. 1990; Torp et al. 1996; Chen et al. 2005), factors which have been linked to LBP.

It is noteworthy that with regard to construction workers and agricultural workers, there was no higher proportion of the disorder in cases than in controls, although these jobs have been reported as high-risk occupations for the development of LBP in the literature (see Table 22.6: Eurofound 2007 cited by Norasteh 2012).

This could be because the workers of these occupations, in Mexico, are very often not affiliated with the MISS, since most of them work independently and are not employees. Companies doing business in Mexico are the entities that enroll their employees in the MISS.

Table 22.6 Percentage reported a backache by sector and gender (Eurofound 2007 cited by Norasteh 2012)

Sector	Men	Women	Total
Agriculture and fishing	43.8	54.4	47.0
Manufacturing and mining	28.0	31.2	29.0
Electricity, gas and water supply	24.7	17.2	23.3
Construction	39.2	17.7	37.0
Wholesale and retail trade	21.0	18.7	19.8
Hotels and restaurants	20.0	24.9	22.2
Transport and communication	31.4	17.5	27.9
Financial intermediation	9.7	14.6	11.9
Real estate and business service	16.6	16.7	16.6
Public administration	19.7	19.7	19.7
Education and health	19.6	22.4	21.7
Other services	21.1	21.2	21.2
Total	27.0	23.6	25.6

22.4.3 Socio-demographic Aspects

As an economic impact indicator, we may consider that the average salary is $202.00 dollars monthly, and, multiplied by the average of workdays lost due to disability (41 days), results in an outlay of $8,282.00 dollars per person.

Distribution by sex was similar in cases and controls, but it seems that there may be a greater natural exposure to heavy labor for men than for women and, therefore, the disorder is more frequently encountered in men.

Figure 22.3 presents a comparison between data obtained in this study and the findings of Walsh et al. (1989; cited by Norasteh 2012). Here, it may be observed that even though the foreign-based data still show greater prevalence of LBP in men, the percentages among women are much higher, probably for the reasons cited in the above paragraph.

The evaluation showed the risk factor of obesity not to be statistically significant, probably due to the small sample of subjects who were classified as obese. However, a slight risk was indicated, with an OR of 1.6 and 1.7 for mild obesity and moderate obesity, respectively. Other authors have encountered greater risk, as high as 5.7 (Ordoñez et al. 2012), relating it to light physical activity and sedentary work.

The significant differences observed in the last category (of scholarship and salary) would suggest that higher education and higher salary constitute protective factors. These data are consistent with reporting by Croft and Rigby in 1994, as well as more recently, by Katz in 2006, in whose work it can be observed that a large proportion of people with LBP have low wages and low educational levels.

The above could be explained on the basis that workers in these socioeconomic strata, in Mexico, are the ones who carry out most of the heavy work involved in the MMH or are occupied as chauffeurs of transport vehicles.

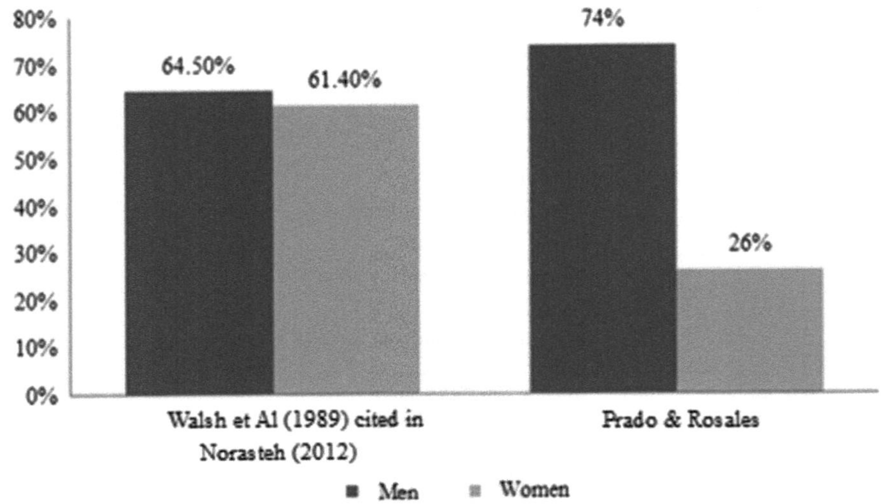

Fig. 22.3 Prevalence rate comparison, according to gender

22.5 Conclusions

The present study provided data that characterize the population suffering from lumbar spondylarthrosis, suggesting possible risk factors such as low educational levels and occupations related to physical overload and vibration.

This data provides us with the necessary knowledge to deepen our understanding of the relationship between LBP and personal and occupational factors, from an ergonomic point of view.

Of particular importance in the context of Mexico: In Mexico, to this date, the disease is considered to be unrelated to the occupation. Furthermore, in addition to the high level of incidence of this disorder, one must consider its socioeconomic impact: the seriousness of the illness results in a large number of days of incapacity, and, finally, total disability.

If the role of occupational factors in the development of LBP were to be recognized, then prevention strategies would also change, focusing on the inherent ergonomic risks of some of the specific tasks performed by workers in these occupations.

There are two approaches to establishing preventive strategies: administrative controls, and engineering. The administrative approach is focused on such measures as job rotation, breaks, the inclusion of exercise routines within the workday, shared activities, etc. As an example of the latter, it is suggested that when tasks require MMH in excess of the permitted weight or with frequent repetition, they can be shared among various persons (see Fig. 22.4); likewise, when it is necessary to mobilize patients (see Fig. 22.5).

Fig. 22.4 When manual lifting or transporting of objects is necessary and the objects are heavy, or the task must be performed frequently, it is safer to do so between two people

Fig. 22.5 Patient handling is also a heavy burden that must be carried out between two people and/or with mechanical aids

The engineering approach refers to changes in the workstation, including the machinery and equipment used to perform work activities. For example, Figs. 22.6, 22.7 and 22.8 show a change in the height of the handle on a transport device, based on an anthropometric adaptation of the equipment, which permits better posture and less effort.

In any case, it is necessary to carry out future studies that provide more data on this disorder and its occupational-ergonomic risk factors in the Mexican population. Most probably, a longitudinal study over a larger population can greatly enrich this

Fig. 22.6 Handle height lower than the user's bent elbow, tiny wheels

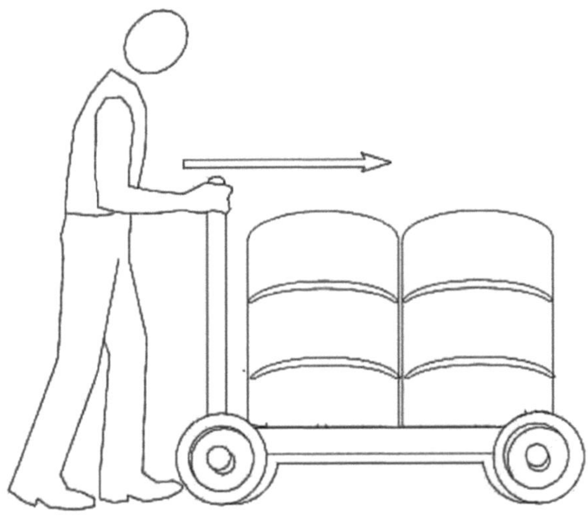

Fig. 22.7 Adequate handle height (for flexed elbows) and large wheels, which require less effort

field of knowledge, since the principal limitations of this study are found in the sample size and in the fact that temporal relationships are difficult to establish in a cross-sectional study.

It bears mentioning that in the present study, although the seated position was addressed in some occupations (for example, accountants and administrative personnel), it was not evaluated in relation to prolonged periods of time in that position. It would, therefore, be of import to launch further studies which investigate the demands of current labor activities, as some occupations require more tasks to be undertaken in a prolonged sitting position. Such is the case in the textile, leather goods, petrochemical, pharmaceutical, health, communications, and transport sectors, among others. According to Maradei-García et al. 2016, even though to date it has not been demonstrated that this posture causes LBP, it is considered to aggravate the condition. Furthermore, the discomfort of LBP increases with

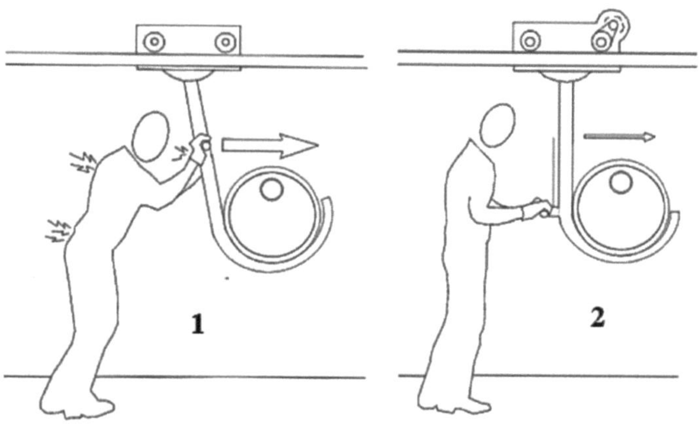

Fig. 22.8 (1) Handles above shoulder height. (2) Ergonomic adaptation to the height of the flexed elbow

exposure to a prolonged sitting position, even in subjects who do not suffer lumbago. For these reasons, it is important to find out more about this posture and its relationship to one of today's most important musculoskeletal disorders.

Finally, it is also necessary to consider other factors that may influence the development of LBP and that have already been reported in other studies, such as psychosocial aspects (which could include job dissatisfaction and stress), either on the job or off work (Hartvigsen et al. 2004; Torp et al. 1999).

References

Alperovitch-Najenson D, Santo Y, Katz-Leurer M et al (2010) Low back pain among professional bus drivers: ergonomic and occupational-psychosocial risk factors. Israel Med Assoc J 12:26–31

Bassols A, Bosch F, Campillo M, Baños J (2003) Back pain in the Catalan population. Preval Charact Ther Behav Gaceta Sanitaria 17:97–107. https://doi.org/10.1016/s0213-9111(03)71706-3 (In Spanish)

Cailliet R (1990) Painful syndromes: inability and pain of soft tissues. Manual Moderno Editorial, Mexico (In Spanish)

Chen JC, Chang WR, Chang W (2005) Occupational factors associated with low back pain in urban taxi drivers. Occup Med 55:535–540

Covarrubias-Gómez A (2010) Low back pain: a public health problem. Revista Mexicana De Anestesiología 33(1):106–109 (In Spanish)

Croft P, Rigby A (1994) Socioeconomic influences on back problems in the community in Britain. J Epidemiol Community Health 48:166–170. https://doi.org/10.1136/jech.48.2.166

Dionne C, Dunn K, Croft P (2006) Does back pain prevalence really decrease with increasing age? A systematic review. Age Ageing 35:229–234. https://doi.org/10.1093/ageing/afj055

Duque-Vera I, Zuluaga-González D, Pinilla-Burgos A (2011) Prevalence of low back pain and risk factors in nurses and auxiliaries of the city of Manizales. Revista Hacia la Promoción de la Salud 16(1):27–38 (In Spanish)

Elders L, Burdof A (2001) Interrelations of risk factors and low back pain in scaffolders. Occup Environ Med 58(9):597–603. https://doi.org/10.1136/oem.58.9.597
Garro-Vargas K (2012) Low back pain. Medicina Legal de Costa Rica 29(2):103–109 (in Spanish)
Hartvigsen J, Lings S, Leboeuf-Y C, Bakketeig LS (2004) Psychosocial factors at work in relation to low back pain and consequences of low back pain; a systematic, critical review of prospective cohort studies. Occup Environ Med 61(1):1–10
Hoy D, Brooks P, Blythc F, Buchbinder R (2010) The epidemiology of low back pain. Best Pract Res Clin Rheumatol 24:769–781
Instituto Mexicano del Seguro Social (2017) Bringing the IMSS to the Citizen. Available at: Imss.gob.mx. http://www.imss.gob.mx/conoce-al-imss/memoria-estadistica-2016 (in Spanish)
Kant I, Notermans J, Borm P (1990) Observations of working postures in garages using the Ovako working posture analysing system (OVVAS) and consequent workload reduction recommendations. Ergonomics 33(2):209–220. https://doi.org/10.1080/00140139008927111
Katz J (2006) Lumbar disc disorders and low-back pain: socioeconomic factors and consequences. J Bone Joint Surg (Am) 88(Supplemet):21–24. https://doi.org/10.2106/jbjs.e.01273
Konz S, Johnson S (2008) Work design: occupational ergonomics, 7th edn. Holcomb Hathaway Publ, Scottsdale
Kuorinka I, Jonsson B, Kilbom A, Vinterberg H, Biering-Sørensen F, Andersson G, Jørgensen K (1987) Standardized Nordic questionnaires for the analysis of musculoskeletal symptoms. Appl Ergon 18:233–237. https://doi.org/10.1016/0003-6870(87)90010-x
Maradei-García F, Quintana-Jiménez L, Barrero L (2016) The relationship between low back pain and movements in prolonged sitting posture. Lit Rev Salud Uninorte 32(1):153–173. 10.14482/sun.32.1.8481 (in Spanish)
Mitchell T, O'Sullivan P, Burnett A, Straker L, Rudd C (2008) Low back pain characteristics from undergraduate student to working nurse in Australia: a cross-sectional survey. Int J Nurs Stud 45(11):1636–1644. https://doi.org/10.1016/j.ijnurstu.2008.03.001
Nahit ES, Macfarlane GJ, Pritchard CM, Cherry NM, Silman AJ (2001) Short term influence of mechanical factors on regional musculoskeletal pain: a study of new workers from 12 occupational groups. Occup Environ Med 58:374–381. https://doi.org/10.1136/oem.58.6.374
Norasteh AA (2012) Low back pain. InTech editorial, Rijeka. https://doi.org/10.5772/3151
Occupational Safety & Health Administration (2004) OSHA Publications|Occupational Safety and Health Administration. In: Osha.gov. http://www.osha.gov/publications/osha3173.pdf
Okunribido O, Shimbles S, Magnusson M, Pope M (2007) City bus driving and low back pain: a study of the exposures to posture demands, manual materials handling and whole-body vibration. Appl Ergon 38(1):29–38. https://doi.org/10.1016/j.apergo.2006.01.006
Ordoñez A, Durán S, Hernández J, Castillejos M (2012) Association between work activity with great demand for physical effort and low back pain. Acta Ortopédica Mexicana 26:21–29 (In Spanish)
Plouvier S, Renahy E, Chastang J, Bonenfant S, Leclerc A (2008) Biomechanical strains and low back disorders: quantifying the effects of the number of years of exposure on various types of pain. Occup Environ Med 65(4):268–274. https://doi.org/10.1136/oem.2007.036095
Putz-Anderson V (1988) Cumulative trauma disorders: a manual for musculoskeletal diseases of the upper limbs. Taylor & Francis
Seguí-Díaz M, Gérvas J (2002) Low back pain. SEMERGEN Medicina De Familia 28(1):21–41. https://doi.org/10.1016/s1138-3593(02)74401-8 (in Spanish)
Smith D, Mihashi M, Adachi Y, Koga H, Ishitake T (2006) A detailed analysis of musculoskeletal disorder risk factors among Japanese nurses. J Saf Res 37(2):195–200. https://doi.org/10.1016/j.jsr.2006.01.004
Sobrino F (2003) Chronic cumulative disease caused by repeated microtrauma: New definition, pathogenesis, general clinical signs and symptoms, risk factors, controversies. Mapfre Med 14:125–133 (in Spanish)
Thelin A, Holmberg S, Thelin N (2008) Functioning in neck and low back pain from a 12-year perspective: a prospective population-based study. J Rehabil Med 40:555–561. https://doi.org/10.2340/16501977-0205

Torp S, Riise T, Moen B (1996) Work-related musculoskeletal symptoms among car mechanics: a descriptive study. Occup Med 46(6):407–413. https://doi.org/10.1093/occmed/46.6.407
Torp S, Riise T, Moen B (1999) How the psychosocial work environment of motor vehicle mechanics may influence coping with musculoskeletal symptoms. Work Stress 13(3):193–203. https://doi.org/10.1080/026783799296011
Wipf J, Deyo R (1995) Low back pain. Med Clin North Am 79(2):231–246. https://doi.org/10.1016/s0025-7125(16)30065-7

Chapter 23
The Knowledge-Based Maintenance: An Approach for Reusing Experiences in Industrial Systems

Luis Alberto López-Ramos, Guillermo Cortés-Robles, Eduardo Roldán-Reyes, Giner Alor-Hernández and Cuauhtémoc Sánchez-Ramírez

Abstract The maintenance process plays a crucial role in any production process. Frequently, the common vision about the maintenance process takes only into account the probability of an event and the risk associated with some expected effects with the goal to reduce the effective cost. Nevertheless, maintenance is not a routine activity; it is a creative process that continually produces new knowledge. The traditional approaches to maintenance do not propose an explicit mechanism to preserve this knowledge. As a consequence, the experts managing the maintenance process face the problem to reinvent past solutions, instead of reusing their experiences to solve new problems. The lack of a learning process in the maintenance process produces the loss of knowledge, increases the dependency on some experts, but more important yet, it makes the evolution of the maintenance process more difficult. This chapter presents a Knowledge-Based Maintenance (KBM) approach that aims to create a framework to manage a maintenance program but also to preserve the valuable experience deployed in the maintenance activities.

Keywords Knowledge capitalization · Maintenance · Knowledge-based maintenance · Case-based reasoning

23.1 Introduction

The purpose of any company competing in the market is to produce profit. In this sense, three generic strategies are possible: (1) To increase the presence of the company in the market through different strategies (i.e., marketing, new product

---

L. A. López-Ramos · G. Cortés-Robles (✉) · E. Roldán-Reyes · G. Alor-Hernández · C. Sánchez-Ramírez
Division of Research and Postgraduate Studies, Tecnologico Nacional de México/Instituto Tecnológico de Orizaba, Av. Oriente 9 no. 852, 94320 Orizaba, Veracruz, Mexico
e-mail: gcortes@ito-depi.edu.mx

J. L. García Alcaraz et al. (eds.), *Best Practices in Manufacturing Processes*, https://doi.org/10.1007/978-3-319-99190-0_23

development, intellectual property strategies, specialization, among thousands of possibilities); (2) To improve the internal operation (i.e., lean manufacturing, zero risk design, ergonomics, to mention just a few), and (3) A combination of both. Each strategy is in itself a technical world. Following this logic, the efficiency of the maintenance process has a direct effect on the internal operation of any industrial system. In the present context, an industrial system is a unit that interacts with other units (similar or not) in a dynamic context that consumes and transforms some tangible or intangible resources into something valuable to the market (Manzini et al. 2010; Morvan 1997; Badiru 2013). An efficient maintenance system, independently of the context where it is deployed (services, at any block of the supply chain or even in nonprofit organizations) has several advantages, which are given as follows:

- It helps to reduce the operation problems and even to minimize the recurrence of problems
- It increases productivity while reduces production costs
- A mature maintenance system also reduces risks and improves quality.

Then, the importance of the maintenance will grow as other characteristics in the industrial and commercial systems evolve such as the complexity of the industrial system, the safety, the pressure in the price reduction, the quality requirements, among other significant features. Under such conditions, the maintenance process will undoubtedly evolve into a more performing process. One of the more relevant tendencies in the evolution of the maintenance process is the assimilation of information technologies and the use or consumption of knowledge (Scherer and White 1989; Lirov and Yue 1991; Pieri et al. 2001; Potes-Ruiz et al. 2014; Geng et al. 2017). This chapter corresponds to this research direction. The objective of this chapter is to conceive and implement a knowledge-based process to assist the industrial maintenance activities. The Case-Based Reasoning (CBR) is a useful approach to manage experience in the industrial maintenance (Potes-Ruiz et al. 2013; Chebel-Morello et al. 2010). Five sections compose the chapter. Section 2 briefly exposes the theoretical background. Section 3, describes the methodological approach. Section 4 contains a case study that demonstrates the objective of this research. Section 5 discusses some findings, results, and limitations. The last section contains the conclusion and the future work.

23.2 Background

The concept of Knowledge-Based Maintenance (KBM), as part of the evolution process in the concept of industrial maintenance, is a dynamic research field. The work of (Clark et al. 1992) explains that the use of the knowledge-based systems can improve the maintenance process. The authors propose to use a knowledge-based system to support the preventive maintenance of buildings. The objective of this system is to improve reliability and reduce the consumption of resources (money, time, replacement parts, and corrective maintenance). Iudica (1987) describes how the industrial planning can benefit from the use and exploitation of knowledge. Toro et al. (2007) move the concept of knowledge-based industrial management to a new resource: portable devices. The authors use augmented reality techniques and ontologies to model experiences and facilitate the usability. Pieri et al. (2001) present a knowledge-based decision support system for the maintenance of chemical plant equipment. The purpose of this work is to combine information from the equipment with the corrosion agents and some crucial factors of each component. Bekkaoui et al. (2015) depict the importance of reuse the human knowledge as an essential element in the continuous improvement of maintenance activities. The authors focus their attention on the problem of selection of experts in maintenance processes. Talamo and Bonanomi (2016) offer a broad perspective on the use of knowledge management and information tools for building maintenance. The authors underline that almost any system produces information about its state or behavior, and hence, this resource, frequently underestimated, can guide the diagnosis and prevention of failures. The use of this information and knowledge extraction reduce the operative cost. Ruschel et al. (2017) explore another application of the knowledge extracted from a system: autonomous equipment predictions and self-diagnosis. The authors underline that both approaches are continually evolving, and hence revealing new uses of the knowledge acquired. The articles in this brief review offer a broad perspective of the value that the use of expertise provides to the industrial maintenance. Among the approaches used to support the maintenance activities one has an extended application: the Case-Based Reasoning approach (CBR) (Kobbacy 2008a, b; Olsson and Funk 2009; Potes-Ruiz et al. 2013).

The Case-Based Reasoning (CBR) suggests that problems are recurrent and maintenance activities are not the exception. Hence, past solutions are the building block to propose new solutions. A case is then a structure that generally contains three components: the description of a problem or situation, the solution, and the result. Once a case is complete, this case is stored in a case memory. The access to this memory and the retrieving mechanism could be an open process or a closed one. An open case memory facilitates collaboration and the knowledge diffusion, but it does not guarantee confidentiality. On the other hand, a close case assures the preservation of knowledge for an organization but does not profit from collaboration (no more than the experts that use the software). Also, a case indexed in the case memory is the subject of a validation process in the industrial context, then it is known with certitude. This condition avoids the problem to deal with uncertainty right after the moment to launch the system. Later, some guidelines or criteria become useful to evaluate a case with the purpose to add it to the memory. In consequence, all cases in the memory are valid in the industrial context. According to Aamodt and Plaza (1994), the reasoning process has four stages: retrieve, reuse, revise, and retain. The CBR process starts describing the input problem. The information in this initial step is then compared to retrieve a problem or set of previously solved problems (cases), stored and indexed in the memory. Then, if one or various stored cases match with the initial problem, the case with the higher

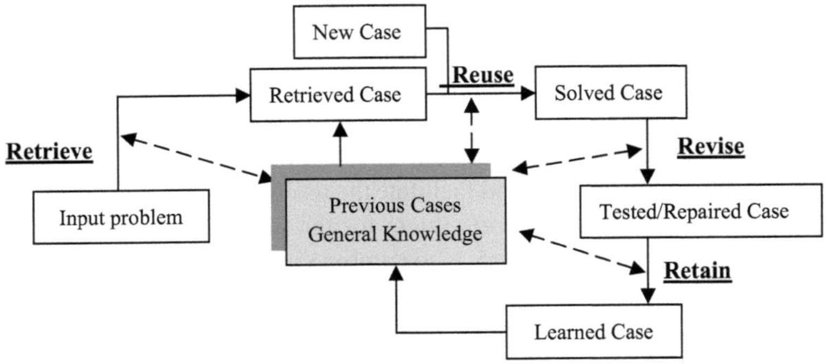

Fig. 23.1 The case-based reasoning process

similarity evaluation is the right candidate to reuse its solution. Subsequently, the expert adapts the known solution to the input problem. The new solution must pass a validation process. Two alternatives emerge at this point: the proposed solution must be revised, tested, and repaired if necessary to reach an objective or the solution is satisfactory. Finally, the new experiences, which comprise a result (failure or success), the strategies to repair and implement the final solutions, among others particular features, are retained for a future utilization and the cases memory is updated. Figure 23.1 depicts this process.

The CBR process offers several benefits, among the most important are as follows:

- Longtime learning is the most valuable product in the CBR process. The capacity to store and index a problem with its associated solution, the implementation of a mechanism to compare and retrieve a problem enables the learning process. According to Leake (1996), the reasoning and learning process have a reliable connection, which makes the CBR process an efficient tool for implementing a Knowledge Management Systems (KMS).
- From a technical point of view, the reuse of experiences in a CBR system is more efficient than a rule-based approach. This advantage is founded in the facility for an individual to understand, utilize, and apply the specific examples provided by a CBR system. This empathy is based on the reasoning process. Both actors—system and reasoner—utilize the same reasoning process for solving problems: the analogical thinking, which is crucial in the CBR process and also, the most common human problem-solving strategy (Aamodt and Plaza 1994; Terninko et al. 1998).
- A CBR manages high volumes of information and the knowledge in the case memory is updated automatically using the system. Hence, time is another determining factor because, despite the technological platform used to implement the CBR, the system becomes more competent over time.

Like any other approach, the CBR also has several disadvantages, which are given as follows:

- The case memory has a specific design objective. Hence, contains only particular problems for a unique context, which is, in fact, an advantage. Nevertheless, problems evolve, new materials, and new technologies emerge continually. Thus, the case memory must take into account this process to preserve the pertinence of the system.
- The automation of the validation process to decide when to increase the cases in the memory is a challenge. Consequently, it is necessary the human involvement in this process.
- The use of cases without an objective and critical examination frequently leads to an unsuccessfully adaptation and revision process. This blind adaptation frequently produces unsatisfactory solutions and uncertainty in the solver about the efficiency of the system.

The use of the CBR to assist maintenance activities has a positive tendency. The number of publications in this field is growing at a considerable pace. One of the first works in this topic is the article by Graham-Jones and Mellor (1995) that proposes the use of the CBR to deal with the challenge of failure analysis. The authors explain that the advantages of CBR, particularly, the elicitation knowledge facilitates the failure analysis. The work of Ganag et al. (2015) follows a similar direction. The authors describe an intelligent fault diagnosis model for power equipment based on CBR. Later, the work of Miyashita and Sycara (1995) describes a CBR process that collects and indexes information about a context-dependent user optimization preferences and tradeoffs. The information in the memory is useful in a reactive schedule management that acts when there are unexpected events. Göker and Roth-Berghofer (1999) explore the CBR drawbacks and develop a case-based help-desk support system that overcomes, even partially some of the CBR shortcomings. As the CBR applications in the industrial maintenance domain continue to evolve, the work of Kobbacy (2008a, b) offers a broad perspective about the use of CBR in crucial maintenance activities (i.e., fault diagnosis and preventive maintenance). Other relevant works about the use of the CBR to support the industrial maintenance process are Nick et al. (2003) that uses the CBR in the e-security to avoid failures. Wang and Hsu (2004) use the CBR process for PC troubleshooting. The authors design a Web-based CBR knowledge management system to support the maintenance service centers of a computer company. Cheng et al. (2008) introduce the CBR process in the reliability-centered maintenance as the central approach to reuse past experiences. The work of Ahmed et al. (2010) uses the CBR approach to reduce mistakes in the medical domain. The work of Goh and Guo (2018) combines the CBR with a Rule-Based Reasoning to design an active fall protection system for the construction industry. The purpose of this article is to improve safety in the construction industry by using past cases.

As a partial conclusion, it is possible to observe that the use of the Case-Based Reasoning approach in the industrial maintenance domain is a well-accepted approach. It produces valuable information, and in most of the cases explored in this section, the CBR creates the conditions to capitalize on past experiences (Rasovska et al. 2008). As a result, the assimilation of the CBR approach in the industrial maintenance process reveals a research direction that is in constant evolution. Despite the usefulness of the CBR to support the industrial maintenance, there are just a few reported articles in Latin America, probably for lack of diffusion. This chapter indicates the use of the CBR to facilitate preventive and corrective maintenance.

23.3 Methodology

The methodology used in this chapter is a synthesis from the INRECA methodology (Bergmann 1998), the work of Cortés-Robles et al. (2006), and Chebel-Morello et al. (2010). The process to develop a KBM system envelops the next stages:

- Identify a group of experts: this initial stage has the purpose to select the expert's panel that will supervise the system design, and that will validate the usefulness of the system.
- Delimit the problem to solve and the maintenance approach: it is crucial to precisely delimit the objective of the knowledge-based maintenance system and the context where it will perform.
- Knowledge extraction and modeling: once it is available a source of experiences (expert panel) and the objective of the system is correlated to the problem to solve, then it is possible to model the problem to explain. The records about the problem are a valuable source of information.
- Experience validation: the problem contains three components, which are a description of the problem, the model of the solution, and the possible results for a problem.
- Design of the case memory: the available information shapes the configuration of the memory. It is important to notice that the memory must implement a mechanism to facilitate the addition or elimination of some variables in the problem description. A flexible memory takes into account possible changes in the problem description.
- Implementation of the reasoning process: once there is a validated case structure and the memory case has a minimal but adaptable configuration, it is possible to deploy the complete process. The JColibri 2.0 application is useful to accomplish this task.
- Verification and desk validation: once there is a connection between the database and the reasoning process, it is crucial to test if all the CBR steps are operational.
- Training and launching: the system stores and indexes all the experiences considered as relevant by the expert panel. These are the initial cases that will be retrieved in the CBR process.
- Return of experiences: The use of the CBR process in the particular problem produces a return of experiences that allows the improvement of the proposed reasoning process. This information is useful to redesign the reasoning process if necessary.

Next section depicts the application of the methodology and describes a case study. Also, next section contains a brief description of a Knowledge-Based Maintenance system (KBM).

23.4 Case Study: A KBM in the Paper Industry

The case study is part of the strategic development of a new approach for preventive and corrective maintenance in a Mexican company with more than 50 years in the market. The company is in constant transformation. A few years ago, the company developed and implemented a new production process to recycle paper. Nowadays, the company is the leader in Latin America, and its exportations are growing at a regular pace. The company sells for almost 1 billion USD and employs more than 10,000 workers. The facilities that develop the KBM system are located in Veracruz, Mexico and have a production capacity of 80,300 tons/year. There are two production lines in this manufacturing center with 110 and 130 tons/day. The production process has six stages (Fig. 23.2).

23.4.1 Experts in the KBM

The company has a maintenance department, which envelops several technical domains called as specialties that perform the maintenance function. The expert panel involves an expert from each area. According to the work of Hanaysha (2016) and Barak et al. (1999), some basic criteria to select the expert were: (1) The individual competencies. In this case, the number of years executing the activity; (2) the capacity to communicate. Thus, criteria was subjectively evaluated by the head responsible for each specialty. (3) The ability to communicate, which again is a subjective appreciation. (4) The problem-solving skills evaluated via the past records in each specialty. Finally, (5) the will to participate in this initiative. The purpose was to create a team where all participants are a voluntary members of the teamwork. Five experts get involved in the teamwork: one from the mechanical specialty, two from the electrical specialty, one from the instruments specialty, and the expert in charge of the maintenance department.

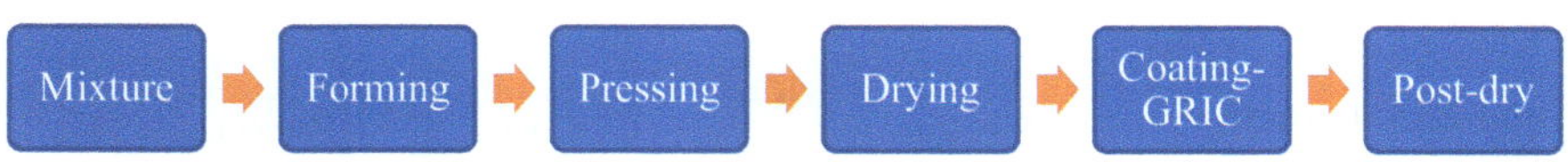

Fig. 23.2 Paper making process

23.4.2 *Problem Delimitation*

It is essential to identify the problem to solve with the KBM system. The right identification of the maintenance problem facilitates the adoption to the final user and diminishes the resistance to change. Three activities are necessary: (1) to evaluate the productivity of each production line to identify the loss of time, (2) to determine the critical area or process, and (3) to identify the component, system or subsystem responsible of this effect.

The first evaluation revealed that the first production line had the more significant loss of time. Then, the team collected information about the availability of each process in this production line by comparing the effective production time with the unproductive time of each process. The resulting percentage indicates the effectiveness of the process (Table 23.1).

$$\text{Effectiveness:}(\text{total available time} - \text{unproductive time}) * 100 \qquad (23.1)$$

It is important to notice that the unproductive time takes into account the time planned for preventive maintenance. The identification of the critical area is the process that generates the more corrective maintenance. The time loss or unproductive time has four components (Table 23.2).

Table 23.2 shows that the activities of corrective maintenance represent the 46.11% of the total unproductive time. Table 23.3 contains the distribution of corrective maintenance time by specialty.

Table 23.1 Distribution of time in the production line no. 1

Concept	Hours	Percentage (%)
Time loss	304.77	5.49
Usage time	5246.43	94.51
Total available time	5551.20	100

Table 23.2 Unproductive time composition

Concept	Hours	Percentage (%)
Planned maintenance	54.5	17.88
Loss of time for setup	65.99	21.65
Corrective maintenance	140.53	46.11
Other concepts	43.75	14.36

Table 23.3 Corrective maintenance by speciality

Specialty	Hours	Percentage (%)
Mechanical	83.19	59
Electrical	49.25	35
Control instruments	8.08	6
Total	140.53	

The mechanical specialty dedicates 39.53 or 28.12% hours to corrective maintenance in one equipment: the Gate Roll Inversion Coater or GRIC. This equipment is then the objective of the KBM system.

23.4.3 Knowledge Extraction and Modeling

The GRIC is a machine that envelops the paper in a layer through the action of some rollers. The GRIC modifies the thickness of the paper according to the customer requirements. The Failure Mode and Effect Analysis (FMEA) are useful in this stage to determine what components are more susceptible to failure (Natarajan 2015). The critical components will be part of the KBM system. Table 23.4 describes the failure mode in the GRIC.

Taking into account the Risk Priority Number it is possible to calculate the relevance of each failure mode. Table 23.5 contains this information.

Table 23.5 shows that the measuring roller and the application roller are the highest relevance in the GRIC. Additionally, the head responsible for the maintenance department included the paper guider, the guider, and the two valves as significant components. The next activity is the identification of the operative variables for each element to configure a case. The panel of experts used a cause–effect diagram to observe the relation of all variables. Table 23.6 lists the state variables that compose the problem description of the case.

For instance, Table 23.7 shows the datasheet for both rollers (problem description).

The variables in Table 23.7 are the input data to calculate the similarity between cases. A case has three components: the problem description, the problem solution, and the result.

23.4.4 Experience Validation

The validation of the case is an activity that occurs during the memory case design. The case design passes through a validation process to assure that it takes into account the most significant variables. This activity also verifies that variables are not overlapping, that the weight of each variable (if there is a hierarchy) is valid, to mention some significant validation steps. All the initial cases stored in the case memory are known with certitude. Theoretically, the case, solution, and result are known. In this way, the entire experience has already happened in the past; hence there is no uncertainty in the data to avoid a lack of trust in the initial stage of the reasoning process. The expert panel validates the case. The main source of information was the failure records stored in the maintenance department.

Table 23.4 Failure mode in the GRIC and the risk priority number

No.	GRIC component	Function	Failure mode	RPN
1	Measuring roller	Applies the layer in the paper sheet	The incorrect supply if the chemical mix that forms the layer	315
2	Application roller	It transfers the thin film to the paper sheet	Nonhomogeneous application of the layer	180
3	Paper guider	Controls the paper movement	The paper sheet gets blocked	105
4	Guider	It controls the direction of the paper sheet through the control	A wrong direction of the paper sheet	90
5	Paper guider motor	Supplies energy to the paper guider	The bearings suffer an uncontrolled wear	60
6	Scanner	A unit that measures all the paper variables	Lack of precision in the unit or bad calibration	50
7	Pulley	Facilitates the guider movement	Blockage in the pulley	48
8	Valve of left connector	Produces the signal to put in action the left connector	A communication error with the left connector	40
9	Left head	To contain the chemical mix of the layer	Inadequate storage	40
10	Right head	Provides the right flow to the measuring roller	Imprecise or variation in the flow	36
11	Monorail	To guide the crane	Excessive wear in the monorail	27
12	Right head valve	Produces the signal to activate the right head	No signal comes out from the valve	24
13	Control	It controls the tension according to the paper speed in the process	A sensible variation between the paper sheet tension and the speed	24
14	Spooner	It keeps the paper on suspension through an air flow	The air supply is not the right	18

23.4.5 *Design of the Case Memory*

The solution for each case has 22 variables (Table 23.8). On the other hand, the result has two sections: the evaluation and justification with six variables.

23.4.6 *Implementation of the Reasoning Process*

In this step, the use of the JColibri 2.0 is fundamental. JColibri 2.0 framework is a platform that assists a user in developing CBR applications. The JColibri architecture is extensible and reusable in a transversal way (Recio-García et al. 2014).

Table 23.5 Failure mode in the GRIC and the risk priority number

No.	GRIC component	RPN events	RPN accumulated	Percentage
1	Measuring roller	315	315	30
2	Application roller	180	495	17
3	Paper guider	105	600	10
4	Guider	90	690	9
5	Paper guider motor	60	750	6
6	Scanner	50	800	5
7	Pulley	48	848	5
8	Valve of left connector	40	888	4
9	Left head	40	928	4
10	Right head	36	964	3
11	Monorail	27	991	3
12	Right head valve	24	1015	3
13	Control	24	1039	3
14	Sponner	18	1057	3

Table 23.6 Variables that compose the problem description

Variable	Measuring roller	Application roller	Paper guider	Guider	Valve
Type	X	X	X	X	X
Position in the GRIC	X	X			X
Diameter	X	X			
Paper density	X	X	X	X	
Air supply pressure					X
Cable connectors					X
Papers sheet speed	X	X	X	X	
Tension	X	X	X	X	
Amperage	X	X	X	X	X
Time in operation	X	X	X	X	X
Time between failures	X	X	X	X	X

Hence, it is possible to reuse an available CBR. Nevertheless, the search for a similar system did not produce a satisfactory result. Consequently, the team developed a new CBR system. The system has a four-layer architecture (Fig. 23.3).

Presentation Layer: in this layer are the components of the user interface. It uses HTML5, which is useful for representing the content of a site web. The Cascading Style Sheets (CSS) adds some functionality to the user interface and produces different styles for the HTML etiquettes. JavaScript executes different events on the site web.

Table 23.7 Variables that compose the problem description in the roller

Variable	Measuring roller	Units
Type	#Value	A/M
Position in the GRIC	#Value	I/D
Diameter	#Value	Mm
Paper density	#Value	Gr/m^2
Paper sheet speed	#Value	RPM
Tension	#Value	N/m^2
Amperage	#Value	Amp
Time in operation	#Value	Hours
Time between failures	#Value	Days

Table 23.8 Variables that compose the solution

Solution variables		
#	Name	Type
1	DriveResp	varchar(10)
2	VelocidadOper	Float
3	tAplicador	Float
4	t4Seccion	Float
5	DiametroMi	Float
6	DiametroMd	Float
7	DiametroAf	Float
8	DiametroAv	Float
9	FrecuenciaMd	Float
10	FrecuenciaMd	Float
11	FrecuenciaAF	Float
12	FrecuenciaAv	Float
13	PBomba	Float
14	FlujoE	Float
15	dFibra	Float
16	IntensidadMi	Float
17	IntensidadMd	Float
18	IntensidadAf	Float
19	IntensidadAv	Float
20	vibracionReductoresIn	Float
21	vibracionRedutoresGe	Float
22	caudalEncolante	Float

Service Layer: the service layer offers all the necessary functions in the CBR process: retrieve, reuse, revise, and retain a case. It is in this layer that the JColibri 2.0 deploys all their capacities and tools. It provides the similarity metrics, an ontology to describe a case, the KNN algorithm to select a case, and several functionalities to maintain the memory case. As a consequence, the user can insert the case description, and the system will recommend the most similar cases in the

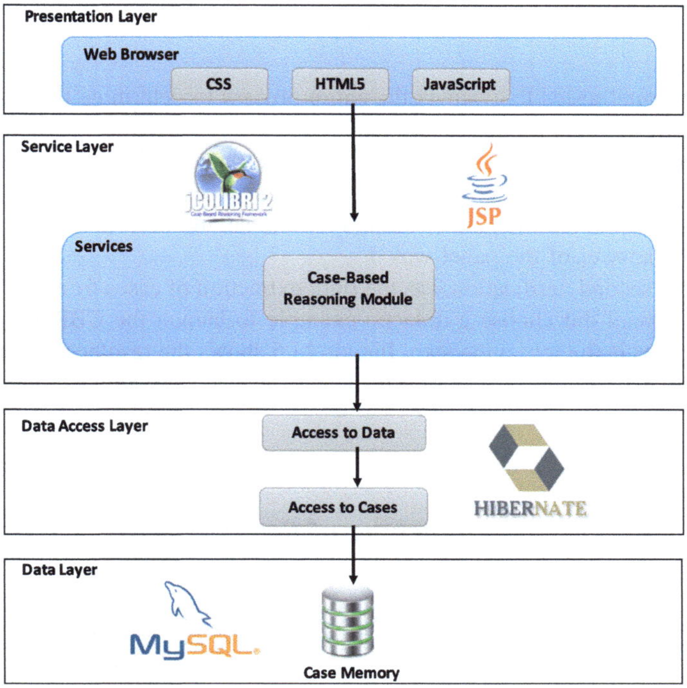

Fig. 23.3 The architecture of the KBM system

memory. The JColibri and the presentation layer provide the structure for reusing a case and storing it the memory. The layer uses Java Server Pages (JSP) in the Web server. It is important to notice that the JColibri 2.0 platform uses Java, which is compatible with the JSP.

Data Access Layer: the scripts are stored in this layer. They describe the logic for managing data and execute the necessary operations in data such as access, delete, add or modify the data. It also contains the query for retrieving some data. This layer also has a connection with the service layer and the databases. The framework Hibernate is then crucial for developing the CBR in the JColibri platform. Hibernate enables a way of mapping some entities with relational objects, which facilitates the access to all data in the system.

Data Layer: the databases are in this layer. The system has access to this data through the scripts. A database management system controls all the operations in the database. The layer uses MySQL, one of the most tested data management systems.

23.4.7 Verification and Desk Validation

The expert panel needs to evaluate the entire process to determine their usefulness and usability. Figure 23.4 schematizes the possible alternatives in the user interface.

The expert panel evaluates subjectively the interface as simple and easy to use. The evaluation uses a Likert scale from 1 to 5, where 1 means very complex interface and difficult to use and 5 means very simple, intuitive and easy to use. The mean in the answer of the panel was 4.

Then, the second verification was the right extraction of cases from the database. The expert panel did choose a didactic example to launch the CBR process. The crucial process is the retrieving step. Figure 23.5 shows the results of the retrieving process in the JColibri 2.0 platform. The expert panel validated these results manually.

23.4.8 Training and Launching of the System

The training does not refer to training the reasoning process. This stage is the objective the desk validation. Before launching the system in the maintenance department, it is necessary that the future user develops some essential competencies and understands the CBR process. To accomplish this objective, a training program, which is part of the system, starts some days before launching the operation. Some didactic cases explain the concept of the KBM with the goal to demonstrate the usefulness of the system. It is important to underline that the training program clarifies that there is some work to add cases to the memory.

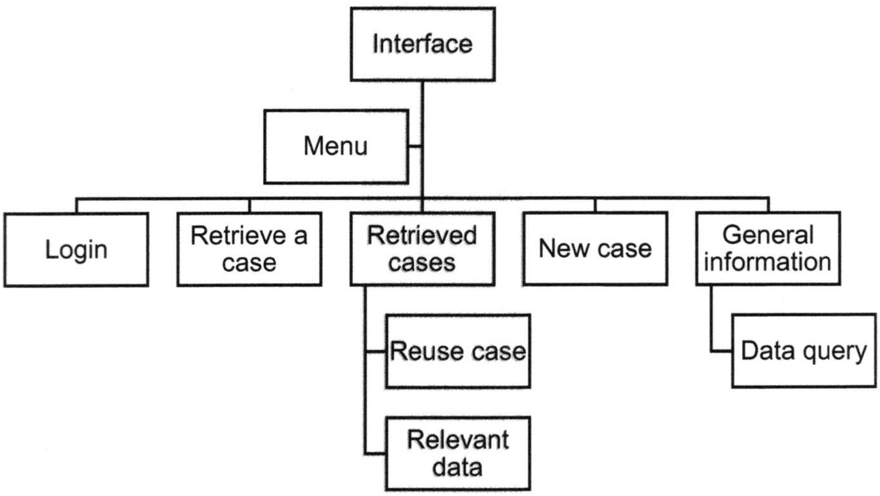

Fig. 23.4 The alternatives in the user interface

```
Equipo  equipo = new Equipo(12.0, 25, "A", 35, 5, 1, 30, 70)

[Description: {11.0, 0.10, B, 40, 1.5, 2.5, 35, 75}][Solution: {ajuste,calibracion}][Sol.Just.: null][Result: null]   -> 0.8
[Description: {11.5, 0.10, A, 37, 1.5, 1.5, 45, 65}][Solution: {mezcla,monitoreo}][Sol.Just.: null][Result: null]   -> 0.55
[Description: {12.0, 0.25, A, 35, 0.5, 1.5, 30, 70}][Solution: {programado,preventivo}][Sol.Just.: null][Result: null] -> 0.47999999999999997
[Description: {12.0, 0.15, C, 50, 1.0, 2.5, 50, 90}][Solution: {limite,correctivo}][Sol.Just.: null][Result: null]   -> 0.47999999999999997
[Description: {13.0, 0.15, B, 35, 0.5, 2.5, 35, 70}][Solution: {mezcla,monitoreo}][Sol.Just.: null][Result: null]   -> 0.25
[Description: {13.5, 0.30, B, 40, 1.0, 2.0, 45, 80}][Solution: {ajuste,calibracion}][Sol.Just.: null][Result: null]   -> 0.25
[Description: {13.5, 0.30, C, 45, 1.5, 1.5, 35, 85}][Solution: {limite,calibracion}][Sol.Just.: null][Result: null]   -> 0.175
```

Fig. 23.5 The alternatives in the user interface

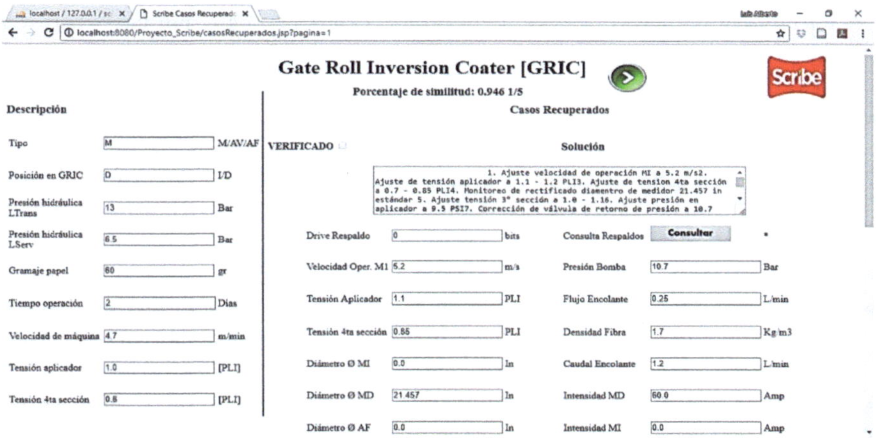

Fig. 23.6 The user interface of the KBM system

The addition of cases means a significant work to the maintenance team, but it pays back by reducing the solving effort while facing the problems covered by the system. Figure 23.6 shows the final user interface after some minor modifications.

23.4.9 *Return of Experiences*

The system is already online on the enterprise intranet. The parameter to evaluate the efficiency of the system is the corrective maintenance time in the GRIC and the mean time to repair. The first test of the system, from June to August 2017 produced some small but positive results. The corrective maintenance time in the GRIC revealed a reduction of 4.2%. It is expected that this percentage increases as the number of cases, and the learning curve of the maintenance department evolve over time.

23.5 Conclusions and Future Work

The KBM system offers several advantages to the maintenance departments. Next points describe the most significant:

- The case memory creates the condition for deploying a longtime learning process. All cases and their evaluation are available for the teamwork. Hence, the experiences are easily shared. This advantage has another benefit: it reduces the dependency on one expert. Sometimes while solving a problem related to the objective of the KBM, we ask the team for a particular case, but the expert was out of shift and unreachable. Consequently, the KBM system becomes support for problem-solving.
- The use of the KBM system is also a training tool. The cases in the memory are useful to train new personnel in the maintenance department. In this case, the problem description of one or a set of cases represents the didactic situation. The answer or strategy proposed by the personnel in training is compared with the solution stored in the case memory to get feedback. In the future, with a vast number of cases, it is even possible to design a simulator.
- The impact of the system on the productivity of the manufacturing process even if for the moment is not remarkable, it is positive. The efficiency of the system will increase over time.
- The KBM system is responsive to mobile devices. The purpose of this capacity is to facilitate collaboration, even if an expert is in another country or out of the enterprise facilities.
- In the case design process, there are two latent variables in the problem description and the solution. Their purpose is to offer some flexibility if the case needs a new variable. If this condition arrives, then the system manager activates the variable, and the system recalculates the similarity index of all cases. Even if this represents some effort, particularly in the adaptation of the user interface, it is easier than redesign the entire case memory.

Despite the advantages that the KBM system provides to the maintenance department, there are also some limitations. Next points describe the most relevant in the present context.

- The effort for defining the case, and in consequence the configuration of the case memory is considerable. In fact, this activity consumes most of the time of the design process. The validation of the case and the selection of the expert panel asks in one hand, a creative effort to propose mechanism or validation strategies, and in the other, criteria well understood and clear for all the team to select the right personnel.
- Another limit or better say, a restriction, in this case, is that the final user not always has all the competencies to operate and maintain the system. This situation leads to dependency on another expert in the process and an increasing workload.

- There is considerable work for adding cases to the memory. The teamwork needs to identify the case, prepare and document the problem. Collect the information of the solution and the results. These activities are executed off-line, condition that increases the workload for the team. This is a restriction because the pressure in the system to accomplish the production goals overcomes the needs of the KBM system.
- In the particular context of the KBM system described in this chapter, there is another limitation: the user, which is the responsible for the production process not always speaks English. This makes difficult to transfer the experiences in the case memory to other countries.

As a future work, the maintenance department considers expanding the objective and goal of the KBM system to cover all the GRIC. With the objective to accomplish this task, it is indispensable to conceive and test a more significant case or to develop independent KBM system for each problem and assign the control of each system to an agent. Also, the enterprise has some similar equipment in other countries. Hence, it is necessary to develop a case memory in English with the most successful cases and also the cases that produced unsatisfactory results. The purpose is to share the cases as a knowledge transfer mechanism. Finally, the KBM system will expand their objective to assists preventive maintenance and add other evaluation parameters to the system.

The conclusion of this chapter is to put into evidence that the Knowledge-Based Maintenance approach consents several advantages, even if the design and implementation of such systems increase the workload temporarily. The KBM approach could assist another perspective about the maintenance as the Condition-Based Maintenance or Reliability-Based Maintenance.

References

Aamodt A, Plaza E (1994) Case-based-reasoning: foundational issues, methodological, variations, and systems approaches. AI Commun 7:39–59 (IOS Press)

Ahmed MU, Begum S, Olsson E, Xiong N, Funk P (2010) Case-based reasoning for medical and industrial decision support systems. In: Montani S, Jain LC (eds) Successful case-based reasoning applications—I. Studies in computational intelligence, vol 305. Springer, Berlin, Heidelberg

Badiru A (ed) (2013) Handbook of industrial and systems engineering, 2nd edn. CRC Press, New York

Barak M, Maymon T, Harel G (1999) Teamwork in modern organizations: implications for technology education. Int J Technol Des Educ 9(1):85–101. https://doi.org/10.1023/A:1008849803984

Bekkaoui M, Karray M, Sari Z (2015) Knowledge formalization for experts' selection into a collaborative maintenance platform. IFAC-PapersOnLine 48(3):1445–1450. https://doi.org/10.1016/j.ifacol.2015.06.290

Bergmann R, Breen S, Fayol E, Göker M, Manago M, Schmitt S, Stahl A, Wess E, Wilke W (1998) Collecting experience on the systematic development of CBR applications using the

INRECA methodology. In: Smyth B, Cunningham P (eds) Advances in case-based reasoning. EWCBR 1998. Lecture notes in computer science, vol 1488. Springer, Berlin, Heidelberg

Chebel-Morello B, Haouchine K, Zerhouni N (2010) A methodology to conceive a case based system of industrial diagnosis. In: Kiritsis D, Emmanouilidis C, Koronios A, Mathew J (eds) Engineering asset lifecycle management. Springer, London

Cheng Z, Jia X, Gao P, Wu S, Wang J (2008) A framework for intelligent reliability centered maintenance analysis. Reliab Eng Syst Saf 93(6):806–814. https://doi.org/10.1016/j.ress.2007.03.037

Clark G, Mehta P, Thomson T (1992) Application of knowledge-based systems to optimised building maintenance management. In: Belli F, Radermacher FJ (eds) Industrial and engineering applications of artificial intelligence and expert systems. IEA/AIE 1992. Lecture notes in computer science (Lecture notes in artificial intelligence), vol 604. Springer, Berlin, Heidelberg

Cortes-Robles G, Negny S, Le Lann JM (2006) Innovation and knowledge management: using the combined approach TRIZ-CBR in process system engineering. Comput Aided Chem Eng 21:2237–2242. https://doi.org/10.1016/S1570-7946(06)80381-0

Gang M, Jiang L, Xu G, Zheng J (2015) A model of intelligent fault diagnosis of power equipment based on CBR. Mathe Probl Eng Article ID 203083 https://doi.org/10.1155/2015/203083

Geng J, Li Y, Wang R, Wang Z, Zhou D (2017) A virtual maintenance-based approach for satellite assembling and troubleshooting assessment. Acta Astronaut 138:434–453. https://doi.org/10.1016/j.actaastro.2017.06.018

Goh YM, Guo B (2018) FPSWizard: a web-based CBR-RBR system for supporting the design of active fall protection systems. Autom Constr 85:40–50. https://doi.org/10.1016/j.autcon.2017.09.020

Göker M, Roth-Berghofer T (1999) Development and utilization of a case-based help-desk support system in a corporate environment. In: Althoff KD, Bergmann R, Branting L (eds) Case-based reasoning research and development. ICCBR 1999. Lecture notes in computer science (Lecture notes in artificial intelligence), vol 1650. Springer, Berlin, Heidelberg

Graham-Jones PJ, Mellor BG (1995) Expert and knowledge-based systems in failure analysis. Eng Fail Anal 2(2):137–149. https://doi.org/10.1016/1350-6307(95)00009-F

Hanaysha J (2016) Examining the effects of employee empowerment, teamwork, and employee training on organizational commitment. Proc Soc Behav Sci 229:298–306. https://doi.org/10.1016/j.sbspro.2016.07.140

Iudica NR (1987) On the requirements for the implementation of knowledge-based industrial planning and configuration systems. In: Menges G, Hövelmanns N, Baur E (eds) Expert systems in production engineering. Lecture notes in engineering, vol 29. Springer, Berlin, Heidelberg

Kobbacy K (2008a) Artificial intelligence in maintenance. In: Complex system maintenance handbook. Springer series in reliability engineering. Springer, London

Kobbacy KAH (2008b) Artificial intelligence in maintenance. In: Complex system maintenance handbook. Springer series in reliability engineering. Springer, London

Leake D (1996) Case-based reasoning: experiences, lessons, and future directions. AAAI Press/MIT Press

Lirov Y, Yue OCJ (1991) Expert maintenance systems in telecommunication networks. J Intell Rob Syst 4:303–309. https://doi.org/10.1007/BF00314937

Manzini R, Regattieri A, Pham H, Ferrari E (eds) (2010) Maintenance for industrial systems. Springer, London

Miyashita K, Sycara K (1995) CABINS: a framework of knowledge acquisition and iterative revision for schedule improvement and reactive repair. Artif Intell 76(1–2):377–426. https://doi.org/10.1016/0004-3702(94)00089-J

Morvan J (1997) Management de la sécurité des systèmes industriels de production. Dunod

Natarajan D (2015) Failure mode and effects analysis. In: Reliable design of electronic equipment. Springer, Cham

Nick M, Snoek B, Willrich T (2003) Supporting the IT security of eServices with CBR-based experience management. In: Ashley KD, Bridge DG (eds) Case-based reasoning research and development. ICCBR 2003. Lecture notes in computer science, vol 2689. Springer, Berlin, Heidelberg

Olsson E, Funk P (2009) Agent-based monitoring using case-based reasoning for experience reuse and improved quality. J Qual Maintenance Eng 15(2):179–192. https://doi.org/10.1108/13552510910961129

Pieri G, Klein MR, Milanese M (2001) A knowledge based system for the maintenance of chemical plants and its implementation using OPTRANS. In: Monostori L, Váncza J, Ali M (eds) Engineering of intelligent systems. IEA/AIE 2001. Lecture notes in computer science, vol 2070. Springer, Berlin, Heidelberg

Potes-Ruiz P, Kamsu-Foguem B, Noyes D (2013) Knowledge reuse integrating the collaboration from experts in industrial maintenance management. Knowl-Based Syst 50:171–186. https://doi.org/10.1016/j.knosys.2013.06.005

Potes-Ruiz P, Kamsu-Foguem B, Noyes D (2014) Generating knowledge in maintenance from experience feedback. Knowl-Based Syst 68:4–20. https://doi.org/10.1016/j.knosys.2014.02.002

Rasovska I, Chebel-Morello B, Zerhouni N (2008) A mix method of knowledge capitalization in maintenance. J Intell Manuf 19:347–359. https://doi.org/10.1007/s10845-008-0087-3

Recio-García J, González-Calero P, Díaz-Agudo B (2014) jcolibri2: a framework for building case-based reasoning systems. Sci Comput Program 79:126–145. https://doi.org/10.1016/j.scico.2012.04.002

Ruschel E, Portela-Santos E, Rocha-Loures E (2017) Industrial maintenance decision-making: a systematic literature review. J Manuf Syst 45:180–194. https://doi.org/10.1016/j.jmsy.2017.09.003

Scherer WT, White CC (1989) A survey of expert systems for equipment maintenance and diagnostics. In: Tzafestas SG (eds) Knowledge-based system diagnosis, supervision, and control. Applied information technology. Springer, Boston

Talamo C, Bonanomi M (2016) Knowledge management and information tools for building maintenance and facility management. Springer

Terninko J, Zusman A, Zotlin B (1998) Systematic innovation: an introduction to TRIZ. St. Lucie Press

Toro C, Sanín C, Vaquero J, Posada J, Szczerbicki E (2007) Knowledge based industrial maintenance using portable devices and augmented reality. In: Apolloni B, Howlett RJ, Jain L (eds) Knowledge-based intelligent information and engineering systems. KES 2007. Lecture notes in computer science, vol 4692. Springer, Berlin, Heidelberg

Wang SL, Hsu SH (2004) A web-based CBR knowledge management system for PC troubleshooting. Int J Adv Manuf Technol 23:532. https://doi.org/10.1007/s00170-003-1676-0

Chapter 24
Industry 4.0 and Engineering Education: An Analysis of Nine Technological Pillars Inclusion in Higher Educational Curriculum

Guadalupe Maribel Hernández-Muñoz, Lizbeth Habib-Mireles, Flor Araceli García-Castillo and Fernando Montemayor-Ibarra

Abstract With the introduction of Industry 4.0, it is necessary to make an analysis from the curriculum perspective and to evaluate how to satisfy the new technological requirements that emerge from this revolution. This work shows an analysis from four universities about their curriculum and the relation with the nine technological pillars of the Industry 4.0; this universities were selected from the most important in the Nuevo Leon State and from the premise that it is very important to Mexico from the industrial perspective. The results show that engineering curriculum have subjects that contribute directly to the pillars of Industry 4.0; but some are missed or not entirely defined, such as: Big data, Cloud, Augmented Virtual Reality, Internet of Things and Additive Manufacturing. As an important conclusion, it is required to establish regulations in Mexico that encourage the actualizations and modifications of educational programs in higher education institutions to meet the new technological knowledge required by the industry in this new digital age.

Keywords Industry 4.0 · Engineering education · Curriculum

G. M. Hernández-Muñoz (✉) · L. Habib-Mireles · F. Montemayor-Ibarra
Universidad Autónoma de Nuevo León, Facultad de Ingeniería Mecánica Y Eléctrica. Centro de Investigación E Innovación En Ingeniería Aeronáutica, Av. Universidad S/N. Ciudad Universitaria, C. P. 66455 San Nicolás de Los Garza, Nuevo León, Mexico
e-mail: guadalupe.hernandezmn@uanl.edu.mx

F. A. García-Castillo
Universidad Autónoma de Nuevo León, Facultad de Ciencias Físico Matemáticas, Av. Universidad S/N. Ciudad Universitaria, C. P. 66451 San Nicolás Los Garza, Nuevo León, Mexico

J. L. García Alcaraz et al. (eds.), *Best Practices in Manufacturing Processes*,
https://doi.org/10.1007/978-3-319-99190-0_24

24.1 Introduction

It is pertinent to establish that the curriculum of the educational programs must be constantly updated or to have what is also known as a curricular redesign, to allow future graduates to be prepared and not only the graduates that are required nowadays, this is why the curricular redesign is more and more important each time, although the Institutions of Higher Education (IES) have an already planned timing to do the planning, which means that you carry out the action with the pressure of time that sometimes implies some limitations (Icarte and Labate 2016). Therefore, the objective of this paper is to analyze ten educational programs of the northeast of Mexico in order to establish how they impact the development of future graduates with characteristics to perform, within the trend that the Industry 4.0 points out. In addition, this paper presents statistical data of the population with an emphasis on the careers focused on the manufacture.

In order to identify the learning units that impact on the training of future graduates and establish which pillar is not contemplated or is not reflected in the curriculum of engineering students, it is intended to make pertinent recommendations for the curricular redesign. All this aligned to the recommendations of the OEI which establishes within its *Reflections on current and critical issues in the curriculum, learning and evaluation*, a space, where they make reference particularly for this research within its factor (v) of the importance of the promotion of sustainable development through the STEM (Science, Technology, Engineering and Mathematics) study programs (Opertti 2017).

Actually, the teachers who participate in the curricular design are involved in diverse problems that they support; however they don't focus to the care of knowledge and skills that is appropriate to support because most of the cases who perform this activity do not have the experience. Therefore, experts are required to guide the development and adaptation of the curriculum and for them to keep company to the teachers from other perspectives and with the vision that the company requires of the future students of each of the redesigned programs (Huizinga et al. 2014).

Where manufacturing is a priority area by being in the process of automation to produce without the need to replace the human hand and reducing time, taking advantage of the knowledge and experience of the human being, who should enrich the final product. For that reason, the training of future engineers, who will face the established innovations by the industry every day, requires from the beginning inside the classroom, to introduce them learning and development of skills and abilities to allow them to be at the forefront and face the challenges that the industry and the society impose on the future graduates.

24.2 Education and Industry Evolution

Manufacturing is the most elementary form of the industry, almost everything we use is the result of this process and nearly everything that is manufactured is made in large factories. The origin of the industry goes back since the beginnings of mankind it has always had the need to transform the resources that nature provided to take advantage of them (Osorio Díaz 2011). Manufacturing is considered a strategic sector for the economic activity of the territories in which it is located.

The world has experienced three industrial and technological revolutions during the last two centuries, we understand by the industrial revolution, the process of economic, social and technological transformation that causes changes (generally abrupt) in existing modes of production and technology (Castillo 2016). Currently, this technique is experiencing a fourth wave of technological advance. Industry 4.0 is part of the global trends of digitalization, whose importance is increasing in all areas of life and the economy (Schroeder 2016).

In the presence of the eminent boost of technology, young people need to develop talents linked to the manipulation of these, which requires a broad spectrum of skills, attitudes and technological experiences that help them focus on the generation of innovation and creativity without losing sight of the social and environmental welfare. A question then arises from this challenge: What is the role that the professional, who will be immersed in this new world full of technologies, needs to play?

The essential components of this new industrial scenario are machines working with humans, the purpose of Industry 4.0 is not to replace people for machines, but to work collaboratively interacting with each other, the new factory 4.0 will adapt to the specific needs of the society or the client at each moment, we will frequently find the autonomous facilities, as well as the fusion between the virtual and the real, this means that the product is designed and tested virtually in order to foresee possible errors before manufacture the final product. Industry 4.0 promotes the automation of manufacturing through the incorporation of information technologies and advanced manufacturing processes, which will give rise to "smart factories" (Gilchrist 2016).

From an academic perspective, the main concepts related to this fourth industrial revolution with regard to the field of engineering are described, finding the following:

The fourth industrial revolution contains the technology of digitalization, information and communications, autonomous learning, robotics and artificial intelligence; this will change the decision making of humans to machines more. The social changes that will occur will have a profound impact on research personal and management sales practices (Syam and Sharma 2018).

Based on the above, the engineering student should be profiled by developing STEM talents and scientific-technological vocations, with programs based on innovation in teaching methods, where institutions and companies connect, in order

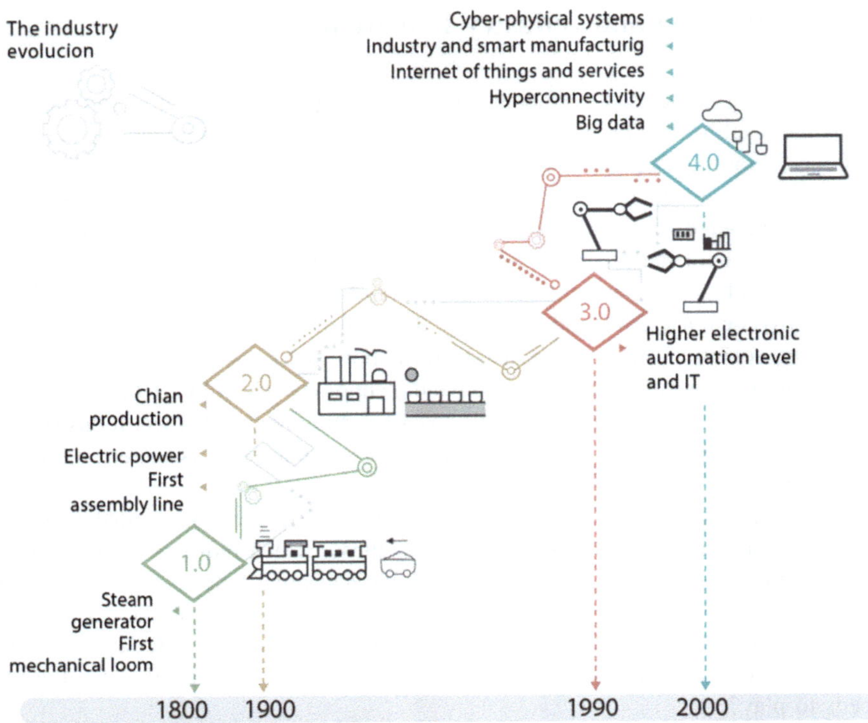

Fig. 24.1 The industry evolution. *Source* Informe Tendencias (2017)

to obtain innovation and creativity competencies that provide engineers a professional future and benefit their employability.

The growth, trends and evolution of manufacturing show us an exponential speed of changes in the market, which is becoming increasingly sophisticated, focused on the automation and connectivity of processes, based on the massive use of artificial intelligence tools (Schwarz 2014).

Throughout history, important technological, cultural and socioeconomic transformations have taken place, which have impacted on our daily life, these revolutions have been given in the following order and are visualized in Fig. 24.1 (Baygin et al. 2016).

- *Industry 1.0*: Began in England in the eighteenth and nineteenth century with the arrival of mechanical production equipment driven by water.
- *Industry 2.0*: In the nineteenth century and near the twentieth century it surpassed with the assembly lines fed with electricity and the division of labor of the second industrial revolution.
- *Industry 3.0*: The automation of production using robots, electronics and IT (Information Technology) took place in the second half of the 20th century, this is often considered as the third industrial revolution in production.

- *Industry 4.0*: The term "Industry 4.0" originates from high-tech strategic project of the German government that promotes the computerization of the manufacturing sector. The objective of Industry 4.0 is to revolutionize production by creating an "intelligent factory" characterized by automated machines and network processes. This industrial infrastructure works through the internet and information in the cloud. It combines the physical integration of machinery and devices connected in network with sensors and software used to predict, control and plan for better business results.

24.2.1 Nine Pillars of Technology of Industry 4.0

The Boston Consulting Group, identifies nine basic areas or also known as pillars of the technological advance of Industry 4.0 (Rüßmann et al. 2015), which are shown in Fig. 24.2 and described below (Bartodziej 2016; Cheng et al. 2016; Gilchrist 2016).

- *Internet of Things (IoT)*: intelligent objects are everyday physical objects that have been improved through small electronic devices that provide the ability to store information and transmit it, communicating in this way with other devices

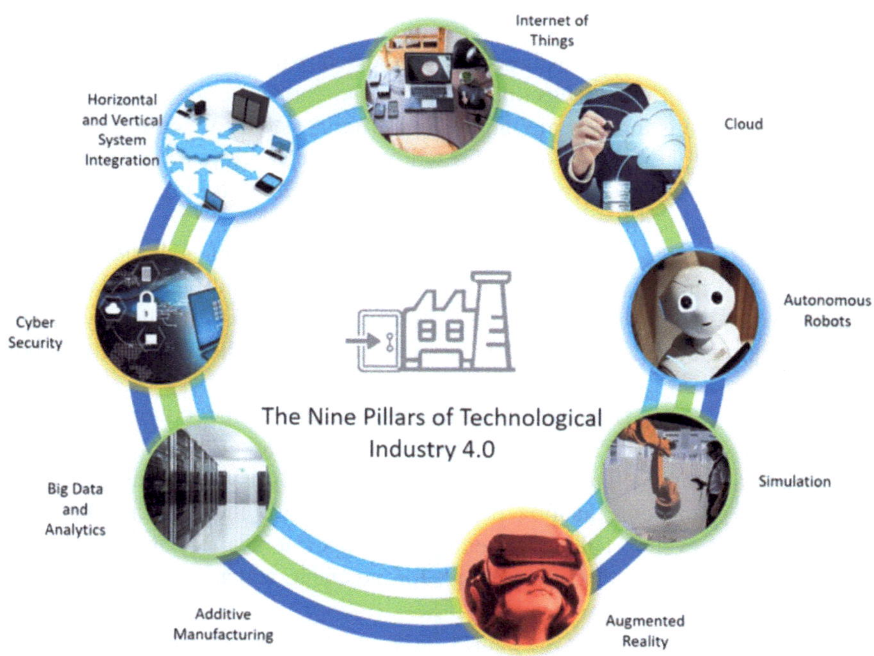

Fig. 24.2 The nine pillars of technological

through cyberspace established by Internet. For this, three elements are necessary: the physical thing itself, the sensor or label that is incorporated in the object to establish a communication link and the computer or device that processes the data.

- *The Cloud (Cloud Computing)*: The cloud comprises applications and infrastructures offered as a service through public or private networks, often in a pay-per-use model. Smart products and systems (CPS and CPPS) will generate huge amounts of data to store and process that must be accessible online from anywhere. The cloud allows this flow of data without borders and eliminates the need for investment in infrastructure to increase capacity, allowing unprecedented flexibility.
- *Autonomous Robots*: The improvement of artificial intelligence together with new sensors has allowed to create increasingly autonomous, flexible and cooperative robots. Over time, they will interact with each other and work safely with humans, learning from them, offering a range of capabilities far superior to those used in today's manufacturing.
- *Simulation*: In the engineering phase, 3-D simulations of products, materials and production processes are already used, but in the future, the simulations will be used more widely in plant operations as well. These simulations take advantage of the data in real time to reflect the physical world in a virtual model, which can include machines, products and human beings. This allows operators to test and optimize machine settings for the next online product in the virtual world before the physical change, boosting this way the machine setup times and increasing quality.
- *Augmented Reality*: Systems based on augmented reality support a wide variety of services, such as the selection of parts in a warehouse and the sending of repair instructions through mobile devices. These systems are currently in their infancy, but in the future, companies will make much wider use of augmented reality to provide workers with real-time information to improve decision making and work procedures.
- *Additive Manufacturing*: Companies have begun to adopt additive manufacturing, 3D printing, which is used to create prototypes and produce individual components. With Industry 4.0, these 3D printing methods are already widely used to produce small batches of customized products that offer construction advantages, such as complex and lightweight designs.
- *Data Analysis (Big Data)*: With an increasing number of products (CPS) and intelligent systems (CPPS) in the factories and the market, the amount of data available to manufacturers will multiply. Their analysis will identify patterns and interdependencies, analyze processes and discover inefficiencies and even predict future events. This will open up new opportunities, not only to improve efficiency, but also to discover services for the client, which will be much better known.
- *Additive Manufacturing*: Companies have begun to adopt additive manufacturing, 3-D printing which is used to create prototypes and produce individual components. With Industry 4.0, these 3-D printing methods are already widely

used to produce small lots of customized products, offering advantages for construction, such as complex and light designs.
- *Data Analysis (Big Data)*: With an increasing number of products (CPS) and intelligent systems (CPPS) in factories and market, the amount of data that manufacturers provide will be multiplied. Their analysis will identify patterns and interdependencies, processes analyze and discover inefficiencies and even predict future events. This will open up new opportunities, not only to improve efficiency, but also to discover client service, which will be much better known.
- *Informatics Security*: With the increase in connectivity and the use of standard communication protocols that come with Industry 4.0, the necessity to protect critical industrial systems and manufacturing lines from cyber security threats increases dramatically. As a result, secure and reliable communications, as well as sophisticated identity management systems and access to machines, and users will be essential.
- *Integration of horizontal and vertical systems*: Now a day most of the IT systems are not totally integrated. Companies, suppliers and customers are rarely closely linked, neither the departments, such as, engineering, production and service. Likewise, the functions of top management, as well as labor section are not fully integrated. But with Industry 4.0, companies and departments will be more cohesive with each other and will allow value chains to truly become automated.

24.3 The Curriculum in Engineering and Industry 4.0

The curriculum for a particular program can be seen as a group of subjects, courses or interdisciplinary projects, which are associated with a number of credits and prerequisites (Phadke and Kulkarni 2018).

In engineering education, curriculum design is an important academic process (Phadke and Kulkarni 2018). Engineering curriculum design must meet the needs of the productive sector, so that graduates have solid and fundamental bases that facilitate their asset into the labor market. Therefore, it is required that the educational plans be analyzed and be able to propose the update of these, based on a labor study and the tendency that makes the requirements of the industry.

In study plans, a key concept is the incorporation of teaching methodologies, such as learning, one of the most presented by a scenario as it will face a graduate is the project-based learning methodology (PBL), which is represented in a curriculum based on the results model, facilitating the student's insertion into the industry or with spirit, business because it develops skills of design, communication, teamwork, creativity, knowledge of customer treatment and investment analysis, among others; having a great set of skills (Irfan et al. 2018). There are other methodologies that are carried out in the classroom so that students can learn in scenarios as close to what they will face as engineers.

In Mexico, it was convened in 2006 to change the traditional teaching approach to a pedagogical trend that focuses on students, so one of the strategies used by teachers is the implementation of projects or projects recognized by others authors such as PBL, however this strategy requires a broad domain of the subject and have well-defined concepts and knowledge of science (Hernández and Ramos 2016). Different sources refer to the advantages of using PBL, by encouraging collaborative learning, through the development creativity, the need to find their own results and conclusions (Bilgin et al. 2015; Boss and Krauss 2014; Savery 2015; Vande Wiele et al. 2017)

Students learn the principles of scientific research when they see the need to define the problem, identify their object of study, choose their sample, plan and execute their project, analyze and present results by sharing their experiences and main learnings. In order to implement the PBL each of the projects must contain the stage of planning, development, communication and evaluation, these broad stages allow students to show their knowledge, skills, teamwork, but without doubt the opportunity to develop a project and know what it entails, a situation that favors its introduction to working life, by familiarizing them with inquiry, communication, learning to learn and the use of creativity.

Technology is the great descending of economic progress, putting people at the center of the digital transformation, with the hard skills that will make possible the digital transformation of companies faster. The talent of managers, technicians and workers is the key to succeed in the digital economy (Lombardero 2015).

Competencies have become the global currency of twenty first century economies. But these "currencies" can be devalued as the requirements of labor markets evolve and individuals lose the skills they do not use. For skills to maintain their value, they must be continuously developed (For Economic Co-operation and Development 2012).

With the emergence of the so-called Industry 4.0, it is pointed out from Germany that the new technologies, work environments, organizational structures and forms of internal and external cooperation of Industry 4.0 have important implications for initial and continuing Vocational Training (VT) at all levels (For the Development of Vocational Training (Cedefop) 2015; Guillén and García 2016), summarizes it in a simple and clear way: for factories 4.0 we are going to need an education also 4.0 (Astigarraga et al. 2017).

World Economic Forum (Soffel 2016) details the new ways of working in digital environments, which must include the competencies that support the evolution of the same (Fig. 24.3), although among the existing lists are: Problem solving, creativity, effective communication, collaboration/teamwork, decision making, initiative and digital and media literacy.

Pérez and Graus (2017), mention two trends in curricular design, the first tendency is to look for a curriculum that broadly interprets the life and experience that allows directing the student towards the processes and development of activities that allow learning, while the second trend refers to the importance of directing the curriculum according to practical requirements, which correspond to the skills and abilities that graduates will require.

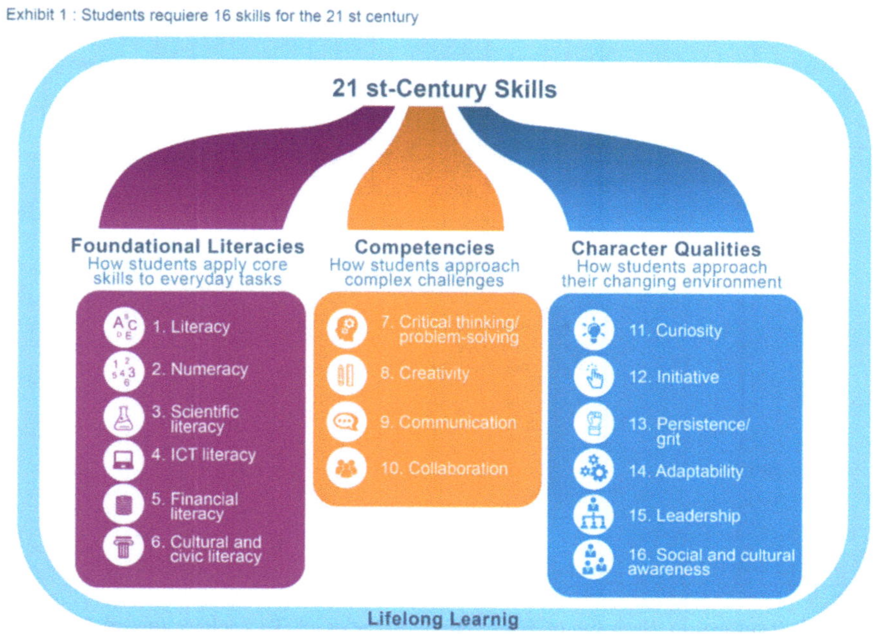

Fig. 24.3 Skills for the 21st century (Soffel 2016)

Table 24.1 Questionnaire proposed by Sackey and Bester for universities (Sackey and Bester 2016)

Questions
How is analytics addressed in the IE curriculum?
Does the IE curriculum cover techniques for human-robot collaboration?
Does the curriculum introduce students to augmented (virtual) reality systems?
Beyond simulation, does the IE curriculum accommodate virtual plant models using real-time data?
How much device data communication/networking knowledge does the IE curriculum offer students?
In the respondent's view, how should IE departments in universities respond to Industry 4.0 through curriculum content enhancements?

Coupled with the above, Nicolai (1993) in his work states: "*The engineering curriculum should allow the student to be an engineer by introducing a problematic situation that forces the student to link the theory of engineering with real-world problems. An original thought, evaluating alternative solutions, taking a decision and defending it*".

With regard to the study of the impact on the curriculum and industry 4.0 (Sackey and Bester 2016), they present a paper in which they propose improvements in the industrial engineering curricula of 10 universities in South Africa, concluding that only one university has adopted the infrastructure of the industry

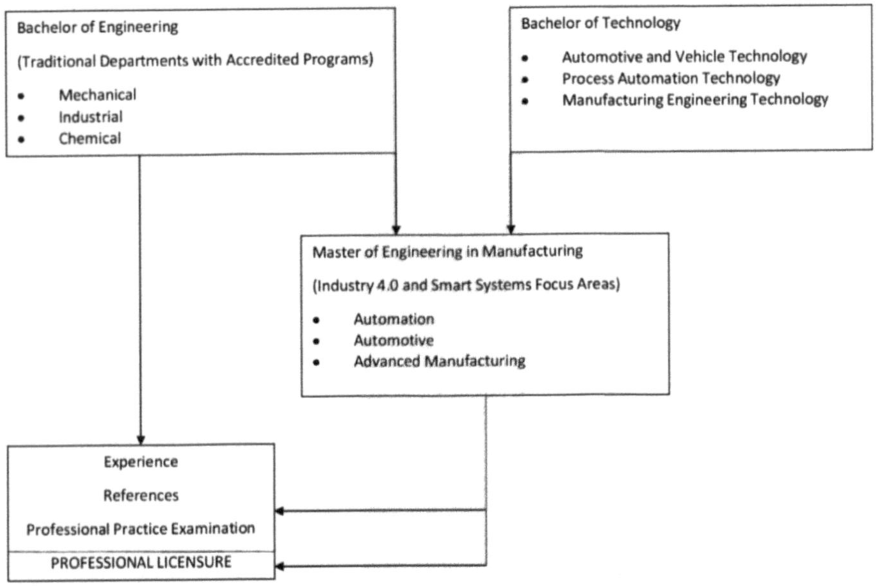

Fig. 24.4 Pathways to Bachelor of Engineering, Bachelor of Technology toward Master of Engineering in Manufacturing for Professional License (Justason et al. 2018)

4.0 and perform a set of curricular enrichment items as a basis for the reform. Table 24.1 shows the questionnaire with the secondary questions that were applied to the universities related to curriculum and industry 4.0, related to some pillars presented by the same and that are detailed below.

Justason et al. (2018), present the design and implementation of new master-level programs that focus on modern approaches in manufacturing; namely Industry 4.0 and Smart Systems. In addition, they relate it to undergraduate studies so that the student count on what is necessary for their professional development (Fig. 24.4). The elements that include cyber physical systems, internet of things and development of intelligent systems which are trends of the industry.

24.4 Analysis of the Context National and State

The government of Mexico and the companies are aware of the important fourth industrial revolution, currently in Mexico they have taken 4 federal programs: Connected Mexico, which proposes to implement free connectivity in public centers across the country; The National Digital Strategy, which fosters adoption of high technology through joint effort carried out by different Ministries; (National Digital Inclusion and Literacy Program), which distributes tablets to education centers so that students can develop digital skills; and finally, The Program for

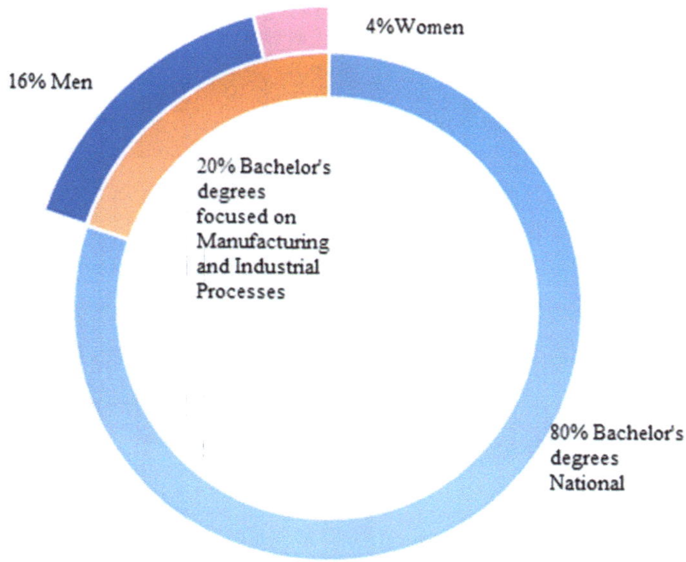

Fig. 24.5 Student population national level-bachelor's degree and bachelor's degree manufacturing and process area

Developing the Software and Innovation Industries, which supports digital entrepreneurship and the IT development of small and medium businesses (México Business Summit 2016), of which, the first three programs are focused directly on education.

The Mexican Institute for Competitiveness (IMCO) in its 2017 report states that the total enrollment at the undergraduate level is 10,088,573 students, of which 4,982,747 (49%) are men and 5,105,826 (51%) are women. In bachelor's degrees focused on Manufacturing and Industrial Processes, the enrollment is 2,004,745, it is only 20% of the total bachelor's degree. With respect to the gender, the participation is of 1,592,422 (80%) men and 412,323 (20%) women. Figure 24.5 clearly shows the participation of the student population in careers directly related to manufacturing.

Performing a focused study based on the statistical data provided by The Mexican Institute for Competitiveness (IMCO), the undergraduate courses focused directly on the 4.0 industry were selected, showing as results only 5 of 66 courses in the Manufacturing and Processes area are directly related to any of the pillars of industry 4.0. Figure 24.6 shows the percentage corresponding to each of the careers, we can see that the largest student population are studying careers focused on information technology and communication and Industrial, mechanical, electronic and technology engineering, multidisciplinary or general programs, 26 and 25%, respectively. Having a considerable deficit in Electronics and automation area, with only 8%.

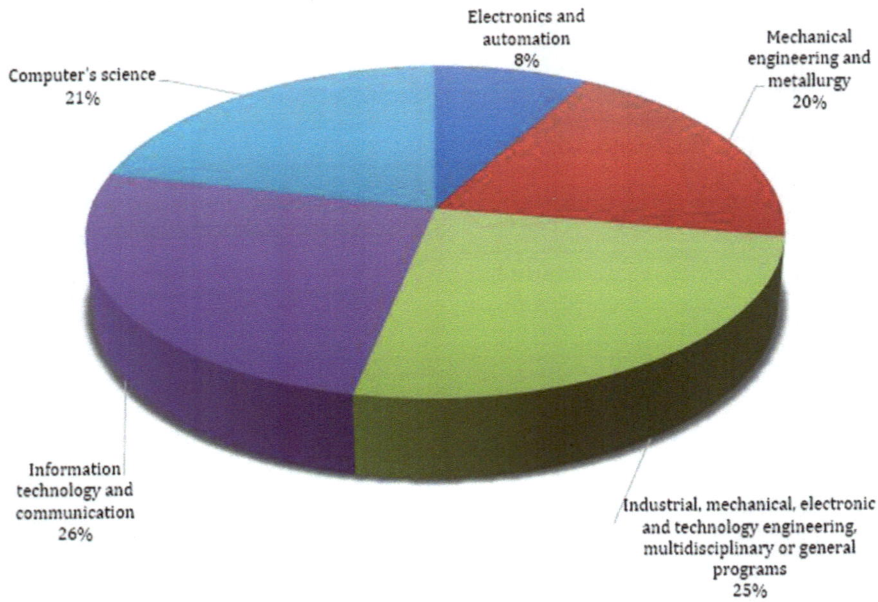

Fig. 24.6 Students studying at bachelor's degrees in Mexico focused on Industry 4.0

The fourth world industrial revolution is advancing really quickly; the government of the state of Nuevo Leon is considering this has undertaken with a fund equivalent to 7.5 million dollars the initiative Nuevo León 4.0 with the purpose of encouraging the industrial development of the state (Gobierno de Nuevo León 2017a).

The support is focused on the innovation projects of companies in the strategic sectors such as, automotive, information technology, education, household appliances, energy, health, aerospace and agrifood (Gobierno de Nuevo León 2017b).

The proposed model defines three actors: government, academia and industry as the innovation managers that drive the economic development of the region. The pillars to be developed in this initiative include the Internet of things, cloud, robotics, artificial intelligence, simulations, additive manufacturing, advanced materials, big data, augmented virtual reality, computer security and software.

Currently enrolled in Bachelor of Manufacturing degrees or Industry-focused areas 4.0 of 196,381 students enrolled in the 2016–2017 school year, having 126,942 (65%) are in public schools and 69,439 (35%) are in private.

The academic actors at the local level are integrated by the main universities of the State of Nuevo Leon, of which three are private, such as the Instituto Tecnológico y Estudios Superiores de Monterrey (Private University A), the Universidad de Monterrey (Private University B) and the Universidad Regiomontana (Private University C). The public university that is in this analysis is the Universidad Autonóma de Nuevo Leon (Public University D). All these

Table 24.2 Educational offer in local universities

University	
Private University A	Digital systems and robotics engineer
	Computer engineer technology
	Industrial and systems engineering
	Mechatronics engineering
Private University B	Computer engineer technology
	Industrial and systems engineering
	Mechatronics engineering
	Electronic technologies and robotics engineer
Private University C	Industrial and systems engineering
	Mechatronics engineering
	Computer engineer technology
Public University D	Bachelor of information technology security
	Bachelor of Information Technology
	Electronic and automation engineering
	Manufacture engineer
	Materials engineer

educational institutions have as their mission to develop knowledge, technology, and specialized human talent with a high added value.

Within the mapping that was made to these universities, the educational offer of engineering careers related to industry 4.0, is shown in Table 24.2, although there are other careers that may not have an industrial approach.

The roles of the academy contemplated in the initiative are:

- Specialized laboratories
- Develops and shares knowledge and technology
- Update the educational offer
- Collaborative work with the industry
- Develops specialized human resources
- International academic link
- Push the frontiers of knowledge

The engineering programs offered by the different universities in their curricular maps must have subjects that propitiate the qualified human resources capable of promoting and consolidating each of the pillars of the 4.0 industry.

Table 24.3 shows the analysis of the relationship between the pillars and the engineering careers. This table was built based on the information by each university web site and with the curricula of each carrier, some careers could not be related to the pillar to which they contribute, this mainly due to the lack of information about the analytical programs study of each subject, this doesn't not mean that this work is not in the pillar in the curricula operation.

Table 24.3 Example of relation between the pillars and some the engineering careers

Pillars	Careers
Private University A	
Internet of things	Digital systems and robotics engineer Computer technologies engineer
Cloud	Digital systems and robotics engineer
Robotics	Mechatronics engineer Digital systems and robotics engineer
Artificial intelligence	Computer technologies engineer
Simulation	–
Additive manufacture	–
Advanced materials	–
Big data	Mechatronics engineer Digital systems and robotics engineer Industrial and systems engineer
Augmented virtual reality	–
Informatics security	Computer technologies engineer
Software	Digital systems and robotics engineer Computer technologies engineer
Private University B	
Internet of things	Computer technologies engineer Electronics technologies and robotics engineer Industrial and systems engineer Mechatronics engineer
Cloud	Computer technologies engineer Electronics technologies and robotic engineer
Robotics	Mechatronics engineer
Artificial intelligence	Computer technologies engineer
Simulation	Industrial and systems engineer
Additive manufacture	–
Advanced materials	–
Big data	Computer technologies engineer
Augmented virtual reality	
Informatics security	Computer technologies engineer
Software	Computer technologies engineer Electronics technologies and robotic engineer Mechatronics engineer Industrial and systems engineer
Private University C	
Internet of things	Computer technologies engineer
Cloud	Computer technologies engineer
Robotics	Computer technologies engineer Mechatronics engineer

(continued)

Table 24.3 (continued)

Pillars	Careers
Artificial intelligence	Computer technologies engineer
Simulation	Mechatronics engineer
Additive manufacture	Mechatronics engineer
Advanced materials	–
Big data	Computer technologies engineer
Augmented virtual reality	–
Informatics reality	Computer technologies engineer
Software	Computer technologies engineer
Public University D	
Internet of things	–
Cloud	–
Robotics	Manufacture engineer Mechatronics engineer Electronic and automation engineer
Artificial intelligence	Mechatronic engineer
Simulation	Manufacture engineer Mechatronic engineer
Additive manufacturing	Mechatronic engineer
Advanced materials	Materials engineer
Big data	–
Augmented virtual reality	–
Informatics security	Computer technologies engineer
Cloud	–
Robotics	Manufacture engineer Mechatronics engineer Electronic and automation engineer

Table 24.4 shows an example the analysis of the relationship between the pillars and the subject of different engineering careers (FIME 2013). It is important to highlight that the thematic contents of the subjects could be adapted or detailed to fulfill the expected scope. Likewise, the pillars of the Inter-net of Thing, Cloud and Big Data are required to be strengthened in the curricular plans of the different careers offered by the universities.

The state of Nuevo León in México has taken seriously to develop projects that favor the best positioning of companies at the international level, where both public and private universities, that private sector and the government work hand in hand within the initiative Nuevo Leon 4.0.

The Software Cluster and the Interactive Media Cluster, integrated by more than 400 companies, work together with the Universities in the development of products for the manufacturing industry, putting into practice pillars of the 4.0 industry implementing augmented reality, virtual reality, internet of things, big data and artificial intelligence.

Table 24.4 Example of relation between the pillars and the subject engineering careers

Pillars	Careers	Subject
Internet things	–	–
Cloud	–	–
Robotics	Manufacture Mechatronics Electronics and automation	Robotics Robot architecture Robots control
Artificial intelligence	Mechatronics	Artificial intelligence and neural networks
Simulation	Manufacture Mechatronics	Finite element analysis/CAE Finite element analysis
Additive manufacturing	Mechatronics	Rapid prototypes
Advanced materials	Materials	
Big data	–	
Augmented virtual reality	–	
Informatics security	Software technology	Information security and cryptography
Software	Systems administrator Software technology	

One of the efforts that are currently being developed is the construction of the first 4.0 Industry laboratory which will be established in a Private University, this laboratory will be called the Center for Industrial Innovation in the Development and Integration of Talent and Technologies, which has as an objective to pro-mote the link between universities and industry members.

The Nuevo Leon's central government strategies for the Industry 4.0, should be improved to build an economic and social systems that can respond to changes in a flexible manner. An in-depth analysis of the government's initiatives, it shows that it is necessary to work in the transformation mechanisms that will help to implement a collaboration ecosystem between Government, Academia, Industry, Entrepreneurs and Investors that makes the difference in the Society, next, we mention the pillars of this strategy:

- *Proactive Liaison*: Invite action, let all Nuevo Leon know the importance of joining this new industrial revolution and how to do it.
- *Talent Development*: Transform/Create curricula for Talent 4.0 needed in this fourth industrial revolution, transform existing talent in companies based on the technologies and key I4.0 capabilities.
- *Technological Infrastructure*: Mapping of Existing Capacities (Infrastructure and Talent) and available for NL4.0, complement of Infrastructure and Talent 4.0 based on the challenges of NL4.0 …
- *Business Models*: Mapping of Business Models of the Iconic Companies of Nuevo Leon and their Cases of Success, Examples for the rest of the NL4.0 community.

- *Public Policy*: Identify and Remove Barriers that prevent NL4.0 from achieving its objectives, Propose Public Policy based on success cases from other countries/regions, Align Incentives/Funds to support NL4.0.
- *Proactive Linking*: Connecting Tractor Company with SMEs and technology-based Startups, Link Tractor Value Chain with Industry 4.0 Solutions in the region.
- *Humanist Ethics*: To deal frontally with the impact of technology (automation) on the human being, Develop the strategy of the "Universal Basic Income" with Purpose of Life, Democratize Industry 4.0 with the Trade Unions.
- *Strategic Alliances*: Identify and Negotiate Favorable Conditions for the NL4.0 Network of Solutions 4.0.
- *Standardization*: Propose guidelines in architecture and communication protocols such as those of the German Initiative (RAMI 4.0).

24.5 Conclusions

In Mexico, the term industry 4.0 is very recent, it represents great challenges that require strategic decisions as well as investments of capital and human talent, and not in the distant future, but in the immediate present. Therefore, the strategy to be followed must be a fundamental axis in the planning and implementation of public policies to conform a country project with an essential derivation of a new industrial policy that puts these potentialities and emerging challenges at the center.

On behalf of Nuevo León, a general strategy has been devised to create an Ecosystem of collaboration between Government, Academia, Industry, Entrepreneurs and Investors that makes a difference in the Society, however, within this initiative, the inclusion of collaborations and shared experiences is absent in the transformation models with the governments of the different countries that already have an advance in the implementation of the most appropriate strategies of the industry 4.0, the same ones that will help in an efficient way to increase the industrial added value and the qualified employment in the sector, stimulation a proper model for the industry of the future.

The development level or scope of the different pillars is not reflected in the government initiative, comparing the different engineering careers offered, subjects related to the requirements of Industry 4.0 can be found. Nevertheless, when the thematic content of the subjects is analyzed, it does not achieve the 100% with the expected scope.

The academy, as one of the main actors of the fourth industrial revolution requires an update of the study programs with topics related to the economic development plan 2016–2021 and aligned with the Nuevo León 4.0 initiative.

It's very important to establish regulations in Mexico that measure the actualizations or modifications of educational programs in higher education institutions; nowadays, those institutions are influenced by external changes to make curricular redesigns or in some cases, create new programs to meet the society and industry demands. However, this process must be a permanent activity in the institutions, without forgetting creating mechanism that permit evaluate if those program adjustments or new creation programs meet the expected labor market profile of the future engineers that are preparing in the classrooms today.

This process can't be immediate, the human factor takes a priority place to convince, train and professionalize the authorities linked with the education that are the main rolls, including the teacher in the classroom and laboratories, academic coordinators, tutors and even the students; in the incorporation of the new technological trends, where all the involved actors should be sensitive to changes and adjustments. The proposal is that educative institutions have a deadline of 5 years to analyze and perform a restructure in the continuous research of engineer's education according to technological evolution.

Therefore, the educational institutions must face the challenge with a holistic change in their study programs, in such a way that they educate the current engineers and update those who are already in the labor world, with the required knowledge based on the technological changes in the new digital age.

References

Astigarraga E, Agirre A, Carrera X (2017) Innovation and change in vocational training in the Basque Country. The ETHAZI model. Revista Iberoamericana de Educación 74:55–82 (in Spanish)

Bartodziej CJ (2016) The concept industry 4.0: an empirical analysis of technologies and applications in production logistics. Springer, Germany. https://doi.org/10.1007/978-3-658-16502-4

Baygin M, Yetis H, Karakose M, Akin E (2016) An effect analysis of industry 4.0 to higher education. In 2016 15th international conference on information technology based higher education and training (ITHET), pp 1–4. https://doi.org/10.1109/ITHET.2016.7760744

Bilgin I, Karakuyu Y, Ay Y (2015) The effects of project based learning on undergraduate students' achievement and self-efficacy beliefs towards science teaching. Eurasia J Math Sci Technol Educ 11(3):469–477. https://doi.org/10.12973/eurasia.2014.1015a

Boss S, Krauss J (2014) Reinventing project-based learning: your field guide to real-world projects in the digital age. International Society for Technology in Education, Arlington

Cheng GJ, Liu LT, Qiang XJ, Liu Y (2016) Industry 4.0 development and application of intelligent manufacturing. In: 2016 international conference on information system and artificial intelligence (ISAI), pp 407–410. https://doi.org/10.1109/ISAI.2016.0092

FIME (2013) Manufacturing engineering curriculum. Retrieved 3 Feb 2018, from http://www.fime.uanl.mx/oferta_educativa/licenciatura/ESP/401/IMF.html

For Economic Co-operation and Development (2012) Better skills, better jobs, better lives: a strategic approach to skills policies. Organization for Economic Co-operation and Development. http://dx.doi.org/10.1787/9789264177338-en

For the Development of Vocational Training (Cedefop) (2015) Innovation and training: partners in change. http://doi.org/10.2801/80843

Gilchrist A (2016) Industry 4.0: the industrial internet of things. Springer, Berlin. http://doi.org/10.1007/978-1-4842-2047-4

Gobierno de Nuevo León (2017a) Technological revolution begins "Nuevo León 4.0. Retrieved 3 Feb 2018, from http://www.nl.gob.mx/noticias/inicia-revolucion-tecnologica-nuevo-leon-40 (in Spanish)

Gobierno de Nuevo León (2017b) Intelligent state. Nuevo León 4.0. Retrieved from https://goo.gl/dtrjyP (in Spanish)
Guillén B, García R (2016) Industry 4.0 demands an education 4.0. [Blog] CincoDías 40. Available at: https://goo.gl/VeEa1j. Accessed 11 May 2018 (in Spanish)
Hernández JLB, Ramos MTG (2016) Works by projects in the secondary science classroom: curricular tensions and teaching resolutions. Revista Mexicana de Investigación Educativa 21 (68):141 [In Spanish]
Huizinga T, Handelzalts A, Nieveen N, Voogt JM (2014) Teacher involvement in curriculum design: need for support to enhance teachers design expertise. J Curric Stud 46(1):33–57. https://doi.org/10.1080/00220272.2013.834077
Icarte GA, Labate HA (2016) Methodology for the revision and updating of a curricular design of a university career incorporating competency-based learning concepts. Formación Universitaria 9:3–16. http://dx.doi.org/10.4067/S0718-50062016000200002 (in Spanish)
Informe Tendencias (2017) New ways of working in virtual environments and evolution of the industry. Retrieved 11 Feb 2017, from https://goo.gl/WC4h9h
Irfan MM, Rajamallaiah A, Ahmad SM (2018) Paradigm shift in the engineering curriculum: design thinking. J Eng Educ Transform 2018:1–5 (Special Issue)
Justason MD, Centea D, Belkhir L (2018) Development of M. Eng. programs with a focus on Industry 4.0 and smart systems. In Auer ME, Zutin DG (eds) Online engineering & internet of things. Springer International Publishing, Cham, pp 68–76. https://doi.org/10.1007/978-3-319-64352-6_7
Lombardero L (2015) Work in the digital age. LID Editorial Empresarial, LID editorial, México
México Business Summit (2016) Industry 4.0 is here to stay, p 1. Retrieved from https://www.cumbredenegocios.com.mx/newsletter5/en/4.html
Nicolai L (1993) An industry view of engineering design education. In: 31st aerospace sciences meeting. Reno NV, USA, p 328
Opertti R (2017) 15 analysis keys to support the 2030 educational agenda. Available at: http://unesdoc.unesco.org/images/0025/002590/259069s.pdf (in Spanish)
Osorio Díaz NJ (2011) Market analysis of the manufacturing industries sector based on ISIC 3 under a focus of economic concentration in the period 2000–2008 in Ecuador. QUITO/EPN/2011
Pérez JJF, Graus MEG (2017) Theoretical aspects of curricular design and its particularities in the sciences. Revista Boletín Redipe 6(3):83–112 (in Spanish)
Phadke AS, Kulkarni SS (2018) Use of network model for analysis of curriculum and its mapping to program outcomes. J Eng Educ Transform 31(3):1–5
Rüßmann M, Lorenz M, Gerbert P, Waldner M, Justus J, Engel P, Harnisch M (2015) Industry 4.0: the future of productivity and growth in manufacturing industries. Boston Consulting Group, p. 9
Sackey SM, Bester A (2016) Industrial engineering curriculum in Industry 4.0 in a South African context. S Afr J Ind Eng 27:01–114. https://doi.org/10.7166/27-4-1579
Savery JR (2015) Overview of problem-based learning: definitions and distinctions. Essent Read Probl Based Learn Explor Ext Leg Howard S. Barrows 9:5–15. https://doi.org/10.7771/1541-5015.1593
Schroeder W (2016) The German strategy Industry 4.0: Rhenish capitalism in the era of digitalization. Friedrich Ebert Stiftung, Madrid. Retrieved from https://goo.gl/pZiB7ySchwarz
Soffel J (2016) What are the 21st-century skills every student needs? [Blog] Digital Lead, Center for the Fourth Industrial Revolution, World Economic Forum. Available at: https://www.weforum.org/agenda/2016/03/21st-century-skills-future-jobs-students/. Accessed 21 May 2018
Syam N, Sharma A (2018) Waiting for a sales renaissance in the fourth industrial revolution: machine learning and artificial intelligence in sales research and practice. Ind Mark Manage 59:135–146. https://doi.org/10.1016/j.indmarman.2017.12.019
Vande Wiele P, Morris D, Ermine JL (2017) Project based learning for professional identity: a case study of collaborative industry projects in marketing. Indep J Teach Learn 12(2):44–63